Frontiers in Sedimentary Geology

K.L. Kleinspehn   C. Paola
Editors

New Perspectives in Basin Analysis

With 225 Illustrations

Indiana University
Library
Northwest

Springer-Verlag
New York Berlin Heidelberg
London Paris Tokyo

K.L. Kleinspehn
C. Paola
Department of Geology and Geophysics
University of Minnesota
Minneapolis, Minnesota 55455, USA

Library of Congress Cataloging-in-Publication Data
New perspectives in basin analysis.
   (Frontiers in sedimentary geology)
   A collection of papers based largely on a symposium
held in Minneapolis, Minn., May 8–9, 1986, in honor
of Francis J. Pettijohn.
   Bibliography: p.
   Includes index.
   1. Sedimentation and deposition—Congresses.
2. Sedimentary structure—Congresses.   I. Kleinspehn,
K.L. (Karen Lee)   II. Paola, C. (Chris)   III. Pettijohn, F. J.
(Francis John), 1904-   .   IV. Series.
QE571.N39   1988   551.3   87-23430

Typeset by Publishers Service, Bozeman, Montana.
Printed and bound by Halliday Lithograph, West Hanover, Massachusetts.
Printed in the United States of America.

9 8 7 6 5 4 3 2 1

ISBN 0-387-96611-0 Springer-Verlag New York Berlin Heidelberg
ISBN 3-540-96611-0 Springer-Verlag Berlin Heidelberg New York

Series Preface

In the extensive field of earth sciences, with its many subdisciplines, the transfer of knowledge is primarily established via personal communication, during meetings, by reading journal articles, or by consulting books. Because more information is available than can be assimilated, it is necessary for the individual to search selectively. Books take more time from the inception of an idea until publication than any of the other means of communication mentioned. As a consequence, their function is somewhat different. Many good books are a compilation of up to date knowledge and serve as reference or instruction manuals. Some books are a collection of previously published papers dealing with a certain topic, while others may basically provide large sets of data or examples.

The *Frontiers in Sedimentary Geology* series was established both for students and practicing earth scientists who wish to either stay abreast of the most recent ideas or developments or to become familiar with an important topic in the field of sedimentary geology. The series attempts to deal with subjects that are in the forefront of both scientific and economic interest. The treatment of a subject in an individual volume should be a combination of topical, regional, and interdisciplinary approaches. Although these three terms can be defined separately, in reality they should flow into each other. A topical treatment should relate to a major category of sedimentary geology. The publication should deal with different geographic areas to avoid competition with excellent local treatises. Most important is the interdisciplinary aspect to stress that the understanding of and the approach to a problem can seldom be handled by one discipline, and that the interaction of many expertises enriches the outcome significantly. As a consequence, most of the publications in this series will be compilations of papers by many authors.

To obtain the highest quality, the editor(s) should strive for the following objectives:

1. To provide a continuous and connected flow of concepts throughout the volume by the use of introductory chapters that outline a topic sufficiently to grasp its problems and to understand the purpose of the chapters that follow. The guidance should not compromise the individuality of a contributor or a specific study.
2. To focus on aspects of sedimentary geology that are at or have returned to the forefront of earth sciences.
3. To select topics that contain a high level of experimental design or new developing concepts.

4. To stress the integration of approaches and the application of both mature methods and new techniques to produce a book that combines state of the art developments with the problems that are being faced.

The *Frontiers of Sedimentary Geology* series thus attempts to present books in such a manner that the reader can become informed in a minimum amount of time and will obtain good background knowledge about that specific topic.

This volume *New Perspectives in Basin Analysis* is the result of a symposium held in the honor of Francis J. Pettijohn. Dr. Pettijohn has been an active leader and student of sedimentary rocks and their position within a basin. Scientifically as well as industrially, modern and ancient basins have received tremendous attention, and several cycles of activity and intensity of basin studies can be discerned. To use Dr. Pettijohn's work as a catalyst and to step back to evaluate where we stand, what we know, and where we should be going are excellent reasons for a symposium.

This volume is divided into four major categories, each preceded by an introduction to that topic, including an indication as to the purpose of the contributions that follow. This publication clearly demonstrates the complexity of basin analyses, the value of some of the older techniques and their updated approaches, and what the near-future trends may be.

Karen Kleinspehn and Chris Paola organized the symposium and made use of the tremendous input provided by Dr. Pettijohn and some of his students. We are pleased that they were interested in accepting the position of Editor and that they requested this publication be part of the *Frontiers in Sedimentary Geology* series.

Arnold H. Bouma
Series Editor
Baton Rouge, Louisiana

Preface

On May 8, 1986, the University of Minnesota awarded an honorary doctorate to Francis J. Pettijohn, Professor Emeritus at the Johns Hopkins University, who received his B.S., M.S., and Ph.D. degrees from Minnesota in 1924, 1925, and 1930, respectively. Present at the ceremony were some 100 researchers from the United States, Canada, and Europe, who had assembled in Minneapolis to honor Professor Pettijohn by celebrating one of his best-known and most far-reaching accomplishments—the development of basin analysis.

The forum for this celebration was a 2-day conference entitled "New Perspectives in Basin Analysis." It was clear from the range of topics discussed at the conference that basin analysis, like any area of vigorous research, has grown a good deal since its inception in the 1940s. It was equally striking that through all this growth and diversification, two of Francis Pettijohn's guiding principles have remained as cornerstones in basin analysis: 1) the importance of an integrated approach to the study of sedimentary basins; and 2) the central importance of careful field work in basin analysis.

This book is a collection of contributions based largely on the Pettijohn symposium. In assembling it, we have adopted what we hope is the spirit of the Pettijohn approach: "Consideration of the sedimentary basin as a whole provides a truly unified approach to the study of sediments" (Potter, P.E. and Pettijohn, F.J., *Paleocurrents and Basin Analysis*. New York: Springer-Verlag, 1977, p. 1). This emphasis on an integrated approach makes our division of the contents of the book into four sections somewhat arbitrary. In general, we have separated reports dealing with the use of basin sediments to infer properties of areas outside the basin (Source-Area Characterization) from those dealing mainly with internal features of basin sediments (Lithostratigraphy and Chronostratigraphy). The broadest category, Tectonics and Sedimentation, includes studies that focus on the relation of tectonic setting, subsidence pattern, and sediment supply to basin sedimentation. Finally, we have included reports on Precambrian Basins as a separate section in recognition of Francis Pettijohn's lifelong interest in the difficult, but fundamentally important, problems of basin analysis in Precambrian rocks.

We offer this book, then, not as a comprehensive survey, but as a testimonial to the breadth, vitality, and innovativeness of research on sedimentary basins today. As a pioneering researcher, Pettijohn has given us not only his own work, but also the foundation for the work that has built upon it. As a teacher, he has given us not only his enormously influential textbooks, but also a set of guiding principles that have remained at the heart of basin analysis through four decades of growth and change. We expect that another such conference 40 years hence would find them there still.

Karen L. Kleinspehn
Chris Paola

Acknowledgments

We are grateful to many members of the University of Minnesota community for helping to organize and run the "New Perspectives in Basin Analysis" symposium on May 8–9, 1986. Lori Graven and Steve Weiland of the Department of Professional Development and Conference Services, and Kathy Ohler and Mary Reinhart of the Department of Geology and Geophysics, helped us organize the symposium. Jack Kohler designed the logo for the symposium, which also appears on the cover of this book. Dick Ojakangas and Dave Southwick arranged a post-meeting field trip to northern Minnesota, which included stops near the site of Francis Pettijohn's first field season. Financial support for the symposium was provided by the following groups within the University of Minnesota: the Office of Continuing Education and Extension, the Office of the Vice President for Academic Affairs, and the Department of Geology and Geophysics.

We thank the following reviewers for their thoughtful comments and suggestions:

E.C. Alexander	T.M. Harrison	N.D. Opdyke
L.D. Ashwal	W. Haxby	P.E. Potter
A.W. Bally	P.L. Heller	H.G. Reading
S.K. Banerjee	K.V. Hodges	F.J. Rich
C. Beaumont	P. Homewood	L.H. Royden
J. Bourgeois	D.G. Howell	J.F. Sarg
G.C. Bond	E. Ito	E.A. Silver
A.H. Bouma	R.V. Ingersoll	B.M. Simonson
J.S. Bridge	T.E. Jordan	R.C. Speed
D.W. Burbank	G.D. Karner	R.J. Steel
B.C. Burchfiel	G. de V. Klein	J.R. Steidtmann
J.L. Cisne	M.J. Kraus	J.H. Stout
M. Covey	T.F. Lawton	C.P. Teyssier
K.D. Crowley	N. Lundberg	J. Thorne
W.R. Dickinson	J.G. McPherson	J.D. Walker
R.H. Dott, Jr.	A.D. Miall	R.G. Walker
G.H. Eisbacher	G.V. Middleton	N. Wells
K.A. Eriksson	R.J. Moiola	H. Williams
J.C. Ferm	J. Namson	R.S. Yeats
J.P. Grotzinger	R.W. Ojakangas	W.L. Zhao

We thank the staff of Springer-Verlag New York Inc., for their help in organizing this book, and we thank Susan Swanson, Gary Powell, and Maureen Mullen for their typing and editorial help. We are also indebted to Arnold Bouma, editor of the Frontiers in Sedimentary Geology series, for invaluable scientific and editorial advice throughout the preparation of this book. We thank Eric Mohring and Edie McEwen for their patience and encouragement during the last 2 years.

Contents

Contributors

James A. Beer, Department of Geological Sciences, Institute for Study of the Continents, Snee Hall, Cornell University, Ithaca, NY 14853-1504, USA

Gerard C. Bond, Lamont-Doherty Geological Observatory, Columbia University, Palisades, NY 10964, USA

Arnold A. Bouma, Department of Geology, Louisiana State University, Baton Rouge, LA 70803, USA

Douglas W. Burbank, Department of Geological Sciences, University of Southern California, Los Angeles, CA 90089-0741, USA

P.F. Cerveny, Department of Geology and Geophysics, University of Wyoming, Laramie, WY 82071, USA

H. Edward Clifton, U.S. Geological Survey, 345 Middlefield Road, Menlo Park, CA 94025, USA

Sierd Cloetingh, Vening Meinesz Laboratory, Institute of Earth Sciences, University of Utrecht, 3508 TA Utrecht, The Netherlands

Elizabeth Dawson-Saunders, Department of Medical Humanities, Southern Illinois School of Medicine, Springfield, IL 62708, USA

William R. Dickinson, Laboratory of Geotectonics, Department of Geosciences, University of Arizona, Tucson, AZ 85721, USA

Rebecca J. Dorsey, Department of Geological and Geophysical Sciences, Princeton University, Princeton, NJ 08544, USA

Kenneth A. Eriksson, Department of Geological Sciences, Virginia Polytechnic Institute, Blacksburg, VA 24061, USA

Peter B. Flemings, Department of Geological Sciences, Institute for Study of the Continents, Snee Hall, Cornell University, Ithaca, NY 14853-1504, USA

Carol D. Frost, Department of Geology and Geophysics, University of Wyoming, Laramie, WY 82071, USA

James V. Gardner, U.S. Geological Survey, 345 Middlefield Road, Menlo Park, CA 94025, USA

John P. Grotzinger, Lamont-Doherty Geological Observatory, Columbia University, Palisades, NY 10964, USA

Paul L. Heller, Department of Geology and Geophysics, University of Wyoming, Laramie, WY 82071, USA

Kenneth J. Hsü, Geological Institute, ETH Zürich, CH-8092 Zürich, Switzerland

Ralph E. Hunter, U.S. Geological Survey, 345 Middlefield Road, Menlo Park, CA 94025, USA

N.M. Johnson, Department of Earth Sciences, Dartmouth College, Hanover, NH 03755, USA (Deceased)

Teresa E. Jordan, Department of Geological Sciences, Institute for Study of the Continents, Snee Hall, Cornell University, Ithaca, NY 14853-1504, USA

William S.F. Kidd, Department of Geological Sciences, State University of New York, Albany, NY 12222, USA

Karen L. Kleinspehn, Department of Geology and Geophysics, University of Minnesota, Minneapolis, MN 55455, USA

Michelle A. Kominz, Lamont-Doherty Geological Observatory, Columbia University, Palisades, NY 10964, USA

Bryan Krapez, Department of Geology, University of Western Australia, Nedlands, Western Australia, Australia 60009

Neil Lundberg, Department of Geological and Geophysical Sciences, Princeton University, Princeton, NJ 08544, USA

Earle F. McBride, Department of Geological Sciences, The University of Texas at Austin, Austin, TX 78712, USA

David S. McCormick, Lamont-Doherty Geological Observatory, Columbia University, Palisades, NY 10964, USA

Lee E. McRae, Department of Earth Sciences, Dartmouth College, Hanover, NH 03755, USA

Andrew D. Miall, Department of Geology, University of Toronto, Toronto, Ontario, Canada M5S 1A1

C.W. Naeser, U.S. Geological Survey, Mail Stop 424, P.O. Box 25046, Federal Center, Denver, CO 80225, USA

N.D. Naeser, U.S. Geological Survey, Mail Stop 424, P.O. Box 25046, Federal Center, Denver, CO 80225, USA

W. Nemec, Geological Institute (A), University of Bergen, Allégate 41, 5007 Bergen, Norway

Richard W. Ojakangas, Department of Geology, University of Minnesota, Duluth, MN 55812, USA

Chris Paola, Department of Geology and Geophysics, University of Minnesota, Minneapolis, MN 55455, USA

Robert G.H. Raynolds, AMOCO Production Company, P.O. Box 3092, Houston, TX 77253, USA

Harold G. Reading, Department of Earth Sciences, University of Oxford, Parks Road, Oxford OX1 3PR, United Kingdom

James G. Schmitt, Department of Earth Sciences, Montana State University, Bozeman, MT 59717, USA

G. Shanmugam, Dallas Research Laboratory, Mobil Research and Development Corporation, Dallas, TX 75381, USA

Khalid A. Sheikh, Branch of Stratigraphy and Paleontology, Geological Survey of Pakistan, Islamabad, Pakistan

James R. Steidtmann, Department of Geology and Geophysics, University of Wyoming, Laramie, WY 82071, USA

F.B. Van Houten, Department of Geological and Geophysical Sciences, Princeton University, Princeton, NJ 08544, USA

J. Douglas Walker, Department of Geology, University of Kansas, Lawrence, KS 66045-2124, USA

P.K. Zeitler, Research School of Earth Sciences, The Australian National University, Canberra, ACT 2601, Australia

FRANCIS J. PETTIJOHN
May 11, 1986

Thomson Dam, Minnesota.
The rocks are turbidites of the Proterozoic Thomson Formation.

Francis Pettijohn as a Teacher: Reflections on a Golden Era

H. Edward Clifton

The late 1950s may have been the best of times to become a student of sedimentary rocks. The focus of sedimentary geology was shifting from the descriptive aspects of texture, grain shape, and mineralogy to processes, facies, and paleogeography. When I began graduate studies at Johns Hopkins in 1959, the term "turbidite" was only 2 years old, and the first applications of crossbedding and other directional structures to regional paleogeography had just been completed. Francis Pettijohn's students had begun the reinterpretation of the sedimentary rocks that compose the Appalachians.

It was a heady environment. I joined, in the years 1959 to 1963, a group of Pettijohn students who would virtually all develop substantial reputations in the fields of teaching, research, or industry: Earl McBride, Norm McIver, Ralph Hunter, Bob Adams, Larry Meckel, Don Swift, Hans-Ulrich Schmincke, Lawrie Hardie, Lynton Land, and others. The intellectual stimulation generated by these Pettijohn students and the other Johns Hopkins students complemented that dispensed by the faculty. Ernst Cloos, Aaron Waters, Hans Eugster, Joe Donnay, and others continuously challenged us to do our best in a wide range of disciplines. Paul Potter arrived via a Guggenheim fellowship, and we watched firsthand as the Pettijohn–Potter team produced the *Atlas and Glossary of Primary Sedimentary Structures* and *Paleocurrents and Basin Analysis*.

The courses taught by Francis Pettijohn, to be honest, seemed unremarkable at the time. They were steeped in traditional approaches to sedimentology, and the 1957 edition of *Sedimentary Rocks* was our bible. We were required to be competent petrographers and spent long hours at the microscope. By recollection, the lectures were rather dry, but superbly organized. I occasionally refer to my class notes from 1959 and 1960 and am perpetually amazed at their content and logical flow. To this day, they summarize a remarkable amount of what is known in the field of sedimentary geology. I do recall that the lectures caught fire whenever we could disengage Francis from his lecture notes. We made concerted efforts to do so and sat spellbound until he would catch himself and return to the loose leaf binder in which he kept his course notes.

Our education hinged strongly on field studies. On field trips, we would have time to examine the rocks, then be subjected to a series of penetrating questions. We were taught to be scientifically conservative; too much conjecture was likely to be met with the comment, "All I know is what I can see in the rocks." Largely because of Pettijohn's interest in paleogeography and paleocurrents, his students developed a field sense of spatial arrangement and orientation that I think remains rare today.

Our thesis guidance was loose, but distinctly present. Francis was always available for consultation. These sessions typically consisted more of his asking questions than giving answers, and we undoubtedly profitted from this approach. Tradition was that the Pettijohn students received little editorial comment on their dissertations, and this seemed generally to be the case. Where editorial criticism was warranted, however, it came in abundance, as I discovered. I received my first real education in scientific writing as I attempted to salvage my disastrous preliminary drafts.

All students at Johns Hopkins at this time were required to be conversant with all the subject matter taught within the Department. The Pettijohn students were expected to be proficient in thermodynamics, crystallography, structural petrofabrics, and the petrogenesis of igneous and metamorphic rocks. It has always been my fancy that we could have, upon graduation, made an adequate living working in an igneous or metamorphic terrain or with any kind of rock. Indeed, the second paper that I published after graduation was on basalt flows in the southern California Coast Ranges, their directional features and paleogeographic implications. The rocks may have been crystalline, but the approach was pure Pettijohn.

One aspect of an education under Pettijohn that seemed to have received little attention was a consideration of modern sediments (nonetheless, Pettijohn students have made some signal advances in the understanding of modern deposits, perhaps in part because they approached them from the perspective of sedimentary rocks). The students themselves assigned a hierarchy to the rocks whereby orthoquartzites and pure carbonates held the highest rank and modern sediments came near the bottom. I don't know that Francis sanctioned the hierarchy, but I suspect that our prejudices derived from his predilection for Precambrian sedimentary rocks. This predoctoral sentiment seems to have been pervasive and long-lasting. I recall an incident years later when Ralph Hunter visited Willapa Bay, Washington, where I was studying modern tidal deposits. He slogged around the mudflats all afternoon, waxing increasingly disdainful. "Mud!" he sniffed. "Might as well study coal."

I like to contemplate how the Pettijohn influence is transferred to successive generations of sedimentary geologists. As we, who were taught and influenced by him, teach and influence younger geologists, it is inconceivable that his standards are not passed through to those whom we contact. I suspect that a large number of young geologists would be surprised to learn how strongly their professional lives have been touched indirectly by Francis Pettijohn.

Without question, I consider myself to have been blessed with the opportunity to study under Pettijohn at an absolutely superb point in time. Yet I strongly suspect that those who studied under him earlier at the University of Minnesota, Oberlin College, the University of Chicago, or in later years at Johns Hopkins, share the same view. If so, the "golden era" must be extended to encompass his entire teaching career.

Part I
Source-Area Characterization: Introduction

EARLE F. MCBRIDE

Source-area characterization is part of the broad topic of provenance and is one aspect of basin analysis. Provenance concerns the location of the source area from which detritus was derived, identification of the rock types exposed in the source area, and an interpretation of the climate and relief of the source. Facies analysis, grain-size trends, and paleocurrent measurements generally provide data to determine the direction from which detritus was derived (although see Heller and Frost in this section for a problematic example), but only detrital grains and fossils can provide information to help solve the other aspects of provenance. The three papers in this section of the book illustrate the remarkable details of source-area characterization that can be obtained using thin sections and certain modern analytical data.

Francis Pettijohn was early to recognize the importance of source-area characterization and basin analysis. The first edition of his classic textbook, *Sedimentary Rocks* (Pettijohn 1949), has chapters treating the detrital composition of terrigenous clastic rocks (as a key to characterizing source areas), weathering, transportation, and deposition. In the second edition (Pettijohn 1957), new chapters treat provenance and dispersal, and in the third edition in 1975 additional chapters were added on sedimentation and tectonics, and sediments and earth history. Thus, although he did not use the terms "source-area characterization" or "basin analysis," the first topic received considerable attention in all editions of *Sedimentary Rocks*, and by the third edition we see all component parts of basin analysis treated from the perspectives of the field geologist and petrographer.

In his 1984 memoirs, Francis Pettijohn expressed apprehension about the potential damage to geology

that might be accomplished through "black box" laboratory studies on samples not tied to careful field studies. All of us can cite horrible examples that verify this prediction. However, the articles in this section document the remarkable advances in source-area characterization and basin analysis that can be made with the proper blend of careful field work, thin-section petrography, and modern analytical techniques.

The paper by Dickinson expands his ideas about the value and power of petrofacies in interpreting paleotectonics and paleogeography. He shows how petrofacies data can help unravel the history of basins that receive detritus from adjacent plates that change position with time, why petrofacies of mixed tectogenetic provenance should be expected in certain basins, and the value of integrating megageomorphology with paleotectonics in order to infer past global patterns of sediment dispersal. Dickinson again demonstrates the remarkably broad interpretations that can be made from thin-section petrography that is tied to both local and regional geology.

Heller and Frost summarize how the thermal histories of detrital grains can be reconstructed using isotopic data on K-Ar, $^{40}Ar/^{39}Ar$, and possibly Rb-Sr, and how the ages of individual grains can be determined using U-Pb, Sm-Nd, and Rb-Sr isotopes. The Eocene Tyee Formation of the Oregon Coast Range is used as a case study to show the value of these methods. The Tyee Formation makes an interesting case study because petrographic and isotopic data are in conflict with facies and paleocurrent data.

Cerveny and his co-authors document the novel use of fission-track dating of individual detrital zircon grains in the Indus river sand, Pakistan, to identify the source of the zircons, and also as a clue to the uplift history of the young mountain range. When zircon cools below its "closure temperature," the fission-track clock starts running. Detrital zircons in the modern Indus River that were derived from rapidly uplifted massifs (5 m/10^3 yr) are dated at 1 to 5 Ma. The Siwalik Group sandstones deposited by the ancestral Indus River over the past 18 million years contain zircons that are only 1 to 5 million years older than the depositional ages of the sandstones. Thus, young zircons have been a consistent component of Himalayan surface exposures for the past 18 million years, implying that the relief of the Himalaya has been essentially constant over this time period and that uplift also has been essentially constant.

The three papers in this section of the book illustrate the trend of current, and I believe much future, research. The first trend is the use of "high tech" analytical techniques that permit remarkably accurate comparisons to be made between the composition and age of detrital minerals in terrigenous rocks and the composition and age of possible source terrains. These techniques are especially powerful if one is able to sample source areas that might have supplied detritus to the sedimentary formation that is being studied. The second and third papers in this section illustrate successful uses of this approach. If potential source areas have been lost by erosion, tectonics, or other plate movements, these techniques can put constraints on interpretations, but they are less helpful.

The second trend in research is the tendency to look at the history of basin filling on a global or megageologic scale. Dickinson's paper addresses this theme. The level of sophistication of rock characterization is improving, and we look at ever larger scale problems. We are interested not only in identifying the source area of terrigenous detritus, but we wish to reconstruct plate-tectonic setting, and place the history of basin development and filling in the perspective of continental evolution.

Both trends in research are creating powerful advances, yet there remain some nagging problems,

both scientific and financial. The authors of these papers are candid in stating the limitations of their respective techniques or approaches, and they report how erroneous conclusions are possible if one is not cognizant of all pertinent geological and geochemical constraints, and all of them stress the need for much additional careful work to broaden our data base and case histories. For example, Dickinson notes that the diagenetic modification of framework grains must be recognized and taken into consideration in all petrofacies studies. Failure to do so in some formations can yield significant errors of interpretation. Heller and Frost stress that not all formations are amenable to isotopic study (e.g., quartz-rich sandstones with even minor amounts of heavy minerals are potentially inappropriate for isotopic study), leakage of daughter products will yield erroneously young radiometric ages, and that isotopic techniques are expensive. People are generally willing to pay $8 to $10 for a thin section for a reconnaissance look at a problematic sample, but this reconnaissance approach is not financially feasible for isotopic studies. Projects of the scope described by Cerveny and his four co-authors generally require team efforts, and the use of expensive and highly specialized equipment. In spite of these drawbacks, investigations of these kinds must be made and will continue to be made in order to continue to improve our understanding of sedimentary basins.

References

PETTIJOHN, F.J. (1949) Sedimentary Rocks. New York: Harper and Brothers, 526 p.

PETTIJOHN, F.J. (1957) Sedimentary Rocks (2nd edition). New York: Harper and Brothers, 718 p.

PETTIJOHN, F.J. (1975) Sedimentary Rocks (3rd edition). New York: Harper and Row, 628 p.

PETTIJOHN, F.J. (1984) Memoirs of an Unrepentent Field Geologist: A Candid Profile of Some Geologists and Their Science, 1921–1981. Chicago: University Chicago Press, 260 p.

1
Provenance and Sediment Dispersal in Relation to Paleotectonics and Paleogeography of Sedimentary Basins

WILLIAM R. DICKINSON

Abstract

Provenance interpretations can be used in conjunction with other evidence to test alternate paleogeographic and paleotectonic reconstructions. Where crustal blocks have moved as parts of mobile lithospheric plates, detritus transported from one block to another may record the times during which the two blocks were adjacent. Where orogenic belts are deeply eroded, sediment shed into nearby basins may record the former existence of rock masses removed by erosion from orogenic highlands.

Sediments derived from different types of provenance terrane display contrasting petrofacies, but petrofacies of mixed provenance are common because dispersal paths connecting sediment sources to basins of deposition may be complex. Consequently, the geodynamic relations of different types of sedimentary basins as revealed by their overall morphology, structural relations, and depositional systems do not predict reliably the nature of the petrofacies that some basins contain. Adequate evaluation of sedimentary linkages between varied provenances and basins requires improved understanding of regional paleogeomorphology and an integrated view of global sediment dispersal.

Sediment dispersal is controlled by distributions of continental blocks and oceanic basins with margins of varying tectonic character, diverse climatic regimes related to paleolatitude and to changing patterns of seas and landmasses, configurations of subduction zones and associated orogenic belts, and locations of large rivers draining highlands and traversing lowlands. Conceptual models that integrate megageomorphology with paleotectonics are thus needed to infer past global patterns of sediment dispersal.

At present, the sinuous world rift system is an interconnected network of linked segments, and the two principal orogenic belts follow portions of two great circles. If the present is a key to the past, these relationships afford potential means to predict the arrangement of continental blocks, oceanic basins, drainage systems, and dispersal paths associated with global paleotectonic regimes. Provenance studies of selected sedimentary assemblages can be used to test reconstructions.

Introduction

Provenance interpretations based on analysis of detrital constituents of sedimentary rocks can be used to address paleogeographic and paleotectonic questions. Detrital modes of clastic strata may preserve unique information about provenance and sediment dispersal in cases where the present geologic setting of a sedimentary basin bears little or no resemblance to its setting when the strata preserved within it were deposited. However, realistic concepts about provenance relations require attention to the variability and complexity of sediment dispersal systems on a dynamic earth. Data on provenance cannot be interpreted properly without improved and more sophisticated models for paleogeomorphology and paleoclimatology on a global scale.

An overall approach to provenance interpretations for sedimentary basins is developed here by discussing basic ways of using provenance data for paleotectonic analysis, the importance and significance of mixed petrofacies in sedimentary basins, complex continental and oceanic paths of sediment dispersal, and the global geometry of orogenic belts and rift systems. As the global distribution of sediment sources and sedimentary basins is fundamentally controlled by tectonic patterns, a fully integrated view of paleogeography, including provenance-basin relationships, must stem from a systematic appraisal of global paleotectonics.

Constraining Paleotectonic Reconstructions

For paleotectonic analysis, data on provenance can be used both to monitor the positions of mobile crustal blocks through time, and to infer the former nature of rock masses that once formed parts of eroded crustal profiles.

The knowledge that many orogenic belts are tectonic collages (Helwig 1974) of suspect terranes (Coney *et al.* 1980), or form along sutures between colliding continents, places a premium on positive evidence for the proximity of presently adjacent crustal blocks at specific times in the past. The transfer of clastic sediment from one block to another is one of the few geologic processes that can provide a permanent record of proximity. For example, Figure 1.1 displays schematically the accretionary history of the Canadian Cordillera.

The sedimentary record shows the times at which clastic sediment was transported from one tectonic belt to another, and thus sets constraints on possible times of accretion or suturing of each belt to its present neighbors.

Where crustal blocks are deeply eroded, high-level rock bodies that existed prior to erosion may be lost forever to direct view. Examination of the sedimentary record of detritus shed from such eroded regions is one of the few ways to establish the original nature of the missing rocks. In simple cases of uplifted blocks, inverted clast stratigraphy can be used to establish the history of unroofing. In more complex cases, where evolving orogenic belts undergo continued erosion, study of derivative detritus may be the only means to establish the nature of an orogenic belt during early phases of its history.

For example, Figure 1.2 displays compositions of the dominant sandstone petrofacies within strata of a late Mesozoic forearc basin that was filled with

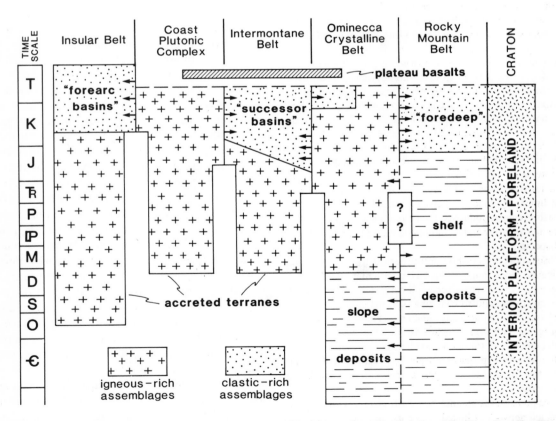

Fig. 1.1. Space and time distribution of lithologic assemblages within major tectonic belts of the Canadian Cordillera (from data of Monger *et al.* 1972, 1982). Arrows denote transport of sedimentary detritus across major tectonic boundaries; spaces between columns indicate times for which no sedimentary linkages between belts are known. Tertiary plateau basalts form a widespread overlap assemblage.

Fig. 1.2. Lower portion of ternary diagram showing mean detrital modes of dominant sandstone petrofacies in formations of different ages within the upper Mesozoic Great Valley Group (or sequence) of northern and central California (data from Ingersoll 1983). Qm = monocrystalline quartz grains; F = total feldspar grains; Lt = all lithic fragments including chert.

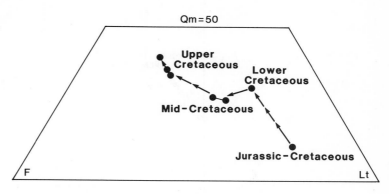

detritus derived mainly from an adjacent magmatic arc. The eroded roots of the arc orogen are now exposed in the Sierra Nevada of California. The changing character of the sandy detritus deposited within different stratigraphic intervals of the basin fill reflects the changing nature of the magmatic arc through time (Dickinson and Rich 1972). The lowest horizons contain volcaniclastic sediment delivered to the forearc basin when the magmatic arc was mantled with volcanogenic materials before the major plutons of the Sierra Nevada had been emplaced. The highest horizons contain arkosic sediment derived from the source terrane after deep erosion had exposed widespread granitic plutons that were comagmatic with the volcanogenic deposits of the arc. Ingersoll (1983) showed that the compositions of transitional petrofacies present at intermediate horizons can be correlated in detail with successive stages in the tectonic evolution of the magmatic arc, or with contrasting longitudinal segments of the arc orogen. Schwab (1981) presented analogous interpretations of the petrofacies succession observed in strata derived from a collision orogen.

Stratigraphic variations in petrofacies can be used to detect other major changes in paleotectonic or paleogeographic setting. For example, the Neogene transition from convergent to transform continental margin in California is reflected by changing petrofacies that record the lateral shift of mobile crustal blocks along the coastal region. The distribution of contrasting petrofacies allows the dating of times at which different terranes were brought into joint proximity as combined sediment sources for sedimentary sequences (Graham *et al.* 1984). In coastal Oregon, stratigraphic variations in Paleogene petrofacies date the time at which a different drainage pattern was established across a contin-

ental block to tap new sediment sources lying farther within the continental interior (Heller and Ryberg 1983).

Interpreting Detrital Modes

Figure 1.3 indicates systematic relationships that have been established between detrital modes of sandy detritus and different generic types of provenance terrane (Dickinson and Suczek 1979; Dickinson and Valloni 1980; Dickinson 1985). Different provenance types distinguished empirically on the basis of petrofacies variations in derived sediment include the following broad tectonic elements: stable cratons of continental blocks, uplifted basement masses or eroded arc plutons, active magmatic arcs of island chains or continental margins, and "recycled orogens," which include uplifted subduction complexes, collision suture belts, and back-arc fold-thrust belts. In broad terms, the five main classes of petrofacies most significant for provenance interpretations, and their most common sources, are the following (see Fig. 1.3 for meaning of symbols for grain types):

1. Quartzose: dominantly Qm with minor Qp and F (K > P); deeply weathered cratonic landmasses or recycled sediments.
2. Volcaniclastic: dominantly Lv (Lv > F) and F (P > K) with low Qm; volcanic fields of active magmatic arcs.
3. Arkosic: dominantly F (variable K/P) and Qm with low Qt; uplifted continental basement or eroded arc plutons.
4. Volcanoplutonic (Dickinson 1982): mixed Qt (Qm > Qp), F (P > K), and L (Lv > Ls); variably dissected magmatic arcs.

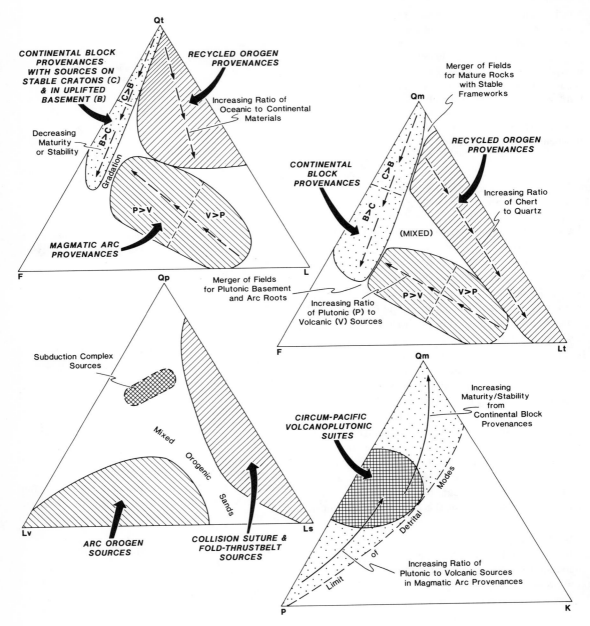

Fig. 1.3. Distributions of mean detrital modes of sandstone suites derived from different generic types of provenance terrane plotted on standard ternary diagrams (after Dickinson 1985). Symbols for grain types: Qt = total quartzose grains (Qm + Qp); Qm = monocrystalline quartz; Qp = polycrystalline quartzose lithic fragments (including chert); F = total monocrystalline feldspar grains (P + K); P = plagioclase; K = K-feldspar; Lt = total polycrystalline lithic fragments (L + Qp); L = unstable lithic fragments (Lv + Ls); Lv = volcanogenic lithic fragments (volcanic, metavolcanic, hypabyssal); Ls = sedimentary and metasedimentary lithic fragments (except chert and metachert). Detrital limeclasts (Lc) are excluded from these plots. Reproduced with permission of Reidel Publishing Company from Volume 148 of NATO ASI Series C.

5. Quartzolithic (Dickinson *et al.* 1986): mixed Qm, Qp, and Ls (variable Qt/L and Qm/Lt ratios) with minor F and Lv; uplifted strata of fold-thrust belts (Mack 1981).

Figure 1.3 places prime emphasis on the role of provenance tectonics in controlling the source and nature of framework grains in sandstones, but the secondary influences of weathering, transport, and diagenesis also affect provenance interpretations. Surficial processes of weathering and sediment transport are largely a function of paleoclimate, whereas subsurface diagenesis is controlled mainly by trends in basin evolution. However, these surficial and subsurface effects both tend to eliminate chemically less stable and physically less resistant grains, and thus to increase the proportion of quartz in the remaining grain population (Mack 1984; McBride 1985).

The tectonic stability of cratonic landmasses typically promotes the degree of weathering or reworking, or both, needed to produce quartz-rich sands. In most cases, weathering in soil horizons concentrates quartz in sand much more effectively than does abrasion or breakage of grains during sediment transport (Dickinson 1985). Although most quartz arenites are probably multicyclic in origin (Suttner *et al.* 1981), they can also be derived from intensely weathered tropical cratons during only one cycle of erosion (Franzinelli and Potter 1983). Conversely, cratons exposed in arid or glaciated regions might conceivably yield quartzo-feldspathic sands to form arkosic petrofacies whose origin could be attributed erroneously to active tectonism.

The quartz content of sands derived from tectonically active highlands is also influenced to a variable but commonly significant degree by the intensity of local weathering. Where relief is great enough, erosion into fresh bedrock may yield quartz-poor sands even where interfluves are deeply weathered (Ruxton 1970). In general, however, sands derived from comparable source rocks are more quartzose in humid regions than in arid regions (Suttner *et al.* 1981). Therefore, any petrofacies differences that involve only variations in quartz content may be caused by changes in paleoclimate rather than paleotectonics. Within an intracratonic rift basin of peninsular India, for example, fluvial sandstones derived from nearby continental basement exhibit stratigraphic variations in petrofacies that alternate between arkosic and quartzose compositions. Reference to Figure 1.3 alone would imply changes in

tectonic setting during sedimentation. In this case, however, the differences in petrofacies reflect changes in climatic regime caused by shifts in paleolatitude (Suttner and Dutta 1986).

During fluvial transport, temporary storage of sediment on exposed bars or floodplains may permit additional weathering after initial erosion but before final burial. For this reason, sandstones derived from the same highland source may display different quartz contents produced by varying degrees of weathering on alluvial plains standing at distinctly lower elevations than the source area (Houseknecht 1980). For provenance interpretations of petrofacies, distinctions between the effects of weathering in the source and weathering in transit from source to basin clearly are difficult to establish.

With allowance for the ambiguities that arise from limited understanding of the complicated interrelations between paleotectonics and paleoclimate, sandstone petrofacies remain a reliable general guide to the overall tectonic settings of most sediment provenances. Although processes of weathering and sediment transport clearly modify the composition of sedimentary detritus, recent analyses imply that the fundamental imprint of provenance tectonics is preserved in the final sedimentary products. For example, sands collected from beaches around the periphery of South America display bulk compositions expected for their respective tectonic settings despite the varied topography and climate of that large and diverse continent (Potter 1984, 1986). Recent studies show, however, that detailed variations in proportions of quartz, feldspar, and rock fragments within sandstone suites derived from the same provenance are sensitive to the combined influence of mechanical abrasion during transport and hydraulic sorting during deposition (Garzanti 1986).

Dissolution or replacement of framework grains during diagenesis of sandstones may produce residual frameworks more quartzose than initial detrital frameworks (McBride 1985). In such instances, detrital modes cannot be established with confidence unless petrographic analysis allows the effects of diagenesis to be reversed during point counting (Dickinson 1970). For example, secondary intragranular porosity must be counted as part of enclosing grains (Shanmugam 1985), and replaced grains must be recorded as they existed before replacement. If preserved replacement textures do not permit valid inferences regarding original framework composition, petrofacies useful for prove-

Table 1.1. Typical sandstone petrofacies for closely linked provenance-basin pairs.

Provenance	Petrofacies	Basin(s)
Craton Basement	Quartzose	Miogeocline, platform
uplift	Arkosic	Rift basin, Aulacogen
Island arc	Volcaniclastic	Forearc basin, Backarc basin
Continental arc	Volcanoplutonic	Forearc basin (or trench)
Fold-thrust belt	Quartzolithic	Foreland basin (or trench)

nance interpretations cannot be defined reliably. For example, McBride (1987) has recently described "diagenetic quartz arenite" in which calcite pseudomorphs (probably after feldspar) form 13% of the observed framework. Nearly all the remaining framework silicate grains are quartz, but the rock was apparently a subarkose prior to diagenesis. If the parent or host of such abundant pseudomorphs is too uncertain to specify, no detrital mode can be calculated and no petrofacies can be assigned.

Mixed Petrofacies

Table 1.1 indicates the types of sedimentary basins in which key petrofacies occur most typically when paths of sediment transport are short and direct from provenance to adjacent or nearby basin (e.g., examples given by Dickinson and Suczek 1979). The simple correlations shown do not encompass transitional tectonic settings (Mack 1984), and are not valid where dispersal systems transporting detritus from provenance to basin are complex and areally extensive (Velbel 1985).

Sandstone petrofacies in many sedimentary basins were derived from multiple sources showing complex paleotectonic and paleogeographic relationships to the basins. For example, the Upper Cretaceous and Tertiary turbidites of the Alps and Apennines display petrofacies that scatter across the whole waist of the QmFLt diagram (Fig. 1.4). Although these strata were deposited coevally in the same overall tectonic setting, the mixture of petrofacies represented cannot be ascribed to any single provenance type. Moreover, many of the sandstone suites contain petrofacies that plot near the center of the QmFLt diagram within a region not typical of detritus derived from any of the generic provenance types of Table 1.1. This region of the diagram is apparently indicative of mixed provenance (Fig. 1.3).

Figure 1.5 indicates how petrofacies similar to many of those represented within the Alpine and Apennine suites can be interpreted as mixtures of detritus from some combination of generic prove-

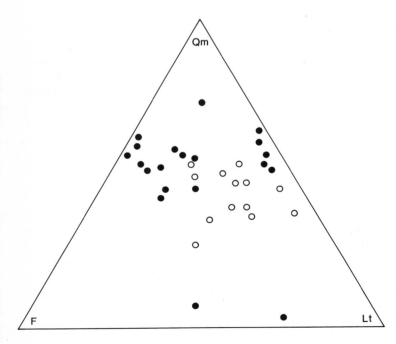

Fig. 1.4. Scatter plot of mean compositions of various sandstone suites from synorogenic turbidite and related sequences of the Alps (data from Hubert 1967) and Apennines (data from Valloni and Zuffa 1984); see Figure 1.3 for meaning of symbols for grain types, except that Lt here includes Lc for about half the Apennine suites plotted.

Fig. 1.5. Diagram to illustrate concept of mixing detritus from different provenance types to produce detrital modes reflecting mixed provenance (data from Dickinson 1985). Typical foreland-basin sand suites were derived from uplifted fold-thrust belts exposing sedimentary and metasedimentary strata. Sands from mixed suture-belt sources were derived from collision orogens of the modern Himalayan system (Ingersoll and Suczek 1979; Suczek and Ingersoll 1985) and the Carboniferous Ouachita-Marathon system (McBride 1966 as interpreted by Graham *et al.* 1975). Data on petrofacies of Hornbrook Formation (Oregon-California) from Golia and Nilsen (1984).

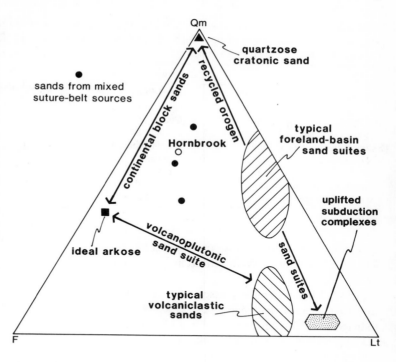

nance types. In effect, the ternary diagram can be viewed as a mixing chart, rather than a merely descriptive display. The three points for sands from mixed suture-belt sources represent suites derived from collision orogens. Along collision sutures, a variety of tectonic elements can be juxtaposed structurally and uplifted jointly: deformed continental basement with overlying miogeoclinal or platform sediments, imbricated deep-sea sediments with underlying oceanic crust of volcanogenic origin, and the partly metamorphosed massifs of evolved magmatic arcs with their volcanic cover and plutonic roots. Such an integrated provenance terrane includes essentially all the generic provenance types of Table 1.1. The points for suture-belt sources may thus reflect detrital mixtures of quartzose, arkosic, quartzolithic, volcanoplutonic, and volcaniclastic components in varying proportions. An array of different mixing models could be constructed for each, but any such models are difficult to constrain without geologic evidence independent of petrographic data.

As some orogenic belts assembled by subduction systems along active continental margins are tectonic collages of accreted terranes, petrofacies of mixed provenance may also be common in successor basins of various types in tectonic settings unrelated to major intercontinental sutures. In coastal Wash-

ington and British Columbia, for example, both marine Cretaceous and nonmarine Paleogene sandstone suites of forearc basins modified by strike-slip faulting include varied petrofacies derived from a range of provenance types (Frizzell 1979; Ward and Stanley 1982; Johnson 1984; Pacht 1984). The complex sources occupy structurally juxtaposed tectonic belts within the orogenic system along the continental margin (Fig. 1.1). Farther south, the Hornbrook Formation exposed along the border between Oregon and California contains a mixed-provenance petrofacies indistinguishable in bulk composition from sands with sources along collision suture belts (Fig. 1.5). Hornbrook detritus was derived mainly from the complex tectonic collage of the nearby Klamath Mountains, and deposited within a southern extension of the Ochoco forearc basin of central Oregon.

Petrofacies of mixed provenance need not be derived exclusively from suture belts or tectonic collages where contrasting source terranes have been placed into close proximity by tectonic processes. Comparable mixing of detrital components from varied sources might be achieved within large drainage systems of broad enough geographic extent to include widely separated tectonic elements of diverse character. The emphasis that Potter (1978a) has placed on the importance of big rivers for

Table 1.2. Possible provenance combinations and mixed petrofacies from sediment dispersal paths in complex drainage networks of big river systems.

Dispersal path	Petrofacies mix
Rift-belt source plus craton transit	Arkosic and Quartzose
Thrust-belt source plus craton transit	Quartzolithic and Quartzose
Sediment transport along suture belt and/or foreland basin	Quartzolithic and Volcano-plutonic
Sediment transport across and/or along accretionary collage	Quartzolithic and Volcaniclastic

sediment dispersal underscores this means of mixing detrital contributions from different provenance types.

Table 1.2 is a preliminary assessment of possible mixed petrofacies that might be produced by the drainage networks of big rivers with large discharges and long courses. Headwater tributaries of big rivers typically drain highlands and uplands of orogenic belts or the uplifted flanks of rift belts. Trunk streams typically flow across cratonic lowlands, along foreland trends parallel to orogenic fronts, or longitudinally within orogenic systems. Along active continental margins and island arcs, short transverse streams that follow steep courses to the sea may carry large sediment loads but tap more restricted sediment sources.

Global Dispersal Systems

An evaluation of the significance of different kinds of drainage networks for provenance studies requires an analysis of sediment dispersal systems on a global scale. The perspective of plate tectonics encourages the study of erosion and sediment transport in the framework of megageomorphology (Gardner and Scoging 1983; Douglas 1985; Baker and Head 1985), a term coined to denote the treatment of landforms from the viewpoint of the planet as a whole. For sedimentary geology, some pioneering thoughts at this scale are already available. For example, Inman and Nordstrom (1971) classified the morphology of continental coasts in terms of tectonic setting, Audley-Charles *et al.* (1977) treated the tectonic settings of major deltas, Potter (1978a) discussed tectonic controls on the locations of major trunk rivers, and Dickinson and Valloni (1980) con-

sidered the sediment sources for different kinds of ocean basins. The following discussions amplify these earlier concepts, with particular attention to implications for sediment dispersal from varied provenance types.

From a tectonic viewpoint, there are four main types of continents (Fig. 1.6):

1. Composite "Eurotype" continents, within which typical drainage systems head in collision orogenic belts (such as the Himalaya and central Asian ranges), which separate multiple cratonic lowlands (e.g., Siberia, India, Arabia in the case of modern Eurasia); major trunk rivers either flow longitudinally within and beside the mountain systems or traverse the lowlands.
2. Asymmetric "Amerotype" continents, within which typical drainage systems (such as the modern Amazon River) head in marginal orogenic belts (such as the modern Andes) and flow across cratonic lowlands toward passive continental margins; shorter rivers with smaller drainage basins flow toward active continental margins from nearby crests of mountain systems.
3. Symmetric "Afrotype" continents, within which typical drainage systems head in rift highlands of the continental interior and flow centrifugally toward passive continental margins (as in modern Africa exclusive of the Saharan region).
4. Low-lying "Austrotype" continents, within which the relief of a dominantly cratonic landmass is insufficient to support many major drainage systems (as in modern Australia).

These four continent types comprise the full spectrum of fundamentally different types possible from the standpoint of plate tectonics, and can be viewed as end members of that spectrum for paleogeographic analysis. Austrotype continents display no active plate boundaries, although the surrounding passive continental margins are inherited from earlier episodes of extensional rifting. Afrotype continents are also surrounded by passive continental margins, but active belts of extensional rifting occur within the continental interiors. Amerotype continents display both active and passive continental margins, whereas composite Eurotype continents contain collisional orogenic systems where active continental margins evolved into suture belts between once separate continental blocks.

Some paleocontinents may have included combinations of these basic end members within the same

Fig. 1.6. Major drainage systems of the four tectonic types of modern continents. Stippled pattern denotes major highlands and mountain belts. Heavy hachured lines locally delimit main parts of continental landmasses (representative of characteristic tectonic settings) from atypical peripheral portions (shown in dotted outline).

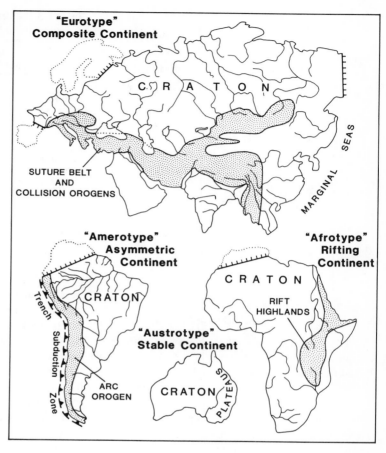

contiguous landmass, but must have been composed of constituent segments resembling various of the four types distinguished here. Indeed, modern Africa can be considered to include both an Afrotype continental segment with an active interior rift belt and a stable Austrotype continental segment; the cratonic Saharan region in the northwestern part of the continent lacks active internal rifts, yet is as large as modern Australia (Fig. 1.6).

Refinements to such a simplified catalog of continent types are clearly possible. For example, modern North America is not a typical Amerotype continent, for the Appalachian chain is formed by the eroded vestiges of an old collisional orogenic system (typical of Eurotype continents); moreover, its marginal orogenic belt has been modified by transform motions along the continental margin and by rifting in the Basin-Range province of the western interior.

Despite the variability of continental drainage networks, the mean elevation of a continent is related in a simple way to continental size, regardless of tectonic character (Fig. 1.7). This relationship confirms the fundamental geologic insight that erosional processes tend to grade any landmass to sea level. Provided their shapes are approximately equant, larger landmasses sustain higher interior elevations to maintain adequate stream gradients from headwater reaches to mouths along the shoreline.

Continental Dispersal Paths

Figure 1.6 shows that most exposed areas of continental blocks, excluding continental shelves, are occupied by drainage basins of big rivers that transport clastic sediment into sedimentary basins. The loads of these and smaller rivers are variable, of course, and no simple correlation exists between water and sediment discharges. However, Milliman and Meade (1983) showed that the 25 modern rivers with the largest sediment discharges supply about

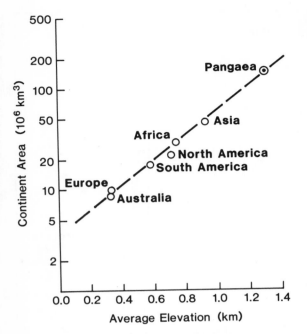

Fig. 1.7. Average continental elevation as a function of continental area (data from Hay *et al.* 1981); elevation of Pangaea inferred from its estimated total area.

Table 1.3. Modern proportions of different types of continental margins.*

Passive margins (by tectonic type of continent)[†]		Active margins (by type of plate boundary)	
Amerotype	20	Interarc	15
Eurotype	15	Arc	15
Austrotype	15	Rift	5
Afrotype	10	Transform	5
	60		40

Adapted from data of Inman and Nordstrom (1971), using terminology adapted from Audley-Charles *et al.* (1977).
*Figures are approximate percentages by cumulative global length.
[†]See Figure 1.6 for continent types.

half the total suspended load that reaches the ocean. This suspended sediment includes sand as well as silt and clay. Bed load composed of coarser sand and gravel forms a significant fraction of the total sediment load of some rivers, and is particularly important in the case of steep rivers that reach the sea along active continental margins. On a global basis, however, the total bed load discharged to the ocean is estimated to be an order of magnitude less than the total suspended load (Milliman and Meade 1983).

Coarse detritus in proximal sedimentary successions containing either marine or nonmarine conglomeratic facies is transported to basins of deposition mainly as bed load in steep streams draining nearby highlands. However, much of the debris in typical sequences of interbedded sandstone and shale forming more distal facies may come from sources far removed from the basin of deposition. As most big rivers rise in highlands distant from their mouths along the flanks of sedimentary basins, little correlation may exist between petrofacies in deltas or submarine fans and the nature of nearby bedrock. At the mouth of the Amazon River, for example, an estimated 82% of the suspended load is derived from mountainous Andean headwaters that occupy only 12% of the total area of the drainage basin (Meade

et al. 1979). The mountainous provenance lies on the opposite side of the continental craton traversed by the trunk stream (Fig. 1.6).

A global assessment of provenance patterns thus requires analysis of the megageomorphology of continental drainage basins. Taking the present as a key to the past, existing relationships can be used as a guide to overall patterns of sediment dispersal on continents. Important insights include the nature of continental margins toward which big rivers flow, and the tectonic settings that are typical for headwaters and trunk streams of big river systems.

Table 1.3 is a summary of relative lengths of different types of modern continental margins. Interarc margins refer to tectonically complex regions where belts of marginal seas occupy interarc or backarc basins separating continental blocks from offshore island arcs. Most modern big-river mouths are located on either passive continental margins or the continental shores of such marginal seas (Fig. 1.6). In this paper, the term marginal sea is used consistently and solely for marine basins of various origins lying between continental landmasses and offshore island arcs with active subduction systems.

Table 1.4 indicates the gross tectonic relations of river systems whose drainage basins are the 25 largest. By analogy with the continent types of Figure 1.6 and using parallel terminology, drainages are classified (as defined in Table 1.4) by the tectonic settings of their headwater reaches and the courses of their trunk streams. Two-thirds to three-quarters of these biggest rivers rise within or along the flanks of major orogenic belts. Nearly all the trunk streams then flow across cratons to reach passive continental margins or the flanks of marginal seas, although a few flow along the axes of foreland basins adjacent to the orogenic belts. Stream courses that traverse cratons are commonly controlled by

Table 1.4. Big river sources and courses: Provenance relations and dispersal systems of 25 largest modern drainage systems.*

Drainage class	Headwaters region	Trunk stream course	
Amerotype	Retroarc orogen	Craton transit	26
Eurotype	Collision orogen	Craton transit	24
Longitudinal	Collision orogen	Suture or foreland	18
Afrotype	Continental rift	Craton transit	18
Austrotype	Craton or plateau	Craton transit	14
			100

Adapted from data of Inman and Nordstrom (1971), using terminology adapted from Audley-Charles *et al.* (1977).
*Figures are percentages of aggregate drainage area.

major geofractures, which guide the big rivers to mouths at marginal aulacogens (Potter 1978a).

The relative paucity of modern big rivers with headwaters on cratons doubtless reflects in part the comparatively low relief characteristic of cratons, but also reflects in part the fact that the cratonic landmasses of Australia and the Saharan region are among the most arid regions of the world. This observation is a reminder that paleoclimate as well as paleotectonics should be taken into account for paleogeographic reconstructions of ancient drainage systems on paleocontinents. Bear in mind also that the transit of cratons by big rivers could not occur, to the same extent as observed now, at times when eustatic high stands submerged large portions of the cratons.

Table 1.4 implies that the mixed petrofacies suggested in Table 1.2 for big river systems are realistic expectations. The orogenic sources for Amerotype rivers are denoted as retroarc orogens, rather than as arc orogens, because such rivers rise typically within uplifted highlands developed along backarc fold-thrust belts, rather than in the volcanic chains themselves. Sources of most big rivers thus lie in recycled orogens, of either retroarc or collisional (suture) type. Consequently, many big river sands studied by Potter (1978b) are dominantly quartz-olithic (Table 1.1). Volcaniclastic and volcanoplutonic sands derived from magmatic arcs tend to be transported to active continental margins by short transverse drainages. Where the continental block behind the arc is narrow enough, however, arc-

derived detritus may reach passive continental margins (Potter 1984, 1986).

Fluvial Sediment Loads

As some big rivers transport much more sediment than others (Milliman and Meade 1983), areas of drainage basins occupying various tectonic settings are not a reliable guide to relative volumes of sediment transported by rivers. Consequently, the relative abundances of different kinds of mixed petrofacies in modern sands transported by big rivers cannot be inferred directly from Table 1.4. As the presence of a large delta is one clear indication that a given river transports large volumes of sediment to the sea, it is instructive to examine the tectonic settings and provenances of major modern deltas.

Table 1.5 indicates the distribution of the 25 largest modern deltas in terms of the types of continental margins where they are located. Nearly half occur along the flanks of marginal seas, and the remainder lie along passive continental margins. The association of large modern deltas with the shores of marginal seas poses important but largely unexplored implications for interpretations of paleogeography (Audley-Charles *et al.* 1977, 1979).

Table 1.6 indicates the overall tectonic relations of the trunk streams whose drainage basins feed the 25 largest modern deltas. Three-quarters of these streams head in collisional orogenic systems within the interior of Eurasia. About half the rivers flow across cratons to reach the sea, but the other half pursue longitudinal courses within orogenic systems or along their foreland flanks. Audley-Charles *et al.* (1977, 1979) also observed the prevalence of drainages parallel to tectonic trends in association with large modern deltas.

Table 1.5. Big delta locations: Types of continental margins where 25 largest modern deltas occur.*

Flank of "marginal sea"	40
Eurotype passive margin	35
Afrotype passive margin	15
Amerotype passive margin	10
	100

Adapted from data of Inman and Nordstrom (1971), using terminology adapted from Audley-Charles *et al.* (1977).
*Figures are percentages by area of exposed subaerial delta platforms.

Table 1.6. Big delta provenance and dispersal relations: Types of rivers feeding 25 largest modern deltas.*

Drainage class	Headwaters region	Trunk stream course	
Longitudinal	Collision orogen	Suture or foreland	45
Eurotype	Collision orogen	Craton transit	30
Afrotype	Continental rift	Craton transit	15
Austrotype	Craton or plateau	Craton transit	5
Amerotype[†]	Retroarc orogen	Craton transit	5
			100

Adapted from data of Inman and Nordstrom (1971), using terminology adapted from Audley-Charles *et al.* (1977).
*Figures are percentages by area of exposed subaerial delta platforms.
[†]Amazon delta is not included in the data set.

Table 1.7. Tectonic relations of 25 modern rivers that discharge the most suspended sediment (sand and finer grain sizes) to the oceans.*

A. Location of mouths	
Flank of "marginal sea"	40
Eurotype passive margin	30
Amerotype passive margin	25
Other tectonic settings	5
	100
B. Nature of drainages	
Longitudinal (within orogen or foreland basin)	50
Eurotype (from collision orogen across craton)	25
Amerotype (from retroarc orogen across craton)	20
Other tectonic settings (including active margins)	5
	100

Adapted from data of Milliman and Meade (1983), using terminology adapted from Audley-Charles *et al.* (1977).
*Figures are percentages of aggregate suspended-sediment discharge.

Sole emphasis on the occurrence of major deltas as indicators of rivers with large sediment loads is potentially misleading. Along the steep coasts of active continental margins, for example, rivers may deliver voluminous sediment almost directly to submarine fans deposited in deep water. In these cases, temporary storage of sediment on coastal plains is not significant, and the size of local delta platforms is not a reliable guide to the relative importance of stream loads. Measurements of suspended load at modern river mouths provide independent data bearing on the problem (Milliman and Meade 1983). The available data set must be used with caution, however, because it is not fully reliable for all big rivers, and may underestimate the contributions of smaller rivers draining steep coasts along active continental margins. Moreover, information on bed load is too fragmentary to be useful on a global scale.

Table 1.7 indicates the tectonic relations of the 25 modern rivers that discharge the most suspended sediment to the ocean. Comparison with Tables 1.5 and 1.6 for large deltas shows that the same principal inferences can be drawn from either data set. Approximately three-quarters of the big rivers with prominent sediment loads head in orogenic highlands and pursue longitudinal or Eurotype courses to passive continental margins or the flanks of marginal seas.

Table 1.8 displays calculated global data for estimated total suspended sediment delivered to the ocean by modern rivers of all sizes. Longitudinal and Eurotype drainages, mostly those of big rivers, are the most important for sediment transport, and

together account for half the total estimated volume. A quarter to a third of the total suspended sediment is apparently contributed by small rivers draining island arcs and active continental margins.

The foregoing appraisal of the tectonic settings and provenance relations of big rivers, large deltas, and sediment-laden rivers support the following inferences:

Table 1.8. Estimated proportions of suspended sediment discharged to modern oceans from different types of drainage basins.*

Longitudinal drainages within orogenic belts or along trends of adjacent foreland basins	30
Archipelagic, isthmian, and penisnsular highlands of island arcs and related tectonically active belts	24
Eurotype drainages reaching passive continental margins or flanks of marginal seas from headwaters in collision orogens	20
Amerotype drainages reaching passive continental margins from headwaters in retroarc orogens	12
Afrotype drainages reaching passive continental margins from headwaters in rift belts	5
Austrotype drainages reaching passive continental margins from headwaters in cratonic areas	5
Transverse drainages reaching active continental margins from headwaters in continental arcs	4
	100

Adapted from data of Milliman and Meade (1983), using terminology adapted from Audley-Charles *et al.* (1977).
*Figures are approximate percentages of global total.

1. Large drainage systems with large sediment loads typically head in orogenic belts and flow toward passive continental margins or the flanks of marginal seas.
2. The trunk streams of such drainage systems are about equally likely to flow across cratons to distant continental margins, or longitudinally within the interiors or along the flanks of the orogenic belts.
3. As the orogenic sources are dominantly recycled orogens, the large drainage systems typically supply dominantly quartzolithic detritus to sedimentary basins.
4. Sources in magmatic arcs are not important for many large drainage systems, because the proximity of continental arc orogens to active continental margins and the insular location of other magmatic arcs encourages sediment dispersal mainly through small local drainage systems.
5. Although available data do not allow an accurate figure to be specified, more than half of all modern fluvial sand is probably of mixed petrofacies.

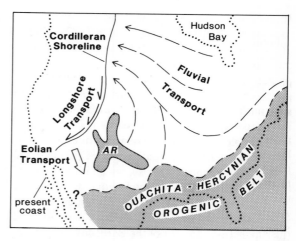

Fig. 1.8. Permian sediment dispersal system dominating the Cordilleran margin of North America, then a part of Pangaea. Adapted from unpublished paleogeographic reconstructions by Jordan (1979) and S.J. Johansen (personal communication, 1986). Compare with more general map in Dott and Batten (1981, p. 335). AR = Ancestral Rockies Province.

Marine Dispersal Paths

A global perspective on sediment dispersal cannot be achieved by attention to continental dispersal paths alone. Processes of sediment transport within marine basins, either along continental shelves or within turbidite systems on the seafloor, form an integral part of the total picture. Oceanic basins can be subdivided into generic compartments (Dickinson and Valloni 1980) that lie adjacent to different kinds of continental margins. Where oceanic crust forms part of the same lithosperic plate that contains an adjacent continent, oceanic dispersal paths can be regarded geometrically as extensions of continental dispersal paths. For example, submarine fans built off passive continental margins should contain petrofacies representative of those delivered to the sea at big river mouths. However, tectonic transport of the seafloor itself can carry some sedimentary assemblages far from their original sites of deposition. In this way, seafloor turbidites can be placed in close juxtaposition with continental blocks or island arcs unrelated to their provenance.

As an illustration of the complexity introduced into provenance analysis by combined continental and marine dispersal paths, Figure 1.8 is a provisional paleogeographic reconstruction of Permian North America. The diagram shows a mixed dispersal system along the Cordilleran margin of the continent. Sediment derived from fold-thrust highlands of the Ouachita and Appalachian segments of the Hercynian orogenic belt, and from basement uplifts of the Ancestral Rockies, was delivered to the continental margin somewhere in the region of the present Northern Rockies. Longshore shelf transport carried sediment southward (Scott 1965), in present coordinates, into the region of the present Colorado Plateau. Winds then piled large volumes of shelf-derived sediment into immense dune fields (Blakey and Middleton 1983), which migrated back toward the Marathon extension of the Ouachita orogenic belt. The distant provenance of the dune sands cannot be determined without understanding the complex dispersal path in the context of regional paleogeography.

Petrofacies of accretionary subduction complexes present special challenges for provenance interpretation. On the one hand, Quaternary turbidites deposited along 2000 km of the modern Chile Trench display a range of volcanoplutonic petrofacies reflecting derivation from various sources in the magmatic arc system that lies along the adjacent active continental margin (Thornburg and Kulm 1987). Many subduction complexes contain

voluminous sandstones composed of similar volcanoplutonic petrofacies (Dickinson 1982). On the
other hand, Velbel (1985) demonstrated that plate
motions transport oceanic turbidites of diverse
provenance into other subduction zones. Where
such turbidites of varied origin are incorporated into
subduction complexes, their petrofacies may be
incongruous with respect to nearby magmatic arcs.
For example, quartzose and quartzolithic sands derived from northeastern South America are dispersed
to an offshore continental rise, which is then carried
tectonically into the subduction zone of the Lesser
Antilles island arc, where the turbidites are off-
scraped from the descending oceanic plate to build
the subduction complex of the Barbados Ridge (Velbel 1985). Recent work indicates that Paleogene turbidites of the Barbados Ridge were derived mainly
from longitudinal drainage of an evolving orogenic
system along tectonic strike in South America
(Kasper and Larue 1986).

Figure 1.9 shows ways in which turbidites can be
delivered to subduction zones from provenances that
lie along tectonic strike by a combination of
sedimentological and tectonic processes. In both
cases, tectonic processes are by far the most important volumetrically. In Figure 1.9A, turbidites of
mixed provenance from a collision orogen (Fig. 1.5)
are dispersed longitudinally along the trench floor.
However, much larger volumes of turbidites from
the same provenance are transported laterally into
the subduction zone by post-depositional tectonic
movement of a large submarine fan, whose flank is
being subducted by continued plate motion. In
Figure 1.9B, lithofeldspathic sands of volcanoplutonic character (Fig. 1.5) are dispersed to the
seafloor from a complex orogenic belt including the
eroded roots of an arc orogen along the continental
margin. Some of the sands that reach the subduction
zone are dispersed longitudinally along the trench
floor by sedimentological processes. However,
much larger volumes are contained in seafloor turbidites that are being carried to the subduction zone by
lateral tectonic transport parallel to the transform
system that bounds the continental block.

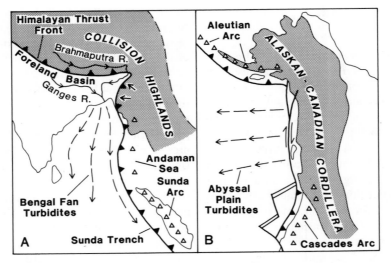

Fig. 1.9. Diagrammatic maps to illustrate tectonic transport of seafloor turbidites of distal provenance into subduction zones associated with magmatic arcs (*cf.* Velbel
1985). In A (left), Bengal Fan turbidites (Ingersoll and
Suczek 1979) derived from the composite provenance of
the Himalayan orogenic belt are transported longitudinally
(dashed arrows) down the axis of a remnant ocean basin in
the Bay of Bengal, and then carried tectonically into the
subduction zone of the Sunda magmatic arc (Graham *et al.*
1975) to be accreted and recycled. In B (right), turbidites
of the Alaskan abyssal plain in the northeast Pacific
(Stewart 1976) derived from the complex collage-like orogenic provenance of the Canadian Cordillera are transported (dashed arrows) into the ocean basin across a
transform bounding the continental block, and then carried tectonically into the subduction zone of the Aleutian
magmatic arc (Dickinson 1982; MacKinnon 1983) to
become incorporated within the continental margin of
southern Alaska. Atlantic seafloor turbidites derived from
South America are carried tectonically into the subduction zone of the Lesser Antilles magmatic arc in comparable fashion (Fig. 3 of Velbel 1985).

Fig. 1.10. Paleotectonic sketch map of southeastern North America in Carboniferous time to illustrate provenance relations of turbidites in "Ouachita flysch" (deformed synorogenic sequence). Heavy arrow indicates net paleocurrent vector for turbidites in exposed overthrust masses of Ouachita Mountains.

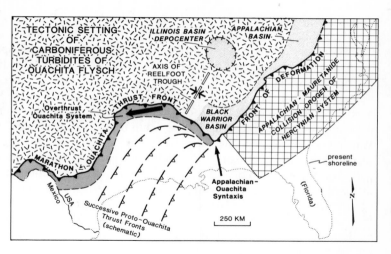

Remnant Ocean Turbidites

As many previous authors have noted, longitudinal dispersal systems of mixed fluvial and marine character are especially important in the evolution of orogenic belts. Plate tectonics explains their significance in terms of the sequential closure of remnant ocean basins by subduction as continental collision propagates along strike (Graham *et al.* 1975). Turbidites derived from longitudinal drainages flowing along evolving suture belts travel down the axial trends of remnant ocean basins to form synorogenic "flysch" successions, which are deformed and incorporated into the growing orogenic belts shortly after their deposition. Figure 1.10 applies this concept to Carboniferous turbidites of the "Ouachita flysch."

Graham *et al.* (1975) suggested that the thick (~ 10 km) Carboniferous turbidite succession of the Ouachita system can be understood as the sedimentary record of large submarine fans deposited during sequential closure of a remnant ocean basin lying south of the North American craton. The collision system involved was the Hercynian orogenic belt developed between Laurasia (or Laurussia) and Gondwanaland as Pangaea was assembled. Longitudinal transport of the turbidites westward within the evolving orogenic system, earlier documented by Cline (1970), was analogous to the longitudinal transport of Bengal Fan turbidites into the modern remnant ocean of the Bay of Bengal (Fig. 1.9A). The distribution of turbidite subfacies within the Ouachita system supports this concept (Moiola and Shanmugam 1984). Moreover, the compositional similarity of quartzolithic petrofacies in the

Ouachita turbidites and coeval fluviodeltaic deposits of the Black Warrior Basin (Fig. 1.10) is compatible with derivation of both sequences from the same orogenic sources lying still farther east (Graham *et al.* 1976). As sequential continental collision proceeded, the Ouachita turbidites were incorporated into thrust sheets that were emplaced upon the southern margin of the continent (Wickham *et al.* 1976).

In proposing a Himalayan-Bengal model (Fig. 1.9A) for the Appalachian-Ouachita system (Fig. 1.10), Graham *et al.* (1975, 1976) placed emphasis on an inferred dispersal system that involved fluvial drainages flowing longitudinally, either within the Appalachian-Mauretanide segment of the Hercynian orogen or along the Appalachian foreland basin. This terrestrial drainage was inferred to pass through a deltaic complex in the Black Warrior Basin into a large turbidite fan that spread Ouachita turbidites westward along the floor of the remnant ocean basin lying south of the continent. Contributions of quartzose sand were assumed to have been added to the quartzolithic sand of orogenic provenance by drainages passing through the Illinois Basin (Fig. 1.10) or across adjacent parts of the craton.

Subsequent detailed analysis of both lithofacies and petrofacies has shown, however, that Carboniferous detritus entered the Black Warrior Basin from the southwest as well as from the northeast (Mack *et al.* 1981, 1983; Thomas and Mack 1982; Owen and Carozzi 1986). This detritus evidently was derived from a proto-Ouachita orogen that began to develop as the collisional orogenic system propagated westward from the Appalachian-Ouachita syntaxis (Fig. 1.10). Recognition of its presence in fluviodeltaic

strata implies that the Black Warrior Basin, which
lies within the cusp of the syntaxis (Fig. 1.10),
received contributions of sand from around its entire
periphery. Much of the sand that bypassed the Black
Warrior deposystems to form the Ouachita turbidites
was thus presumably a product of mixed provenance. Additional contributions of sediment from
the rising proto-Ouachita orogen lying south of the
remnant ocean basin doubtless augmented the sand
supply to the closing trough, as did sediment transported off the craton to the north. The appreciation
that proto-Ouachita as well as Appalachian sources
contributed detritus to the Black Warrior Basin further enhances the suggested analogy with the
modern Himalayan-Bengal example. Sediment
sources for the Bengal Fan include the Indo-Burman
Ranges east of peninsular India, as well as the main
Himalayan chain (Fig. 1.9A).

Global Provenance Geometry

The overall geometry of orogenic belts produced by
subduction, and of midocean ridges produced by
rifting, is fundamental for provenance analysis on

a global scale. The variability of continent types,
the geometry of sediment dispersal paths, and the
tectonic transport of seafloor are all controlled ultimately by patterns of plate tectonics. The distribution of different types of continents is controlled
mainly by the locations of the subduction zones and
orogenic systems in relation to continental blocks.
Furthermore, the distribution of arc orogens and
collision orogens is controlled by the changing positions of continental blocks with respect to evolving
subduction zones. Sediment dispersal paths are controlled largely by the locations of ocean basins with
respect to the positions of orogenic belts. The
development of ocean basins is controlled in turn by
the global pattern of rift systems and oceanic spreading centers in relation to subduction zones where
oceanic plates are consumed.

Effective reconstruction of global paleogeography
thus depends upon the recognition of systematic
relationships among various kinds of plate boundaries. Where complex dispersal paths are involved in
sediment transport, important questions about provenance cannot be addressed by analysis of local relations alone. On the contrary, understanding the
megageomorphology of whole continents and ocean

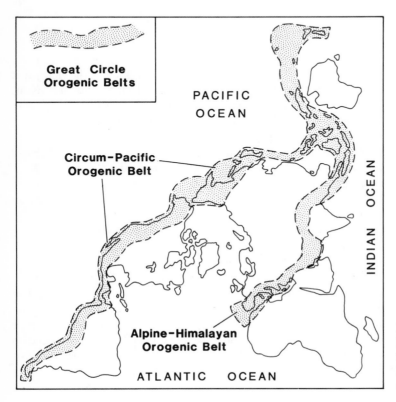

Fig. 1.11. Circum-Pacific and Alpine-
Himalayan orogenic belts depicted on
modified circular projection to high-
light their linear continuity as segments
of two true great circles on the globe.

Fig. 1.12. World rift system of oceanic spreading centers plotted on standard Mercator projection to illustrate its longitudinal continuity and relationship to major orogenic belts.

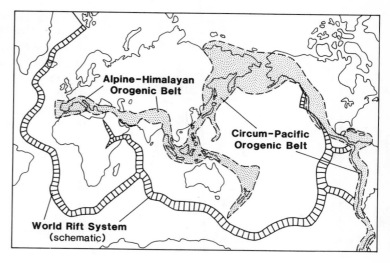

basins may be required to specify a realistic set of alternate working hypotheses for the provenance of a local sedimentary assemblage.

The two great orogenic belts of the modern world (Fig. 1.11) are sinuous in detail, but each is aligned along a true great circle on the globe (de Montessus de Ballore 1903; Wilson 1954; Dickinson *et al.* 1986). The two intersect in the tectonically complex Indonesian region (Hamilton 1979). The world rift system (Heezen 1960) is composed of branching segments that are linked together into a continuous network that interlocks geometrically with the two great orogenic belts (Fig. 1.12). Major arms of the world rift system evidently propagate longitudinally, as the Atlantic spreading system is now propagating into the Arctic region, and the Indian Ocean system into the Red Sea and East Africa. Major plate boundaries today thus form a remarkably systematic array.

The geodynamic processes that control the formation of extensional backarc basins occupied by marginal seas remain uncertain (Taylor and Karner 1983). The question is important for provenance studies because so many large deltas and the mouths of so many big rivers are presently located along the continental flanks of marginal seas. Backarc spreading to separate fringing island arcs from the edges of continental blocks and backarc thrusting to develop fold-thrust belts and foreland basins between cratons and continental-margin arcs are opposite styles of geodynamic behavior (Chase 1978; Dickinson 1978; Molnar and Atwater 1978; Uyeda and Kanamori 1979; Uyeda 1982; Cross and Pilger 1982). Backarc thrusting reflects structural coupling between descending and overriding plates to induce net con-

traction between the continental interior and the subduction zone. Backarc spreading reflects a degree of decoupling to allow the arc-trench system to migrate away from the continental block in the overriding plate.

Nelson and Temple (1972) observed that backarc spreading is most common for westward-underthrust, eastward-facing arc-trench systems, whereas the reverse is true for backarc thrusting. They ascribed this asymmetric geodynamic behavior of modern orogenic systems to systematic east-west differential movement between lithosphere and asthenosphere. Figure 1.13 illustrates the remarkably asymmetric distribution of the two types of orogenic systems in the modern world. Figure 1.14 depicts the hypothesis that net westward drift of lithosphere with respect to asthenosphere is responsible for the contrasting behavior of arc-trench systems with different geographic orientations.

The diagram is based upon the supposition that slabs of lithosphere inserted into the asthenosphere acts as anchors to retard net westward drift of lithosphere with respect to underlying asthenosphere. The overriding plates in west-facing systems are able to advance on "trench hinges," where slabs bend to descend into the mantle, and structural coupling promotes compressional tectonics. The overriding plates in east-facing systems pull away from anchored "trench hinges," and marginal seas occupy backarc basins formed by consequent spreading. Systematic advance and retreat of the overriding plates give rise to an asymmetric ocean, analogous to the modern Pacific, in which the

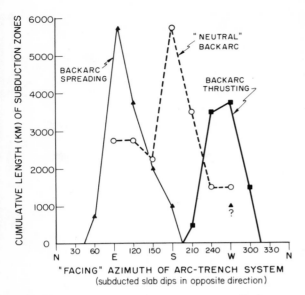

Fig. 1.13. Plot showing relationship between Cenozoic backarc tectonics and geographic orientation of associated arc-trench system (after Dickinson 1980); "facing azimuth" is given by a line drawn normal to the subduction zone pointing toward the plate being subducted (i.e., arc "faces" toward trench). Queried point represents Andaman Sea behind west-facing arc, but reflects atypical transtensional opening by strike slip. Reproduced with permission from Geological Association of Canada Special Paper 20.

spreading center is displaced from a mid-oceanic position. Younger oceanic lithosphere on the eastern flank of such an ocean may then be buoyant enough to subduct at a shallow angle that enhances the degree of coupling with the overriding plate, whereas older oceanic lithosphere on the western flank of the same ocean may subduct at a comparatively steep angle that further facilitates decoupling and backarc spreading.

If the present is a key to the past, perhaps global paleotectonic and paleogeographic reconstructions should be based on the actualistic hypotheses that major orogenic belts follow great circle trends, that major rifts propagate longitudinally from preexisting branches of the world rift system, that backarc spreading occurs where arc orogens face east, and that backarc thrusting occurs where arc orogens face west. If so, positions of past subduction zones, continental rifts, and arc-trench systems of varying character cannot be inferred arbitrarily apart from their global context. To date, however, such global considerations have played little role in constraining paleotectonic and paleogeographic reconstructions for local regions. Consequently, important alternatives for provenance interpretations may have been ignored.

The methodology to undertake more systematic analysis of global paleotectonics is available. Shifting positions of the continental blocks can be inferred from the geometry of seafloor magnetic anomalies and apparent paleowander paths for the

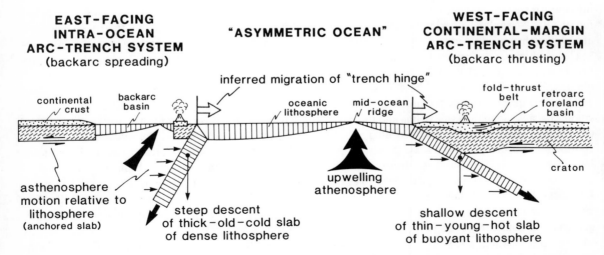

Fig. 1.14. Diagram to illustrate inferred systematic asymmetry in tectonic behavior of east-facing and west-facing arc-trench systems in response to hypothetical net westward drift (to the left in the figure) of lithosphere with respect to asthenosphere.

continents (Smith and Briden 1977; Kanasewich *et al.* 1978; Scotese *et al.* 1979; Smith *et al.* 1981). Although specific paleopoles provide direct information about paleolatitude and azimuthal orientation only, sophisticated analyses of apparent paleowander paths yield insight into paleolongitude as well (Gordon *et al.* 1984; Livermore *et al.* 1986). Alternate paleocontinental reconstructions can be evaluated by cladistic methods using branching diagrams to represent possible sequences of continental fragmentation and assembly (Young 1986). The morphology of various paleogeographic elements can be treated in terms of a discrete number of paleorelief categories (Ziegler *et al.* 1985). The new discipline of paleoceanography permits specific inferences about the paleobathymetry of past ocean basins (Sclater *et al.* 1977). The capability of modern computer graphics provides a tool potentially able to coordinate multiple data sets in routine ways.

Research programs designed to test the provenance implications of diverse paleotectonic and paleogeographic reconstructions may produce several benefits. Comparative study of petrofacies derived from the same generic types of provenance in different paleolatitudes and paleoclimatic settings could improve understanding of the influence of weathering and sediment transport on detrital modes. Detrital modes of sandstones at different stratigraphic horizons in sediment deposited along passive continental margins or within adjacent ocean basins might be used to monitor tectonic phases or the progress of erosion in orogenic belts along distant active continental margins and suture belts. Investigation of stratigraphic variations in petrofacies within appropriate sedimentary basins might help document successive stages in the assembly of a supercontinent such as Pangaea. During Cenozoic time, for example, Eurasia probably evolved from a generally Amerotype configuration to its present Eurotype configuration as continental collisions along its southern margin converted arc orogens to collision orogens, and backarc spreading along its eastern margin converted continental-margin arcs to island arcs.

Conclusions

Consideration of global patterns of sediment dispersal leads to the following general conclusions impor-

tant for paleogeographic and paleotectonic analysis of sedimentary basins:

1. Where sediment dispersal paths between orogenic provenances and adjacent sedimentary basins are short and direct, stratigraphic variations in sandstone petrofacies offer an effective means to monitor the tectonic evolution of the orogens through time.

2. Although major differences in sandstone petrofacies most commonly reflect provenance tectonics, the effects of weathering, sediment transport, and diagenesis also influence framework composition to varying degrees.

3. Petrofacies of mixed provenance may reflect derivation from complex orogens where varied source terranes are juxtaposed tectonically, but also may be produced by mixing of sediment from varied sources exposed within large drainage systems of big rivers.

4. Depending upon the tectonic framework of different types of continents, big rivers typically head in orogenic belts or the uplifted flanks of rift belts, and flow either parallel to major tectonic trends or across cratons to the sea.

5. A majority of modern big rivers head in collision or retroarc orogens, and transport dominantly quartzolithic detritus to passive continental margins or the continental flanks of marginal seas.

6. Sediment-laden rivers that have built the largest modern deltas and deliver the most suspended sediment to the ocean typically head in collision orogens and pursue longitudinal courses, either within orogenic highlands or along the axes of adjacent foreland basins.

7. Although many subduction complexes contain mainly volcanoplutonic petrofacies derived from adjacent magmatic arcs, others are composed of sedimentary assemblages derived from more diverse sources and transported tectonically to subduction zones by plate motions involving underlying oceanic crust.

8. Deformed sedimentary assemblages exposed in many collision orogens include voluminous remnant-ocean turbidites composed of detritus transported longitudinally from orogenic highlands that developed along tectonic strike within the evolving orogenic systems.

9. Improved reconstructions of global paleotectonic patterns that controlled paths of sediment

dispersal can be achieved by using the distribution of different kinds of modern plate boundaries as an actualistic guide for geometric analysis.

10. Provenance interpretations for sedimentary assemblages can be addressed most effectively in the context of global paleogeographic patterns inferred from paleotectonic reconstructions, and can be used to test such reconstructions.

Acknowledgments. The work of F.J. Pettijohn and his students has long inspired my thinking about sediment dispersal, P.E. Potter turned my attention to dispersal systems as viewed on a grand scale, and V.R. Baker introduced me to the concept of megageomorphology. Joint work with P.L. Heller and M.A. Klute improved my thoughts on mixed provenance, as did discussions about the sedimentary tectonics of the Apennines with Gian Luca Ferrini, Eduardo Garzanti, Fabio Lentini, Franco Ricci Lucchi, Renzo Valloni, and Gian Gaspare Zuffa. A preprint by S.J. Johansen improved my understanding of lateral dispersal paths along continental shelves. I thank Karen Kleinspehn and Chris Paola for encouragement to complete the manuscript, which was prepared with support from the Laboratory of Geotectonics in the Department of Geosciences at the University of Arizona. All figures were drafted by Rick Brokaw. Reviews by J.A. Bourgeois and G.deV. Klein improved the text.

References

AUDLEY-CHARLES, M.G., CURRAY, J.R., and EVANS, G. (1977) Location of major deltas. Geology 5:341–344.

AUDLEY-CHARLES, M.G., CURRAY, J.R., and EVANS, G. (1979) Significance and origin of big rivers: A discussion. Journal Geology 87:122–123.

BAKER, V.R. and HEAD, J.W., III (1985) Global megageomorphology. In: Hayden, R.S. (ed) Global Mega-Geomorphology. National Aeronautics Space Administration Conference Publication 2312, pp. 113–120.

BLAKEY, R.C. and MIDDLETON, L.T. (1983) Permian shoreline eolian complex in central Arizona. In: Brookfield, M.E. and Ahlbrandt, T.S. (eds) Eolian Sediments and Processes. Amsterdam: Elsevier, Developments in Sedimentology 38, pp. 551–581.

CHASE, C.G. (1978) Extension behind island arcs and motions relative to hot spots. Journal Geophysical Research 83:5385–5387.

CLINE, L.M. (1970) Sedimentary features of late Paleozoic flysch, Ouachita Mountains, Oklahoma. In: Lajoie, J. (ed) Flysch Sedimentology in North America, Geological Association Canada Special Paper 7, pp. 85–101.

CONEY, P.J., JONES, D.L., and MONGER, J.W.H. (1980) Cordilleran suspect terranes. Nature 288: 329–333.

CROSS, T.A. and PILGER, R.H., JR. (1982) Controls of subduction geometry, location of magmatic arcs, and tectonics of arc and back-arc regions. Geological Society America Bulletin 93:545–562.

DE MONTESSUS DE BALLORE, F. (1903) Sur l'existence de deux grands cercles d'instabilité sismique maxima. Academie Sciences France Compte Rendu 136: 1707–1709.

DICKINSON, W.R. (1970) Interpreting detrital modes of graywacke and arkose. Journal Sedimentary Petrology 40:695–707.

DICKINSON, W.R. (1978) Plate tectonic evolution of north Pacific rim. Journal Physics Earth 26:S1–S19.

DICKINSON, W.R. (1980) Plate tectonics and key petrologic associations. In: Strangway, D.W. (ed) The Continental Crust and its Mineral Deposits. Geological Association Canada Special Paper 20, pp. 341–360.

DICKINSON, W.R. (1982) Compositions of sandstones in circum-Pacific subduction complexes and fore-arc basins. American Association Petroleum Geologists Bulletin 66:121–137.

DICKINSON, W.R. (1985) Interpreting provenance relations from detrital modes of sandstones. In: Zuffa, G.G. (ed) Provenance of Arenites. Dordrecht, Holland: Reidel, pp. 333–361.

DICKINSON, W.R., LAWTON, T.F., and INMAN, K.F. (1986) Sandstone detrital modes, central Utah foreland: Stratigraphic record of Cretaceous-Paleogene tectonic evolution. Journal Sedimentary Petrology 56:276–293.

DICKINSON, W.R. and RICH, E.I. (1972) Petrologic intervals and petrofacies in the Great Valley Sequence, Sacramento Valley, California. Geological Society America Bulletin 83:3007–3024.

DICKINSON, W.R. and SUCZEK, C.A. (1979) Plate tectonics and sandstone compositions. American Association Petroleum Geologists Bulletin 63:2164–2182.

DICKINSON, W.R., SWIFT, P.N., and CONEY, P.J. (1986) Tectonic strip maps of Alpine-Himalayan and Circum-Pacific orogenic belts (great circle projections). Geological Society America Map Chart Series MC-58, 1:20,000,000.

DICKINSON, W.R. and VALLONI, R. (1980) Plate settings and provenance of sands in modern ocean basins. Geology 8:82–86.

DOTT, R.H., JR. and BATTEN, R.L. (1981) Evolution of the Earth (3rd edition). New York: McGraw-Hill, 573 p.

DOUGLAS, I. (1985) Global megageomorphology. In: Hayden, R.S. (ed) Global Mega-Geomorphology. National Aeronautics Space Administration Conference Publication 2312, pp. 10–17.

FRANZINELLI, E. and POTTER, P.E. (1983) Petrology, chemistry, and texture of modern river sands, Amazon River system. Journal Geology 91:23–40.

FRIZZELL, V.A., JR. (1979) Point count data and sample locations for selected samples from Paleogene nonmarine sandstones, Washington. U.S. Geological Survey Open-File Report, 79–293, 30 p.

GARDNER, R. and SCOGING, H. (1983) Mega-Geomorphology. Oxford, England: Clarendon Press, 240 p.

GARZANTI, E. (1986) Source rock versus sedimentary control on the mineralogy of deltaic volcanic arenites (Upper Triassic, northern Italy). Journal Sedimentary Petrology 56:267–275.

GOLIA, R.T. and NILSEN, T.H. (1984) Sandstone petrography of the Hornbrook Formation, Oregon and California. In: Nilsen, T.H. (ed) Geology of the Upper Cretaceous Hornbrook Formation, Oregon and California. Society Economic Paleontologists Mineralogists, Pacific Section, Book 46, pp. 99–109.

GORDON, R.G., COX, A., and O'HARE, S. (1984) Paleomagnetic Euler poles and the apparent polar wander and absolute motion of North America since the Carboniferous. Tectonics 3:499–537.

GRAHAM, S.A., DICKINSON, W.R., and INGERSOLL, R.V. (1975) Himalayan-Bengal model for flysch dispersal in the Appalachian-Ouachita system. Geological Society America Bulletin 86:273–286.

GRAHAM, S.A., INGERSOLL, R.V., and DICKINSON, W.R. (1976) Common provenance for lithic grains in Carboniferous sandstones from the Ouachita Mountains and Black Warrior Basin. Journal Sedimentary Petrology 46:620–632.

GRAHAM, S.A., McCLOY, C., HITZMAN, M., WARD, R., and TURNER, R. (1984) Basin evolution during change from convergent to transform continental margin in central California. American Association Petroleum Geologists Bulletin 68:233–248.

HAMILTON, W. (1979) Tectonics of the Indonesian Region. United States Geological Survey Professional Paper 1078, 345 p.

HAY, W.W., BARRON, E.J., and SLOAN, J.L., II (1981) Continental drift and the global pattern of sedimentation. Geologische Rundschau 70:302–315.

HEEZEN, B.C. (1960) The rift in the ocean floor. Scientific American 203:98–110.

HELLER, P.L. and RYBERG, P.T. (1983) Sedimentary record of subduction to forearc transition in the rotated Eocene basin of western Oregon. Geology 11:380–383.

HELWIG, J. (1974) Eugeosynclinal basement and a collage concept of orogenic belts. In: Dott, R.H., Jr. and Shaver, R.R. (eds) Modern and Ancient Geosynclinal

Sedimentation. Society Economic Paleontologists Mineralogists Special Publication 19, pp. 359–376.

HOUSEKNECHT, D.W. (1980) Comparative anatomy of a Pottsville lithic arenite and quartz arenite of the Pocahontas Basin, southern West Virginia: Petrogenetic, depositional, and stratigraphic implications. Journal Sedimentary Petrology 50:3–20.

HUBERT, J.F. (1967) Sedimentology of Prealpine flysch sequences, Switzerland. Journal Sedimentary Petrology 37:885–907.

INGERSOLL, R.V. (1983) Petrofacies and provenance of late Mesozoic forearc basin, northern and central California. American Association Petroleum Geologists Bulletin 67:1125–1142.

INGERSOLL, R.V. and SUCZEK, C.A. (1979) Petrology and provenance of Neogene sand from Nicobar and Bengal fans, DSDP sites 211 and 218. Journal Sedimentary Petrology 49:1217–1228.

INMAN, D.L. and NORDSTROM, C.E. (1971) On the tectonic and morphologic classification of coasts. Journal Geology 79:1–21.

JOHNSON, S.Y. (1984) Stratigraphy, age, and paleogeography of the Eocene Chuckanut Formation, northwest Washington. Canadian Journal Earth Sciences 21:92–106.

JORDAN, T.E. (1979) Evolution of the late Pennsylvanian and early Permian western Oquirrh Basin, Utah (Ph.D. dissertation). Stanford, California: Stanford University, 253 p.

KANASEWICH, E.R., HAVSKOV, J., and EVANS, M.E. (1978) Plate tectonics in the Phanerozoic. Canadian Journal Earth Sciences 15:919–955.

KASPER, D.C. and LARUE, D.K. (1986) Paleogeographic and tectonic implications of quartzose sandstones of Barbados. Tectonics 5:837–854.

LIVERMORE, R.A., SMITH, A.G., and VINE, F.J. (1986) Late Palaeozoic to early Mesozoic evolution of Pangaea. Nature 322:162–165.

MACK, G.H. (1981) Composition of modern stream sand in a humid climate derived from a low-grade metamorphic and sedimentary foreland fold-thrust belt of north Georgia. Journal Sedimentary Petrology 51:1247–1258.

MACK, G.H. (1984) Exceptions to the relationship between plate tectonics and sandstone composition. Journal Sedimentary Petrology 54:212–220.

MACK, G.H., JAMES, W.C., and THOMAS, W.A. (1981) Orogenic provenance of Mississippian sandstones associated with southern Appalachian-Ouachita orogen. American Association Petroleum Geologists Bulletin 65:1444–1456.

MACK, G.H., THOMAS, W.A., and HORSEY, C.A. (1983) Composition of Carboniferous sandstones and tectonic framework of southern Appalachian-Ouachita orogen. Journal Sedimentary Petrology 53:931–946.

MacKINNON, T.C. (1983) Origin of the Torlesse terrane and coeval rocks, South Island, New Zealand. Geological Society America Bulletin 94:967–985.

McBRIDE, E.F. (1966) Sedimentary petrology and history of the Haymond Formation (Pennsylvanian), Marathon Basin, Texas. Texas Bureau Economic Geology Report Investigations 57, 110 p.

McBRIDE, E.F. (1985) Diagenetic processes that affect provenance determinations in sandstone. In: Zuffa, G.G. (ed) Provenance of Arenites. Dordrecht, Holland: Reidel, pp. 95–113.

McBRIDE, E.F. (1987) Diagenesis of the Maxon Sandstone (Early Cretaceous), Marathon region, Texas: A diagenetic quartzarenite. Journal Sedimentary Petrology 57:98–107.

MEADE, R.H., NORDIN, C.F., JR., CURTIS, W.F., COSTA RODRIGUEZ, F.M., DO VALE, C.M., and EDMOND, J.M. (1979) Sediment loads in the Amazon River. Nature 278:161–163.

MILLIMAN, J.D. and MEADE, R.H. (1983) World-wide delivery of river sediment to the oceans. Journal Geology 91:1–21.

MOIOLA, R.J. and SHANMUGAM, G. (1984) Facies analysis of upper Jackfork Formation (Pennsylvanian), DeGray Dam, Arkansas [abstract]. American Association Petroleum Geologists Bulletin 68:509.

MOLNAR, P. and ATWATER, T. (1978) Interarc spreading and Cordilleran tectonics as alternates related to the age of subducted oceanic lithosphere. Earth Planetary Science Letters 41:330–340.

MONGER, J.W.H., PRICE, R.A., and TEMPLEMAN-KLUIT, D.J. (1982) Tectonic accretion and the origin of the two major metamorphic and plutonic welts in the Canadian Cordillera. Geology 10:70–75.

MONGER, J.W.H., SOUTHER, J.G., and GABRIELSE, H. (1972) Evolution of the Canadian Cordillera: A plate-tectonic model. American Journal Science 272:577–602.

NELSON, T.H. and TEMPLE, P.G. (1972) Mainstream mantle convection: A geologic analysis of plate motion. American Association Petroleum Geologists Bulletin 56:226–246.

OWEN, M.R. and CAROZZI, A.V. (1986) Southern provenance of upper Jackfork Sandstone, southern Ouachita Mountains: Cathodoluminescence petrography. Geological Society America Bulletin 97:110–115.

PACHT, J.A. (1984) Petrologic evolution and paleogeography of the Late Cretaceous Nanaimo Basin, Washington and British Columbia: Implications for Cretaceous tectonics. Geological Society America Bulletin 95:766–778.

POTTER, P.E. (1978a) Significance and origin of big rivers. Journal Geology 86:13–33.

POTTER, P.E. (1978b) Petrology and chemistry of modern big river sands. Journal Geology 86:423–449.

POTTER, P.E. (1984) South African [sic] modern beach sand and plate tectonics. Nature 311:645–648 ("African" in title should read "American").

POTTER, P.E. (1986) South America and a few grains of sand: Part I, beach sands. Journal Geology 94:301–319.

RUXTON, B.P. (1970) Labile quartz-poor sediments from young mountain ranges in northeast Papua. Journal Sedimentary Petrology 40:1262–1270.

SCHWAB, F.L. (1981) Evolution of the western continental margin, French-Italian Alps: Sandstone mineralogy as an index of plate setting. Journal Geology 89:349–368.

SCLATER, J.G., HELLINGER, S., and TAPSCOTT, C. (1977) The paleobathymetry of the Atlantic Ocean from the Jurassic to the present. Journal Geology 85:509–522.

SCOTESE, C.R., BAMBACH, R.K., BARTON, C., VAN DER VOO, R., and ZIEGLER, A.M. (1979) Paleozoic base maps. Journal Geology 87:217–277.

SCOTT, G.L. (1965) Heavy mineral evidence for source of some Permian quartzose sandstones, Colorado Plateau. Journal Sedimentary Petrology 35:391–400.

SHANMUGAM, G. (1985) Types of porosity in sandstones and their significance in interpreting provenance. In: Zuffa, G.G. (ed) Provenance of Arenites. Dordrecht, Holland: Reidel, pp. 115–137.

SMITH, A.G. and BRIDEN, J.C. (1977) Mesozoic and Cenozoic Paleocontinental Maps. Cambridge, England: Cambridge University Press, 63 p.

SMITH, A.G., HURLEY, A.M., and BRIDEN, J.C. (1981) Phanerozoic Paleocontinental World Maps. Cambridge, England: Cambridge University Press, 102 p.

STEWART, R.J. (1976) Turbidites of the Aleutian abyssal plain: Mineralogy, provenance, and constraints for Cenozoic motion of the Pacific plate. Geological Society America Bulletin 87:793–808.

SUCZEK, C.A. and INGERSOLL, R.V. (1985) Petrology and provenance of Cenozoic sand from the Indus Cone and the Arabian Basin, DSDP sites 221, 222, and 224. Journal Sedimentary Petrology 55:340–346.

SUTTNER, L.J., BASU, A., and MACK, G.H. (1981) Climate and the origin of quartz arenites. Journal Sedimentary Petrology 51:1235–1246.

SUTTNER, L.J. and DUTTA, P.K. (1986) Alluvial sandstone compositions and paleoclimate, I. Framework mineralogy. Journal Sedimentary Petrology 56:329–345.

TAYLOR, B. and KARNER, G.D. (1983) On the evolution of marginal basins. Reviews Geophysics Space Physics 21:1727–1741.

THOMAS, W.A. and MACK, G.H. (1982) Paleogeographic relationships of a Mississippian barrier-island and shelf-bar system (Hartselle Sandstone) in Alabama to the Appalachian-Ouachita orogenic belt. Geological Society America Bulletin 93:6–19.

THORNBURG, T.M. and KULM, L.D. (1987) Sedimentation in the Chile Trench: Petrofacies and provenance. Journal Sedimentary Petrology 57:55–74.

UYEDA, S. (1982) Subduction zones: An introduction to comparative subductology. Tectonophysics 81:133–159.

UYEDA, S. and KANAMORI, H. (1979) Back-arc opening and the mode of subduction. Journal Geophysical Research 84:1049–1061.

VALLONI, R. and ZUFFA, G.G. (1984) Provenance changes for arenaceous formations of the northern Apennines, Italy. Geological Society America Bulletin 95:1035–1039.

VELBEL, M.A. (1985) Mineralogically mature sandstones in accretionary prisms. Journal Sedimentary Petrology 55:685–690.

WARD, P. and STANLEY, K.O. (1982) The Haslam Formation: A late Santonian-early Campanian forearc basin deposit in the insular belt of southwestern British Columbia and adjacent Washington. Journal Sedimentary Petrology 52:975–990.

WICKHAM, J., ROEDER, D., and BRIGGS, G. (1976) Plate tectonics models for the Ouachita foldbelt. Geology 4:173–176.

WILSON, J.T. (1954) The development and structure of the crust. In: Kuiper, G.P. (ed) The Earth as a Planet. Chicago: University Chicago Press, pp. 138–214.

YOUNG, G.C. (1986) Cladistic methods in Paleozoic continental reconstruction. Journal Geology 94:523–537.

ZIEGLER, A.M., ROWLEY, D.B., LOTTES, A.L., SAHAGIAN, D.L., HULVER, M.L., and GIERLOWSKI, T.C. (1985) Paleogeographic interpretation, with an example from the mid-Cretaceous. Annual Reviews Earth Planetary Sciences 13:385–425.

2

Isotopic Provenance of Clastic Deposits: Application of Geochemistry to Sedimentary Provenance Studies

PAUL L. HELLER and CAROL D. FROST

Abstract

Determining the source areas of sandstone suites typically involves documenting areal and compositional trends across a sedimentary basin, as in paleocurrent and point-count studies. Although traditional basin-analysis techniques suffice in settings where there are few potential source areas, they may prove inadequate in more complex tectonic settings. In these situations isotopic analyses of whole–rock samples and mineral separates from clastic deposits may help identify source areas where mineralogic studies alone cannot. Even minor geochemical variations between source areas of otherwise similar composition can be discerned by the isotopic–provenance technique. When various isotopic systems are brought to bear on whole-rock samples or mineral separates from compositionally immature sandstones, a detailed tectono-thermal history of the source region may be obtained and compared with known histories of specific potential source areas. Isotopic study of mudstones, on the other hand, provides a better indication of the average composition of regional source areas. Although the isotopic provenance technique is not universally applicable, it is a powerful analytical tool that, when used in conjunction with other basin-analysis techniques, can provide a detailed portrait of the source area not available by any other method.

Introduction

One of the goals of basin analysis is to determine the source areas from which clastic deposits were derived. Provenance information of this type can be used to reconstruct the original basin configuration, its depositional systems, and regional paleogeography. Furthermore, provenance studies can aid in reconstructing uplift histories of source areas and, where parent rocks have been removed by erosion,

the basinal sediments may provide the only preserved record of local crustal evolution in the source area.

Determining sandstone provenance typically involves documenting areal and compositional trends across a sedimentary basin. Traditional methods include measuring changes in paleocurrent directions, facies patterns, grain-size trends, and gravel or sandstone compositions (Miall 1984). These traditional approaches are not without problems. Deformation of a basin or its wholesale displacement subsequent to the time of deposition can make provenance data difficult to interpret. In addition, to use paleocurrent and facies data effectively, adequate exposures and detailed chronostratigraphic correlations are needed because several different depositional systems may be found in the basin over time. Grain-size information alone may be very difficult to interpret: Transport in fluvial systems is primarily a function of stream power, which may vary, and does not necessarily reflect distance of transport; and in marine systems, deposits may be resedimented, therefore the distribution of marine clastics may be process-related and not necessarily an indicator of proximity to source area.

Sandstone provenance studies have been used to show that sediments derived from similar tectonic settings have similar overall compositional characteristics (Dickinson and Suczek 1979; Dickinson 1985). Although this method is useful in interpreting regional tectonic settings, it is difficult to differentiate between source areas of similar tectonic setting. For example, Dickinson (1982) has shown that forearc–basin and subduction–zone sandstones from throughout the Pacific Basin are strikingly similar. In areas of complex tectonic history or where source-area compositions evolve through time, sandstone compositions may be very hard to interpret.

Therefore, while traditional provenance techniques work best in simple settings where it is easy to discriminate between distinctive source areas, they may prove inadequate in more complex tectonic settings. For example, along convergent plate margins multiple source areas of generally similar composition may develop. The composition of a single source area may also evolve through time or, in the case of displaced terranes, the sedimentary deposit may be far removed from its source. In these situations, other techniques may also be necessary to identify source areas. Various techniques have been added to the petrologists' arsenal to refine the interpretation of sedimentary provenance. Heavy-mineral studies have been used to interpret source areas by describing accessory minerals derived from the parent rock (Van Andel 1959). However, the source area for the heavy minerals might not be the same as that supplying the bulk of the sediment. Heavy minerals are not the hydrodynamic equivalent of similarly sized light minerals and so become concentrated as placer deposits (Dickinson 1970). The abundance of heavy minerals and, hence, the inferred composition of the source area may be artificially generated, giving a distorted picture of the source area. For example, heavy minerals from a small exotic metamorphic source may locally mask the major source area that supplies most of the basinal sediments. Furthermore, reliance on heavy-mineral composition and abundance may also be undermined by the selective breakdown and removal of certain unstable minerals in the sedimentary environment (Folk 1974).

Another approach to provenance studies involves analyzing the bulk chemical composition of sediments, including trace elements (Bhatia 1983; Hiscott 1984; Roser and Korsch 1986; Bhatia and Crook 1986). This approach provides useful results, especially where later metamorphism destroys original mineralogy (Van de Kamp and Leake 1985). If original mineralogy is preserved, however, point counting will provide as much, or more, information than is interpretable from geochemistry alone because whole-rock chemical composition is dependent on mineralogy. More information can be provided from this technique if separates of mineral and lithic fragments can be analyzed and compared with known compositions from potential source rocks.

A related approach is the determination of ratios of various isotopes from whole rocks and mineral separates of sedimentary deposits to determine source area. By dealing with isotopic ratios, this approach avoids many of the pitfalls of merely looking at total elemental abundances in deposits. Isotope studies have been used extensively in sedimentary deposits to determine mechanisms of cementation and diagenesis (Land 1983) and age of deposition (Cordani et al. 1978; Odin 1982). Their use as a tool for provenance study of clastic deposits, however, has not been fully exploited. Several isotopic systems are relatively insensitive to alteration effects and so can be used on rocks that are unsuitable for point counting. The technique can be quite expensive, but it is potentially a powerful tool in basin analysis when used in conjunction with more traditional methods.

The purpose of this paper is to review isotopic methods which can be used in sedimentary provenance studies. We stress the advantages and disadvantages of various techniques and describe ideal situations in which the isotope provenance approach would be most useful.

Isotopic Provenance Method

Isotopic studies can provide a wealth of information about source areas, such as age of formation, uplift and thermal history, crustal composition and evolution, and secondary alteration of the parent rocks. These types of information paint a detailed picture of the source area from which the sediment was derived. If similar information is available for the potential source areas, these "fingerprints" can be matched and the provenance determined. Clearly, it is essential to select an appropriate sedimentary sequence for study. Potential source areas must be isotopically distinguishable on the basis of age of formation, thermal history, and/or petrogenesis. If these characteristics cannot be used to discriminate among source areas, then the isotopic fingerprint approach is inappropriate and would yield ambiguous results. Different isotopic techniques provide different types of information about source areas, and are briefly described with examples below.

Fission-Track Dating Method

The fission-track dating method (Fleischer et al. 1975; Naeser 1978) has had broad application to a range of sedimentary problems including dating and determining uplift, burial and diagenetic histories. However, relatively few fission-track provenance studies have been undertaken. Nonetheless, fission-track age, that is the time since the material cooled below its blocking temperature permitting track retention (Naeser 1978), may be very useful in provenance studies. Fission-track ages from uranium-bearing minerals in sedimentary rocks may identify source terranes if the sediment has not been subjected to temperatures approaching the mineral blocking temperature (generally less than a couple of hundred degrees Celsius), and if the possible source areas are of distinctly different ages.

Few fission-track provenance studies have been reported (e.g., Zeitler *et al.* 1982; Yim *et al.* 1985; Cerveny *et al.* 1988: this volume). Recently, Baldwin *et al.* (1986) successfully applied fission track methods to the problem of sources for the accreted sandstones of Barbados, Lesser Antilles. In their study, Baldwin *et al.* found that fission–track ages for individual zircon grains from the accreted sediments fell in three age groups: 25–80 Ma, 200–350 Ma, and greater than 500 Ma, which could be correlated to specific source areas from young volcanic centers to the old South American craton.

The fission-track-age method of provenance determination has the advantage that individual grains as well as populations can be dated, so mixed populations can be identified. The minerals used (sphene, apatite, zircon) are common accessory minerals, at least one of which is usually present in sand or sandstone. On the other hand, correctly counting fission tracks requires experience, and the low annealing temperature means that burial can erase the fission-track record. In addition, reliance on heavy minerals may yield results that are not representative of the source area for the bulk of the sedimentary grains (as discussed earlier). Nevertheless, the fission–track approach may be valuable, especially for young deposits that have not been deeply buried.

K-Ar and ^{40}Ar/^{39}Ar Isotopic Method

Potassium-bearing minerals can be dated by the K-Ar and ^{40}Ar/^{39}Ar methods, which record the age at which mineral crystals cool below blocking temperatures (Turner 1968; Albarede 1982; Faure 1986). In the case of some minerals, such as biotite, these temperatures can be quite low, a few hundred degrees Celsius (Dodson and McClelland-Brown

1985). Therefore, ages determined by the K-Ar and ^{40}Ar/^{39}Ar methods may represent either the initial time of mineral formation or the last time of cooling through its blocking temperature. This may be accomplished by uplift following burial diagenesis, igneous heating events, or related mechanisms. In addition, the elemental abundances of potassium and argon can change during diagenesis and weathering, altering the calculated age of the mineral grains. Since all of these processes may take place within the depositional basin any application of these dating techniques to sedimentary provenance studies must be done with caution.

Of the two methods, the ^{40}Ar/^{39}Ar incremental heating method yields potentially more information than K-Ar age dating. Radiogenic argon loss during thermal events may be detected in the ^{40}Ar/^{39}Ar release pattern, whereas such loss cannot be easily identified using conventional K-Ar methods. Although most studies using the ^{40}Ar/^{39}Ar method concentrate on interpreting the thermal history of sedimentary basins, Harrison and Bé (1983) were also able to determine the cooling age of the source area that shed microcline grains into the southern San Joaquin Basin. Their incremental ^{40}Ar/^{39}Ar age spectrum of detrital microclines recorded both the slow cooling during uplift of the source terrane (probably the Sierra Nevada) between about 85 and 60 Ma and a recent thermal event that affected the detrital microclines while they were buried in the basin.

Utilizing K-Ar and ^{40}Ar/^{39}Ar information from detrital minerals has the advantage that datable potassium-bearing minerals are common in most sedimentary rocks (e.g., micas, amphiboles). In addition, as shown in the Harrison and Bé (1983) study, details of the thermal history of the source area may be obtained by age spectra produced by the ^{40}Ar/^{39}Ar incremental heating method.

A disadvantage of the K-Ar and ^{40}Ar/^{39}Ar techniques is that artificially young ages are produced in many minerals by the loss of argon through the crystal lattice during weathering, transport, and burial. Feldspars are especially susceptible to this process. Low blocking temperatures for some minerals mean that calculated ages most likely represent uplift or cooling ages, not the age of original mineral formation. Finally, the samples may include unwanted mixtures of minerals from different source areas. In K-Ar dating this problem is compounded by the fact that the potassium and argon are measured from separate aliquots that may not have identical compositions (Albarede 1982). The resultant age from such samples is, therefore, a mean age of the minerals from the various source areas weighted by the relative contribution from each source area. Such data can be difficult to interpret.

Rb-Sr Isotopic Method

Rb-Sr studies of sedimentary rocks to determine provenance have dealt chiefly with mudstones (e.g., Biscaye and Dasch 1971; Boger and Faure 1974; Clauer 1979) but have seen application to sandstones (e.g., Peterman *et al.* 1967, 1981; Condie *et al.* 1970; Peterman and Whetten 1972; Faure *et al.* 1983; Heller *et al.* 1985). By the isochron dating technique, as described by Faure (1986), samples that crystallize with the same strontium-isotopic composition, such as comagmatic igneous rocks or authigenic minerals, plot as points on a straight line in coordinates of $^{87}Sr/^{86}Sr$ and $^{87}Rb/^{86}Sr$. Linear arrays on isochron diagrams may also be produced by mixtures of two sources with distinct $^{87}Sr/^{86}Sr$

and $^{87}Rb/^{86}Sr$ characteristics. Thus, isochron plots of sedimentary samples may provide either depositional or detrital-mineral ages, or evidence of mixing of sediment from multiple source areas.

In a study of Eocene deposits from the southern part of the Oregon Coast Range, Peterman *et al.* (1981) analyzed rubidium and strontium from whole-rock graywacke samples. They found a surprisingly consistent colinear relationship between $^{87}Sr/^{86}Sr$ and $^{87}Rb/^{86}Sr$ (Fig. 2.1). This "pseudo-isochron" was found to be consistent with the results that would be expected from mixing available Mesozoic source rocks from the adjacent Klamath Mountains. They inferred from their results that the Eocene deposits were derived from this nearby source area. Other Eocene deposits in the Oregon Coast Range (Fig. 2.2) had significantly different isotopic compositions, and therefore, suggested a different source area in the Mesozoic plutonic belt found much farther to the east (Heller *et al.* 1985).

When using the Rb-Sr approach, as with other isotopic systems, one must be aware of difficulties that may arise when it is applied to problems of sedimen-

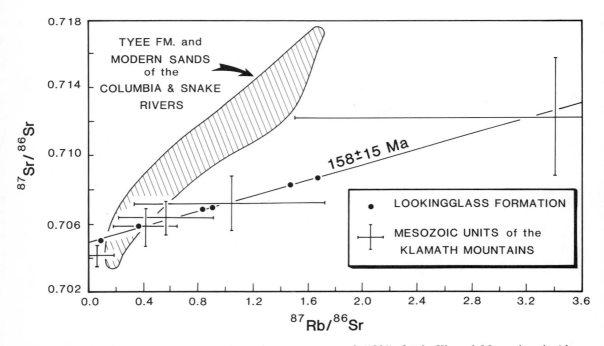

Fig. 2.1. Rb-Sr systematics for whole-rock sandstones from the Lookingglass Formation and the Tyee Formation in the southern Oregon Coast Range, Mesozoic units in the Klamath Mountains, and modern sands from the Columbia and Snake Rivers. See Figure 2.2 for locations. Data for the Lookingglass Formation recalculated from Peter-

man *et al.* (1981); for the Klamath Mountain units (showing mean and standard deviation) from Coleman (1972) and Peterman *et al.* (1981); for the Tyee Formation and modern river sands from Peterman and Whetten (1972) and Heller *et al.* (1985).

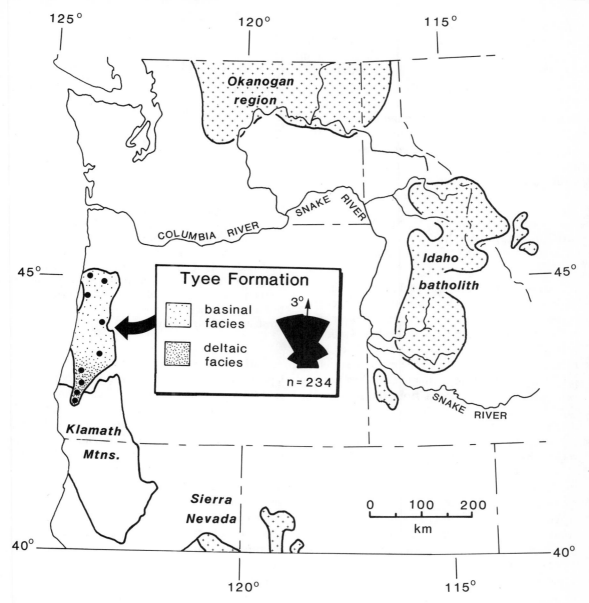

Fig. 2.2. Location of the Eocene Tyee Formation in the Oregon Coast Range and its potential source areas in the U.S. Pacific Northwest. Location of samples collected from the Tyee Formation for isotopic study shown by dots. General facies trends and paleocurrent information from Heller and Ryberg (1983) and Heller and Dickinson (1985). Plutonic source areas shown with a pattern of small plus signs. Modified from Heller *et al.* (1985).

tary provenance, because of differential mobility of rubidium and strontium during weathering, transport, sedimentation, and diagenesis. These problems may be significant for whole-rock analyses of mature clastic rocks (e.g., Nelson and DePaolo, in press), but the method can be used with success on compositionally immature clastics (graywackes) for reasons discussed later. Even if whole-rock analyses are too scattered to fit an isochron accurately, the general range of Rb-Sr values can be compared with values from possible source rocks to determine if derivation from a proposed source area is likely (Heller *et al.* 1985).

Study of the Rb-Sr character of individual minerals in sandstones avoids potential problems associated with whole-rock analyses. Although rubidium and strontium mobility remains a potential problem even for mineral separates, a model mineral age may be obtained given measured Rb/Sr and $^{87}Sr/^{86}Sr$ ratios and an assumed initial ratio. This model age can help identify source areas (e.g., Heller *et al.* 1985). Many common rock-forming minerals contain sufficient rubidium for model ages to be obtained, and the age of Precambrian to Tertiary materials can be estimated. The model-age approach, however, may be misleading if the assumed initial strontium-ratio is significantly in error and the enrichment in ^{87}Sr is not great.

Sm-Nd Isotopic Method

The relative immobility of rare earth elements relative to one another during crustal processes, including weathering, sediment transport, and diagenesis, makes the radiometric parent-daughter pair ^{147}Sm-^{143}Nd potentially useful in studies of sediment provenance (DePaolo and Wasserburg 1976, 1977; O'Nions *et al.* 1983). Because the major partitioning of samarium from neodymium accompanies the generation of crustal melts from the mantle, the Sm/Nd ratio and $^{143}Nd/^{144}Nd$ ratio can be used to determine the "crustal-residence age" or estimated time since the precursors to the sedimentary source rocks were first extracted from the mantle (O'Nions *et al.* 1983). The crustal-residence age represents the age of formation of the crust from which source rocks were later derived. This calculation assumes that there is no appreciable partitioning of samarium from neodymium during erosion, sediment transport, and deposition. It is not equivalent to ages determined by other radiometric dating techniques, which represent time since the source rock passed through the blocking temperature of the radiometric system of interest.

Clastic deposits derived from source areas that were extracted from the mantle at different times have different neodymium-isotopic crustal-residence ages, as demonstrated by O'Nions *et al.* (1983) in their study of sedimentary rocks from the British Isles. Deposits from the Dalradian Supergroup of Late Precambrian depositional age (*ca.* 700 Ma) have crustal residence ages of 1.8 to 2.7 Ga, similar to crustal residence ages of Lewisian and Torridonian deposits (Fig. 2.3). Hence, all these materials may have been derived from Precambrian basement in northwest Scotland and adjacent North

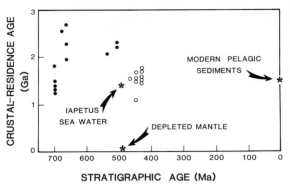

Fig. 2.3. Comparison of stratigraphic age and estimated crustal-residence ages for clastic rocks from the Dalradian Supergroup (dots) and deposits of the Southern Uplands of Scotland (open circles). Crustal-residence ages, during the Ordovician, of Iapetus ocean sea water and depleted mantle are shown for comparison, as is the average crustal-residence age of modern pelagic sediments. Modified from O'Nions *et al.* (1983) with additional data from Frost and O'Nions (1985).

Atlantic cratonic areas. In contrast, Southern Upland deposits have crustal residence ages of 1.43 to 1.74 Ga. Young mantle-derived material combined with older Dalradian-like sedimentary rocks would produce mixtures of sediment with these crustal-residence ages. The variability of crustal residence ages for a formation, such as the Dalradian Supergroup, can be related to incomplete homogenization of source materials of different ages. Moreover, clastics with crustal-residence ages similar to their depositional ages, such as the Lewisian metasedimentary sequence, must have formed from new continental crust and may be first-cycle deposits.

The Sm-Nd isotope approach has the advantage that effects of weathering and metamorphism on the system are apparently minimal; hence, sedimentary rocks that are unsuited to other provenance methods may be analyzed. The Sm-Nd method has been successfully applied to sedimentary sequences of Archean to Recent age (O'Nions *et al.* 1983; Frost and O'Nions 1984; Michard *et al.* 1985; Nelson and DePaolo, in press). A disadvantage of the Sm-Nd approach is that source areas of the same crustal-residence age cannot be distinguished. This problem is compounded by observations that suggest that areas of similar crustal-residence age may cover very large parts of a continent (Nelson and Depaolo 1985). Another problem in some instances is that accessory minerals with high rare-earth contents skew the sediment Sm-Nd composition (Frost and Winston 1987).

U-Pb Isotopic Methods

Use of U-Pb isotope methods to relate whole-rock sediment samples to source area is complicated by the mobility of uranium in oxidized, near-surface environments. However, if the uranium loss is recent, the lead-isotopic ratios of the sedimentary rock reflect the antiquity of the source (Doe 1970). For example, the classic study by Chow and Patterson (1962) showed that the lead-isotopic compositions of pelagic sediments from the Atlantic Ocean are more radiogenic than those from the Pacific Ocean, presumably reflecting a source terrain around the Atlantic that is older, on average, than found around the Pacific Basin.

Recent work applying the U-Pb system to sediments has focused upon mineral separates, zircons in particular (e.g., Froude et al. 1983), although feldspars have also been used (Patterson and Tatsumoto 1964). These minerals may be analyzed as populations or as single grains and ages determined (Faure 1986). The ages may be related to the crystallization age of the source rock. If the population includes grains from different source areas, the resultant age will be intermediate between the source-rock ages. Analyses of single grains can help overcome this difficulty. Single grains have been analyzed both by ion microprobe and by conventional mass spectrometry. By the former method, Froude et al. (1983) dated zircon grains in an Archean quartzite as old as 4.1 Ga along with others in the 3.5 to 3.8 Ga age range. Using mass spectrometry, Schärer and Allègre (1985) found zircons in the same sandstone unit ranging in age from 3.3 to 3.8 Ga, but did not analyze any zircons as old as 4.1 or 4.2 Ga.

Although the interpretation of mineral ages is not always straightforward, particularly when multiple lead-loss events have occurred, U-Pb analyses of mineral separates from sedimentary rocks may be a very useful, but presently underused, technique for provenance studies.

Stable Isotopic Methods

Studies that make use of stable isotopes to determine provenance of sandstones are rare, in part because many stable isotope studies emphasize the thermodynamic and isotopic equilibrium between two coexisting minerals in igneous or metamorphic rocks (Hoefs 1980). In sandstones, minerals that coexist may have come from two different sources, and therefore, are likely not in thermodynamic equilibrium. Conversely, if two coexisting minerals in a sedimentary rock *are* in thermodynamic equilibrium, it allows the possibility that they were derived from the same source rock (Heller et al. 1985). However, thermodynamic disequilibrium does not preclude that the minerals came from different rocks in the same source area or have undergone subsequent differential exchange. Nonetheless, the use of stable isotopes, primarily oxygen and deuterium, may prove to be a powerful tool in provenance studies. If isotopic values in minerals from possible source rocks are known and are distinctive for each source area, then values from the same minerals in the sedimentary rocks can be compared successfully.

Heller et al. (1985) used $\delta^{18}O$ values from white micas (mainly muscovite) in Eocene sedimentary rocks of the Oregon Coast Range, in conjunction with other isotopic data summarized below, to help determine provenance. Microscopic observation of the micas showed them to be unaltered, and consistent $\delta^{18}O$ values from micas collected from throughout the basin also suggested that local alteration did not occur to any significant degree. Values for $\delta^{18}O$ of approximately +9.5 per mil (relative to Standard Mean Ocean Water) were lighter than values from white micas in metamorphic rocks of the adjacent Klamath Mountains, which was considered to be a likely source area. The observed $\delta^{18}O$ values were nearly identical with those of white micas collected in the Idaho Batholith, suggesting a source area in the Mesozoic plutonic belt of the U.S. Cordillera. The results of this multi-isotopic study are discussed in greater detail below.

One problem in dealing with stable isotopes in sandstones is that the original isotopic composition may have been altered in either the source rocks or the sediment by subsequent hydrothermal alteration or other metamorphic changes. Often, it is simple to determine if the minerals have been altered by petrographic examination of thin sections. Although diagenetic events can be inferred from the stable-isotopic composition of cements within the clastic sequence (Land 1983), it is difficult to alter the isotopic composition of many detrital minerals during low-grade diagenesis.

Effects of Compositional Maturity and Grain Size

In contrast to igneous and metamorphic rocks, there are several factors that must be carefully considered when applying isotopic techniques to clastic sedimentary sequences. Generally these factors can be considered either as effects due to compositional maturity or effects due to grain size.

Compositional Maturity

Just as sandstone mineralogy reflects not only source rock composition but subsequent weathering, erosion, transport, and burial (Cameron and Blatt 1971; Galloway 1974; Mack 1978; Suttner et al. 1981), so too do these processes affect the geochemical composition of the sediment. Partitioning, or changing the relative abundances, of radiometric parent and daughter elements by selectively removing one or the other from the geochemical system during sedimentary processes will, with time, affect the isotopic ratios of the daughter element. It is important to evaluate this effect when interpreting isotopic data. Element partitioning may be recognized by examining the compositional maturity of the deposit.

Different elements preferentially collect in specific minerals, for example strontium in carbonates and rubidium in clay. As sedimentary processes selectively remove minerals from the deposits the isotopic composition of the resulting sedimentary whole-rock sample changes. These processes include chemical and physical breakdown during weathering, grain transport and diagenesis. Hydrodynamic processes may tend selectively to concentrate heavy minerals, rich in certain elements, in the stratigraphic sequence as placer deposits. Authigenic mineral growth in sedimentary rocks may affect the bulk isotopic composition (Dasch et al. 1966). Lastly, some elements, such as strontium, are more easily mobilized than others, such as rubidium, in the sedimentary environment. As a result of these processes, whole-rock analyses on clastic sequences may show either an enrichment or depletion of certain parent or daughter elements. With time, this leads to isotopic ratios that differ

from those of the source areas. Hence, whole-rock isotopic analyses are, in the extreme case, unrepresentative of the source-area composition, and may be more closely related to the degree of weathering and diagenesis. At minimum, whole-rock isotopic analyses may produce variable results depending on the specific mineral composition of the sample being analyzed.

Potential problems associated with natural sedimentary processes are reflected in the compositional maturity of the sandstones. Most changes that bring about geochemical partitioning also result in removal of unstable grains in the clastic sediments. Hence, with increasing compositional maturity there is a greater chance that the whole-rock geochemistry is less representative of source-area composition and is more reflective of weathering, transport, and burial history. Furthermore, mature sandstones tend to yield more inconsistent isotopic results. For this reason, quartz sandstones with minor amounts of heavy minerals are potentially inappropriate for isotopic study. Quartz is depleted in all elements except silica and oxygen. Therefore, minor changes in the abundance of heavy minerals, which are enriched in certain trace elements, will produce widely variable concentrations of those elements. Frost and Winston (1987) found that quartz-rich sandstones from the Belt-Purcell Supergroup of Montana and adjacent parts of Canada had variable Sm/Nd ratios that did not reflect variations in source area, but rather were the result of differential accumulation of rare-earth element-bearing heavy mineral suites. Immature sandstones, on the other hand, contain more varied grain types, many of which are relatively rich in trace elements. Minor changes in heavy-mineral content in immature sandstones, therefore, have a less significant effect on overall elemental abundance of the sandstone sample, leading to more consistent isotopic ratios. Peterman et al. (1967, 1981) and Peterman and Whetten (1972) have shown in sedimentary deposits of the Pacific Northwest, that whole-rock Rb-Sr results for immature sandstones primarily reflect source-area composition.

In a similar fashion, the addition of authigenic minerals during deposition and diagenesis can change the geochemical balance of a sedimentary rock. Carbonates, enriched in strontium, could dominate the $^{87}Sr/^{86}Sr$ ratio of a sedimentary rock if present in significant amounts (Perry and Turekian 1974). During diagenesis strontium in carbonates and feldspars may reequilibrate with strontium from other sources unrelated to the source rocks, causing the $^{87}Sr/^{86}Sr$ ratio of the source–rock strontium to be shifted toward the $^{87}Sr/^{86}Sr$ ratio of the foreign strontium. Furthermore, the rubidium and strontium concentrations and, as a result, strontium-isotopic ratios, can be modified by mineralogic changes, dewatering, and the addition of calcareous tests in the marine environment (Peterman *et al.* 1981). However, additions such as these might be detected by petrographic analysis and carbonates, in particular, can be removed by acid leaching.

Due to the effects described above, and because of possible inhomogeneity of isotopic composition related to the amounts of different detrital components, it is difficult to establish strict guidelines for the application of whole-rock isotopic methods to sedimentary rocks (Clauer 1982). At best, whole-rock analyses yield average isotopic compositions for the source areas, weighted by the relative contribution from each source area and by the relative mineral composition of the bulk sample. How representative a sample is of the true source–area composition depends upon the compositional maturity of the sedimentary rock. The more mature the clastic rock is, the more opportunity for parent-daughter element partitioning. Immature lithic sandstones ("graywackes") can yield important information regarding integrated sources because the poor sorting and preservation of lithic fragments associated with these types of rocks implies short times between uplift, erosion, burial, and lithification (Folk 1974).

Grain Size

Another important factor in determining which clastic rocks are suitable for isotopic study is grain size. To a large extent, the choice of fine-grained or coarse-grained clastics depends upon the goals of the provenance study. If *specific* source areas are to be identified, coarser-grained sediments containing detrital grains that can be matched to different source areas are most useful. If, on the other hand, an *average* source-area composition of a large area is desired, then fine-grained mudstones and siltstones that effectively homogenize material from many source areas during cycles of sedimentary transport and deposition are suitable. Furthermore, the ratio of shale to sandstone in a sedimentary system tends to be large. Thus, the volumetric abundance of fine-grained rocks suggests that they are more representative of the *average* source-area composition than are related sandstone deposits.

As stated above, when quartzose sandstones accumulate significant quantities of heavy minerals, they also become enriched in trace elements that are concentrated in those minerals. The remaining finer-grained sedimentary fraction should be left with a complementary depletion of heavy minerals and, therefore, should contain lower trace-element concentrations. However, the overall abundances of most elements in siltstones and mudstones are high compared to quartz sands, so that trace-element concentrations are little affected by the lack of heavy-mineral contribution. For example, while the neodymium contained in a few grains of apatite, sphene and zircon can easily dominate the neodymium content of quartz sand, the absence of those mineral grains from a shale, which contains a significant amount of neodymium in clay minerals, will not appreciably affect its overall content of this element (Frost and Winston 1987). The abundance of immobile elements in fine-grained clastic deposits, therefore, tends to produce more consistent isotopic results than coarser-grained fractions. For this reason, the abundance in shales of elements that are not partitioned during sedimentary processes, such as the rare-earth elements, is considered representative of their true crustal abundance (Taylor and McLennan 1985). However, many elements, including most alkalis and alkali earths, are mobilized during weathering and, hence, their abundances in shales are not necessarily representative of the crust from which they were eroded. Therefore, isotopic studies of only *immobile* elements in shales can be used in characterizing the average composition of continental areas exposed to erosion (e.g., Goldstein *et al.* 1984).

Summary

In consideration of these problems, the best samples to analyze, depending on the purpose of the study, may be either compositionally immature sandstones, that is those rocks that have not undergone extensive alteration, or fine-grained clastics (mudstones). Immature sandstones are best used to determine specific source areas, and mudstones to determine average compositions of large regions. Another approach to determining the provenance of sand-sized material is to restrict the study to specific minerals, in cases where the deposit is not significantly altered and, therefore, contains primary detrital minerals. Although whole-rock analysis has the advantage of giving a bulk composition of the average of all source areas, mineral separates may be more useful in yielding details of the isotopic composition of specific minerals in the source area. Isotopic analyses of specific minerals may provide information about the character of the source area or areas, especially when a specific mineral is derived from a unique source area or at least only those source areas that contain that mineral. Therefore, results from a specific mineral type are less likely to show the average of many sources found in the region. In addition, use of coarse-mineral separates reduces the likelihood of including fine-grained authigenic minerals that may give spurious results (Dasch *et al.* 1966).

Suggested Methodology

Several recommendations can be made on a methodology for determining the provenance of clastic deposits by isotopic methods. Isotopic studies are relatively time consuming and expensive, and should be attempted only after preliminary study indicates that this technique will produce useful results. A fundamental, although not trivial, aspect of such a study is for a sedimentary geologist to find an isotope geochemist with whom to associate. Ideally, the cross-disciplinary approach advocated here benefits by collaborative research.

Setting Up the Problem

The first step in setting up an isotopic provenance study is to identify all potential source areas and determine how they can be differentiated. If source mineralogy is distinctive, then a mineralogic study of the sandstone may be sufficient. If not, then source areas should be examined for differences in age, thermal history, and petrogenesis. This requires a review of both geologic and isotopic literature that is available for possible source terrains. The distinguishing features between possible source areas are then matched to appropriate isotopic methods; for example, different thermal histories may be discerned through fission-track, K-Ar, ^{40}Ar/^{39}Ar, and possibly Rb-Sr studies, whereas age differences may be detected from U-Pb, Sm-Nd, or Rb-Sr analyses.

Suitability of Samples

The next critical step is to determine whether the rocks are suitable for the intended isotopic methods. As reviewed above, different isotopic systems have differing tolerances to secondary processes. For example, Sm-Nd systematics are established primarily during formation of continental crust, whereas fission–track ages can be reset at relatively low temperatures. For these reasons, immature and/or fine-grained whole-rock samples or mineral fractions may be the most useful. Of course, the ideal sample for any isotopic study is as pristine and free of alteration as possible.

Regardless of the sandstone type to be studied, petrographic examination is critical to determine: 1) the degree of secondary alteration and authigenesis that might affect the geochemical composition of the sample; 2) the presence of minerals, such as a disproportionate amount of heavy minerals, that might make results from whole-rock samples difficult to interpret; and 3) the degree to which the sandstones to be studied are of uniform composition. Heterogeneous mineral assemblages indicate the influence of multiple source areas. Isotopic study of these nonuniform samples may still be useful in determining provenance, although to characterize accurately the range of isotopic values on such deposits and then correlate these to their multiple source areas can quickly prove to be a very expensive proposition.

Analysis

Isotopic analyses are expensive when compared with the cost of a thin section. Certainly, sampling should be done in a parsimonious manner; on the other hand, as much information as possible is needed from the samples, including an adequate characterization of the variability of the deposits. Therefore, using the isotopic technique in conjunction with other basin-analysis techniques may allow isotopic results to be extrapolated across a sedimentary basin. It is important to study those deposits that are most representative of source area and that provide the most information about the origin and history of a sedimentary basin.

Studying only those minerals that will give valuable information about the source area can greatly simplify a project. For example, zircons may be the only mineral requiring analysis in fission-track or conventional U-Pb studies. On the other hand, depending on the aim of the study, heavy minerals may not make a good choice because they may represent only a minor source area that did not provide most of the deposit. Conversely, it may be worthwhile to avoid exceedingly common minerals that could easily come from various sources and so might produce nonunique results. However, if different sources yield, for example, feldspars of different isotopic ratios, and only one source area was being tapped, then feldspar would be the logical mineral to study. If possible, iron-rich minerals should be avoided because they weather quickly and may not preserve their original isotopic compositions. Heller *et al.* (1985) found consistent isotopic results from white mica in arkoses and lithic sandstones; these had not suffered any isotopic modification.

Ideally, an isotopic provenance study would use a broad spectrum of mineral species and a wide variety of isotopic systems to characterize or fingerprint the sedimentary sequence. Often the availability of isotopic data from potential source areas is uneven. It is desirable to generate a data set for the sedimentary deposits that is compatible with what is known for the source areas. At the same time, applying various isotopic systems to a single mineral type may yield detailed thermo-tectonic information about the source area. This information can be compared with the known tectonic history of a source area, even if its specific isotopic age data are not available. Hence, we emphasize that appropriate isotopic techniques may vary from one provenance study to another and that sufficient effort must be spent in planning and choosing the most suitable method.

Example of the Isotopic Provenance Approach: Case Study from the Tyee Formation, Oregon Coast Range

An example of the multiple-system approach advocated here is presented in Heller *et al.* (1985). That study attempted to determine the source area for the Tyee Formation of Eocene age in the Oregon Coast Range (Fig. 2.2). Previous studies had suggested that the deposits were derived from the adjacent Klamath Mountains to the south (Fig. 2.2) (Snavely *et al.* 1964; Dott 1966; Lovell 1969; Chan and Dott 1983). The basis for these interpretations was vari-

ous traditional basin-analysis techniques including: 1) paleocurrent indicators, which showed a very tight south-to-north paleoflow for the Tyee Formation; 2) changes in lithofacies of the Tyee Formation from deep basinal turbidite deposits in the northern part of the Oregon Coast Range to deltaic and fluvial facies toward the south; and 3) sandstone compositions, which include grain types that indicate a likely Klamath Mountains derivation, including volcanic lithic grains, potassium feldspar, and coarse white-mica grains. This broad spectrum of basin-analysis data suggested that the Klamath Mountains were the sole, or at least dominant, source area.

Heller *et al.* (1985), on the other hand, argued on the basis of isotopic compositions of mineral grains within the Tyee Formation that the Klamath Mountains were not the primary source area for these deposits. Instead, they were derived primarily from the Mesozoic plutonic belt presently found far to the east of the Oregon Coast Range, as had been suggested by Heller and Ryberg (1983). However, the results of previous basin analyses clearly indicate that streams that transported the detritus to the Oregon Coast Range must have flowed *through* the Klamath Mountains.

Heller *et al.* studied various isotopic systems in whole-rock samples and in potassium-feldspar and white-mica (mostly muscovite) separates in order to determine the isotopic provenance of the deposits. The neodymium- and strontium-isotopic analyses of whole-rock samples allowed the determination of the parent material from which the source rocks were derived. An enrichment of $^{87}Sr/^{86}Sr$ coupled with a depletion of $^{143}Nd/^{144}Nd$ relative to bulk-earth uniform reservoir (DePaolo and Wasserburg 1976) and a depleted-mantle reservoir (Goldstein *et al.* 1984), respectively, indicated that the source rocks were most likely derived from continental crust and suggested that the parent material formed in Proterozoic time (Fig. 2.4). These results suggested that the Klamath Mountains were not the major source area for the Tyee Formation because they consist, for the most part, of Phanerozoic oceanic crust and not Precambrian continental crust. Furthermore, the Rb-Sr isotopic ratios of whole-rock samples of the Tyee Formation differed significantly from similar analyses done on underlying clastic units (the Lookingglass Formation) (Peterman *et al.* 1981; Assad 1982) whose mineralogy and Rb-Sr isotopic ratios are consistent with derivation from the Klamath Mountains (Fig. 2.1).

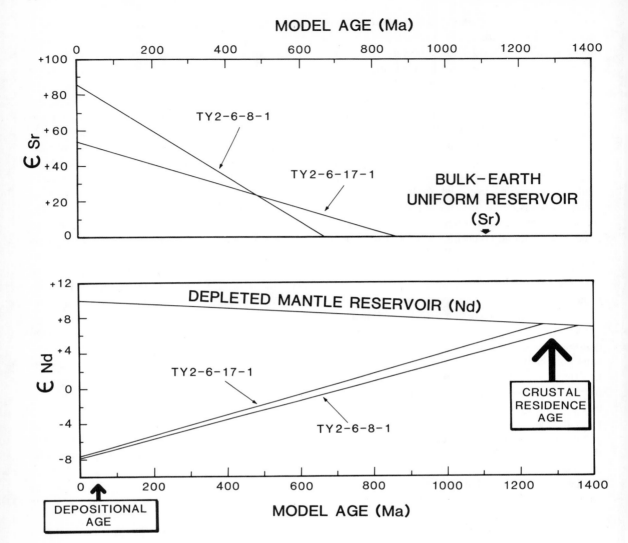

Fig. 2.4. Strontium- and neodymium-evolution diagrams for whole-rock sandstone samples from the Tyee Formation. Data from Heller *et al.* (1985). Depleted mantle reservoir model for neodymium from Goldstein *et al.* (1984). ε is defined as parts per 10,000 deviation from a bulk-earth uniform reservoir for strontium, and a chon-dritic uniform reservoir for neodymium (DePaolo and Wasserburg 1976). Calculated neodymium crustal-residence age of the parent material from which the source rocks for the Tyee Formation were derived range from about 1270 to 1370 Ma.

The Rb-Sr data from the Tyee Formation formed a cloud of points that overlaps with results from modern detritus collected from the Columbia and Snake Rivers in Washington (Fig. 2.1), rivers that drain part of the Mesozoic plutonic belt, including the Idaho Batholith, far to the east of the Oregon Coast Range (Fig. 2.2).

Rb-Sr analyses were also done on potassium-feldspar and white-mica separates from sandstones of the Tyee Formation. The potassium-feldspar values, although having some scatter, were shown not to have been derived from the Klamath Mountains. If Jurassic plutonic bodies in the Klamath region acted as the source for the potassium feldspar grains, then the plutons would have had to have been significantly older than they are, or have a much higher initial $^{87}Sr/^{86}Sr$ ratio than they do. Therefore, Heller *et al.* (1985) concluded that the potassium

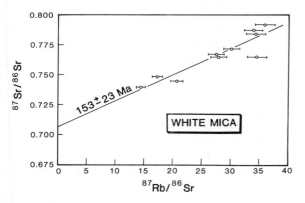

Fig. 2.5. Rb-Sr systematics for white-mica separates from the Tyee Formation. Solid line represents "pseudo-isochron" calculated from data in Heller *et al.* (1985).

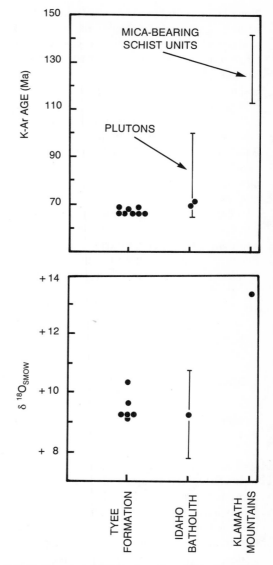

feldspars were probably not derived from the Klamath Mountains, as previously thought (e.g., Dott 1966).

White micas provided the most consistent data set from the Tyee Formation, allowing a detailed tectono-thermal history to be constructed by using a variety of isotope systems. The Rb-Sr data from the micas are relatively colinear, yielding a "pseudo-isochron" of Late Jurassic age (Fig. 2.5). The meaning of this pseudo-isochron is uncertain because it may result from a mixture of white micas that may have been derived from various sources of differing age or reflect a crystallization age, and scatter in the data places large uncertainties on the calculated age. However, Heller *et al.* interpreted the data to mean that the source area was of middle to late Mesozoic age. These same white-mica separates analyzed by the K-Ar method produced consistent ages of about 67 Ma (Fig. 2.6), suggesting that the source area underwent a tectono-thermal event in Late Cretaceous time.

Oxygen-isotope values from the white micas in the Tyee Formation yielded consistent $\delta^{18}O$ values at around +9.5 parts per mil (Fig. 2.6), whereas potassium-feldspar values from the same samples varied between +8.7 and +11.7 parts per mil. The fact that white mica and coexisting potassium feldspars had oxygen values that were not in thermodynamic equilibrium (see Heller *et al.* 1985 for discussion) indicated that these two minerals could not have been derived from the same source rock unless earlier alteration had taken the minerals out of isotopic equilibrium (Criss and Taylor 1983);

Fig. 2.6. K-Ar ages and $\delta^{18}O$ values for white micas from sandstones of the Tyee Formation compared to range of values reported for mica-bearing rock bodies and white-mica separates from the Idaho Batholith and Klamath Mountains. Modified from Heller *et al.* (1985).

however, they could have been derived from two different source rocks in the same source area.

When comparing these data with results reported in the literature for white micas from likely source areas, it became apparent that there were many more isotopic data available on the clastic deposits than had been reported from potential sources.

Nonetheless, available data made it clear that the white-mica values, like those of the whole-rock samples and potassium-feldspar separates, were dissimilar to those obtained from the Klamath Mountains, but were similar in many respects to those from the Idaho Batholith. Although a perfect match for the isotopic fingerprint of the Tyee Formation could not be found, when all the data from the whole rock and mineral separates are taken into account, the Idaho Batholith or similar rocks elsewhere in the Mesozoic plutonic belt of the U.S. Cordillera proves to be the most likely source area. Since rivers tapping this plutonic source must have flowed through the Klamath region to deposit the sediments in the Oregon Coast Range, the Klamath Mountains must have had relatively low relief during this part of middle Eocene time in order not to contribute significantly to the composition of the Tyee Formation. This discovery of a primary source area far to the east of the present Oregon Coast Range allowed previously proposed tectonic models of terrane displacement and rotation to be evaluated.

Conclusions

Isotopic studies of clastic deposits can be used to delineate source areas to a degree not always possible using other basin-analysis techniques. Both specific source areas as well as average source-area composition can be determined depending on the compositional maturity and grain size of the sedimentary rock used. Immature sandstones are best used when specific source areas are to be ascertained, and fine-grained clastics are better for determining the average composition of large regions.

The isotopic-provenance technique is particularly well suited for complex tectonic settings where traditional basin-analysis techniques alone may yield equivocal results. This tool is particularly useful because even if potential source areas have broadly similar mineralogies and so produce sandstones of like composition, even minor isotopic differences in those minerals between source areas can be used to determine specific provenance. A wide variety of isotopic systems can be studied simultaneously on both whole-rock and separate minerals in a clastic deposit, yielding a detailed tectono-thermal history that is more likely to be unique for each source area. However, this approach can be very time-consuming and labor intensive, and therefore, should be used

only in cases where likely source areas cannot be discriminated with certainty by any other means. Even then, the approach is best used in conjunction with other basin-analysis techniques in order to yield the most complete interpretation of the history of the sedimentary basin and the provenance of its deposits. If the application of the isotopic method is well thought out so that only those systems that are likely to produce unequivocal provenance information are used, then the isotopic-provenance technique may prove to be a very powerful tool in the arsenal of sedimentary basin analysts.

Acknowledgments. This manuscript has greatly benefitted by collaborative research over the past several years among Jim O'Neil, Zell Peterman, Muhammad Shafiqullah, and Paul Heller. Reviews by E. Ito, R.V. Ingersoll, A.H. Bouma, two anonymous reviewers, and, especially, Z.E. Peterman are much appreciated. Funding for isotopic provenance studies has been provided by NSF (EAR82-08759, EAR84-07117) to P. . Heller and NSF (EAR84-08357) to C.D. Frost.

References

ALBAREDE, F. (1982) The $^{39}Ar/^{40}Ar$ technique of dating. In: Odin, G.S. (ed) Numerical Dating in Stratigraphy. Chichester: John Wiley and Sons, pt. 1, pp. 181–197.

ASSAD, R. (1982) Comment on provenance of Eocene graywackes of the Flournoy Formation near Agness, Oregon—A geochemical approach. Geology 10:333–334.

BALDWIN, S.L., HARRISON, M.T., and BURKE, K. (1986) Fission track evidence for the source of accreted sandstones, Barbados. Tectonics 5:457–468.

BHATIA, M.R. (1983) Plate tectonics and geochemical composition of sandstones. Journal Geology 91:611–628.

BHATIA, M.R. and CROOK, K.A.W. (1986) Trace element characteristics of graywackes and tectonic setting discrimination of sedimentary basins. Contributions Mineralogy Petrology 92:181–193.

BISCAYE, P.E. and DASCH, E.J. (1971) The rubidium, strontium, strontium isotope system in deep-sea sediments: Argentina Basin. Journal Geophysical Research 76:5087–5096.

BOGER, P.D. and FAURE, G. (1974) Strontium-isotope stratigraphy of a Red Sea core. Geology 2:181–183.

CAMERON, K.L. and BLATT, H. (1971) Durabilities of sand size schist and "volcanic" rock fragments during fluvial transport, Elk Creek, Black Hills, South Dakota. Journal Sedimentary Petrology 41:565–576.

CERVENY, P.F., NAESER, N.D., ZEITLER, P.K., NAESER, C.W., and JOHNSON, N.M. (1988) History of uplift and relief of the Himalaya during the past 18 million years: Evidence from fission-track ages of detrital zircons from sandstones of the Siwalik Group. In: Kleinspehn, K.L. and Paola, C. (eds) New Perspectives in Basin Analysis. New York: Springer-Verlag, pp. 42–61.

CHAN, M.A. and DOTT, R.H., JR. (1983) Shelf and deep-sea sedimentation in Eocene forearc basin, western Oregon—fan or non-fan? American Association Petroleum Geologists Bulletin 67:2100–2116.

CHOW, T.J. and PATTERSON, C.C. (1962) The occurrence and significance of lead isotopes in pelagic sediments. Geochemica Cosmochemica Acta 26:263–308.

CLAUER, N. (1979) Relationship between the isotopic composition of strontium in newly formed continental clays and their source material. Chemical Geology 27:115–124.

CLAUER, N. (1982) The rubidium-strontium method applied to sediments: Certitudes and uncertainties. In: Odin, G.S. (ed) Numerical Dating in Stratigraphy. Chichester: John Wiley and Sons, pt. 1, pp. 245–276.

COLEMAN, R.G. (1972) The Colebrooke Schist of Southwestern Oregon and its Relation to the Tectonic Evolution of the Region. United States Geological Survey Bulletin 1339, 61 p.

CONDIE, K.C., MACKIE, J.E., and REIMER, T.O. (1970) Petrology and geochemistry of early Precambrian graywackes from the Fig Tree Group, South Africa. Geological Society America Bulletin 81:2759–2776.

CORDANI, V.G., KAWASHITA, K., and FILHO, A.T. (1978) Applicability of the rubidium-strontium method to shales and related rocks. In: Cohee, G.V., Glaessner, M.F., and Hedberg, H.D. (eds) Contributions to the Geologic Time Scale. American Association Petroleum Geologists Studies Geology 6, pp. 93–117.

CRISS, R.E. and TAYLOR, H.P., JR. (1983) An $^{18}O/^{16}O$ and D/H study of Tertiary hydrothermal systems in the southern half of the Idaho Batholith. Geological Society America Bulletin 94:640–663.

DASCH, E.J., HILLS, F.A., and TUREKIAN, K.K. (1966) Strontium isotopes in deep-sea sediments. Science 153:295–297.

DEPAOLO, D.J. and WASSERBURG, G.J. (1976) Nd isotopic variations and petrogenetic models. Geophysical Research Letters 3:249–252.

DEPAOLO, D.J. and WASSERBURG, G.J. (1977) The sources of island arcs as indicated by Nd and Sr isotopic studies. Geophysical Research Letters 4:465–468.

DICKINSON, W.R. (1970) Interpreting detrital modes of graywacke and arkose. Journal Sedimentary Petrology 40:695–707.

DICKINSON, W.R. (1982) Compositions of sandstones in circum-Pacific subduction complexes and fore-arc basins. American Association Petroleum Geologists Bulletin 66:121–137.

DICKINSON, W.R. (1985) Interpreting provenance relations from detrital modes of sandstones. In: Zuffa, G.G. (ed) Provenance of Arenites. Dordrecht: Reidel Publishing Co., pp. 333–361.

DICKINSON, W.R. and SUCZEK, C.A. (1979) Plate tectonics and sandstone compositions. American Association Petroleum Geologists Bulletin 63:2164–2182.

DODSON, M.H. and McCLELLAND-BROWN, E. (1985) Isotopic and paleomagnetic evidence for rates of cooling, uplift and erosion. In: Snelling, N.J. (ed) The Chronology of the Geological Record. Geological Society London Memoir 10, pp. 315–325.

DOE, B.R. (1970) Lead Isotopes. New York: Springer-Verlag, 137 p.

DOTT, R.H., JR. (1966) Eocene deltaic sedimentation at Coos Bay, Oregon. Journal Geology 74:373–420.

FAURE, G. (1986) Principles of Isotope Geology (2nd edition). New York: John Wiley and Sons, 589 p.

FAURE, G., TAYLOR, K.S., and MERCER, J.H. (1983) Rb-Sr provenance dates of feldspar in glacial deposits of the Wisconsin Range, Transantarctic Mountains. Geological Society America Bulletin 94:1275–1280.

FLEISCHER, R.L., BUFORD, P. and WALKER, R.M. (1975) Nuclear Tracks in Solids: Principles and Applications. Berkeley: University of California, 605 p.

FOLK, R.L. (1974) Petrology of Sedimentary Rocks. Austin: Hemphill Publishing Co., 182 p.

FROST, C.D. and O'NIONS, R.K. (1984) Nd evidence for Proterozoic crustal development in the Belt-Purcell Supergroup. Nature 312:53–56.

FROST, C.D. and O'NIONS, R.K. (1985) Caledonian magma genesis and crustal recycling. Journal Petrology 26:515–544.

FROST, C.D. and WINSTON, D. (1987) Nd Isotope systematics of coarse- and fine-grained sediments: Examples from the Middle Proterozoic Belt-Purcell Supergroup. Journal Geology 95:309–327.

FROUDE, D.O., IRELAND, T.R., KINNEY, P.D., WILLIAMS, I.S., COMPSTON, W., WILLIAMS, I.R., and MYERS, J.S. (1983) Ion microprobe identification of 4,100–4,200 Myr-old terrestrial zircons. Nature 304:616–618.

GALLOWAY, W.E. (1974) Deposition and diagenetic alteration of sandstone in northeast Pacific arc-related basins: Implications for graywacke genesis. Geological Society America Bulletin 85:379–390.

GOLDSTEIN, S.L., O'NIONS, R.K., and HAMILTON, P.J. (1984) A Sm-Nd isotopic study of atmospheric dusts and particulates from major river systems. Earth Planetary Science Letters 70:221–236.

HARRISON, T.M. and BÉ, K. (1983) $^{40}Ar/^{39}Ar$ age spectrum analysis of detrital microclines from the southern

San Joaquin Basin, California: An approach to determining the thermal evolution of sedimentary basins. Earth Planetary Science Letters 64:244–256.

HELLER, P.L. and DICKINSON, W.R. (1985) Submarine ramp facies model for delta-fed, sand-rich turbidite systems. American Association Petroleum Geologists Bulletin 69:960–976.

HELLER, P.L., PETERMAN, Z.E., O'NEIL, J.R., and SHAFIQULLAH, M. (1985) Isotopic provenance of sandstones from the Eocene Tyee Formation, Oregon Coast Range. Geological Society America Bulletin 96:770–780.

HELLER, P.L. and RYBERG, P.T. (1983) Sedimentary record of subduction to forearc transition in the rotated Eocene basin of western Oregon. Geology 11:380–383.

HISCOTT, R.N. (1984) Ophiolitic source rocks for Taconic-age flysch: Trace element evidence. Geological Society America Bulletin 95:1261–1267.

HOEFS, J. (1980) Stable Isotope Geochemistry (2nd edition). Berlin: Springer-Verlag, 208 p.

LAND, L.S. (1983) The application of stable isotopes to studies of the origin of dolomite and to problems of diagenesis of clastic sediments. Society Economic Paleontologists Mineralogists Short Course 10, pp. 4-1–4-22.

LOVELL, J.P.B. (1969) Tyee Formation: Undeformed turbidites and their lateral equivalents: Mineralogy and paleogeography. Geological Society America Bulletin 80:9–22.

MACK, G.H. (1978) The survivability of labile light mineral grains in fluvial, aeolian, and littoral marine environments: The Permian Cutler and Cedar Mesa Formation, Moab, Utah. Sedimentology 25:587–606.

MIALL, A.D. (1984) Principles of Sedimentary Basin Analysis. New York: Springer-Verlag, 490 p.

MICHARD, A., GURRIET, P., SOUDANT, M., and ALBAREDE, F. (1985) Nd isotopes in French Phanerozoic shales: External vs. internal aspects of crustal evolution. Geochemica Cosmochemica Acta 49:601–610.

NAESER, C.W. (1978) Fission track dating. U.S. Geological Survey Open-file Report 76–190, revised Jan. 1978.

NELSON, B.K. and DEPAOLO, D.J. (1985) Rapid production of continental crust 1.7 to 1.9 by ago: Nd isotopic evidence from the basement of the North American mid-continent. Geological Society America Bulletin 96:746–754.

NELSON, B.K. and DEPAOLO, D.J. (in press) Application of Sm-Nd and Rb-Sr isotope systematics to studies of provenance and basin analysis. Journal Sedimentary Petrology.

ODIN, G.S. (ed) (1982) Numerical Dating in Stratigraphy. Chichester: John Wiley and Sons, pt. 1, 630 p.

O'NIONS, R.K., HAMILTON, P.J., and HOOKER, P.J. (1983) A Nd isotope investigation of sediments related to crustal development in the British Isles. Earth Planetary Science Letters 63:229–240.

PATTERSON, C. and TATSUMOTO, M. (1964) The significance of lead isotopes in detrital feldspar with respect to chemical differentiation within the earth's mantle. Geochemica Cosmochemica Acta 28:1–22.

PERRY, E.A., JR. and TUREKIAN, K.K. (1974) The effects of diagenesis on the redistribution of strontium isotopes in shales. Geochemica Cosmochemica Acta 38:929–935.

PETERMAN, Z.E., COLEMAN, R.G., and BUNKER, C.M. (1981) Provenance of Eocene graywackes of the Flournoy Formation near Agness, Oregon—A geochemical approach. Geology 9:81–86.

PETERMAN, Z.E., HEDGE, C.E., COLEMAN, R.G., and SNAVELY, P.D., JR. (1967) $^{87}Sr/^{86}Sr$ ratios in some eugeosynclinal sedimentary rocks and their bearing on the origin of granitic magma in orogenic belts. Earth Planetary Science Letters 2:433–439.

PETERMAN, Z.E. and WHETTEN, J.T. (1972) Sr^{87}/Sr^{86} variation in Columbia River bottom sediments as a function of provenance. Geological Society America Memoir 135, pp. 29–36.

ROSER, B.P. and KORSCH, R.J. (1986) Determination of tectonic setting of sandstone-mudstone sites using SiO_2 content and K_2O/Na_2O ratio. Journal Geology 94:635–650.

SCHÄRER, U. and ALLÈGRE, C.J. (1985) Determination of the age of the Australian continent by single grain zircon analysis of the Mt. Narryer metaquartzite. Nature 315:52–55.

SNAVELY, P.D., JR., WAGNER, H.C., and MACLEOD, N.S. (1964) Rhythmic-bedded eugeosynclinal deposits of the Tyee Formation, Oregon Coast Range. Kansas Geological Survey Bulletin 169:461–480.

SUTTNER, L.J., BASU, A., and MACK, G.H. (1981) Climate and the origin of quartz arenites. Journal Sedimentary Petrology 51:1235–1246.

TAYLOR, S.R. and McLENNAN, S.H. (1985) The Continental Crust: Its Composition and Evolution. Oxford: Blackwell Scientific Publications, 312 p.

TURNER, G. (1968) The distribution of potassium and argon in chondrites. In: Ahrens, L.H. (ed) Origin and Distribution of the Elements. London: Pergamon Press, pp. 387–398.

VAN ANDEL, T.H. (1959) Reflections on the interpretation of heavy mineral analyses. Journal Sedimentary Petrology 29:153–163.

VAN DE KAMP, P.C. and LEAKE, B.E. (1985) Petrography and geochemistry of feldspathic and mafic sediments of the northeastern Pacific margin. Transactions Royal Society Edinburgh 76:411–449.

YIM, W.W.-S., GLEADOW, A.J.W., and VAN MOORT, J.C. (1985) Fission track dating of alluvial zircons and heavy mineral provenance in northeast Tasmania. Journal Geological Society London 142:351–356.

ZEITLER, P.K., JOHNSON, N.M., NAESER, C.W., and TAHIRKHELI, R.A.K. (1982) Fission-track evidence for Quaternary uplift of the Nanga Parbat region, Pakistan. Nature 298:255–257.

3

History of Uplift and Relief of the Himalaya During the Past 18 Million Years: Evidence from Fission-Track Ages of Detrital Zircons from Sandstones of the Siwalik Group

P.F. Cerveny, N.D. Naeser, P.K. Zeitler, C.W. Naeser, and N.M. Johnson

Abstract

Fission-track dating of individual detrital zircon grains can be used to characterize both ancient and modern sedimentary provenance. Ages of zircons in the modern Indus River drainage system of northern Pakistan are controlled dominantly by uplift rates of the source rocks in the Himalaya. Young detrital zircons come from rapidly rising terrain, whereas old zircon ages imply slow or negligible uplift. Modern Indus River sands contain a distinctive population of young, 1 to 5 Ma, zircons that are derived from the Nanga Parbat-Haramosh Massif, an area of rapid uplift (5 m/10³ yr). Sandstones of the Siwalik Group deposited by the ancestral Indus River over the past 18 million years contain zircons that are only 1 to 5 million years older than the depositional age of the sandstones. Therefore, young zircons have been a consistent component of Himalayan surface rocks for the past 18 million years. These ages imply that a series of uplifted blocks or "massifs," analogous to the contemporary Nanga Parbat area, have been continually present in the Himalaya since 18 Ma, and that over that time the elevation and relief of the Himalaya, on a broad scale, have been essentially constant.

Introduction

This paper describes the relief and uplift in the upper Indus River watershed during the past 18 million years, based on fission-track ages of detrital zircons derived from the sediments of the present Indus River of northern Pakistan and from sandstones of the Siwalik Group, which originated from sediments deposited by the ancestral Indus River during Neogene time. The data analysis represents a new methodology in the study of source terrain. Previous studies have used fission-track dating of detrital zir-

cons to identify source area based on the ages of the detrital grains (e.g., Hurford et al. 1984; Johnson 1984; Yim et al. 1985; Baldwin et al. 1986). However, none of these studies was directly concerned in detail with using the ages of the detrital grains in sedimentary rocks to interpret the *tectonic history* of the source terrain. Detailed interpretation of the tectonic history is the central purpose of this paper.

Zeitler (1985) and Zeitler et al. (1982b,c) have demonstrated that in the Himalaya, areas characterized by rapid uplift rates have zircons with young fission-track ages (Fig. 3.1). Zeitler et al. (1982a) showed further that young zircons are present in the detrital zircon suite currently being eroded from the Himalaya. These data suggest that by dating detrital zircons separated from Siwalik Group sandstones of known stratigraphic age, it would be possible to determine the ages of zircon that were being eroded from the basement source terrain in northern Pakistan at selected intervals during Neogene time, and that the detrital zircons might thus be a vehicle through which uplift rates from the geologic past could be assessed (Zeitler et al. 1982a). This possibility is explored in this paper. The primary objective of the present study is to determine if zircons with young ages, which are indicative of high uplift rates, have been present in the Himalaya over the past 18 million years.

Geography and Geologic Setting

The study area in the Punjab region of northern Pakistan (Fig. 3.2) contains the type sections for most of the formations of the Siwalik Group (Fatmai 1974), which consists of sandstones derived from

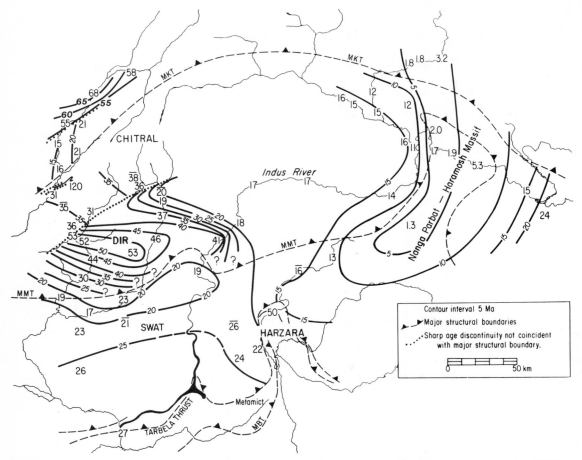

Fig. 3.1. Contour map of zircon fission-track ages from surface rocks in northern Pakistan (Zeitler 1985). Note the 50-fold variation in zircon age. MMT = Main Mantle Thrust; MKT = Main Karakoram Thrust; MBT = Main Boundary Thrust. The geographic location of this map may be identified in Figure 3.2 by the configuration of the Indus River, notably its right-angle, eastward bend in northern Pakistan.

Neogene fluvial sediments eroded from the Himalaya. The Siwalik Group is composed of fluvial cycles (Allen 1965), which represent the migrations of a major trunk river (the ancestral Indus) over the Punjab region. Over the past 18 million years, the return period of the major river in the Punjab region has been 10^4 to 10^5 yr (Johnson *et al.* 1985).

The ages of the Siwalik Group sandstones have been established by magnetic-polarity stratigraphy and fission-track dating techniques (Opdyke *et al.* 1979; Johnson *et al.* 1982, 1985). In the vicinity of Chinji Village and in the Trans Indus (Fig. 3.2), the Siwalik sequence begins at a basal contact with Eocene limestone and continues stratigraphically upward through the Lower, Middle, and, in the case of the Trans Indus section, Upper Siwalik

sequences. The Chinji Village section is approximately 2 km thick and contains the Kamlial (oldest), Chinji, Nagri, and Dhok Pathan Formations (Fig. 3.3). The Trans Indus section is composed of approximately 4.2 km of sedimentary rocks and contains the Chinji, Nagri, Dhok Pathan, and Soan Formations.

The rocks of the Himalaya, which were the source of the sediments forming the rocks of the Siwalik Group, are today being eroded and transported by the contemporary Indus River system. The distribution of zircon fission-track ages in the bedrock presently exposed in the Indus River watershed (Zeitler 1985) is illustrated in Figure 3.1. The Nanga Parbat-Haramosh Massif defines an area of 1.3 to 3.2 Ma zircons, which is surrounded by an area

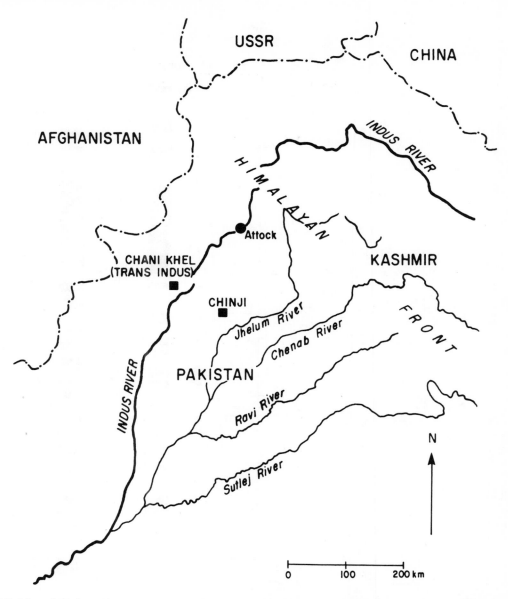

Fig. 3.2. Map of northern Pakistan showing: 1) location of Siwalik Group stratigraphic sections near Chinji Village and Chani Khel (Trans Indus), and 2) collection site for the present Indus River sand (Attock).

where ≈ 5 Ma zircons occur. Moving farther outward from the Nanga Parbat-Haramosh Massif, zircon ages range from 11 to 24 Ma. In the westernmost area, in the Dir-Chitral region, ages are concentrated in the 30 to 55 Ma range, with ages in outlying areas up to 120 Ma. However, essentially all the rocks in the Himalaya and the Lesser Himalaya of northern Pakistan are Mesozoic or older in age. Thus, as shown in Figure 3.1, the zircons from these rocks commonly yield fission-track ages much younger than their true ages. These age differences are explained by the fact that the rocks now at the surface were once at depth, at temperatures sufficiently high to reset their fission-track ages (see below). Thus, the ages of zircons at the surface today are recording the time when the rocks were uplifted through their closure temperature and started to retain fission tracks. In the Nanga Parbat-Haramosh Massif area, where the closure temperature for zircon is assumed to occur at a depth of ≈ 6 km (see

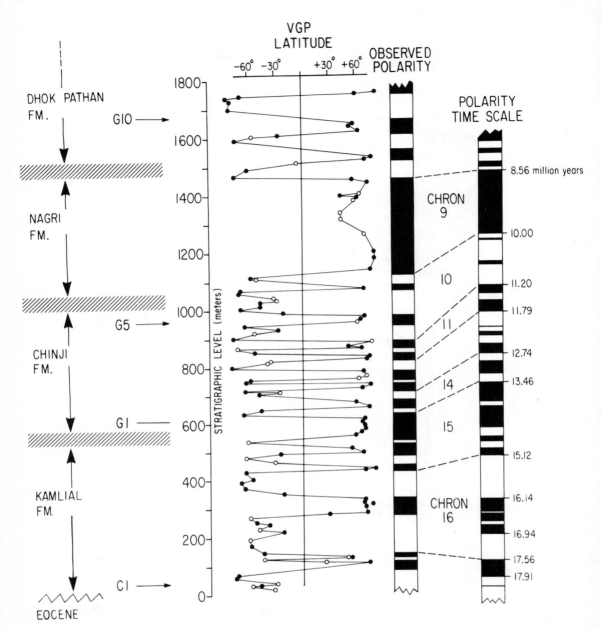

Fig. 3.3. Magnetic polarity stratigraphy of the Siwalik Group sequence near Chinji Village (modified from Johnson *et al.* 1985), showing the stratigraphic level of sand samples (C1, G1, G5, and G10) collected for this study. Solid and open circles in the magnetic data indicate class A data and class B data, respectively.

below), the 1.3 to 3.2 Ma zircon ages indicate that the mean uplift rate for this area at the present time is on the order of 5 m/10^3 yr; the area has been uplifted 10 km in the last 2 million years (Zeitler 1985).

Fission-Track Dating

Theory

Fission tracks can be thought of as analogous to a trace fossil found in uranium-bearing minerals such as apatite, zircon, and sphene. Each fission track records the decay by spontaneous fission of a single nucleus of ^{238}U into subequal halves, with the track being a tubular zone of damage within the host mineral's lattice. Chemical etching can make these damaged zones visible by means of an ordinary optical microscope (etched fission tracks are on the order of 10 to 20 μm in length and 1 to 3 μm in diameter). The fission-track method of dating is not one of high precision, because tracks tend to be relatively few in number (the probability of a given uranium atom decaying by fission in any given year is less than 1 in 10^{16}), but the method is extremely sensitive, with each track corresponding to a single fissioned uranium nucleus.

The most direct geological application of fission tracks is in geochronology. Irradiation with "slow" or "thermal" neutrons in a nuclear reactor induces fission of some of the ^{235}U found in the mineral (the isotopic composition of uranium in nature is constant, and the present-day value is $^{238}U/^{235}U = 137.88$). Together with the use of neutron-fluence monitors and standards of known ages, uranium-bearing minerals can be dated by counting, per unit area, the number of spontaneous or "fossil" fission tracks as well as the number of induced fission tracks created in the reactor. The density of spontaneous tracks is a function of age and uranium content, and the density of the induced tracks is in effect a measurement of uranium content by neutron activation. It is important to note that fission-track ages are very different than those derived using the conventional U-Pb method, which is based on the decay of uranium to lead through a series of intermediate daughters. Details about fission-track dating procedures can be found in Naeser (1976, 1979), Naeser et al. (in press a), and Hurford and Green (1982). The book by Fleischer et al. (1975) remains an invaluable introduction to the early history and fundamentals of fission-track dating and charged-particle tracks in general.

Annealing of Fission Tracks

Despite the great value of the fission-track method in dating young rocks, its most important contribution to geology lies in the determination of the thermal histories of rocks. At temperatures of several hundreds of degrees Celsius (about 100°C for apatite, 200°C for zircon, and 300°C for sphene), over geological time spans, fission tracks will fade or anneal. In response to increasing temperatures, observed track densities are reduced because tracks shorten; this shortening reflects repair of the host crystal's lattice as a consequence of the enhanced rate of solid-state diffusion that occurs at higher temperatures. Like vitrinite reflectance, track fading is a time-dependent and temperature-dependent process (Fleischer et al. 1965), and equivalent effects can be achieved by heating at low temperatures for prolonged periods or at high temperatures for shorter periods. Examples of studies demonstrating the way in which fission-track dating can be used to constrain thermal events include those by Briggs et al. (1981), Gleadow and Duddy (1984), and Naeser et al. (in press a). Gleadow et al. (1983) and Green et al. (in press) include a review of the use of track-length distribution as a further constraint on the thermal history of a sample. At high enough temperatures, fission tracks fade approximately as fast as they form. After sufficient cooling, a point will be reached where no annealing occurs and all tracks are fully retained. The temperature at which full retention begins is known as the closure temperature (Dodson 1979), and varies from mineral to mineral. Thus, when a mineral cools through its closure temperature, the fission-track clock begins to run. The closure temperature for cooling has a small but significant dependence on cooling rate (about a 10% increase per order-of-magnitude increase in cooling rate). Fixed values for the closure temperature of each mineral simply do not exist. Two examples of studies employing the closure behavior of fission tracks in minerals are those by Wagner et al. (77) in the Alps and Zimmermann et al. (1975) in the northern Appalachians.

In the present study, it is the thermally sensitive nature of fission tracks that is exploited. Most of the bedrock exposed in the northwestern Himalaya today has cooled by well over 200°C during Tertiary time in direct response to uplift and erosion. Thus, zircons entrained in these rocks yield young ages that record the time that they passed through 200°C (which is, roughly, the closure temperature for zircon). In an active mountain belt, an attempt to fix rigorously the location of a paleo-isotherm in the crust would require sophisticated modelling (Parrish 1982). However, as a rough approximation we can say that uplift from the 200°C isotherm represents approximately 6 km of uplift and erosion (less in areas of fast and sustained uplift). Clearly, uplift rate is inversely proportional to the fission-track age, and in a setting like the northwestern Himalaya, younger ages imply more rapid uplift.

Detrital Fission-Track Ages

Zircons deposited in a sandstone at a given time represent a composite of the zircons that were being eroded from the source area and deposited at that time. A detrital zircon acquires some fraction of its present age during its residence time in the uplift, transport, and post-depositional portions of its history.

Figure 3.4 is a schematic representation of the life history of a detrital zircon grain in the uplift-erosion-depositional cycle of the Himalayan-Siwalik system, where (in years before present)

$T1$ = time the zircon passes through its closure temperature and begins to accumulate fission tracks

$T2$ = time the zircon reaches the surface (through uplift and erosion) and enters the sedimentary cycle

$T3$ = time of deposition in the sedimentary basin

Thus,

$T_u = T1 - T2$ = elapsed time zircon required to reach surface and enter sedimentary cycle

$T_t = T2 - T3$ = elapsed time spent in transport

$T_s = T3$ − present = elapsed time since final deposition, i.e., the stratigraphic age of the sediment, and

$$\text{Observed Age} = T1 = T_u + T_t + T_s \qquad (1)$$

For zircon ages measured in the bedrock of the Himalaya today (i.e., Fig. 3.1), $T_t = T_s = 0$ and, therefore, $T1 = T_u$. Zeitler (1985) was thus able to evaluate observed ages ($T1$) directly as uplift ages (T_u).

To obtain uplift ages from detrital zircons, however, it is necessary to know or constrain T_t and T_s. Stratigraphic ages (T_s) in the Himalayan-Siwalik system have already been established in detail by magnetic-polarity stratigraphy and fission-track dating of the Siwalik Group (Opdyke *et al.* 1979; Johnson *et al.* 1982, 1985). Data from modern Indus River sand allow us to evaluate detritus transport time, T_t (see below). By thus fixing T_s and T_t, we can effectively isolate T_u from the observed ages of detrital grains in the Siwalik sandstones and gain insight into the uplift history of the northwestern Himalaya over the last 18 million years, during the time of deposition of the Siwalik Group.

Methods

Sampling

Four Siwalik Group sandstone samples collected in the vicinity of Chinji Village (Figs. 3.2, 3.3) were dated. The stratigraphic ages of these sandstones range from 18 to 7 Ma (Johnson *et al.* 1985). Four additional Siwalik sandstone samples from the Trans Indus area 120 km to the west near Chani Khel (Fig. 3.2) were dated. The ages of these sandstones were estimated from their stratigraphic position and from the known ages of the formations elsewhere in Pakistan (Opdyke *et al.* 1979; Johnson *et al.* 1982, 1985). One sample collected from a sandbar of the modern Indus River was also dated.

In both stratigraphic sections, samples were collected from the base of the local Siwalik Group sequence above the contact with Eocene limestone. In the Chinji Village section, samples were collected from the basal Kamlial Formation (sample number C1), the Chinji Formation (G1 and G5), and the lower Dhok Pathan Formation (G10; Fig. 3.3). The Trans Indus section, where the Kamlial is not present, included samples from the Chinji (K-7 and CK-11), Nagri (CK-10), and Dhok Pathan Formations (CK-5).

Analysis of the heavy-mineral content from the nine dated samples and from 35 additional samples collected in the study area is given elsewhere (Cerveny 1986; Cerveny *et al.* in press).

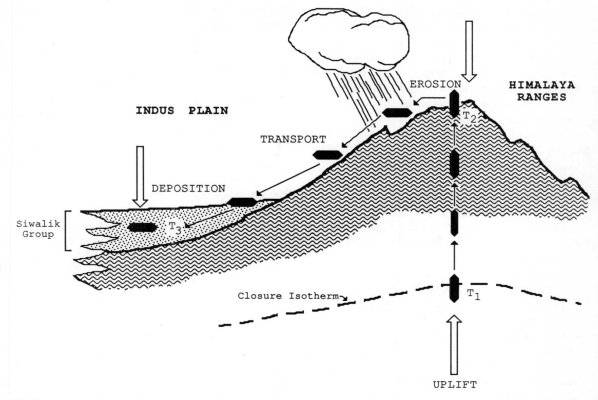

Fig. 3.4. Schematic representation of the life history of a zircon grain (solid symbol) in the Himalaya-Indus-Siwalik system. Dashed line = approximate position of closure isotherm for retention of fission tracks in zircon (see text). For explanation of T1, T2, T3, see text.

Laboratory Procedures

The external-detector method (Naeser 1976, 1979) was used to determine the age of each of the detrital zircons. In the external detector method, the fossil tracks are counted in a single grain and the induced tracks are counted in a muscovite detector that covered that grain during the neutron irradiation. This procedure permits the determination of an age for each individual grain in the mount. Zircon crystals must be dated individually because: 1) uranium is distributed inhomogeneously within and between zircon crystals; and 2) in the case of detrital grains in sediments, each zircon crystal may have a different age.

Thermal neutron fluence was determined from a calibrated muscovite detector covering National Bureau of Standards glass standard SRM 962 (37.38 ± 0.08 ppm U) placed at the top and bottom of each irradiation tube. The fluences were calibrated against the Cu value determined at the National Bureau of Standards (Carpenter and Reimer 1974). The fluence for each sample was calculated by interpolation between the values determined for the standards. This method of fluence determination, when used in conjunction with a value of 7.03×10^{-17} yr^{-1} (Roberts et al. 1968) for the spontaneous fission decay constant and with the laboratory procedures followed in the United States Geological Survey fission-track laboratory, consistently yields fission-track ages that are concordant with K-Ar ages of co-existing phases in rapidly cooled (volcanic and hypabyssal) rocks (e.g., Naeser et al. 1977).

The methods used to prepare zircons from a sample of unimodal age for fission-track dating are well known (Naeser 1976; Naeser and Naeser 1984). However, in a detrital zircon suite the conventional procedure must be modified to allow grains from a wide range of ages to be properly etched. Generally, the time required to etch fission tracks properly in a zircon grain is inversely proportional to fission-track density, and fission-track density is roughly proportional to the age of the grain. Thus it is usually not possible to develop tracks properly in all grains in a detrital suite with a single etch. A double-etch procedure (Naeser *et al.* in press b; see also Baldwin *et al.* 1986) was used to overcome this problem in the detrital zircon samples.

In the double-etch method, two mounts are made of the detrital zircon suite from a given sandstone. One of the mounts (long etch) is etched in a eutectic KOH-NaOH melt (Gleadow *et al.* 1976) to the point where there are no grains that are visibly under-etched. At this point the grains characterized by low track density (generally the younger grains) are fully etched, whereas most of the grains with high track density (generally the older grains) are over-etched or completely dissolved away (in the case of meta-mict grains). Grains with intermediate track density can range from fully etched to over-etched.

A second mount (short etch) is etched for roughly half the time of the first, until the grains characterized by high track density are properly etched. In this mount, the grains with low track density will be under-etched and grains with intermediate track density (generally the intermediate ages) will usually range from under-etched to fully etched.

The end result is that young grains will usually be properly etched in the long-etch mount and old grains will be properly etched in the short-etch mount, whereas grains with intermediate ages will be properly etched in one or the other or both of the two mounts. Therefore, by using the double-etch method, we hope to have properly etched grains available for counting from the full range of ages in the detrital zircon suite. Data from the short-etch and long-etch mounts were not combined when plotting the ages of individual grains in a sample (see below) because such a combined plot would over-emphasize the grains from age populations that are properly etched and countable in both mounts (usually the intermediate-age grains). The following discussion of the detrital zircon ages will concentrate on data from the long-etch mounts because the primary interest in this study is determining the ages of the *youngest* zircons present in the detrital suites.

In a sample with a single age, it is usually sufficient to count fission tracks in only 6 to 12 grains (Naeser 1976). However, in most detrital suites there is a wide range of ages and 6 to 12 grains are unlikely to be representative of the sample. Experience suggests that dating 40 grains per mount (80 grains per sample) is sufficient to provide a representative cross section of the major age populations present in the sample.

Only fully etched grains were counted, and no grain that was fully etched and countable was omitted. The long-etch mount was generally easier to count because by design it contained no under-etched grains. By contrast, the short-etch mount was more difficult because there were many varying stages of etch from grain-to-grain that made judgement of the quality of etch in each grain more subjective. As an operational rule, if we were uncertain of the degree of the etch of a grain, the grain was not counted. The counting techniques used were those of Naeser (1976).

The standard error of the mean for each zircon age was estimated from the equations of McGee *et al.* (1985). The specific data for each sample are given in Cerveny (1986).

Data Plotting

Once the zircons from a detrital suite have been dated, a method of combining the individual grain ages is needed. The simplest method is to plot a histogram of the individual calculated ages. However, a more realistic representation of the age populations in a detrital suite must take into account the analytical uncertainty in each individual grain age. A probability density distribution of age for each grain may be determined from its calculated age (A_i) and standard error of the age (s_i):

$$f(a) = s_i^{-1}(2\pi)^{-1/2} \exp\text{-}(a\text{-}A_i)^2/2s_i^2 \qquad (2)$$

where a is any given age.

The probability density distribution for all the zircons dated in a detrital suite is formed by combining the probability distributions of the individual grains:

$$f'(a) = \sum_{i=1}^{n} s_i^{-1}(2\pi)^{-1/2} \exp\text{-}(a\text{-}A_i)^2/2s_i^2 \qquad (3)$$

This summation produces a plot, or age spectrum, of relative probable frequency versus age for the detrital suite (e.g., Fig. 3.5).

This plotting method has been used in several previous studies to identify the age populations present in a detrital suite (Hurford *et al.* 1984; Kowallis *et al.* 1986). Naeser *et al.* (in press b) have demonstrated, by working with a "detrital suite" manufactured from zircon populations of known age, that the age modes, or peaks, in a spectrum do

Fig. 3.5. Expected distribution of detrital zircon ages when the age spectrum from the contemporary Indus River sands (long-etch mounts) is projected into the future (see text).

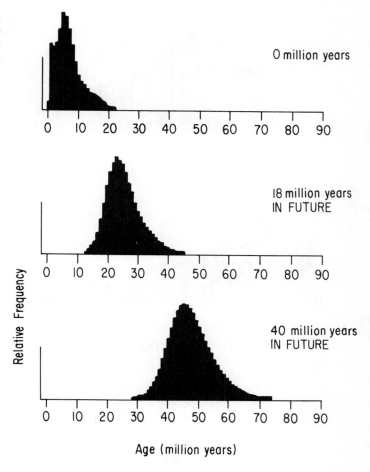

accurately portray the age populations present in the suite. The peaks are particularly well defined for the younger age populations.

To compare the age spectra of detrital grains from different samples, it is useful to standardize them to equivalent geologic conditions, for example, as they existed at the time of deposition of the sediment. This requires a transformation of each age spectrum as it is observed today into what it was at the time of deposition. This transformation requires that the number of spontaneous fission tracks (N_s) used to calculate the age and standard error of the age for each grain be changed as a function of time. The number of neutron-induced fission tracks is set by the particular neutron dose used in the experiment; it is independent of the age of the zircon and is used only to estimate the uranium concentration in the grain being dated. Because of the stochastic nature of the fission process, it is not possible to predict exactly what number of spontaneous fission events has occurred in the time interval concerned. It is

possible, however, to predict the *most likely* number of spontaneous fission events (ΔN_s) that would have occurred in a given time interval (Δt) for each grain (McGee *et al.* 1985):

$$\Delta N_s = (2/3) \, \lambda_F \, {}^{238}U \, (\Delta t) \qquad (4)$$

where:

λ_F = fission constant of U^{238} (7.03×10^{-17} yr^{-1})

${}^{238}U$ = ppm of common uranium $\times$ ($^{238}U/U$)

The 2/3 factor accounts for the probability of a fission track from the cubical volume [(range of fission fragment)^3] around the fissioning nucleus reaching the polished surface on which tracks are counted. Equation 4 may be recognized as the basic radioactive decay law for conditions where Δt is short compared to the half-life.

To restore an age spectrum to its original state at the time of deposition, we must subtract, grain by grain, the number of spontaneous tracks that have

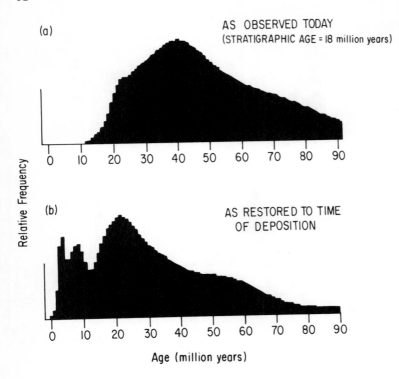

Fig. 3.6. Comparison of detrital zircon-age spectrum as (a) observed today and (b) after restoration to the time of deposition (18 million years ago). Sample is C1 from the basal Kamlial Formation, Chinji Village section (Fig. 3.3).

formed since the sand was deposited (determined from Equation 4). The resulting adjusted age and standard error for each grain are used to generate the restored spectrum (using Equation 3). The simulated de-ageing process, or its converse the ageing process, as derived from Equation 4 is completely reversible and a synthetically derived spectrum could be restored intact to its original state. Equation 4 is not capable of predicting exactly what happened, but it is capable of predicting what would most likely have happened.

By way of illustration, Figure 3.5 shows how the zircon-age spectrum determined from the long-etch mount of the modern Indus River sand sample will most likely appear at various times in the future. As this particular age spectrum becomes older in the future, the peaks broaden and become less distinct. In the opposite sense, Figure 3.6 shows the result of moving an age spectrum backward in time. Figure 3.6a is the age spectrum as it appears today for a sample whose stratigraphic age is 18 Ma. Figure 3.6b shows the result when the age spectrum of this sample is restored to its likely appearance at the time of deposition (i.e., 18 million years ago) using Equation 4. Note that this transformation has sharpened the peaks, transforming an otherwise smooth age distribution into at least three distinct age compo-

nents. This transformation thus has the effect of defining more clearly zircon age populations that have been obscured by time. Thus, adding time tends to smooth an age spectrum, whereas subtracting time tends to sharpen it. The transformation in shape of an age spectrum is differential over time because as the age of a given grain increases, so does the standard error of the age (s_i in Equations 2 and 3) in years, and vice versa.

Detrital Zircon Fission-Track Data

The age spectra for the long-etch and short-etch mounts of the detrital zircons collected from a sandbar in the modern Indus River are shown in Figure 3.7. The range of ages in this sample corresponds well with the range of ages present in the surface rocks of the Indus watershed (Fig. 3.1). In fact, essentially all of the surface zircon ages as mapped by Zeitler (1985) in the Indus watershed (Fig. 3.1) are represented in either the long-etch or the short-etch spectrum of the Indus River sand (Fig. 3.7). The only ages not represented in modern Indus River detrital zircons are those greater than 100 Ma. We conclude, therefore, that the sands of the modern

Fig. 3.7. Zircon-age spectra, contemporary Indus River sand. Note that most of the surface-rock ages from northern Pakistan (Fig. 3.1) are represented. Sample location is shown in Figure 3.2.

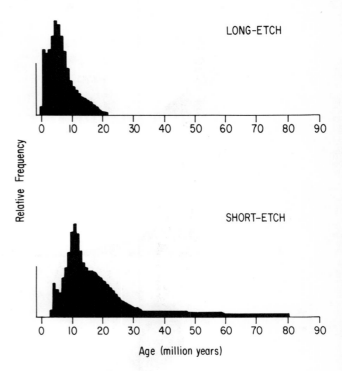

Indus River are representative of the rocks exposed in its contemporary watershed.

A significant population of 1 to 5 Ma zircons occurs in modern Indus River sand, suggesting that only 1 to 5 million years were required for these zircons to pass through their closure temperature and subsequently be uplifted, eroded, and become Indus River detritus. The important implication of this observation and of the observed general close correspondence between the ages of zircons present in the surface rocks of the Indus River watershed and those present in the modern Indus River is that the transport time (T_t) for detritus in the modern Indus River is quite short on a geologic scale, probably less than 10^6 yr, and for purposes of our analysis can be ignored. We assume that this relatively brief transport time has also been true of the Indus River system over the past 18 million years.

In the Siwalik Group sandstones in the Chinji Village section (Figs. 3.3, 3.8), the oldest sample (C1) is from the basal Kamlial Formation and has a stratigraphic age of 18 Ma (Johnson *et al.* 1985). The youngest peak in the zircon-age spectrum (long-etch mount) for this sample occurs at ≈ 22 Ma. Moving up section, the youngest peak in the spectrum of sample G1 from the lower Chinji Formation (14 Ma) occurs at ≈ 18 Ma. Sample G5 from the uppermost

Chinji (10.8 Ma) has a peak at 14 Ma and sample G10 from the middle Dhok Pathan Formation (7.9 Ma) has a distinct 11 Ma peak. The Siwalik sandstones at Chinji Village, therefore, consistently contain zircons that are only ≈ 3 to 4 million years older than their depositional age, a situation comparable to that found in the modern Indus River (Fig. 3.7).

In the Trans Indus section (Fig. 3.9), sample K-7 is from the basal Siwalik sandstone unit overlying the Eocene strata. Based on lithologic similarity with the Chinji-Kamlial boundary, the stratigraphic age of K-7 is estimated to be approximately 14 Ma. The youngest peak in the age spectrum for this sample occurs at 17 Ma (Fig. 3.9). Moving up section to sample CK-11 (upper Chinji, stratigraphic age ≈ 11 Ma), a subtle peak or shoulder occurs at 14 Ma. CK-10 (lower Nagri, ≈ 10 Ma) shows a definite peak at 11 Ma, and CK-5 (upper Dhok Pathan, ≈ 4 Ma) has a peak at 6 Ma. The stratigraphic ages of these samples have been estimated by assuming that formation ages in the Trans Indus area are identical to those in the Potwar Plateau (Johnson *et al.* 1982, 1985). These age estimates are therefore not as precise as those specified for the Chinji Village area, having perhaps an error of a million years or so. Nevertheless, there seems to be a 1 to 3 million-year age difference between the youngest zircons present

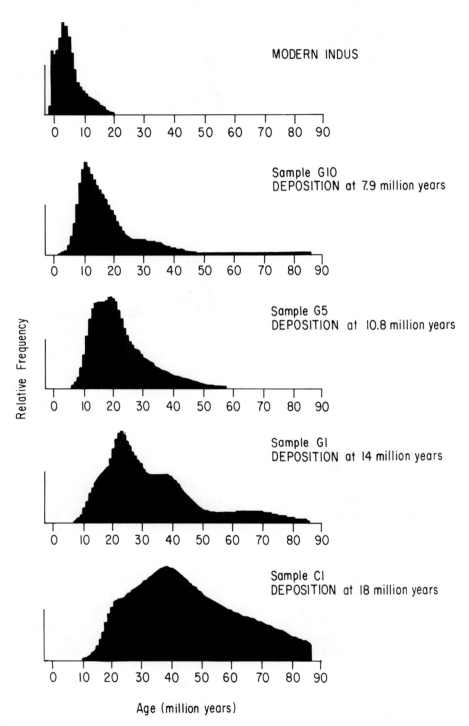

Fig. 3.8. Zircon-age spectra (long-etch mounts) for the Chinji Village section. Spectrum from the modern Indus River sand sample is included for reference. The presence of a youthful tail in the spectra, which may be younger than the stratigraphic age, is an artifact of the statistical assumptions used in generating the age spectrum (Equation 3).

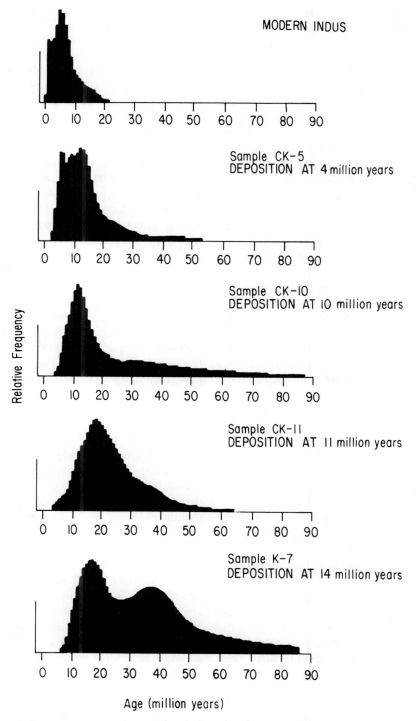

Fig. 3.9. Zircon-age spectra (long-etch mounts) for the Trans Indus section. Note spectrum from modern Indus River sand for reference.

and the estimated stratigraphic age of the sandstones. This relationship reinforces that found in the Chinji Village section (Fig. 3.8).

The relationship is even more obvious when the detrital zircon-age spectra are restored to their likely appearance at the time of deposition. As discussed previously, this transformation helps to distinguish and identify underlying age populations in the spectra. Figure 3.10 shows the age spectra from samples in the Chinji Village section restored to time of deposition. The spectrum of the modern Indus River sample is included for reference. Figure 3.11 shows the comparable results from the Trans Indus section. These results (Figs. 3.10, 3.11) clearly show the presence of zircons that are only 1 to 5 million years older than the stratigraphic age of the sandstones.

In general, the Chinji Village and Trans Indus detrital zircon-age spectra (Figs. 3.8–3.11) broaden with increasing stratigraphic age, as anticipated (Fig. 3.5). It is also clear that as we go back in time, the older part of the spectra becomes decidedly more conspicuous. That is, the older components of the age spectra become relatively more abundant with increasing stratigraphic age. A similar trend, i.e., enhanced older components with increasing stratigraphic age, also occurs in the age spectra from the short-etch mounts of these samples (Cerveny 1986).

Interpreting the Detrital Zircon-Age Spectra

The age spectra of detrital zircons from modern Indus River sand and Siwalik Group sandstones have established that 1 to 5 million-year-old zircons have been a significant component of Indus River detritus for the past 18 million years. These young zircons cannot be explained as reworked grains from older Siwalik units. Any such reworked grains would necessarily be older than the stratigraphic age of the unit from which they were eroded because none of the Siwalik sandstones has undergone burial deep enough to reset zircon ages. Thus, reworked and recycled zircons would be expected to contribute to the older end of the age spectra, rather than to the younger end. It is also unlikely that the presence of young detrital zircons in Siwalik sandstones can be explained by continuous volcanism during the past 18 million years. Very little evidence exists for volcanism of any sort affecting northern Pakistan during Neogene time. The Siwalik Group contains only

three minor volcanic ash zones in the Chinji Village area, and no ash has been observed in the Trans Indus area. The volume of volcanic rocks that would be necessary to contribute a significant component of young zircons into the Indus River system is absent in northern Pakistan. We conclude, therefore, that the young zircons in the modern Indus River sand and in Siwalik sandstones were derived directly from the Himalaya by erosion and transport in the Indus River system.

The rapid uplift observed today in the Nanga Parbat-Haramosh Massif area ($\approx$ 5 m/10^3 yr) is attributed to the collision of Asia with India, producing locally high uplift rates. The area of rapid uplift around Nanga Parbat is characterized by young zircons at the surface (Fig. 3.1). Because this area is within the Indus watershed, it is the most likely source area for the young detrital zircon seen in the modern Indus River sands (Fig. 3.7). The fact that throughout the Siwalik Group sequence there are zircons whose ages were only 1 to 5 Ma at the time of deposition (Figs. 3.8–3.11) suggests that these high rates of uplift must have been a characteristic feature of portions of the Himalaya for at least the past 18 million years. The Nanga Parbat-Haramosh Massif exemplifies the kind of source area that would be necessary to supply such young zircons to the sandstones of the Siwalik Group.

A series of "massifs" analogous to Nanga Parbat but varying spatially throughout the Indus watershed over time therefore seems likely. Where might these massifs have been located in the past? It is possible that they were in an area north of Nanga Parbat, and that the evidence for their existence has since been eroded away. Areas with surface zircons presently ranging in age from 10 to 25 Ma also might be former Nanga Parbat analogs. The broad region between Dir-Chitral and Nanga Parbat shows zircons 12 to 20 Ma in age and this region may have been the site of an ancestral mountain front, tectonically similar to that of the present-day Nanga Parbat-Haramosh Massif (Fig. 3.1). This area may have served as the source for young zircons to the ancestral Indus, until the focus of uplift shifted to its present position to the east. In another scenario it is possible that the present site of the Nanga Parbat-Haramosh Massif has been the locus of long-sustained, rapid uplift. This would, of course, require a constant replenishment of crustal rock to the area to compensate for the large amount of erosion that would have taken place. From the opposite point of view, we can say with some confidence that the

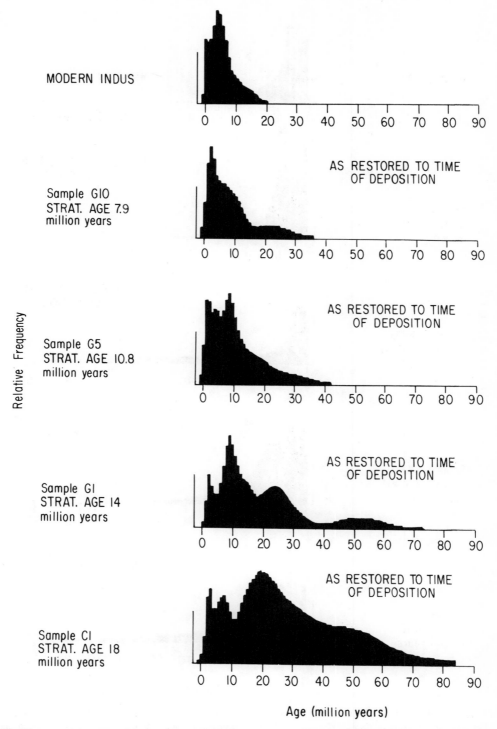

Fig. 3.10. Zircon-age spectra (long-etch mounts) for Chinji Village section, *restored to time of deposition*. Note zircon ages <5 Ma in all samples. Negative ages are the result of statistical dispersion inherent in the de-aging transformation (Equation 4).

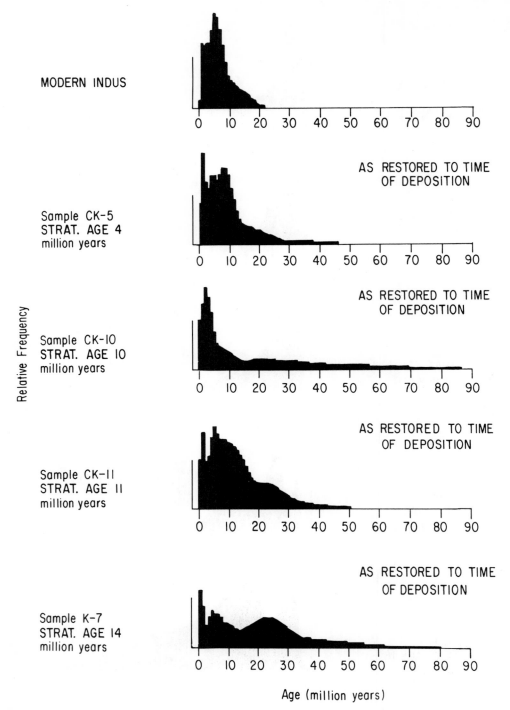

Fig. 3.11. Zircon-age spectra (long-etch mounts) for Trans Indus section, *restored to time of deposition*. Note zircon ages <5 Ma in all samples. Negative ages (see Fig. 3.10 caption) have been compiled as 0–1 Ma dates, which tends to enhance or produce a peak in this interval.

Dir-Chitral region to the west of Nanga Parbat could *not* have been the source of the young zircons found in Siwalik sandstones because most of the surface rocks in that region have zircons with old (≥ 30 Ma) ages (Fig. 3.1).

Older zircons are decidedly less abundant today in Indus River sand than they were in the past (Figs. 3.7–3.11). As seen in Figure 3.7, the detrital-age spectra of the modern Indus River are dominated by zircons between 1 and 30 Ma in age. Zircons older than this are present, but not in significant amounts. In contrast, detrital zircons with ages greater than 30 Ma were prevalent in the Indus River in the past (Figs. 3.10, 3.11). This suggests that in the past larger areas of "old" terrain were present in the Himalaya. In this scenario we envision that for some time after continental docking, rocks with "old" zircons provided the main source of detritus to the Indus River. These rocks were not buried to great depths so their zircons were not annealed. As these "old" rocks were systematically consumed by uplift and erosion, successively deeper and partially annealed zircons, and then fully annealed zircons, made their way to the surface.

The detrital zircon-age spectra suggest that lofty Himalayan peaks and ranges, much like those present today, must have been present in northern Pakistan for the past 18 million years. To the extent that the high Himalaya of today have a profound effect on the climate of central Asia, we suggest that this orographic effect has been a long-term (18 million year) feature of central Asia.

Summary and Conclusions

This study has established that the contemporary Indus River contains young, 1 to 5 Ma, zircons as a major component of its detrital zircon suite. These zircons have not aged significantly since they were released from bedrock by erosion. The ancestral Indus River sands (Siwalik Group sandstones) also contain zircons whose ages were only 1 to 5 Ma at the time they were deposited. "Old" zircons become systematically more abundant going stratigraphically downward.

These observations lead to several important conclusions, the most general being that for at least the last 18 million years there have been areas in the Indus River watershed with uplift rates, relief, and erosion rates comparable to those observed in the

Nanga Parbat region today. In effect, the contemporary Himalayan landscape, on a broad scale, has been a relatively steady-state feature for at least the past 18 million years. Over this time period there has always been an area of rapid uplift, similar to the region around Nanga Parbat, supplying 1 to 5 Ma zircons to the fluvial system and delivering them to the Indo-Gangetic plain.

At present the Indian plate is being thrust under the Asian plate. At first, this would necessitate that the upper crustal plate, the Asian plate, would be the first eroded. Rocks from this upper plate would have older zircon ages because they were never buried deeply enough and long enough to reset their zircon ages. Today, however, rocks of the lower plate are being eroded and evidently these rocks have been buried deeply enough for a long enough time to anneal the fission tracks in their contained zircons. This sequence of events would require older zircons to be eroded first and then, with the passage of time, to be replaced by progressively younger zircons. This scenario is supported by the observed data from the Siwalik record.

In studying paleo-climates it is essential to know what topographical constraints were placed on weather systems in the past. We infer from our detrital zircon data that the Himalaya have continuously maintained their presence and relief during the past 18 million years. To the extent that the monsoon system of Asia may be affected by the modern high Himalaya, we may infer that this condition has existed for the past 18 million years. Furthermore, it is quite likely that the high Himalaya have been present for even longer than this, but we lack data below the 18-Ma stratigraphic level. The critical evidence prior to 18 Ma is lacking as there is a conspicuous absence of Oligocene and older fluvial deposits suitable for study in the Punjab area.

In summary, the case study presented here suggests that detrital zircon fission-track ages are a powerful tool for the analysis of sedimentary provenance. The concepts and technology behind this tool are rather simple, straightforward, and readily available.

Acknowledgments. We are indebted to R. Tahirkheli, C. Gronseth, I. Khan, Q. Jan, and N. Bonis for their direct contributions to this study. Financial support was provided by National Science Foundation grants INT-8308069 and EAR-8206184

and by the United States Geological Survey Evolution of Sedimentary Basins Program.

References

ALLEN, J.R.L. (1965) Review of the origin and characteristics of recent alluvial sediments. Sedimentology 5:89–191.

BALDWIN, S.L., HARRISON, T.M., and BURKE, K. (1986) Fission track evidence for the source of accreted sandstones, Barbados. Tectonics 5:457–468.

BRIGGS, N.D., NAESER, C.W., and McCULLOH, T.H. (1981) Thermal history of sedimentary basins by fission-track dating (Abstract). Nuclear Tracks 5:235–237.

CARPENTER, B.S. and REIMER, G.M. (1974) Standard Reference Materials: Calibrated Glass Standards for Fission Track Use. National Bureau Standards Special Publication 260–49, 16 p.

CERVENY, P.F. (1986) Uplift and erosion of the Himalaya over the past 18 million years: Evidence from fission track dating of detrital zircons and heavy mineral analysis. Unpublished M.S. thesis, Hanover, New Hampshire: Dartmouth College, 198 p.

CERVENY, P.F., JOHNSON, N.M., TAHIRKHELI, R.A.K., and BONIS, N.R. (in press) Tectonic and geomorphic implications of Siwalik Group heavy minerals, Potwar Plateau, Pakistan. In: Malinconico, L. and Lillie, R. (eds) Tectonics and Geophysics of the Western Himalaya. Geological Society America Special Paper.

DODSON, M.H. (1979) Theory of cooling ages. In: Jäger, E. and Hunziker, J.C. (eds) Lectures in Isotope Geology. New York: Springer-Verlag, pp. 194–202.

FATMAI, A.N. (1974) Lithostratigraphic Units of Kohat Potwar Province, Indus Basin, Pakistan. Geological Survey Pakistan Memoir 10, 80 p.

FLEISCHER, R.L., PRICE, P.B., and WALKER, R.M. (1965) Effects of temperature, pressure, and ionization on the formation and stability of fission tracks in minerals and glasses. Journal Geophysical Research 70:1497–1502.

FLEISCHER, R.L., PRICE, P.B., and WALKER, R.M. (1975) Nuclear Tracks in Solids – Principles and Applications. Berkeley: University California Press, 605 p.

GLEADOW, A.J.W. and DUDDY, I.R. (1984) Fission track dating and thermal history analysis of apatites from wells in the north-west Canning Basin. In: Purcell P.G. (ed) The Canning Basin. Perth: Geological Society Australia and Petroleum Exploration Society Australia, pp. 377–387.

GLEADOW, A.J.W., DUDDY, I.R., and LOVERING, J.F. (1983) Fission-track analysis: A new tool for the evaluation of thermal histories and hydrocarbon potential. Australian Petroleum Exploration Association Journal 23:93–102.

GLEADOW, A.J.W., HURFORD, A.J., and QUAIFE, R.D. (1976) Fission track dating of zircon: Improved etching techniques. Earth Planetary Science Letters 33:273–276.

GREEN, P.F., DUDDY, I.R., GLEADOW, A.J.W., and LOVERING, J.F. (in press) Apatite fission-track analysis as a paleotemperature indicator for hydrocarbon exploration. In: Naeser, N.D. and McCulloh, T.H. (eds) Thermal History of Sedimentary Basins – Methods and Case Histories. New York: Springer-Verlag.

HURFORD, A.J. and GREEN, P.F. (1982) A user's guide to fission-track dating. Earth Planetary Science Letters 59:343–354.

HURFORD, A.J., FITCH, F.J., and CLARKE, A. (1984) Resolution of the age structure of the detrital zircon populations of two Lower Cretaceous sandstones from the Weald of England by fission track dating. Geological Magazine 121:269–277.

JOHNSON, N.M., OPDYKE, N.D., JOHNSON, G.D., LINDSAY, E.H., and TAHIRKHELI, R.A.K. (1982) Magnetic polarity stratigraphy and ages of Siwalik Group rocks of the Potwar Plateau, Pakistan. Palaeogeography, Palaeoclimatology, Palaeoecology 37:17–42.

JOHNSON, N.M., STIX, J., TAUXE, L., CERVENY, P.F., and TAHIRKHELI, R.A.K. (1985) Paleomagnetic chronology, fluvial processes, and tectonic implications of the Siwalik deposits near Chinji Village, Pakistan. Journal Geology 93:27–40.

JOHNSON, S.Y. (1984) Stratigraphy, age, and paleogeography of the Eocene Chuckanut Formation, northwest Washington. Canadian Journal Earth Sciences 21:92–106.

KOWALLIS, B.J., HEATON, J.S., and BRINGHURST, K. (1986) Fission-track dating of volcanically derived sedimentary rocks. Geology 14:19–22.

McGEE, V.E., JOHNSON, N.M., and NAESER, C.W. (1985) Simulated fissioning of uranium and testing of the fission track dating method. Nuclear Tracks 10:365–379.

NAESER, C.W. (1976) Fission Track Dating. United States Geological Survey Open-File Report 76–190, 65 p.

NAESER, C.W. (1979) Fission-track dating and geologic annealing of fission tracks. In: Jäger, E. and Hunziker, J.C. (eds) Lectures in Isotope Geology. New York: Springer-Verlag, pp. 154–169.

NAESER, C.W., HURFORD, A.J., and GLEADOW, A.J.W. (1977) Fission-track dating of pumice from the KBS Tuff, East Rudolf, Kenya. Nature 267:649.

NAESER, N.D. and NAESER, C.W. (1984) Fission-track dating. In: Mahaney, W.C. (ed) Quaternary Dating Methods. Amsterdam: Elsevier, pp. 87–100.

NAESER, N.D., NAESER, C.W., and McCULLOH, T.H. (in press a) The application of fission-track dating to the depositional and thermal history of rocks in sedimentary basins. In: Naeser, N.D. and McCulloh, T.H. (eds) Thermal History of Sedimentary Basins — Methods and Case Histories. New York: Springer-Verlag.

NAESER, N.D., ZEITLER, P.K., NAESER, C.W., and CERVENY, P.F. (in press b) Provenance studies by fission track dating — Etching and counting procedures. Nuclear Tracks.

OPDYKE, N.D., LINDSAY, E., JOHNSON, G.D., JOHNSON, N.M., TAHIRKHELI, R.A.K., and MIZRA, M.A. (1979) Magnetic polarity stratigraphy and vertebrate palaeontology of the Upper Siwalik Subgroup of northern Pakistan. Palaeogeography, Palaeoclimatology, Palaeoecology 27:1–34.

PARRISH, R.R. (1982) Cenozoic thermal and tectonic history of the Coast Mountains of British Columbia as revealed by fission-track and geological data and quantitative thermal models. Unpublished Ph.D. thesis, Vancouver: University of British Columbia, 166 p.

ROBERTS, J.A., GOLD, R., and ARMANI, R.J. (1968) Spontaneous-fission decay constant of ^{238}U. Physical Review 174:1482–1484.

WAGNER, G.A., REIMER, G.M., and JÄGER, E. (1977) Cooling Ages Derived by Apatite Fission-track, Mica Rb-Sr and K-Ar Dating: The Uplift and Cooling History of the Central Alps. Memorie degli Istituti Geologia Mineralogia Universita Padova 30, 28p.

YIM, W.W.-S., GLEADOW, A.J.W., and VAN MOORT, J.C. (1985) Fission track dating of alluvial zircons and heavy mineral provenance in northeast Tasmania. Journal Geological Society London 142:351–356.

ZEITLER, P.K. (1985) Cooling history of the NW Himalaya, Pakistan. Tectonics 4:127–151.

ZEITLER, P.K., JOHNSON, N.M., BRIGGS, N.D., and NAESER, C.W. (1982a) Uplift history of the NW Himalaya as recorded by fission-track ages on detrital Siwalik zircons (Abstract). Program, Symposium Mesozoic and Cenozoic Geology (60th Anniversary Meeting, Geological Society China), Bedaike, China, p. 109.

ZEITLER, P.K., JOHNSON, N.M., NAESER, C.W., and TAHIRKHELI, R.A.K. (1982b) Fission track evidence for Quaternary uplift of the Nanga Parbat region, Pakistan. Nature 298:255–257.

ZEITLER, P.K., TAHIRKHELI, R.A.K., NAESER, C.W., and JOHNSON, N.M. (1982c) Unroofing history of a suture zone in the Himalaya of Pakistan by means of fission track annealing ages. Earth Planetary Science Letters 57:227–240.

ZIMMERMANN, R.A., REIMER, G.M., FOLAND, K.A., and FAUL, H. (1975) Cretaceous fission-track dates of apatites from northern New England. Earth Planetary Science Letters 28:181–188.

Part II
Lithostratigraphy and Chronostratigraphy: Introduction

F.B. VAN HOUTEN

Lithostratigraphy and chronostratigraphy have long been central concerns in the study of sedimentary basins. In the past, however, the focus was commonly on local problems or on conventional codes for cataloging global geology. During the past several decades a rapid rise of interest in basin analysis, generated in part by petroleum exploration and stimulated by the plate-tectonics paradigm, has led to "the greening of stratigraphy" (Sloss 1984). With it has come an expansion of pertinent techniques and methods, recently reviewed by Miall (1984), who points out that stratigraphy now requires a synthesis of many sorts of geologic data. Concurrently, the principal kinds of basins have been delineated in a plate-tectonics framework (Bally and Snelson 1980; Mitchell and Reading 1986). This has increased the role of detailed description, accurate correlation, and precise interpretation of their basin fill because that is the main source of information about the subsidence history.

Early in the "greening" Francis Pettijohn recognized the importance of careful study of the sedimentary rock record—a focus that was a persisting theme throughout his constructive career. In particular, Pettijohn's extensive work on sedimentary structures, as well as that of his many students, is reflected in the *Atlas and Glossary of Primary Sedimentary Structures* (Pettijohn and Potter 1964) and in *Paleocurrents and Basin Analysis* (Potter and Pettijohn 1977). This, in turn, contributed immensely to the development of facies models. Use of these two-dimensional vertical profiles (Visher 1965; Walker 1979) in reporting and analyzing sedimentary sequences has helped alert students to the variety of successions of facies, and encouraged them to try to find some order in the lithostratigraphic record.

Among the many advances in stratigraphic analysis several have been especially effective. Recognition of regionally extensive unconformity-bound sequences (Sloss 1963), corroborated by seismic stratigraphy (Vail *et al.* 1977), emphasized the need for reliable means of regional correlation, and it increased the significance of unconformities in basin analysis. This has led to a more general interest, the meaning of hiatuses in basin history and clues to lapses in normal sedimentation. Appreciation that sedimentation may be essentially discontinuous, consisting of increments and gaps (Sadler 1981), has improved our understanding of "rates of accumulation." Development of the graphic correlation technique (Shaw 1964) provided a reliable way to discover changes and differences in subsidence history. In the past this method was applied mostly to successions with rigorous biostratigraphic control; its broader value has not yet been fully appreciated. A backstripping method of analyzing stratigraphic sections can produce tectonic subsidence curves of passive margins which reflect thermal contraction of the heated lithosphere (Sleep 1971). A continuing complementary concern has been the identification and interpretation of transgressive and regressive alternations and the role of global rise and fall of sea level. Construction of oxygen, carbon, and sulfur isotope curves has contributed to the correlation of sea-level changes as well as to the interpretation of the marine environment. Identification of vertical patterns of magnetic reversals in sedimentary rocks affords still another new source of chronostratigraphic markers, especially when they have been calibrated with the detailed radiometric time scale (Berggren *et al.* 1985). These methods of correlation are especially useful today when there is an enthusiastic interest in finding allocyclic control expressed extensively in the stratigraphic record. With this trend has come an increased awareness of the effect of Milankovitch orbital cycles in producing some of the patterned sedimentary sequences.

Each of the papers assembled in this section elaborates a particular method or technique of lithostratigraphic or chronostratigraphic analysis that contributes to the more general study of sedimentary basins and will have an increasingly important role in the future. Miall has reviewed changing ideas about detrital facies architecture, and he concludes that in many situations two-dimensional vertical profiles are inadequate for accurate understanding of complex three-dimensional sedimentary deposits. In an effort to generate some new unifying concepts, he emphasizes the importance of recognizing a hierarchy of physical scales and three-dimensional architectural elements, or depositional units, characterized by their lithofacies composition, and external and internal geometry. In this approach his focus is on a six-fold hierarchy of bounding surfaces, or bedding contacts, as a source of detailed information about the relation between duration of events and the scale and geometry of the product.

Shanmugam returns to an old stratigraphic theme — unconformities. He identifies erosional unconformities of local to global scale, discusses the origin of subaerial and submarine varieties, and reviews criteria for recognizing them. In emphasizing their role in basin analysis Shanmugam points out some of the effects of eustatic and tectonic controls of sea level, as well as problems of distinguishing between them, especially in seismic stratigraphic records. In petroleum exploration, erosional unconformities are important, especially where they mark the upper boundary of increased porosity, separate reservoir and source rocks, or provide avenues for migration of hydrocarbons.

Clifton, Hunter, and Gardner have analyzed a thick late Cenozoic shallow marine and shoreline deposit, with the aim of isolating the major factors controlling its transgressions and regressions. For this purpose ten depositional facies, identified on the basis of biota, sedimentary structures, and textures, have been reconstructed in a succession of recurring transgressions and regressions. The consistent pattern of alternations compares closely with a revised Pleistocene eustatic sea-level curve generated by these authors. The study also reveals that the basin subsided throughout its history at an average

rate of 1 metre per 1,000 years, and that a significant change in the rate of sedimentation recorded within the deposit probably resulted from a major change in provenance.

Bond, Kominz, and Grotzinger have analyzed the lower Paleozoic stratigraphic record of the Cordilleran and Appalachian orogens to test whether multiple orders of sea-level change prevailed then. They have constructed subsidence curves by the backstripping method which show that in general the subsidence decayed exponentially with time in response to slow cooling of the crust. Significantly, however, the curves also reveal systematic deflections. These suggest that the long-term subsidence of the Paleozoic passive margins was modified by 40–60 m.y. and 2–10 m.y. patterns of eustatic rise and fall of sea level that were similar to patterns that prevailed during Jurassic time.

Nemec demonstrates that sequences of coal measures provide important clues to short-term intrabasinal differential subsidence not accounted for in general basin modelling. Noting that beds of coal commonly accumulate during basin-wide hiatuses in normal sedimentation, he regards them as essentially chronostratigraphic markers. Then he uses the coal measures as "biological" events and analyzes the data with the graphic correlation technique (Shaw 1964). By this means he identifies changing rates of tectonic basin subsidence and sediment accumulation in a foreland basin, an intracratonic basin, and two extensional basins. The results suggest that other factors, such as compaction and sediment loading, have relatively little effect on subsidence.

Johnson and his colleagues describe a method for calculating stratigraphic variability based on the magnetic-reversal time scale. Use of these time lines for lateral tracing and correlation of Miocene fluvial deposits in Pakistan reveals both lateral stratigraphic nonuniformity between sections and local unsteadiness, or vertical variation in sedimentation, within sections. The authors address the question of how much of the observed variability is in the stratigraphic record and how much may be an artifact of sampling. In spite of recognized complications, the identified variation in sedimentation from place to place and from time to time is a measure of the dynamic processes at work. With continued refinement and extension the magnetic-reversal time scale will become an increasingly useful chronostratigraphic tool.

References

BALLY, A.W. and SNELSON, S. (1980) Realms of subsidence. In: Miall, A.D. (ed) Facts and Principles of World Petroleum Occurrence. Canadian Society Petroleum Geologists Memoir 6, pp. 9–94.

BERGGREN, W.A., KENT, D.V., FLYNN, J.J., and VAN COUVERING, J.A. (1985) Cenozoic geochronology. Geological Society American Bulletin 96:1407–1418.

MIALL, A.D. (1984) Principles of Sedimentary Basin Analysis. New York: Springer-Verlag, 490 p.

MITCHELL, A.H.G. and READING, H.G. (1986) Sedimentation and tectonics. In: Reading, H.G. (ed) Sedimentary Environments and Facies. Boston: Blackwell, pp. 471–519.

PETTIJOHN, F.J. and POTTER, P.E. (1964) Atlas and Glossary of Primary Sedimentary Structures. New York: Springer-Verlag, 380 p.

POTTER, P.E. and PETTIJOHN, F.J. (1977) Paleocurrents and Basin Analysis. New York: Springer-Verlag, 425 p.

SADLER, P.M. (1981) Sediment accumulation rates and the completeness of stratigraphic sections. Journal Geology 89:569–584.

SHAW, A.B. (1964) Time in Stratigraphy. New York: McGraw-Hill, 365 p.

SLEEP, N.H. (1971) Thermal effects of the formation of Atlantic continental margins by continental breakup. Geophysical Journal Royal Astronomical Society 24:325–350.

SLOSS, L.L. (1963) Sequences in the cratonic interior of North America. Geological Society America Bulletin 74:93–113.

SLOSS, L.L. (1984) The greening of stratigraphy 1933–1983. Annual Review Earth Planetary Sciences 12:1–10.

VAIL, P.R., MITCHUM, R.M., and THOMPSON III, S. (1977) Seismic stratigraphy and global changes of sea level, Part 4: Global cycles of relative changes of sea level. In: Payton, C.E. (ed) Seismic Stratigraphy—Applications to Hydrocarbon Exploration. American Association Petroleum Geologists Memoir 26:83–97.

VISHER, G.S. (1965) Use of vertical profile in environmental reconstruction. American Association Petroleum Geologists Bulletin 49:41–61.

WALKER, R.G. (ed) (1979) Facies Models. Geoscience Canada Reprint Series 1, 211 p.

4
Facies Architecture in Clastic Sedimentary Basins

Andrew D. Miall

Abstract

Clastic deposits can be subdivided into microforms, meso-forms, and macroforms [using Jackson's (1975) terminology], reflecting a range of physical scales of deposit and the time scale during which they form. A formal three-dimensional subdivision of some types of deposit, which clarifies these physical and temporal scales, can now be attempted using the concept of a hierarchy of internal bounding surfaces. Brookfield's (1977) three-fold classification of eolian bounding surfaces is widely accepted. Allen's (1983) comparable attempt to classify fluvial bounding surfaces is also useful but, it is suggested here, should be expanded to a six-fold hierarchy, mainly to facilitate the definition of macroforms.

The three-dimensional depositional units so identified can be classified using standard facies–analysis techniques. To emphasize their three-dimensional nature they are termed "architectural elements." The lithofacies composition, external shape, and internal geometry of these elements are characteristic of various suites of processes in depositional systems; for example, the well known lateral–accretion deposit represents the accretion of bank-attached bars inside meander bends of channels in rivers, deltas, tidal creeks, and submarine fans.

Analysis of vertical profiles, in particular their cyclicity, was formerly a vital component of facies analysis methodology. However, in many situations, especially where the architectural elements have complex three-dimensional shapes, such analyses have become much less important. For example, subsurface mapping techniques can make use of petrophysical log correlations in well developed oil and gas fields, and recent advances in three-dimensional seismic data-display techniques permit many elements to be mapped using horizontal seismic sections.

Most of the ideas discussed in this paper evolved from studies of nonmarine deposits, but it is suggested that there is considerable potential for further development in the field of facies analysis methodology by the application of these concepts to marine deposits, including those in which chemical sediments are deposited by clastic processes.

Introduction

One of the fundamental objectives of stratigraphy is to describe and interpret the three-dimensional architecture of the sediments filling sedimentary basins. On the largest scale this involves the disciplines of lithostratigraphy and biostratigraphy, the reconstruction of major depositional sequences, perhaps with the aid of regional seismic sections, and an assessment of the roles of sea-level change and the various mechanisms of basin subsidence (e.g., Bond *et al.* 1988; Cloetingh 1988; Jordan *et al.* 1988; Paola 1988: all this volume). At an intermediate to small scale, basin architecture depends on the interplay between subsidence rates and the autocyclic processes governing the distribution and accumulation of sediments within a particular range of depositional systems. Exploration for stratigraphic petroleum traps and for many types of stratabound ore bodies requires close attention to basin architecture at the intermediate scale. For example, the search for and the interpretation of subsurface "trends" in potential reservoir rocks is a major preoccupation of petroleum geologists. Since the late 1950s many improvements have been made in techniques of basin analysis that facilitate this task.

For some time we have been quite good at predicting the location and trend of such major depositional entities as alluvial fans, barrier reefs, various types of deltas, and barrier-island sand bodies, but one of the outcomes of modern sedimentological research on depositional processes has been a much improved understanding of the composition and architecture of the various facies assemblages formed *within* each type of depositional system. We now have the ability, therefore, to construct much better models of the heterogeneities that exist within clastic reservoirs.

The purposes of this paper are to review changing ideas about facies architectures at the intermediate to small scale, and to suggest ways in which research may be directed. The focus is on detrital facies, and their distribution and accumulation by air and water currents. Many of the ideas are applicable to chemical sediments because carbonate and evaporite particles are commonly redistributed as clastic detritus prior to final sedimentation.

The paper focuses on two interrelated ideas:

1. The concept of architectural *scale*. Deposits consist of assemblages of lithofacies and structures over a wide range of physical scales, from the individual small-scale ripple mark to the assemblage produced by an entire depositional system. Recent work, particularly in eolian and fluvial environments, suggests that it is possible to formalize a hierarchy of scales. Depositional units at each size scale originate in response to processes occurring over a particular time scale, and are physically separable from each other by a hierarchy of internal bounding surfaces.
2. The concept of the *architectural element*. An architectural element is a lithosome characterized by its geometry, facies composition and scale, and represents a particular process or suite of processes occurring within a depositional system.

Architectural Scale and the Bounding–Surface Concept

As noted by Allen (1983, p. 249):

The idea that sandstone bodies are divisible internally into "packets" of genetically related strata by an hierarchically ordered set of bedding contacts has been exploited sedimentologically for many years, although not always in an explicit manner. For example, McKee and Weir (1953) distinguished the hierarchy of the stratum, the set of strata, and the coset of sets of strata, bedding contacts being used implicitly to separate these entities.

Allen (1966) showed that flow fields in such environments as rivers and deltas could be classified

into a hierarchical order. His hierarchy was designed as an aid to the interpretation of variance in paleocurrent data collected over various areal scales, from the individual bed to large outcrops or outcrop groups. The hierarchy consists of five categories: small-scale ripples, large-scale ripples, dunes, channels, and the "integrated system," meaning the sum of the variances over the four scales. Miall (1974) added the scale of the entire river system to this idea, and compiled some data illustrating the validity of the concept.

Jackson (1975), following earlier Russian work, showed that bedforms could be grouped into three orders of time scale and physical scale: *Microforms* are such structures as small-scale ripples, and with a time scale of variation in the order of seconds to hours. *Mesoforms* are larger (dm-m) scale deposits, many of which are active mainly during what Jackson termed "dynamic events," such as hurricanes, seasonal floods, spring tides, or eolian sandstorms, when disproportionately large volumes of sediment are moved in geologically instantaneous time periods. The system may remain virtually unchanged between dynamic events. Examples of mesoforms are subaqueous dunes and sand waves, which occur in a variety of environments, sandy linguoid and transverse "bars" and gravel longitudinal "bars" in rivers, and spillover lobes in shoreline and shallow shelf sand bodies. These features are volumetrically at least an order of magnitude larger than microforms, which has implications for the significance of the currents that formed them. *Macroforms* represent the long-term accumulation of sediment in response to major tectonic, geomorphic, and climatic controls, and typically consist of the superimposed deposits of many microforms and mesoforms. Examples include fluvial point bars, eolian draas, tidal deltas, and shelf sand ridges. Macroforms in channels are comparable in height to channel depth, and comparable in length to channel width. They are at least an order of magnitude larger than mesoforms and may have complex three-dimensional geometries but, because of the limitations of outcrop and subsurface data, a great deal remains to be learned about them.

It is suggested that a long-standing confusion in the literature about the meaning of the term "bar" could be clarified by restricting its use to macroforms, a useage that would return us to the definition proposed by the American Society of Civil Engineers (1966). Mesoforms, many of which have been termed "bars" in the literature (see previous paragraph), consist mainly of periodic or quasi-periodic deposits that are better termed "bedforms."

The study of macroforms, as they are preserved in the ancient record, has the potential to make substantial contributions to the science of geomorphology. As Hickin (1983) has noted, geomorphologists have great difficulty documenting geomorphic processes at time scales intermediate between the human scale (about the last 100 yr) and geological time ($> 10^5$ yr). Macroforms represent the action of depositional and erosional processes acting over a 10^2 to 10^3 yr time scale, which goes a considerable way toward filling this gap.

Brookfield (1977) discussed the concept of an eolian bedform hierarchy, and tabulated the characteristics of four orders of "eolian bedform elements": draas, dunes, aerodynamic ripples, and impact ripples. These four orders occur simultaneously, superimposed on each other. Brookfield showed that this superimposition resulted in the formation of three types of internal bounding surfaces. His first-order surfaces are major, laterally extensive, flat-lying or convex-up bedding planes between draas (macroforms). Second-order surfaces are low to moderately dipping surfaces bounding sets of cross strata formed by the passage of dunes across draas (mesoforms). Third-order surfaces are reactivation surfaces bounding bundles of laminae within cross-bedded sets, and are caused by localized changes in wind direction or velocity (mesoforms to micro-forms).

Brookfield's (1977) development of the relationship between the time duration of a depositional

event, the physical scale of the depositional product, and the geometry of the resulting lithosome, was a major step forward that has been of considerable use in the analysis of eolian deposits. Brookfield (1977), Gradzinski *et al.* (1979), and Kocurek (1981) showed how these ideas could be applied to the interpretation of ancient eolian deposits (Fig. 4.1). Kocurek (personal communication, 1986) found that first order surfaces include two types of surface, the most laterally extensive of which he terms "super-surfaces." Characterization of eolian deposits therefore now requires a four-fold hierarchy of bounding surfaces.

Bounding Surfaces in Fluvial Deposits

Several workers have attempted to develop a breakdown of the range of physical scales present in fluvial deposits. Williams and Rust (1969) proposed an ordering of the scales of channels and bars in the modern Donjek River, Yukon. Campbell (1976), in an analysis of the Westwater Canyon Member of the Morrison Formation in New Mexico, recognized several scales of fluvial sequence that occur in tabular channel-fill sandstone bodies of a range of dimensions. Jones and McCabe (1980) described three types of reactivation surfaces occurring within sets of giant crossbedding, and related them to changes in bedform orientation and to stage changes in the river. A similar type of analysis was performed by Haszeldine (1983a,b) on the bounding surfaces within a Carboniferous sand-flat deposit. Bridge and

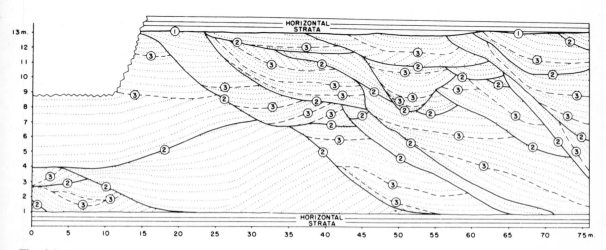

Fig. 4.1. Interpretation of an outcrop of the Entrada Sandstone (Jurassic), in which the crossbed sets have been subdivided using Brookfield's (1977) three-fold bounding surface hierarchy. Reprinted from Kocurek 1981, with permission.

Diemer (1983) and Bridge and Gordon (1985) referred to "major" and "minor" bounding surfaces within Paleozoic fluvial sequences. The major surfaces are typically horizontal and planar or slightly concave up, and enclose tabular sheets representing channel-fill successions. Friend *et al.* (1979) called these sand sheets "storeys."

Allen's (1983) study of the Devonian Brownstones of the Welsh Borders represents the most explicit attempt to formalize the concept of a hierarchy of bounding surfaces in fluvial deposits, and makes reference to Brookfield's work in eolian strata as a point of comparison. Allen described three types of bounding surface. He reversed the order of numbering from that used by Brookfield (1977), such that the surfaces with the highest number are the most laterally extensive. No reason was offered for this reversal, but the result is an open-ended numbering scheme that can readily accommodate developments in our understanding of larger scale depositional units, as discussed below. *First-order contacts*, in Allen's scheme, are set boundaries, in the sense of McKee and Weir (1953). *Second-order contacts* "bound clusters of sedimentation units of the kinds delineated by first-order contacts." They are comparable to the coset boundaries of McKee and Weir (1953) except that more than one type of lithofacies may comprise a cluster. Allen (1983) stated that "these groupings, here termed complexes, comprise sedimentation units that are genetically related by facies and/or paleocurrent direction." Many of the complexes in the Brownstones are produced by macroforms, in the sense defined by Jackson (1975). *Third-order* surfaces are comparable to the major surfaces of Bridge and Diemer (1983). No direct relationship is implied between Allen's three orders of surfaces and those of Brookfield, because of the different hydraulic behavior and depositional patterns of eolian and aqueous currents.

I found it useful to expand Allen's classification to a six-fold hierarchy, to facilitate the definition of fluvial macroform architecture, and to include the largest, basin-scale heterogeneities in the classification (A.D. Miall, L.T. Middleton, and C.E. Turner-Peterson, unpublished data from Jurassic-Cretaceous fluvial units of the Colorado Plateau).

First-Order Surfaces

First-order and second-order surfaces record boundaries within microform and mesoform deposits. The definition of *first-order surfaces* is unchanged from Allen (1983). They represent crossbed–set bounding surfaces (Figs. 4.2, 4.3). Little or no internal erosion is apparent at these boundaries, and they represent the virtually continuous sedimentation of a train of bedforms of similar type. Subtle modifications in attitude, with minor erosion, may be caused by reactivation following stage changes (Collinson 1970), or may be the result of changes in bedform orientation (Haszeldine 1983b). In core these surfaces may not be very prominent, but the presence of a reactivation surface can be recognized by truncation or wedge-out of crossbed foresets above the base of the set (Fig. 4.2).

Second-Order Surfaces

These are simple coset bounding surfaces, in the sense of McKee and Weir (1953). They define groups of microforms or mesoforms and indicate changes in flow conditions, or a change in flow direction, but no significant time break (Figs. 4.2, 4.3). Lithofacies above and below the surface are different, but the surface is usually not marked by significant bedding truncations or other evidence of erosion, except for the same kinds of minor modification that occur on first-order surfaces, as noted above. In core these surfaces may be distinguished from first-order surfaces by a change in lithofacies.

Third-Order Surfaces

Third-order and fourth-order surfaces are defined when architectural reconstruction indicates the presence of macroforms, including lateral–accretion deposits and downstream–accreting macroforms (elements LA and DA, in a classification modified slightly from that of Miall 1985: see below). Individual depositional units ("storeys" or "architectural elements") are bounded by surfaces of fourth-order or higher rank.

These are cross-cutting erosion surfaces within macroforms that dip at a low angle (normally < 15 degrees) and truncate underlying crossbedding at a low angle. They may cut through more than one crossbed set. They are commonly draped with intraclast breccia. Facies assemblages above and below the surface are similar.

Third-order surfaces may also develop at the top of minor bar or bedform sequences, and are draped by mudstone or siltstone, indicating lower discharge. Succeeding strata commonly contain a basal

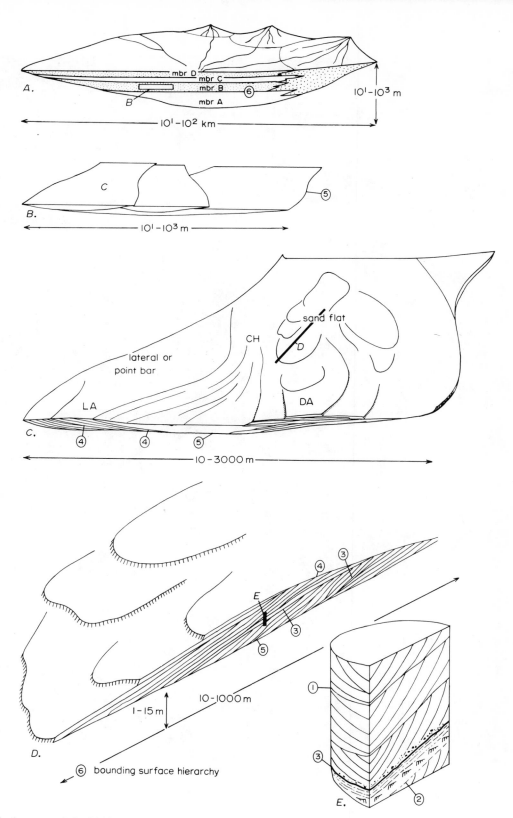

Fig. 4.2. A proposed six-fold bounding-surface hierarchy for fluvial deposits. Diagrams A to E represent successive enlargements of part of a fluvial unit, in which bounding surfaces of six distinct types may be recognized. See text for discussion. Diagram A shows basin-scale bounding surfaces, whereas diagrams B–E represent smaller scale macroforms and mesoforms.

Fig. 4.3. Illustrations of the fluvial bounding surface hierarchy in outcrop. The rank of the surfaces is indicated by numerals. The examples are discussed in the text. Locations: A = Cretaceous, Banks Island, Arctic Canada; B = Eocene, Axel Heiberg Island, Arctic Canada; C = Morrison Formation (Jurassic), New Mexico; D = Kayenta Formation (Jurassic), Colorado; E = Carboniferous, Alabama; F = Cretaceous, Gamtoos Basin, South Africa; G = Permian, Poland; H = Morrison Formation (Jurassic), New Mexico.

intraclast breccia composed of rip-up clasts of the draping fine-grained sediments. These characteristics are readily recognized in core.

These surfaces indicate stage changes, or changes in bedform orientation within the macroform (Fig. 4.3). They indicate a form of large-scale "reactivation," to adapt Collinson's (1970) term. In core these surfaces may be recognized by their gentle dip. Correlation of third-order and fourth-order surfaces in the subsurface could theoretically be possible with very close well spacing, but probably only if core were available, as the units defined by third-order surfaces are estimated to have dimensions of less than 0.1 km² (1 km² = 100 ha ≈ 247 acres).

Fourth-Order Surfaces

These surfaces represent the upper bounding surfaces of macroforms. They typically are flat to convex-upward. Underlying bedding surfaces and first to third order bounding surfaces are truncated at a low angle or may be locally parallel to the upper bounding surface, indicating that they are surfaces of lateral or downstream accretion. The form of this surface is commonly echoed by that of internal third-order surfaces within the underlying macroform element. Mud drapes on the element underlying the surface are common.

It may be convenient to define a second type of fourth-order surface: the basal scour surface of minor channels, such as chute channels. Where major channels are present, they would be bounded by higher order surfaces.

In the subsurface the best clue to the presence of third-order and fourth-order surfaces is their low depositional dip (Fig. 4.3). This should be recognizable in core and may also be apparent from dipmeter records or the traces produced by such modern devices as the Schlumberger MicroScanner tool. Third-order and fourth-order surfaces may be very difficult to distinguish from each other in small outcrops or individual cores. The low dips and draping breccias or mud lenses are similar. The best way to distinguish these surfaces from each other is if the lithofacies assemblages above and below the surface are different, indicating a change in element (macroform) type.

Second-, third-, and fourth-order surfaces, in this classification, were all included in the second-order category of Allen (1983). Third-order and fourth-

order surfaces correspond to the "minor" surfaces of Bridge and Diemer (1983).

Fifth-Order Surfaces

These are surfaces bounding major sand sheets, such as channel-fill complexes (Figs. 4.2, 4.3). They are generally flat to slightly concave-upward, but may be marked by local cut-and-fill relief and by basal lag gravels (third-order surfaces of Allen 1983; "major" surfaces of Bridge and Diemer 1983).

Sixth-Order Surfaces

These are surfaces defining groups of channels, or paleovalleys. Mappable stratigraphic units such as members or submembers are bounded by sixth-order surfaces (Figs. 4.2, 4.3; not defined by Allen 1983).

Fifth-order and sixth-order surfaces are potentially the easiest to map in the subsurface because of their wide lateral extent and essentially simple, flat or gently curved, channelized geometry. Many examples of such mapping have, in fact, been reported in the literature (e.g., Busch 1974). Considerable potential now exists for mapping these surfaces with three-dimensional seismic data, as described by Brown (1986). Fourth-, fifth-, and sixth-order surfaces may appear very similar to third-order surfaces in core. They are best differentiated by careful stratigraphic correlation between closely spaced cores, an objective that is best achieved in the most intensively developed fields, where well spacing may be a few hundred metres, or less.

Identification and correlation of these various bounding surfaces clearly can make a major contribution to the unravelling of the complexities of a fluvial depositional system. They are likely to be particularly useful in the recognition and documentation of macroforms, about which much remains to be learned.

Even in excellent outcrop the correct classification of bounding surfaces is not always easy. Three useful rules may make the task easier: 1) a surface of any given order may be truncated by a surface of equal or higher order, but not by one of lower order; 2) because bounding surfaces commonly record erosional events they may sometimes be more logically defined on the basis of what follows them rather than what precedes them. For example, the top of a macroform is defined by a fourth-order

Table 4.1. The range of scales of depositional units in fluvial sandstones, as illustrated by the Westwater Canyon Member, Morrison Formation, near Gallup, New Mexico.*

Rank of bounding surface	Lateral extent of unit	Thickness of unit	Area of unit (ha)	Origin	Subsurface mapping methods
6	200×200 km	0–30 m	4×10^7	Members or submembers, subtle tectonic control	Regional correlation of wireline logs
5	1×10 km	10–20 m	10^4	Sheet sandstone of channel origin	Intra-field correlation of wireline logs, 3-D seismic
5	0.25×10 km	10–20 m	2500	Ribbon channel sandstone	Mapping very difficult except with very close well spacing, 3-D seismic
4	200×200 m	3–10 m	40	Macroform elements (elements LA,DA)	Dip of 4th- and 3rd-order surfaces may be recognizable in core
3	100×100 m	3–10 m	10	Reactivation of macroforms	Dip of 4th- and 3rd-order surfaces may be recognizable in core
2	100×100 m	5 m	10	Cosets of similar crossbed facies	Facies analysis of core
1	100×100 m	2 m	10	Individual crossbed sets	Facies analysis of core

Data from A.D. Miall and C.E. Turner-Peterson, unpublished data.
*In columns 2, 3, and 4, approximate maximum dimensions are indicated. Element codes are discussed in text.

surface, except where it has been cut into by a major channel, the base of which typically constitutes a fifth-order surface; 3) minor surfaces may change rank laterally. For example, the upper fourth-order bounding surface of a macroform may merge into a second-order surface in the floor of an adjacent channel.

Some examples of these surfaces are illustrated in Figure 4.3. In Figure 4.3A there are several first-order surfaces between planar crossbed sets. Two second-order surfaces are shown in Figure 4.3B, enclosing a unit of parallel-laminated sandstone, which shows gradational contacts above and below with climbing ripple cross-lamination. In Figure 4.3C a third-order surface occurs within a homogeneous, sandstone-dominated lateral–accretion deposit that accreted to the right. The macroform rests on a fifth-order surface above a siltstone unit (outcrop in foreground is about 8-m high). A downstream-accreted macroform is shown in Figure 4.3D. It is bounded at the top by a convex-up fourth-order surface. One of several internal third-order surfaces is also indicated, and the macroform rests on a fifth-order surface. Figure 4.3E shows a lateral–accretion deposit 4 m thick, resting on a major fifth-order surface and capped by a grada-

tional fourth-order surface, above which is the floodplain sequence. Several internal third-order erosion surfaces are present (arrowed). Figure 4.3F illustrates a macroform, probably of lateral-accretion type, with several internal third-order erosion surfaces indicating changes in accretion direction. One of these is indicated. The close-up of a fifth-order surface shown in Figure 4.3G is at the base of a major sandstone sheet. Note the minor downcutting and intraclast breccia. There is little to distinguish this surface from others of third- to sixth-order in small outcrops or core. Two major sixth-order surfaces divide the unit shown in Figure 4.3H into three submembers, which can be traced for about 200 km (Turner-Peterson 1986). The middle submember contains a channel about 200 m wide, floored by a concave-up fifth-order surface.

It may be possible to document the range in scales and the potential mappability of the hierarchy of depositional units and their bounding surfaces in any given stratigraphic unit. For example, Table 4.1 illustrates the results of work in progress to document the fluvial styles of the Westwater Canyon Member of the Morrison Formation (Jurassic) in northern New Mexico. This type of subdivision is of importance in reservoir development because each level of the hierarchy will be characterized by its own porosity and permeability architecture, which can be built into computer models of fluid flow. However, hierarchies of this type do not have a fixed range of scales. For example, in parts of the Donjek River, Yukon, large compound bars (the first-order bars of Williams and Rust 1969, using an earlier classification; fourth-order macroforms in the present scheme) are in the order of 200 to 400 m in downstream length. Comparable features in the Brahmaputra River, Bangladesh, are several kilometres in length (Coleman 1969) – an order of magnitude larger.

Bounding Surfaces in Other Detrital Deposits

To my knowledge a comparable classification of depositional scales and their bounding surfaces has not been attempted for other clastic environments. The potential would seem to be there, however, and I submit that this may prove to be a useful unifying concept in facies analysis. For example, submarine fans contain depositional lobes, major and minor channels, point bars, all of macroform rank; minor bars and channels and individual sediment–gravity flow events of mesoform rank, and small–scale structures (e.g., the ripples of Bouma turbidite division C) of microform rank (see many examples in Bouma *et al.* 1985). Again, however, the scale of the specific depositional system must be taken into account. For example, in the upper reaches of the small modern Navy Fan, off southern California, the main feeder channel is less than 20 m deep and 1 km wide. Compare this with the main channel in the upper Bengal Fan, off India, which is more than 300 m deep and 15 km wide (Bouma *et al.* 1985).

Other channelized environments, including delta distributaries, tidal creeks in areas of lagoonal, estuarine, or sabkha tidal flats, tidal channels through barrier islands and carbonate reefs, and submarine canyons, are all likely to contain hierarchies of microforms, mesoforms, and macroforms comparable to those of fluvial origin discussed above (examples were discussed by Jackson 1975).

Other kinds of environments may require a different approach. Recent high-resolution seismic and sonar surveys of modern shelves reveal a wealth of small to giant bars and bedforms (e.g., Martin and Flemming 1986; Luternauer 1986; Swift *et al.* 1986), for which many architectural details remain to be worked out. In basin plains, and in many shelf settings and lacustrine environments away from any clastic shoreline, the stratigraphy is controlled by basinwide "events" such as storms, sediment gravity flows, or changes in base level (sea level or lake level). Stratigraphic geometries are typically tabular and individual lithosomes may extend for many kilometres (e.g., the coarsening-upward sequences in the Cardium Formation of Alberta, as described by Plint *et al.* 1986). A much simpler classification of bounding surfaces may suffice, unless a long-term cyclicity is revealed by careful stratigraphic work.

The erection of a depositional-unit hierarchy and a bounding-surface hierarchy for each of these types of deposit would be a useful aid to the analysis of various autocyclic and allocyclic depositional controls, and would provide valuable data on the scale of potential heterogeneities in reservoir rocks formed in these environments.

Architectural Elements

Application of the bounding-surface concept permits the subdivision of a clastic succession into a hierarchy of three-dimensional rock units. This facilitates description, and also makes it easier to reconstruct the appropriate physical extent and time duration of the processes that controlled sedimentation at each level of the hierarchy. Rock description involves the definition of lithofacies and the recognition of facies assemblages. Vertical profile analysis has been a primary descriptive and analytical tool since the classic work of Visher (1965). As discussed in a later section, vertical profiles have little more to offer as research tools, and the time has come to focus more on the three-dimensional architecture of rock bodies.

Many workers have studied two-dimensional and three-dimensional outcrops in an attempt to reconstruct the internal geometry or architecture of a deposit. In the case of fluvial deposits good examples include studies by Nami and Leeder (1978), Nijman and Puigdefabregas (1978), Puigdefabregas and Van Vliet (1978), Friend *et al.* (1979), Jones and McCabe (1980), Allen (1983), Bridge and Diemer (1983), Friend (1983), Haszeldine (1983a,b), Kirk (1983), Ramos and Sopeña (1983), Blakey and Gubitosa (1984), Bridge and Gordon (1985), and Ramos *et al.* (1986). Few fluvial deposits have simple tabular geometries because of the presence of numerous channel scours and large bar deposits.

Many of the studies quoted above have confirmed the value of Jackson's (1975) recognition of macroforms as a specific class of large-scale deposit. In fact, some of these studies, plus my ongoing work with L. Middleton and C.E. Turner-Peterson in the Colorado Plateau, demonstrate that many fluvial deposits consist largely of macroforms several metres thick and tens to a few hundred metres in length and width.

Most deposits may be subdivided into several types of three-dimensional bodies characterized by distinctive lithofacies assemblages, external geometries, and orientations (many of which are macroforms). Allen (1983) coined the term "architectural element" for these depositional units, and Miall (1985) attempted a summary and classification of the current state of knowledge of these elements as they occur in fluvial rocks.

Two interpretive processes are involved simultaneously in the analysis of outcrops that contain a range of scales of depositional units and bounding surfaces: 1) the definition of the various types and scales of bounding surfaces; and 2) the subdivision of the succession into its constituent lithofacies assemblages, with the recognition and definition of macroforms and any other large features that may be present.

In general, the most distinctive characteristic of a macroform is that it consists of genetically related lithofacies, with sedimentary structures showing similar orientations, and internal minor bounding surfaces (first to third order of the classification given above) that extend from the top to the bottom of the element, indicating that it developed by long-term lateral, oblique, or downstream accretion. A macroform is comparable in height to the depth of the channel in which it formed and in width and length is of similar order of magnitude to the width of the channel. However, independent confirmation of these dimensions is difficult in multi-storey sandstone bodies, where channel margins are rarely preserved and the storeys commonly have erosional relationships with each other.

The definition of a macroform in any given outcrop is in part an interpretive process. Some types of macroform, such as the lateral accretion deposits that comprise the typical point bar, are by now so well known that their recognition in outcrop would be classified by some workers as a "descriptive" rather than an "interpretive" exercise. Other types of macroform are less well known. The recognition of a macroform may depend in part on the type of bounding surface that encloses it. Conversely, the appropriate classification of a bounding surface may depend on a description of the lithofacies assemblage and geometry of the beds above and below it. For these reasons description, classification, and interpretation cannot always be completely separate exercises.

A gradation exists between lateral, oblique, and downstream accretion. Oblique accretion may characterize alternate bars (e.g., Jones and McCabe 1980; Okolo 1983). Downstream accretion occurs in the large mid-channel macroforms described by Allen (1983) and Haszeldine (1983a,b). Some elements are likely to contain an internal gradation in accretion directions in response to channel migration, and there may, of course, be other types of deposit that have yet to be recognized. To distinguish these kinds of depositional elements requires a careful examination and interpretation of the geometry and orientation of internal bounding surfaces and constituent sedimentary structures. Lateral accretion is characterized by crossbed dip orientations parallel to the strike of the accretion surface, whereas downstream accretion results in crossbed dip orientation parallel to the dip of the accretion surfaces. These relationships must be determined from measurements within the macroform itself, and not based on comparisons between local and regional paleocurrent patterns, because the local channel orientation may be at a high angle to regional paleoslope.

The range of element types in fluvial rocks is illustrated in Figure 4.4, and an example of an outcrop interpreted using the bounding-surface hierarchy and the architectural-element classification presented here is shown in Figure 4.5, which can be found on the photographic insert on the following page of this book. In this diagram elements bounded by surfaces of fourth order or higher rank are numbered in depositional sequence, and are interpreted using the two-letter element code shown in Figure 4.4. A three-dimensional interpretation of facies architecture can be made from outcrops like this by the careful tracing and classification of the bounding surfaces, and by the documentation of orientations of crossbed sets and the dip of the bounding surfaces. For example, surfaces A and D are tentatively interpreted as fourth-order, because they are convex-up and are, in part, parallel to underlying stratification, suggesting that they are accretion surfaces (see also the fourth-order surface in Figure 4.3D, which shows an adjacent outcrop). The other major bound-

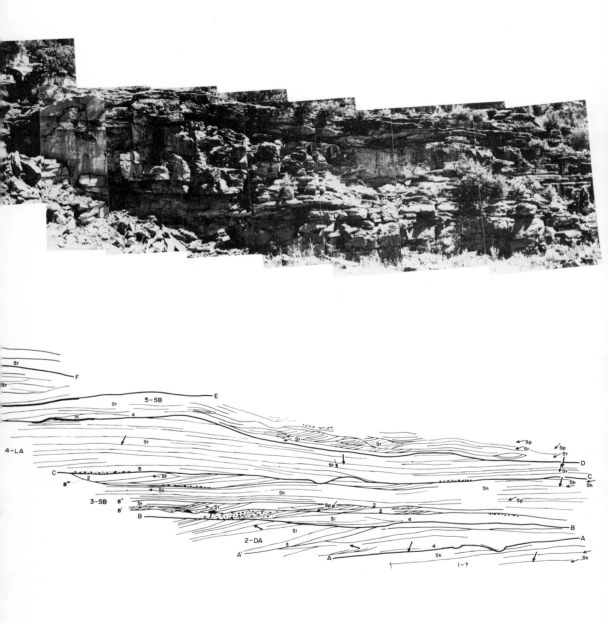

Fig. 4.5. An example of two-dimensional outcrop analysis, showing the classification of the bounding surfaces and the constituent architectural elements. Kayenta Formation (Jurassic), near Dove Creek, Colorado. See text for explanation, and Miall (in press) for additional details.

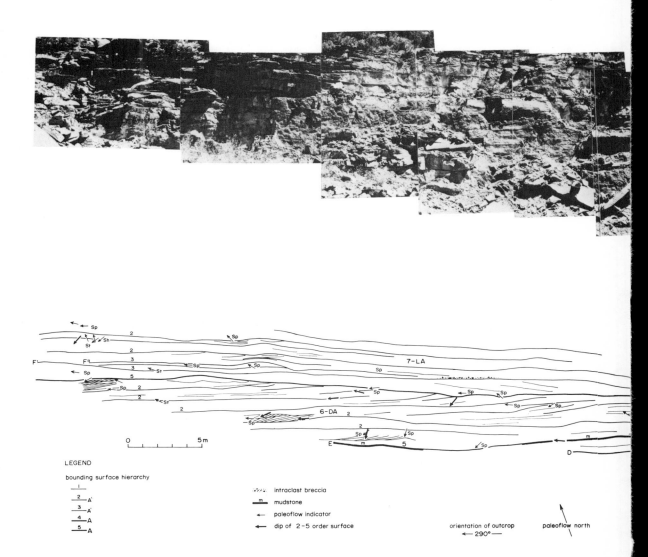

Fig. 4.4. The range of architectural elements in fluvial deposits. Lithofacies codes are those of Miall (1978) (slightly modified from Miall 1985).

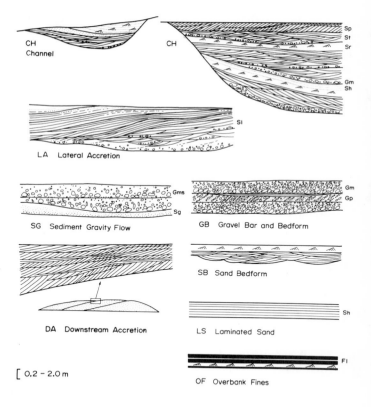

ing surfaces are probably at the base of channel-fill sequences and are of fifth-order. In elements 4 and 7, second-order and third-order bounding surfaces dip out of the outcrop (S to SW), whereas individual crossbed sets are oriented parallel to the face (W), a configuration characteristic of lateral accretion. In element 6, dip orientation of bounding surfaces and crossbed sets is similar (W), which suggests downstream accretion.

As suggested in the previous section, other types of channelized environments probably could be characterized by a suite of bounding surfaces and architectural elements similar to those that are beginning to be understood in the fluvial environment, whereas nonchannelized clastic environments (beaches, shelves, basin plains, etc.) may require a quite different type of classification.

Analysis of Facies Architecture in the Subsurface

Up to this point the discussion has focused on the study of outcrops, where it may be a relatively simple matter to make three-dimensional reconstruc-

tions. Can these ideas be applied to the well and seismic data available from the subsurface?

The vertical-profile model has had a long life in sedimentary geology, partly because of its usefulness in interpreting petrophysical logs and cores (e.g., Fisher *et al.* 1969; Walker 1984). Where possible, subsurface geologists have also constructed stratigraphic sections and fence diagrams from available well data, and the results have been used to amplify environmental reconstructions and provide data for the reservoir engineer.

Central to many interpretations of vertical profiles in clastic successions is the recognition of one of two characteristic motifs: coarsening-and-thickening-upward (c-thk-u) or fining-and-thinning-upward (f-thn-u). The first is characteristic of active progradation of alluvial fans, deltas and submarine fans, and may also occur as a result of the reoccupation of abandoned channels. The second motif is a product of progressive aggradation and abandonment of a channelized environment, by lateral fill of the same channel, or by progressive abandonment of an active depositional area on a delta or fan (subaerial or submarine). Autocyclic processes, such as channel migration and avulsion, and delta or fan lobe switching can produce all these effects, although tectonic control may also be important locally. However, many successions show c-thk-u or f-thn-u cycles of more than one scale (thickness) in the same stratigraphic succession. Problems with the interpretation of cycles include the following:

1. In a given environment there may be several short-term and long-term autocyclic processes that can simultaneously produce vertical profiles of widely varying scales. Sequences (cycles) of a range of thicknesses and types may therefore be nested within each other. Assigning a correct interpretation to each type and scale of cyclicity may be facilitated by information on lateral variability.
2. Different autocyclic mechanisms in the same overall environment can produce very similar profiles (Miall 1980 discussed this with reference to the fluvial environment).
3. Allocyclic processes, particularly local tectonics, may produce very similar vertical sequences, which may or may not be in phase with those resulting from autocyclic energy distribution (Miall 1980). Again, mapping of lateral variability is essential for correct interpretation.
4. A one-dimensional profile would be adequate to describe a succession of facies having simple tabular geometries. However, as the discussion above has shown, many environments are characterized by lithosome geometries with complex three-dimensional forms. This problem is particularly acute in channelized environments where both erosional and constructional depositional relief may be in the order of tens to hundreds of metres.
5. The definition of cyclicity in any given succession may be a rather subjective process (Zeller 1964), and the results can be influenced by the analytical technique (Hiscott 1981a).

How do the various scales and types of cyclicity relate to the architectural elements discussed earlier? Allocycles, such as those produced by tectonic effects and sea-level changes (e.g., Heward 1978; Plint et al. 1986), are bounded by sixth-order surfaces. Cycles of progradation in fans and deltas are of fifth and sixth order, depending on their scale. Channel-fill cycles are typically of fifth order, although cycles produced by dynamic events, such as flood cycles in ephemeral rivers (mesoforms in Jackson's 1975 classification) compare in temporal

significance to the accretionary increments in macroforms that are bounded by third-order surfaces. Some macroforms may be internally cyclic, such as the fining-upward cycles of some point bars. These cycles are bounded by fourth-order surfaces. However, many macroforms are not cyclic, and are not expected to have characteristic vertical profiles. Statistical techniques, such as Markov–chain analysis, which have been used to impart an aura of quantitative rigor to vertical profile analysis (Miall 1973; Hiscott 1981b; Harper 1984), now seem somewhat superfluous, particularly in view of the varying scales of sedimentary cycles and the wide range of interpretations that can be made from them.

Most of the problems relating to the interpretation of vertical profiles can be circumvented by careful mapping of lateral facies variability. The discussion below addresses this point, examining the kinds of architectural variability at various scales, and the kinds of mapping technique appropriate to those scales. Reservoir modellers may be able to break a reservoir unit down into its component elements, and make use of the bounding–surface hierarchy to examine each component separately, in its correct scale context. Table 4.1 summarizes heterogeneity data for the Westwater Canyon Member of the Morrison Formation, based on outcrop studies by the writer and C. E. Turner-Peterson (unpublished data). These data are used in the paragraphs below to discuss subsurface approaches to the use of architectural data.

Mapping of units defined by fifth-order and sixth-order surfaces can be based on conventional methods of subsurface wireline log correlation, provided well spacing is adequate. In most mature basins, with a well spacing of one to a dozen wells per township (wells about 3 to 10 km apart), such correlation should be possible down to the submember level, although local problems may arise as a result of internal erosion and the resulting amalgamation of sandsheet units. For example, Santos and Turner-Peterson (1986) reported local difficulties in subdividing the Morrison Formation into its constituent members, particularly in the subsurface, and Turner-Peterson (1986) relied on outcrop studies of pebble suites and paleocurrent trends to distinguish the three submembers of the Westwater Canyon Member.

Mapping of ribbon sandstone bodies in the subsurface can present particularly difficult problems because the width of the sandstones is usually less than well spacing, except in highly developed fields, where the well spacing may range from 0.32 to 0.65 km² (32–65 ha; 80–160 acres; 8–4 wells per 1 mile section, equivalent to a well spacing of about 600–800 m). For example, anastomosed ribbon sandstone bodies were reported in the subsurface Mannville Formation of Alberta and Saskatchewan by Putnam (1982a,b), but his analyses were challenged by Wightman *et al.* (1981) on several grounds, including the difficulty of correlating sandstones of different thickness and possibly at different stratigraphic levels.

Three-dimensional seismic data promises to yield much valuable information on depositional architecture. Conventional two-dimensional seismic sections have been difficult to interpret sedimentologically because of the limited vertical resolution of the sections relative to the vertical scale of most depositional features. However, on horizontal seismic sections the resolution is more than adequate to reveal major channels and other features bounded by fifth- and sixth-order surfaces (Brown 1986). It remains to be seen whether individual macroforms will be mappable using seismic data.

Based on experience with the Westwater Canyon Member (Table 4.1) macroform bar units, bounded by fourth-order surfaces, and the subunits within them that are defined by third-order surfaces, would require well spacing of 0.32 km² (32 ha; 80 acres; 8 wells per section; wells about 600 m apart) or less if their geometry is to be reconstructed reliably from wireline log and core data. In larger river systems the macroforms are correspondingly more extensive, and a wider well spacing might be adequate, although Busch (1974) mapped a 6-km diameter point bar using a 0.16 km² spacing (16 ha; 40 acres; 16 wells per section; wells 400 m apart). Berg (1968) described another example. The recognition of third-order and fourth-order surfaces in core and dipmeter data was referred to earlier. The necessary data should be collected during pilot studies carried out at the commencement of field development or enhanced recovery projects.

The analysis of individual crossbed sets and cosets in core provides information on those lithosome units bounded by first-order and second-order surfaces. In the Westwater Canyon Member these have a maximum expected areal extent of about 0.1 km² (10 ha; 25 acres; 10⁵ m²), usually much less, and typically cannot be correlated from well to well.

Conclusions

Has facies analysis been "done"?

The continued outpouring of studies on modern sedimentary processes in ever more inaccessible places has hopelessly overloaded the simple facies-model concepts with which most of us were content approximately ten years ago.

The ideas presented here are an attempt to build on this vast (but still very incomplete) data base and to generate some new unifying concepts. The ideas about physical scale and bounding surfaces allow us to separate the products of several overlapping time scales. The ideas about architectural elements represent an attempt to apply facies-model approaches to the *components* of depositional systems and to suggest that some of the various kinds of components may occur in a variety of environments.

So, no, I do not think that facies analysis has been done. We just need to put on better 3-D glasses and peer through a more discriminating zoom lens.

References

ALLEN, J.R.L. (1966) On bed forms and paleocurrents. Sedimentology 6:153–190.

ALLEN, J.R.L. (1983) Studies in fluviatile sedimentation: Bars, bar complexes and sandstone sheets (low-sinuosity braided streams) in the Brownstones (L. Devonian), Welsh Borders. Sedimentary Geology 33:237–293.

AMERICAN SOCIETY OF CIVIL ENGINEERS, Task Force on Bed Forms in Alluvial Channels (1966) Nomenclature for bed forms in alluvial channels. Proceedings American Society Civil Engineers, Journal Hydraulics Division 92(HY3):51–64.

BERG, R.R. (1968) Point bar origin of Fall River Sandstone reservoirs, northeastern Wyoming. American Association Petroleum Geologists Bulletin 52:2116–2122.

BLAKEY, R.C. and GUBITOSA, R. (1984) Controls of sandstone body geometry and architecture in the Chinle Formation (Upper Triassic), Colorado Plateau. Sedimentary Geology 38:51–86.

BOND, G.C., KOMINZ, M.A., and GROTZINGER, J.P. (1988) Cambro-Ordovician eustasy: Evidence from geophysical modelling of subsidence in Cordilleran and Appalachian passive margins. In: Kleinspehn, K.L. and Paola, C. (eds) New Perspectives in Basin Analysis. New York: Springer-Verlag, pp. 129–160.

BOUMA, A.H., NORMARK, W.R., and BARNES, N.E. (eds) (1985) Submarine Fans and Related Turbidite Systems. New York: Springer-Verlag, 351 p.

BRIDGE, J.S. and DIEMER, J.A. (1983) Quantitative interpretation of an evolving ancient river system. Sedimentology 30:599–623.

BRIDGE, J.S. and GORDON, E.A. (1985) The Catskill magnafacies of New York State. In: Flores, R.M. and Harvey, M. (eds) Field Guide to Modern and Ancient Fluvial Systems in the United States, Third International Sedimentology Conference, pp. 3–17.

BROOKFIELD, M.E. (1977) The origin of bounding surfaces in ancient aeolian sandstones. Sedimentology 24:303–332.

BROWN, A.R. (1986) Interpretation of three-dimensional seismic data. American Association Petroleum Geologists Memoir 42, 194 p.

BUSCH, D.A. (1974) Stratigraphic traps in sandstones — Exploration techniques. American Association Petroleum Geologists Memoir 21, 174 p.

CAMPBELL, C.V. (1976) Reservoir geometry of a fluvial sheet sandstone. American Association Petroleum Geologists Bulletin 60:1009–1020.

CLOETINGH, S. (1988) Intraplate stresses: A new element in basin analysis. In: Kleinspehn, K.L. and Paola, C. (eds) New Perspectives in Basin Analysis. New York: Springer-Verlag, pp. 205–230.

COLEMAN, J.M. (1969) Brahmaputra River: Channel processes and sedimentation. Sedimentary Geology 3:129–239.

COLLINSON, J.D. (1970) Bedforms of the Tana River, Norway. Geografiska Annaler 52A:31–55.

FISHER, W.L., BROWN, L.F., SCOTT, A.J., and McGOWEN, J.H. (1969) Delta Systems in the Exploration for Oil and Gas. Austin, Texas: Bureau of Economic Geology, 78 p.

FRIEND, P.F. (1983) Towards the field classification of alluvial architecture or sequence. In: Collinson, J.D. and Lewin, J. (eds) Modern and Ancient Fluvial Systems. International Association Sedimentologists Special Publication 6, pp. 345–354.

FRIEND, P.F., SLATER, M.J., and WILLIAMS, R.C. (1979) Vertical and lateral building of river sandstone bodies, Ebro Basin, Spain. Journal Geological Society London 136:39–46.

GRADZINSKI, R., GAGOL, J., and SLACZKA, A. (1979) The Tumlin Sandstone (Holy Cross Mts., central Poland): Lower Triassic deposits of aeolian dunes and interdune areas. Acta Geologica Polonica 29:151–175.

HARPER, C.W., JR. (1984) Improved methods of facies sequence analysis. In: Walker, R.G. (ed) Facies Models, Second edition, Geoscience Canada Reprint Series 1, pp. 11–13.

HASZELDINE, R.S. (1983a) Fluvial bars reconstructed from a deep, straight channel, Upper Carboniferous coalfield of northeast England. Journal Sedimentary Petrology 53:1233–1248.

HASZELDINE, R.S. (1983b) Descending tabular cross-bed sets and bounding surfaces from a fluvial channel in the Upper Carboniferous coalfield of north-east England. In: Collinson, J.D. and Lewin, J. (eds) Modern and Ancient Fluvial Systems, International Association Sedimentologists Special Publication 6, pp. 449–456.

HEWARD, A.P. (1978) Alluvial fan sequence and megasequence models: With examples from Westphalian D-Stephanian B coalfields, northern Spain. In: Miall, A.D. (ed) Fluvial Sedimentology, Canadian Society of Petroleum Geologists Memoir 5, pp. 669–702.

HICKIN, E.J. (1983) River channel changes: Retrospect and prospect. In: Collinson, J.D. and Lewin, J. (eds) Modern and Ancient Fluvial Systems. International Association Sedimentologists Special Publication 6, pp. 61–83.

HISCOTT, R.M. (1981a) Deep-sea fan deposits in the Macigno Formation (Middle-Upper Oligocene) of the Gordana Valley, northern Apennines, Italy: Discussion. Journal Sedimentary Petrology 51:1015–1021.

HISCOTT, R.M. (1981b) Chi-square tests for Markov chain analysis. Journal International Association Mathematical Geologists 13:53–68.

JACKSON, R.G., II (1975) Hierarchical attributes and a unifying model of bed forms composed of cohesionless material and produced by shearing flow. Geological Society America Bulletin 86:1523–1533.

JONES, C.M. and McCABE, P.J. (1980) Erosion surfaces within giant fluvial cross-beds of the Carboniferous in northern England. Journal Sedimentary Petrology 50:613–620.

JORDAN, T.E., FLEMINGS, P.B., and BEER, J.A. (1988) Dating thrust-fault activity by use of foreland-basin strata. In: Kleinspehn, K.L. and Paola, C. (eds) New Perspectives in Basin Analysis. New York: Springer-Verlag, pp. 307–330.

KIRK, M. (1983) Bar developments in a fluvial sandstone (Westphalian "A"), Scotland. Sedimentology 30:727–742.

KOCUREK, G. (1981) Significance of interdune deposits and bounding surfaces in aeolian dune sands. Sedimentology 28:753–780.

LUTERNAUER, J.L. (1986) Character and setting of sand and gravel bed forms on the open continental shelf off western Canada. In: Knight, R.J. and McLean, J.R. (eds) Shelf Sands and Sandstones. Canadian Society Petroleum Geologists Memoir 11, pp. 45–55.

MARTIN, A.K. and FLEMMING, B.W. (1986) The Holocene shelf sediment wedge off the south and east coast of South Africa. In: Knight, R.J. and McLean, J.R. (eds) Shelf Sands and Sandstones. Canadian Society Petroleum Geologists Memoir 11, pp. 27–44.

McKEE, E.D. and WEIR, G.W. (1953) Terminology for stratification and cross-stratification in sedimentary rocks. Geological Society America Bulletin 64:381–389.

MIALL, A.D. (1973) Markov chain analysis applied to an ancient alluvial plain succession. Sedimentology 20:347–364.

MIALL, A.D. (1974) Paleocurrent analysis of alluvial

sediments: A discussion of directional variance and vector magnitude. Journal Sedimentary Petrology 44:1174–1185.

MIALL, A.D. (1978) Lithofacies types and vertical profile models in braided river deposits, a summary. In: Miall, A.D. (ed) Fluvial Sedimentology. Canadian Society Petroleum Geologists Memoir 5, pp. 597–604.

MIALL, A.D. (1980) Cyclicity and the facies model concept in geology. Bulletin Canadian Petroleum Geology 28:59–80.

MIALL, A.D. (1985) Architectural-element analysis: A new method of facies analysis applied to fluvial deposits. Earth Science Reviews 22:261–308.

MIALL, A.D. (in press) Architectural elements and bounding surfaces in fluvial deposits: Anatomy of the Kayenta Formation (Lower Jurassic), southwest Colorado. Sedimentary Geology.

NAMI, M. and LEEDER, M.R. (1978) Changing channel morphology and magnitude in the Scalby Formation (M. Jurassic) of Yorkshire, England. In: Miall, A.D. (ed) Fluvial Sedimentology. Canadian Society Petroleum Geologists Memoir 5, pp. 431–440.

NIJMAN, W. and PUIGDEFABREGAS, C. (1978) Coarse-grained point bar structure in a molasse-type fluvial system, Eocene Castisent Sandstone Formation, South Pyrenean Basin. In: Miall, A.D. (ed) Fluvial Sedimentology. Canadian Society Petroleum Geologists Memoir 5, pp. 487–510.

OKOLO, S.A. (1983) Fluvial distributary channels in the Fletcher Bank Grit (Namurian R2b), at Ramsbottom, Lancashire, England. In: Collinson, J.D. and Lewin, J. (eds) Modern and Ancient Fluvial Systems. International Association of Sedimentologists Special Publication 6, pp. 421–433.

PAOLA, C. (1988) Subsidence and gravel transport in alluvial basins. In: Kleinspehn, K.L. and Paola, C. (eds) New Perspectives in Basin Analysis. New York: Springer-Verlag, pp. 231–243.

PLINT, A.G., WALKER, R.G., and BERGMAN, K.M. (1986) Cardium Formation 6, Stratigraphic framework of the Cardium in subsurface. Bulletin Canadian Petroleum Geology 34:213–225.

PUIGDEFABREGAS, C. and VAN VLIET, A. (1978) Meandering stream deposits from the Tertiary of the southern Pyrenees. In: Miall, A.D. (ed) Fluvial Sedimentology. Canadian Society Petroleum Geologists Memoir 5, pp. 469–485.

PUTNAM, P.E. (1982a) Fluvial channel sandstones within upper Mannville (Albian) of Lloydminster area, Canada—Geometry, petrography, and paleogeographic implications. American Association Petroleum Geologists Bulletin 66:436–459.

PUTNAM, P.E. (1982b) Aspects of the petroleum geology of the Lloydminster heavy oil fields, Alberta and Saskatchewan. Bulletin Canadian Petroleum Geology 30:81–111.

RAMOS, A. and SOPEÑA, A. (1983) Gravel bars in low-sinuosity streams (Permian and Triassic, central Spain). In: Collinson, J.D. and Lewin, J. (eds) Modern and Ancient Fluvial Systems, International Association Sedimentologists Special Publication 6, pp. 301–312.

RAMOS, A., SOPEÑA, A., and PEREZ-ARLUCEA, M. (1986) Evolution of Buntsandstein fluvial sedimentation in the northwest Iberian Ranges (Central Spain). Journal Sedimentary Petrology 56:862–875.

SANTOS, E.S. and TURNER-PETERSON, C.E. (1986) Tectonic setting of the San Juan Basin in the Jurassic. In: Turner-Peterson, C.E., Santos, E.S., and Fishman, N.S. (eds) A Basin Analysis Case Study—The Morrison Formation, Grants Uranium Region, New Mexico. American Association Petroleum Geologists Studies Geology 22:27–33.

SWIFT, D.J.P., THORNE, J., and OERTEL, G.F. (1986) Fluid processes and sea-floor response on a modern storm-dominated shelf: Middle Atlantic shelf off North America. Part II: Response of the shelf floor. In: Knight, R.J. and McLean, J.R. (eds) Shelf Sands and Sandstones. Canadian Society Petroleum Geologists Memoir 11, pp. 191–211.

TURNER-PETERSON, C.E. (1986) Fluvial sedimentology of a major uranium-bearing sandstone—A study of the Westwater Canyon Member of the Morrison Formation, San Juan Basin, New Mexico. In: Turner-Peterson, C.E., Santos, E.S., and Fishman, N.S. (eds) A Basin Analysis Case Study—The Morrison Formation, Grants Uranium Region, New Mexico. American Association Petroleum Geologists Studies Geology 22:47–75.

VISHER, G.S. (1965) Use of vertical profile in environmental reconstruction. American Association Petroleum Geologists Bulletin 49:41–61.

WALKER, R.G. (ed) (1984) Facies Models, Second Edition. Geoscience Canada Reprint Series 1, Geological Association Canada, 317 p.

WIGHTMAN, D.M., TILLEY, B.J., and LAST, B.M. (1981) Stratigraphic traps in channels sandstones in the upper Mannville (Albian) of east-central Alberta: Discussion. Bulletin Canadian Petroleum Geology 29:622–625.

WILLIAMS, P.F. and RUST, B.R. (1969) The sedimentology of a braided river. Journal Sedimentary Petrology 39:649–679.

ZELLER, E.J. (1964) Cycles and psychology. Kansas Geological Survey Bulletin 169 2:631–636.

5
Origin, Recognition, and Importance of Erosional Unconformities in Sedimentary Basins

G. SHANMUGAM

Abstract

Erosional unconformities of different scales (local to global) are an ubiquitous element of all sedimentary basins. Erosional unconformities of subaerial origin are believed to have been caused by tectonic uplifts and by eustatic sea-level fall. Erosional unconformities of submarine origin may be related to transgression, mass movements, turbidity currents, thermohaline currents, carbonate dissolution, storms, and clastic influx on carbonate shelves. Important criteria for recognizing subaerial unconformities include discordance of dip, karst facies, basal conglomerate, and a major gap in the fossil record. Paleosol horizons, duricrust, and continental deposits, indicative of subaerial exposure, can also be used to define surfaces of potential subaerial unconformities. Submarine unconformities may be recognized by mass-movement deposits, glauconitic minerals, and manganese nodules.

Recognition of unconformities is useful for subdividing stratigraphic units, determining the timing of tectonic activity, interpreting lateral facies relationships, constructing burial and uplift curves, correlating certain stratigraphic boundaries, interpreting sea-level changes, and for reconstructing paleogeography. Erosional unconformities may be important to exploration because they can be used to predict deep-sea turbidite reservoir facies; they can mark upper boundaries of zones of increased porosity (e.g., Statfjord Field, North Sea); they can provide an ideal juxtaposition of reservoir and source rocks (e.g., Prudhoe Bay Field, Alaska); they can act as avenues of hydrocarbon migration (e.g., Maracaibo Basin, Venezuela); they can generate hydrocarbon traps (e.g., Messla Field, Libya); and they can be favorable sites for mineralization (e.g., uranium, aluminum, phosphates, and gold).

Introduction

James Hutton first perceived the meaning of unconformity and temporal breaks in the stratigraphic record in 1788, when he observed the Devonian Old Red Sandstone resting on steeply dipping Silurian strata at Siccar Point, Scotland (Fig. 5.1). Charles Darwin (1859) was first to point out the magnitude of breaks in the geologic record by arguing that, probably, more of geologic time is represented by stratigraphic breaks than by preserved beds. Recognition of these breaks and their magnitude, which may range from a short span of time between storms to hundreds of millions of years, is important in many facets of basin analysis. The purpose of this contribution is to discuss erosional unconformities in terms of their origin, recognition, and importance.

Earlier studies of unconformities were concerned primarily with unconformity traps for hydrocarbons (Levorsen 1934, 1936; Chenworth 1972), and with criteria for their recognition (Krumbein 1942). American Association of Petroleum Geologists Memoirs 16 (King 1972), 32 (Halbouty 1982), and 36 (Schlee 1984) dealt with various aspects of unconformities such as methodology for recognizing unconformity traps, case histories, and the relationship between sea-level changes and unconformity development. This paper is not an exhaustive review of literature on unconformities, and it differs from previous studies in the following ways: 1) Exploration and classical stratigraphic aspects of unconformities are integrated; 2) differing scales of unconformities are emphasized using seismic, outcrop, and core examples; 3) the origin of unconformities in subaerial and submarine environments is discussed using modern and ancient examples; 4) a balanced treatment of siliciclastic and carbonate sequences is attempted; and 5) causes of porosity enhancement beneath erosional unconformities are explored.

Fig. 5.1. The great unconformity at Siccar Point, Berwickshire coast, Scotland, first interpreted by James Hutton (1788), showing discordance of dip between the Upper Old Red Sandstone (Devonian) and underlying vertical beds of Silurian graywackes. (Photo: K.F. Keller).

Unconformities and Condensed Sections

Unconformities

The American Geological Institute Glossary of Geology (Bates and Jackson 1980) defines an unconformity as "the structural relationship between rock strata in contact, characterized by a lack of continuity in deposition, and corresponding to a period of nondeposition, weathering, or especially erosion (either subaerial or subaqueous) prior to the deposition of the younger beds and often (but not always) marked by the absence of parallelism between strata." In this chapter, the term unconformity is used for a surface of erosion representing a significant temporal break in the stratigraphic record.

On the basis of structural relations between unconformable units, three major types of erosional unconformities can be recognized (Dunbar and Rodgers 1957; Krumbein and Sloss 1963). They are: 1) nonconformity (erosional surface between igne-ous or metamorphic rocks below and sedimentary rocks above); 2) angular unconformity (angular discordance separating two units of strata along the erosional surface); and 3) disconformity (two units of strata with the same orientation, but separated by an uneven erosional surface). In general, angular unconformities and nonconformities represent major temporal breaks, and disconformities reflect variable temporal breaks. Degree of angularity, however, is not a measure of magnitude of temporal breaks.

Barrell (1917) introduced the term "diastem" to refer to minor breaks in the geologic record. His primary intention was to distinguish major breaks (unconformities) from minor breaks (diastems). Unconformities represent an overall change in the environment, whereas diastems result from expected random variation in sedimentation rate without any basic change in the environment. Without faunal evidence, however, distinguishing between disconformities and diastems can be difficult.

Condensed Sections

Thin deep-marine intervals characterized by very slow depositional rates (<0.01 m/10^3 yr) are considered to be condensed sections and are regarded as marine hiatuses by seismic stratigraphers (Vail *et al.* 1984). This suggests that condensed sections represent temporal breaks. In reality, intervals of slow deposition are not true hiatuses because there is continuous accumulation of muddy or biogenic sediment even during the periods of "marine hiatus." For example, pelagic mudstone intervals deposited between turbidite sandstones can be interpreted as condensed sections. One could also consider mud layers deposited during slack-water periods in subtidal facies (Visser 1980) as condensed sections relative to the migration of the sand wave. The usage of the term "condensed section," implying temporal breaks, for pelagic intervals of continuous but slow deposition is thus misleading.

Areas of true nondeposition, in contrast with condensed sections, are said to have reached an equilibrium state (i.e., depositional base level), in which neither deposition nor erosion takes place. The transport of sediments across areas of nondeposition, known as sedimentary bypassing, is considered to be a fundamental process in the formation of marine unconformities (Krumbein and Sloss 1963). Manganese nodules, phosphates, glauconitic minerals, hardgrounds, burrowed zones, and borings are considered to be common indicators of nondeposition. Manganese nodules (Watkins and Kennett 1972) and glauconitic minerals (Odin 1985), however, are also associated with areas of vigorous bottom currents in the deep sea. If winnowing (erosion) of fine sediment by strong bottom currents occurs in an area of hardgrounds and chemical precipitates, physical evidence for erosion along these hard surfaces may be lacking. Thus it is conceivable that most, if not all, surfaces routinely identified as episodes of nondeposition in the rock record may actually be due to erosion. Although nondepositional surfaces generally reflect a break in coarse clastic sedimentation, they may not truly represent major temporal breaks in the stratigraphic record because the chemical sedimentary record (e.g., manganese nodules) is usually well preserved along these surfaces. The major differences between unconformities and condensed sections formed by slow deposition are: 1) unconformities are surfaces, whereas condensed sections are intervals; and 2) unconformities represent major breaks in the strati-

graphic record, whereas condensed sections comprise a complete stratigraphic record. Condensed sections, however, may contain many relatively minor breaks (diastems).

Origin of Erosional Unconformities

Erosional unconformities are created by both allocyclic (tectonic and eustatic controls) and autocyclic (sedimentary control) processes in subaerial and submarine environments.

Subaerial Erosion

Tectonic Uplift

Erosional unconformities developed within subaerial deposits generally result from tectonic uplift and subsequent erosion. Erosion of uplifted and tilted strata will produce an easily identifiable truncation unconformity (Fig. 5.2A). Vail *et al.* (1984) ascribed the origin of truncation unconformities to simultaneous eustatic sea-level fall and tectonic uplift; however, tectonism alone, unaccompanied by eustatic sea-level fall, can create such erosional surfaces. Depending on the magnitude of uplift events, these unconformities may extend from a few kilometres to hundreds of kilometres.

Sloss (1984) suggested that interregional unconformities can be a product of globally synchronous episodes of tectonic uplifts. Schwan (1980) compiled evidence for global orogenic activity during Late Jurassic to Late Tertiary time. These data suggest that periods of formation of angular unconformities by orogeny closely correlate with periods of significant discontinuities of sea-floor spreading at 148, 115–110, 80–75, 63, 53, 42–38, 17, and 10–9 Ma. Most of these global tectonic events correlate with periods of rising sea level (Miall 1984). In other words, global unconformities can develop during periods of tectonic uplift and high sea level. Global unconformities created during uplifts pose a real problem for a basic tenet of seismic stratigraphy: the assumption that all global unconformities were created during periods of low sea level (Fig. 5.2B).

The Cimmerian unconformity in the North Sea has been interpreted to represent tectonic uplift by some authors (Sommer 1978; Selley 1984; Cloetingh 1986) and eustatic sea-level fall by others (Vail and Todd 1981; Vail *et al.* 1984). Structural features, such as discordance of dip and truncation

A. EROSION DURING UPLIFT

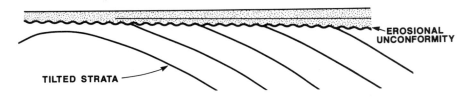

B. EROSION DURING SEA–LEVEL FALL

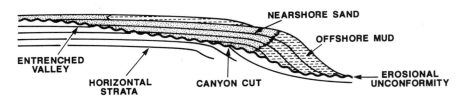

Fig. 5.2. Erosional unconformities formed during tectonic uplift (A) and eustatic sea-level fall (B); note pronounced structural discordance beneath tectonically generated unconformities. Modified after Vail *et al.* (1984). Published with permission of the American Association of Petroleum Geologists.

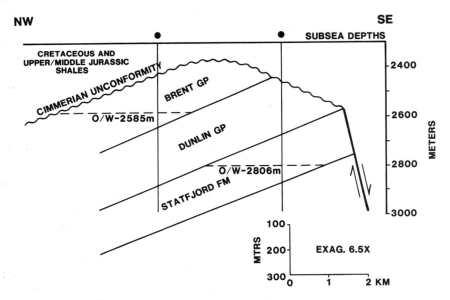

Fig. 5.3. A generalized cross section showing the position of Cimmerian unconformity immediately above the Middle Jurassic Brent Group, Statfjord Field, North Sea (modified after Kirk 1980). Discordance of dip and truncation suggest uplift-related erosion. O/W = oil-water contact. Published with permission of the American Association of Petroleum Geologists.

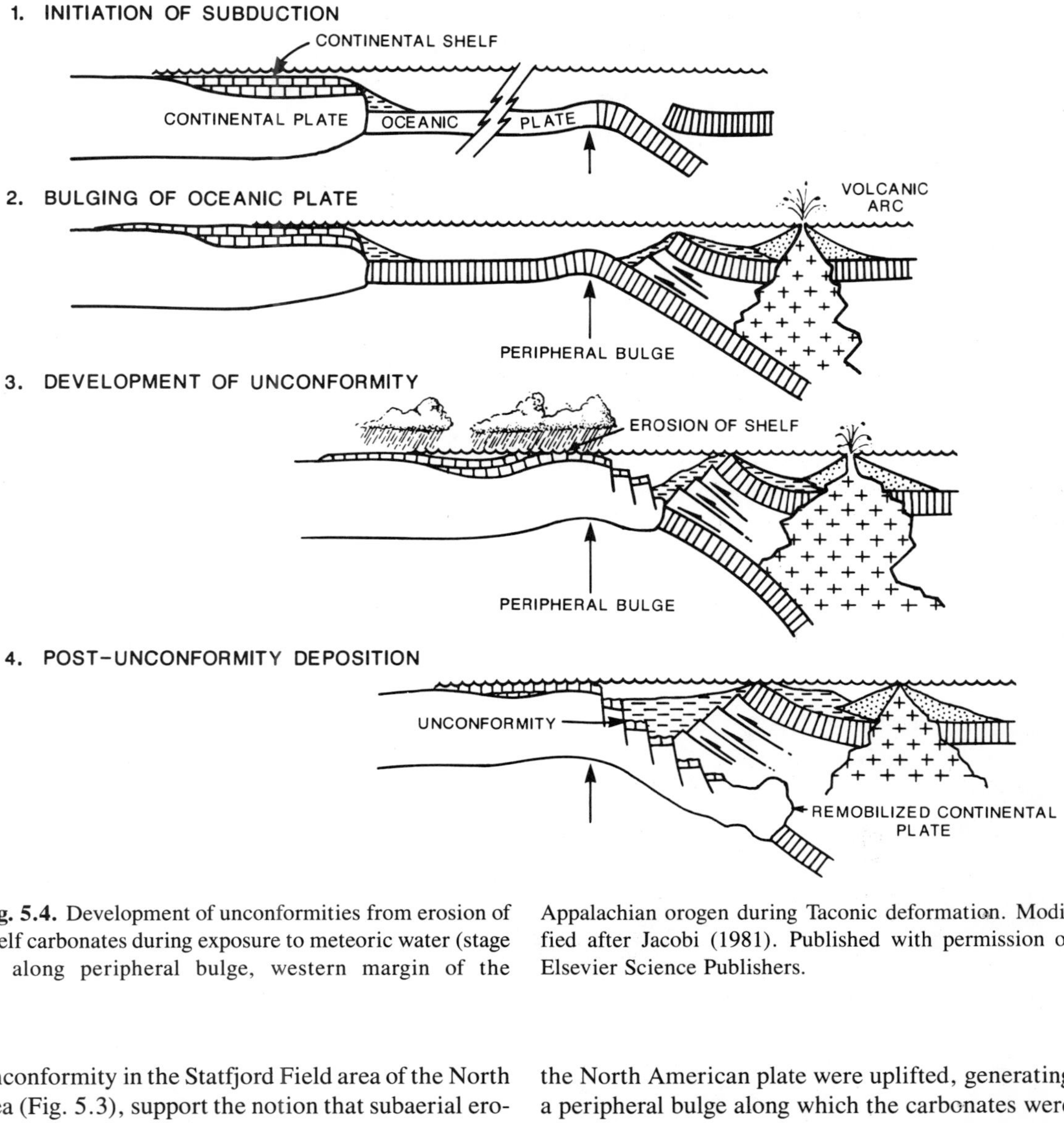

Fig. 5.4. Development of unconformities from erosion of shelf carbonates during exposure to meteoric water (stage 3) along peripheral bulge, western margin of the Appalachian orogen during Taconic deformation. Modified after Jacobi (1981). Published with permission of Elsevier Science Publishers.

unconformity in the Statfjord Field area of the North Sea (Fig. 5.3), support the notion that subaerial erosion during Cimmerian time was uplift-related. Regional uplifts have created erosional unconformities not only in the North Sea (Ziegler 1975) but also in the North Slope (Morgridge and Smith 1972; Jones and Speers 1976) and in other parts of the world (Halbouty 1982).

Jacobi (1981) proposed a plate-tectonic model to explain the origin of the Early/Middle Ordovician erosional unconformity observed along much of the western margin of the Appalachian orogen. According to Jacobi's model, the unconformity was caused by upwarping that occurred as the Ordovician continental margin drifted east into a trench. The Ordovician shelf carbonates on the continental margin of

the North American plate were uplifted, generating a peripheral bulge along which the carbonates were exposed to meteoric water (Fig. 5.4). This exposure resulted in the formation of an erosional unconformity and karst topography in the carbonates. Shanmugam and Lash (1982) explained the origin of this Ordovician unconformity in the Appalachians by upward flexure of the Ordovician shelf carbonates and related erosion during the Taconic orogeny. This is an example of a regional disconformity formed by tectonism.

Eustatic Sea-Level Fall

The eustatic origin of erosional unconformities is discussed by Vail *et al.* (1984); they suggested that eustatic sea-level fall creates global unconformities, whereas tectonism creates local unconformities. Although the above notion is true in most cases, episodes of tectonism associated with changing spreading rates may also create global unconformities (Pitman 1978).

Submarine Erosion

Transgression

Transgressive or onlap facies is usually preceded by an erosional unconformity. Although initial erosion beneath transgressive facies might have taken place in subaerial conditions, the final phase of erosion usually occurs under submarine conditions. Such unconformities tend to produce a sharp and "clean" erosional surface. An excellent example of this type is a nonconformity that separates the Early Cambrian Cape Granite Suite from the overlying Ordovician Graafwater Formation (lower Table Mountain Group) of marine origin from the Cape

Peninsula, South Africa (Fig. 5.5). This type of "marine-swept" regional unconformity is common in the geologic record because of favorable conditions for preservation.

Mass Movement

Mass-movement deposits such as slumps, slides, and debris flows are common in the deep-sea environment. In the Northwest Pacific Basin, an erosional unconformity approximately 25 km in extent separates mass-movement deposits above undisturbed, well stratified sediments below (Fig. 5.6; Damuth *et al.* 1983). Submarine unconformities, created by major slumps (10–50 m thick), have been reported from the continental slope of North Island, New Zealand (Lewis 1971). These slumps occurred on slopes of 1°–4°. The area of slumps ranges from several square kilometres to hundreds of square kilometres. Off Cape Turnagain a curved unconformity at the shelf edge has been recognized on continuous seismic profiles (Lewis 1971). This

Fig. 5.5. A sharp nonconformity contact (arrow) between the Early Cambrian Cape Granite Suite (white) and the overlying Ordovician Graafwater Formation (lower Table Mountain Group), Chapman's Peak, Cape Town, South

Africa. The final phase of erosion and sweeping of the erosion surface is interpreted to be related to marine processes. (Photo: J.G. McPherson).

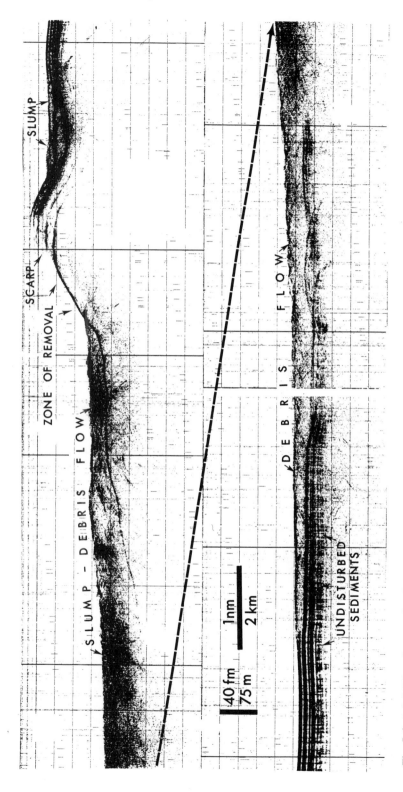

Fig. 5.6. A 3.5 kHz echogram showing mass-movement deposits lying unconformably above well stratified sediments; note slump scarp and zone of removal (Damuth *et al.* 1983). Reproduced with permission of the Geological Society of America Bulletin and the authors.

unconformity represents a buried erosion surface that was formed by slumping. These erosional unconformities may range from regional to local in scale; however, they may not represent significant intervals of time.

Turbidity Currents

Turbidity currents are powerful agents of erosion. In submarine-fan environments, major channels are commonly associated with upper and middle fan deposits. Major deep-sea channels are usually cut by density currents. Modern submarine channels reach a length of up to 3,000 km (e.g., the Bengal fan). Dimensions of selected modern and ancient submarine channels are summarized by Shanmugam et al. (1985a). Submarine channels (Fig. 5.7) may be interpreted as disconformities in terms of stratigraphic breaks.

Thermohaline Currents

The velocity of modern deep-sea bottom currents (contour currents) usually ranges from 1–20 cm/sec (Hollister and Heezen 1972; Grant et al. 1985); however, a velocity of more than 100 cm/sec has been recorded from Iceland-Faroes Ridge (Crease 1965). A comprehensive list of velocity measurements of bottom currents is given by Stow and Lovell (1979). Erosional zones of regional scale (hundreds of kilometres wide), and covering a considerable depth interval (up to 80 m) have been observed on the abyssal sea floor off South Africa (Tucholke and Embley 1984). The erosional zones were created by the northward flow of the Antarctic Bottom Water. Regional disconformities were developed during Cenozoic time by bottom currents between Australia and Antarctica (Watkins and Kennett 1972).

Carbonate Dissolution

In deep-sea environments, widespread hiatuses are created by carbonate dissolution (chemical erosion). In regions of low sediment influx, dissolution hiatuses may represent several millions of years (Keller et al. 1987).

Storms

In clastic shelf environments, minor erosional surfaces may be created by storm and hurricane events (Sloss 1984). These local erosional surfaces would be most likely interpreted as diastems in the rock record.

Clastic Influx

In carbonate shelf environments, introduction of terrigenous clastics reduces penetration of light and related production of nutrients. This could lead to local "erosional" surfaces (e.g., Walker et al. 1983) that are interpreted as surfaces of erosion simply due to the change in sediment type.

Because unconformities of different scales originate from allocyclic and autocyclic processes, they are an integral component of all sedimentary basins. In current practice, many unconformities with the exception of major ones are either unrecognized or underemphasized in basin analysis. For this reason, the following section is devoted to criteria for recognizing unconformities.

Recognition of Erosional Unconformities

Krumbein (1942) first compiled a comprehensive list of 42 criteria by which to recognize unconformities; he divided these criteria into three categories, namely, sedimentary, paleontologic, and structural. Krumbein emphasized that a single criterion by itself may not indicate an unconformity; however, the association of several criteria greatly increases the chance for identification of an unconformity. In this paper, some new criteria have been added to selected criteria from Krumbein's original list.

A distinction between subaerial unconformity and subaerial exposure has been made because subaerial unconformities presuppose prolonged subaerial exposure; however, evidence for subaerial exposure, by itself, does not imply a significant temporal break.

Evidence for Subaerial Unconformities

Discordance of Dip

The angular unconformity invariably develops where older tilted strata have been eroded and have subsequently been overlain by younger strata. An excellent example of this angular relationship is exhibited by the great unconformity at Siccar Point in Scotland (Fig. 5.1). The Silurian beds were intensely folded and eroded during the Caldeonian orogeny. Discordance of dip is best observed in outcrops and in seismic profiles. Unlike the truncation unconformities formed during tectonic uplifts (Fig.

Fig. 5.7. A submarine fan channel showing erosional contact (unfilled arrow) between channel-fill turbidite sandstone and underlying shale, Hecho Group, Eocene, Spain. This type of erosional surface may be interpreted as a disconformity. Black arrow marks a sandstone bed 15-cm thick for scale.

5.2A), unconformities created during eustatic sea-level fall generally exhibit low structural discordance (Fig. 5.2B). If no tectonic uplift is involved, the truncation during periods of low sea level should be greatest at the shelf edge and should progress upshelf by headward erosion and downcutting of river valleys during sea-level fall (Fig. 5.2B). In the subsurface, data from several wells are required to establish this feature. The angular unconformity is the most reliable criterion for erosional unconformities of subaerial origin.

Erosional Surface

When a unit is subjected to subaerial exposure and erosion, an undulatory to even erosional surface results. Undulatory surfaces are particularly well developed in carbonate sequences where karst topography develops due to dissolution. Karst facies, a diagnostic criterion for detecting subaerial unconformities, is recognized by solution features such as vugs, leached fossils, sink holes, caves, and collapse breccia. A detailed description of karst facies is given by Esteban and Klappa (1983). A well developed modern karst from southern China is shown in Figure 5.8.

Ancient erosional surfaces are well developed along the Canadian-Chazyan unconformable boundary in east Tennessee. In the lower part of the Middle Ordovician Sevier Basin in east Tennessee, sharp and even erosional surfaces have been observed to show a brecciated contact between underlying laminated dolostone and overlying lime mudstone (Fig. 5.9; Shanmugam 1978). Scalloped and planar erosional surfaces of karst affinity have been reported from the Middle Ordovician limestones in Virginia (Read and Grover 1977).

Regional Truncation

Regional truncation is best recognized in seismic profiles and in stratigraphic maps of various types

Fig. 5.8. A classic example of modern tower karst topography in the Devonian to Upper Carboniferous limestones near Guilin, southern China. The limestones forming the pinnacles are dark gray in color, massively bedded, extensively fractured, and contain cavernous porosity (arrow). This erosional topography, if preserved, would become a future surface of unconformity. Tower karsts have also been reported to develop in quartzose sandstones (Young 1987).

Fig. 5.9. Erosional surface showing even and sharp brecciated contact between the underlying laminated dolostone and overlying lime mudstone in the lower part of the Middle Ordovician sequence in east Tennessee; note angular dolostone clasts above contact (Shanmugam 1978).

(e.g., lithofacies and isopach; Krumbein and Sloss 1963). Subsurface data from numerous wells are essential to establishing regional truncation of a formation. When established, it is a good indicator of erosional unconformities. A well known example of regional truncation is the truncation of the Permo-Triassic Sadlerochit Group by the Neocomian unconformity near Prudhoe Bay area in Alaska.

Change in Degree of Deformation

If the lower unit is more severely deformed than the upper unit, the inference is that the lower one was deformed and eroded before the upper one was deposited. This is also a reliable criterion for recognizing erosional unconformities. However, the style of deformation could change across an unconformity merely because of differences in rock properties (lithology) on either side.

Fluvial Valleys

Valley-fill fluvial sequences can be used to infer subaerial erosion. Siever (1951) documented a Mississippian-Pennsylvanian unconformity surface that is incised by deep fluvial channels in southern Illinois. Fluvial sequences usually contain widespread basal conglomerate.

Basal Conglomerate

A conglomeratic unit that contains pebbles from an underlying extrabasinal formation is considered to

be a basal conglomerate. Basal conglomerates are characterized by pebbles that are angular and weathered. Pennsylvanian detrital chert (Fig. 5.10) that occurs immediately above the Pre-Pennsylvanian unconformity in the Permian Basin of Texas is an example of basal conglomerate. In this case, the detrital chert was derived from underlying Devonian chert. A counterpart of a basal conglomerate in carbonate rocks is shown in Figure 5.9. Thin layers of conglomerate with bones and teeth may also occur above an erosional unconformity as basal conglomerate.

Weathered Chert

Weathered chert, known as tripolitic chert, is chalky white in color and extremely porous. In contrast, fresh chert is usually dark colored and dense. The porous nature of such chert is easily recognized in thin sections of chert (using blue-dyed epoxy). White weathering rims are common in tripolitic cherts (Fig. 5.10). Weathered chert can occur both above and below erosional unconformities, and is a good indicator of proximity to erosional unconformities of subaerial origin. However, caution must be exercised in distinguishing between transported weathered chert and in-situ weathered chert.

Zones of Enhanced Porosity

When carbonates are exposed subaerially to meteoric water, cavernous porosity and karst topography develop from dissolution. Similarly, when silici-

Fig. 5.10. Photograph of a core showing Pennsylvanian detrital chert in a basal conglomerate; note white weathering rims (arrows) surrounding black chert grains, Permian Basin, Texas.

clastic sequences are exposed to acidic meteoric water during uplift and erosion, unusually high porosity and permeability may develop beneath erosional unconformities from the dissolution of unstable framework grains and cements. This criterion, however, has limitations because unusually high porosity may also develop in zones unrelated to erosional unconformities.

Abrupt Faunal Break

Major temporal breaks along unconformities are recognized primarily by dating rocks above and below an unconformity using fossils. Unless the faunas differ appreciably in age; however, an abrupt faunal break is not evidence of a major temporal break, for it may be due to a change in environmental conditions (Dunbar and Rodgers 1957). Fossils are the most commonly employed tool for estimating the length of temporal breaks.

Seismic Reflection Patterns

Angular relationships are readily observable on high-quality seismic reflection profiles and are indicative of erosional unconformities. The seismic expression of angular unconformities is a function of both the difference in dip of the strata above and/or below the unconformity and the impedance contrasts at the unconformity (Vail and Todd 1981). Angular relationships of the Base Cretaceous unconformity, which is a regional seismic marker in the Viking Graben, Norwegian North Sea, are shown in Figure 5.11.

Log Patterns

Petrophysical wireline logs may be useful in interpreting abrupt changes in shale density/velocity gradients, lithology, porosity, and radioactivity. Because these changes are commonly associated with unconformities, correlating such changes from well to well in a basin may determine the existence of regional unconformities. Dipmeter motifs are useful in recognizing discordance of dips caused either by a change in depositional facies and/or by unconformities. Gilreath and Maricelli (1964) reported that zones of weathering frequently exhibit high-angle dips below subaerial unconformities. Log patterns, however, are not direct indicators of erosional unconformities.

Stable Isotopes

Stable-isotope geochemistry of certain elements may be used to identify the meteoric origin of cements and formation waters. The meteoric origin, in turn, may be used as indirect evidence for erosional unconformities where influx of meteoric waters commonly occurs. Isotopic evidence for meteoric waters in a sequence does not necessarily indicate proximity to an unconformity. The following ranges of deuterium, as well as oxygen isotopes, suggest meteoric waters (Hoefs 1980):

1. δD in $\permil$ (relative SMOW): 0 to -350.
2. $\delta^{18}O$ in $\permil$ (relative SMOW): 0 to -45.

Oil Seeps

Oil seeps are known to occur along some erosional unconformities, but oil seeps may also occur along fractures and faults.

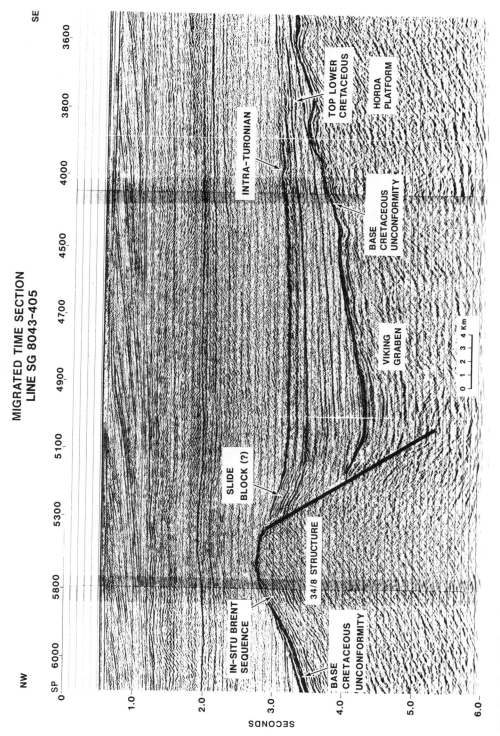

Fig. 5.11. A seismic reflection profile showing the Cimmerian unconformity (Base Cretaceous unconformity) in the Viking Graben, Norwegian North Sea; note change in reflection pattern above and below the unconformity surface (Alhilali and Damuth 1987). Reproduced with permission of the authors.

Fig. 5.12. A massive silcrete zone with a well developed upper crust (bracketed), Grahamstown, South Africa. (Photo: J.G. McPherson).

Evidence for Subaerial Exposure

Features indicative of subaerial exposure do not necessarily represent significant temporal breaks and, therefore, are not conclusive evidence for subaerial unconformities. However, evidence for subaerial exposure can be used to define potential surfaces of subaerial unconformities.

Paleosol Horizons

Humus layer, land plants, vertical rootlets, zones of oxidation and reduction, clay-rich layers, nodules, and vein networks (McPherson 1979; Wright 1986) are suggestive of paleosol horizons. These horizons are commonly difficult to recognize in ancient sequences, and this is especially true for lower Paleozoic weathered regoliths that developed before the advent of land plants. Buried soil horizons are definite indicators of subaerial exposure.

Duricrust

A general term for a hard crust on the surface of a soil horizon in a semiarid climate. The three major types of duricrust are silcrete, calcrete, and ferri-crete. Silcrete represents hard crusts of silicified sand and gravel developed through subaerial weathering (Fig. 5.12). Thick and extensive silcrete develops where there is some equilibrium between wet and dry seasonal periods (Thiry and Millot 1987). Calcrete develops at or near the surface in arid and semiarid regions, and it is composed of calcium carbonate crusts in addition to gravel, sand, and clay (Goudie 1973; Reeves 1976). Rootlets in calcrete horizons are common (Fig. 5.13). Calcretes developed in highly calcareous soils do not necessarily suggest long periods of exposure. The term ferri-crete is used for surficial sand and gravel cemented into a hard mass by iron oxide derived from oxidation by percolating solutions. Red bed zones caused by surface-related weathering may reach a thickness of 200 m.

Abrupt Change in Lithology

Lithologic changes caused by the occurrence of marine sand over fluvial conglomerate, for example, denote subaerial exposure and possibly a major stratigraphic break.

Fig. 5.14. SEM photograph showing authigenic kaolinite with well crystallized hexagonal platelets, Ivishak Formation (Triassic), 2,737 m, Alaska.

Fig. 5.13. A modern calcrete zone showing roots in growth position (arrow), Cape Province, South Africa. (Photo: J.G. McPherson).

Continental Deposits

Eolian dunes, alluvial fans, fluvial channels, and continental glacial deposits are commonly developed on a major subaerial erosional surface.

Tidal Flat Facies

Carbonate tidal flat features such as mudcracks, birdseye vugs, and algal structures are indicative of subaerial exposure (Shinn 1983).

Volcanic Ash

Recognition of subaerially deposited volcanic ash or lava may help define surfaces of subaerial exposure.

Kaolinite

The common association of authigenic kaolinite with surfaces that were subaerially exposed has been studied by Al-Gailani (1981). Kaolinite is a common

diagenetic phase in sandstones (Fig. 5.14) recharged by meteoric water (Longstaffe 1984). Because kaolinite also forms during burial diagenesis unrelated to subaerial exposure, caution must be exercised in using this criterion to recognize erosional unconformities of subaerial origin.

Evidence for Submarine Unconformities

Submarine Canyons and Channels

Submarine canyons and channels are indicative of submarine erosion. Submarine channels are usually filled with turbidite sandstones (Fig. 5.7), slumps, debris flows, hemipelagic mudstones, or a combination of these deposits.

Glauconitic Minerals

These are marine authigenic green minerals ranging in composition from glauconitic smectite to glauconitic mica. They tend to develop in water depths of 50–500 m in an open marine environment where there is very slow deposition. Occasionally, glauconitic minerals occur immediately above a major hiatus produced by contour currents at water depths of 1600–2500 m (Odin 1985). Glauconitic minerals in deep-marine facies may suggest erosion, but their occurrence in shallow-marine facies may indicate periods of slow deposition.

Phosphatic Pellets

Phosphates, in conjunction with glauconitic minerals and manganese nodules, can be used to infer submarine unconformities. For example, Pettijohn (1926) reported phosphatic pellets associated with a disconformity.

Manganese Nodules

Manganese nodules are commonly associated with erosional surfaces produced by strong bottom currents in the deep sea. The strong currents keep the manganese nodules from being covered with sediment and thus allow the nodules to grow (Watkins and Kennett 1972). Manganese "hardgrounds" in the rock record may be used to recognize submarine erosional surfaces created by bottom currents (Tucholke and Embley 1984). Manganese nodules also grow in deep-sea areas of low sedimentation rates and volcanic activity (Cronan 1977).

Abrupt Faunal Break

As for subaerial unconformities, major temporal breaks along submarine unconformities are recognized primarily by fossil evidence.

Seismic Reflection Profiles

Submarine canyons and channels are recognized in seismic reflection profiles by their erosional and depositional morphology. Associated slump deposits usually exhibit chaotic reflections.

Importance of Erosional Unconformities in Basin Analysis

General Significance

Erosional unconformities of different scales are an ubiquitous element of sedimentary basins. In practice, only demonstrable stratigraphic breaks representing a significant period of time are formally recognized as unconformities. Minor unconformities in the rock record are currently not emphasized, even if they span more time than that represented by preserved sediments in a sequence (Dott and Batten

1971). Even a regional unconformity cannot be fully appreciated from a single outcrop or a well; regional mapping of the unconformity covering a wide area is required. Size of outcrops often limit our ability to recognize large-scale erosional features such as submarine channels (Shanmugam *et al.* 1985a). These constraints, however, do not minimize the importance of information concerning time and sediment represented along stratigraphic breaks. In general, unconformities are useful in subdividing geologic time and stratigraphic units, and in determining the timing of tectonic activity (Chenworth 1967). The significance of unconformities in specific areas of basin analysis is discussed below.

Facies Relationships

Sedimentologists routinely use Walther's Law of Facies for interpreting lateral facies relationships using vertical arrangement of facies. Walther's Law states that facies occurring in a conformable vertical succession were formed in laterally adjacent environments. Because Walther's Law applies only to conformable successions (Middleton 1973), recognition of even minor stratigraphic breaks is essential to interpreting lateral facies relationships.

Minor stratigraphic breaks are often ignored in facies analysis. Submarine canyon and channel sequences in the rock record, for example, are customarily interpreted in terms of slope and fan facies. Alternatively, submarine canyons and channels can be interpreted as marine hiatuses or disconformities. The time represented along these stratigraphic breaks is of critical importance in subsidence analysis.

Burial and Uplift Analysis

Accuracy in thickness estimation of eroded sediment along unconformities ultimately dictates the reliability of burial and uplift analysis. Conventionally, the thickness of eroded sediment along unconformities is estimated by comparing the unconformable sequence with the type section in which the entire sequence is preserved. However, this method may not be accurate because lateral facies changes are unknown in missing intervals. In fact, the error in using such a method may be substantial. For example, if the missing interval in an alluvial stratigraphic section were proximal, it could represent either an alluvial fan sequence of several kilometres in thickness, or an alluvial plain sequence of only a few hundred metres in thickness. Additional information from depth-sensitive mineral assemblages (e.g., zeolites), isotopic data, and vitrinite reflectance is needed to establish the depth of burial, which may aid in inferring the thickness of missing sediment. The task becomes even more complex when multiple erosional unconformities are present in a sequence.

Correlation

Unconformities are frequently considered to be time lines and are widely used for the correlation of stratigraphic boundaries (Vail *et al.* 1984). However, when different parts of an unconformity surface are exposed and reflooded at different times, rocks above and below a particular unconformity (even a precisely traceable one) may be of different ages in different places. Thus unconformities may not be accurate time lines. It is advisable for correlation to select a good marker horizon beneath the unconformity rather than the unconformity surface itself.

Depending on depositional slope and rates of transgression and regression, the amount of time and sediment lost by erosion may vary more drastically in a dip direction than in a strike direction. This is important for correlation because unconformities are likely to be more reliable as time lines in the strike direction than in the dip direction.

Stratigraphic correlation using unconformities is meaningful only when unconformities of similar origin are compared. For example, when an unconformity, such as the Cimmerian unconformity in the North Sea, is interpreted to be created by tectonic uplift, direct correlation with unconformities created by eustatic sea-level fall should be avoided.

Sea-Level Changes

The seismic-stratigraphic approach, which assumes that global unconformities are caused by lowstands of sea level (Vail *et al.* 1977), has limitations (e.g., Miall 1984, 1986). Available data suggest that not all global unconformities are caused by fall in sea level. As discussed earlier, some global unconformities seem to have been created by tectonism (Schwan 1980; Sloss 1984). Abyssal bottom currents generate both regional (Tucholke and Embley 1984) and global (Keller *et al.* 1987) unconformities. In the rock record, many bottom-current deposits correlate with global sea-level fall (Shanmugam and Moiola 1982a). This suggests that vigorous bottom currents (thermohaline contour currents) occurred on a global scale during lowstands of sea level.

Locally, drowning and burial of carbonate platforms by siliciclastics may produce drowned unconformities that will simulate a fall in sea level with seaward shift in onlap (Schlager and Camber 1986). Because unconformities of different scales are created by different processes during different stands of sea level, a correct genetic interpretation of unconformities is critical in basin analysis.

Sedimentary Environments

Recognition of the type of unconformity is useful in deciphering sedimentary environments. Karst facies and soil horizons associated with erosional unconformities imply subaerial exposure, and mass-movement deposits usually indicate submarine slope and channel environments.

Paleotopography

Erosional unconformities can be used to infer subaerial paleotopography that existed during the final phase of erosion. Erosional topography with a relief of more than 1800 m has been detected in the Renqiu Field of the North China Basin (Guangming and Quanheng 1982). In the Renqiu Field area, karst development during Tertiary time had developed multi-peaked mountains with caverns and fractures in the Precambrian carbonate sequence.

Economic Significance

Prediction of Turbidite Reservoir Facies

During periods of low sea level, most continental shelves are emergent and are subjected to subaerial erosion. This allows rivers to discharge their sediment loads directly into the heads of submarine canyons at or near the shelf break (Shanmugam and Moiola 1982a,b; Shanmugam *et al.* 1985b). Consequently, major turbidite packages and associated submarine fans generally develop in the deep sea during lowstands of sea level. By recognizing erosional unconformities on the shelf and a regressive seismic facies, the downdip occurrence of coeval deep-sea reservoir facies may be predicted.

Porosity Enhancement

Erosional unconformities can enhance porosity in siliciclastic, carbonate, and even in igneous rocks that occur beneath erosional unconformities. The importance of erosional unconformities in connection with the process of porosity enhancement was

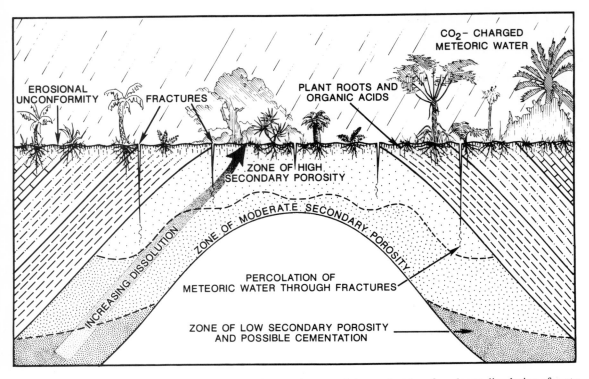

Fig. 5.15. A conceptual model showing an increase in porosity toward the erosional surface due to dissolution of unstable mineral constituents by acidic meteoric waters and organic acids.

first emphasized by Levorsen (1934) and has recently been discussed, for example, by Selley (1985) and Shanmugam (1985a,b). During uplift and erosion of a sequence, unstable cements (e.g., calcite) and framework grains (e.g., feldspar and lithic fragments) are susceptible to dissolution by acidic meteoric waters. The dissolution process may be accelerated along unconformities where soils are formed and organic acids are contributed to the system by plant roots (Fig. 5.15).

Giles and Marshall (1986), on the basis of theoretical considerations, concluded that leaching of minerals by meteoric water is the most viable mechanism for creating significant volumes of secondary porosity in the shallow subsurface. Unlike other popular mechanisms advocated for secondary porosity (e.g., dissolution by acidic fluids and carboxylic acids generated during the thermal maturation of organic matter), meteoric-water leaching does not suffer from a shortage of dissolving fluids. Meteoric-water leaching can occur under two different scenarios: 1) at subaerial depositional sites where sands are exposed to acidic meteoric water during and immediately after deposition; and

2) along unconformity surfaces where uplifted sandstones are subjected to erosion by meteoric water. All sandstones deposited in subaerial environments undergo meteoric-water leaching to some degree. However, when some of these sandstones are exposed to meteoric water during later uplift and related erosion, they may gain additional porosity. Erosional unconformities of subaerial origin provide a number of favorable conditions for dissolution. These conditions include:

1. Availability of undersaturated acidic meteoric water;
2. Unique ability of the soil zone to increase the acidity of percolating meteoric water (Freeze and Cherry 1979);
3. Availability of organic acids from plant roots, which are powerful agents of dissolution in the near-surface environments (Huang and Keller 1970);
4. Sandstones are in direct contact with meteoric water;
5. In humid regions, abundant supply of meteoric water facilitates a net movement of fluid

through the exposed sandstone, which is necessary for pervasive dissolution. Karstic solution features, characteristic of carbonates, have been reported to occur in Proterozoic and Paleozoic quartzose sandstones in the east Kimberley region of northwestern Australia (Young 1987). Extensive dissolution of quartz and the loss of silica have been attributed to acidic, organic-rich solutions percolating through the sandstone. The karst features in sandstones were developed during Late Mesozoic to Late Tertiary time when northwestern Australia was under a warm humid climate (Young 1987);

6. Fractures and faults, commonly associated with uplifted sandstones, serve as conduits through which meteoric waters percolate into the subsurface;

7. Meteoric water may extend to depths approaching 3,000 metres (Galloway and Hobday 1983), however, the zone of intensive dissolution is concentrated above the water table. The zone of weathering is commonly 100 m thick (Rose *et al.* 1979);

8. Continued erosion exposes new surfaces and wide areas for leaching;

9. Erosional unconformities are part of open systems in which dissolved material is constantly transported away from the site of dissolution, resulting in a net loss of dissolved constituents;

10. Along unconformities equilibrium is either attained slowly or not attained at all. Such disequilibrium conditions are conducive to transport-controlled dissolution (Berner 1978).

Because of these favorable conditions, many hydrocarbon reservoirs throughout the world exhibit porosity enhancement beneath erosional unconformities (Table 5.1). Two selected examples (North Sea and Alaskan North Slope) are discussed below.

In the Statfjord Field, North Sea, the middle Jurassic sandstones occur beneath the Cimmerian unconformity (Fig. 5.3). These sandstones exhibit extensive dissolution porosity. Solution channels caused by dissolution of adjacent framework grains are common. Solution channels are primarily responsible for developing effective porosity (Shanmugam 1985a). There is a positive correlation between dissolution porosity and core permeability (Fig. 5.16), suggesting that solution channels, perhaps, enhanced the permeability of sandstones. More importantly, there is a general increase in core permeability toward the unconformity (Fig. 5.16).

Table 5.1. Reservoir quality of sandstones and carbonates that occur beneath erosional unconformities.

Reservoir	Maximum core permeability (md)	Dissolved constituents (secondary porosity)
Latrobe, Cret.-Eocene, Gippsland Basin, Australia (Bodard *et al.* 1984)	3,000	Dolomite, feldspar, rock fragments
Sarir, Cretaceous, Sirte Basin, Libya (Sanford 1970; Hea 1971; Al-Shaieb *et al.* 1981)	3,000	Quartz, feldspar, rock fragments
Halten, Jurassic, Norwegian Sea	22,900	Feldspar
Brent, Jurassic, Statfjord Field, North Sea (Sommer 1978)	6,400	Feldspar
Lunde, Triassic North Sea	8,600	Feldspar
Ivishak, Triassic, Prudhoe Bay, Alaska	3,000	Chert, carbonates
Triassic, DeWijk gas field, The Netherlands (Gdula 1983)	100	Anhydrite
Mesozoic and Paleozoic, Buried Hill Pools, North China Basin (Guangming and Quanheng 1982)	No data (cavernous porosity in karst facies)	Carbonates
Minnelusa, Permian, Powder River Basin, Wyoming and Montana	3,200	Anhydrite, carbonates
Kekiktuk, Mississippian, Mikkelsen Bay, Alaska	12,800	Chert, quartz
Devonian, Ector, Crane, and Pecos counties, Texas (David 1946; Hanson 1985)	200	Chert, carbonates

These trends suggest that meteoric waters percolating from the erosional surface were responsible for dissolving plagioclase feldspars and for enhancing reservoir quality near the unconformity. The dissolution of feldspars in the Jurassic sandstones of the North Sea has been attributed to meteoric waters related to the Cimmerian uplifts (Hancock and Tay-

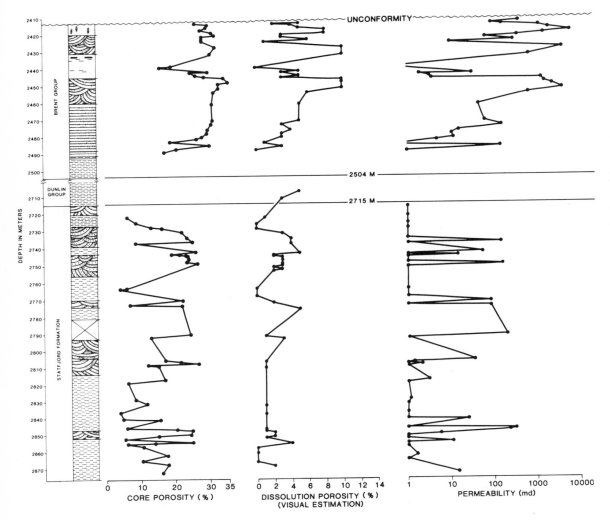

Fig. 5.16. Distribution of porosity and permeability beneath the Cimmerian unconformity; note a general increase in reservoir quality toward the unconformity, Statfjord Field, North Sea.

lor 1978; Selley 1984; Sommer 1978). According to another hypothesis, leaching of feldspars by meteoric water in the Brent Sandstone (Middle Jurassic), Statfjord Field, occurred relatively early during deposition and was later repeated during the Cimmerian uplifts (Bjørlykke 1983; Bjørlykke and Brendsdal 1986).

In the Prudhoe Bay Field, North Slope, Alaska, the Permo-Triassic reservoir (Ivishak Formation of the Sadlerochit Group) is truncated by the Neoco-

mian unconformity. In this reservoir, dissolution of chert is the dominant cause of porosity development. Chert dissolution is believed to have been caused by percolating acidic meteoric waters related to the Neocomian unconformity (Shanmugam 1985a; Shanmugam and Higgins 1987). The uppermost fluvial interval (Zone 4) of the Ivishak Formation exhibits increasing core porosity with increasing stratigraphic proximity to the unconformity (Fig. 5.17). This trend is a clear indication of

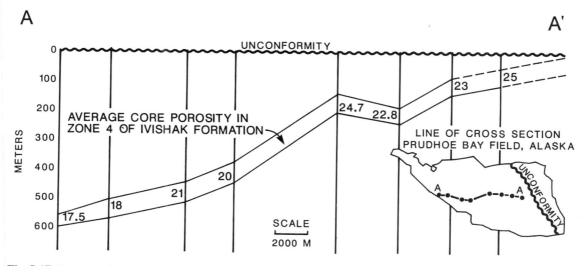

Fig. 5.17. Increase of average core porosity in Zone 4 (fluvial facies) of the Ivishak Formation (Triassic) with increasing proximity to the unconformity, Prudhoe Bay Field, Alaska.

unconformity-induced porosity enhancement. Such trends can be used to predict porosity in frontier areas using subcrop maps.

Porosity Reduction

Although porosity enhancement beneath erosional unconformities is common, extensive zones of cementation can also develop beneath erosional unconformities. Such zones may act as a diagenetic trap for hydrocarbons. An example of dolomite cementation beneath an unconformity is the Parkman Field (Fig. 5.18), Saskatchewan (Miller 1972).

Contact with Source Rocks

During marine transgression over an erosional unconformity, deposition of shale rich in organic matter can provide an ideal juxtaposition of reservoir and source rocks (Levorsen 1934, 1954). For example, in the Prudhoe Bay Field, North Slope, Alaska, the Triassic reservoir rocks (e.g., the Ivishak Formation) are in contact with the Cretaceous source rocks (8.4% TOC; Bushnell 1981) along the Neocomian unconformity.

Avenues of Hydrocarbon Migration

Erosional unconformities can mark the top of a zone of increased porosity and permeability. Therefore unconformities can act as avenues of migration for oil and gas. A classic example of oil migration along

an unconformity surface is the Maracaibo Basin in Venezuela (Dickey and Hunt 1972), where a spectacular series of oil seeps occur within a few hundred metres of the pre-Miocene unconformity.

Unconformity Traps

Unconformity traps are common in the geologic record (Levorsen 1934, 1936, 1954; Halbouty 1982; Faerseth et al. 1986). Unconformity-related traps may occur both above and below an unconformity. Selected examples of unconformity traps are given in Table 5.2.

Economic Mineral Deposits

Erosional unconformities are favorable sites for the concentration of economic mineral deposits because porous zones beneath unconformities serve as conduits for mineralizing solutions (Bateman 1942). Karstification is an important process in localizing some of the world's major lead and zinc deposits (e.g., Mississippi Valley type) in the carbonate strata (Kyle 1983).

The occurrence of uranium deposits in association with unconformities has been discussed by Bowie (1979). Residual concentrations of aluminum, iron, and manganese are common along erosional unconformities. Other economic mineral deposits associated with unconformities include: cobalt, clay, phosphates, tripoli, zinc, tin, and gold (Jensen and Bateman 1981).

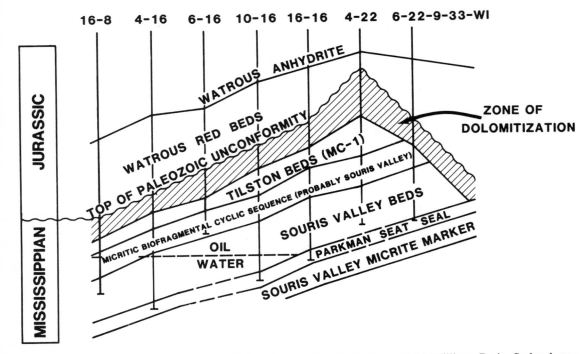

Fig. 5.18. Zone of dolomitization beneath the Paleozoic unconformity, Parkman Field, Williston Basin, Saskatchewan (Miller 1972). Reproduced with permission of the American Association of Petroleum Geologists.

Table 5.2. Selected examples of unconformity traps.

Field	Trap
Antelope Hills, California (Levorsen 1954)	Above Miocene unconformity
Gippsland, Australia (Shanmugam 1985c)	Below Oligocene unconformity
Renqiu, North China Basin (Guangming and Quanheng 1982)	Below Tertiary unconformity
Messla, Libya (Clifford *et al.* 1980)	Below Upper Cretaceous unconformity
Cutbank, Montana (Shelton 1967; Selley 1985)	Above Lower Cretaceous unconformity
Prudhoe Bay, Alaska (Jones and Speers 1976)	Below Neocomian unconformity
Statfjord, Norwegian North Sea (Faerseth *et al.* 1986)	Below Base Cretaceous unconformity

Summary

1. Unconformities are surfaces of erosion, and they represent significant temporal breaks in the stratigraphic record. Unconformities of different scales (local to global) are present in all sedimentary basins. Erosional unconformities of subaerial origin are believed to be caused by tectonic uplifts and by eustatic sea-level fall. Submarine erosional unconformities are considered to be due to transgression, mass movements, density currents, carbonate dissolution, and clastic influx in carbonate shelves.

2. A distinction between subaerial unconformity and subaerial exposure has been made because subaerial unconformities require prolonged subaerial exposure; however, evidence for subaerial exposure, by itself, does not imply a significant temporal break. Criteria for recognizing subaerial unconformities include discordance of dip, karst facies, basal conglomerate, and a major gap in the fossil record. Evidences of subaerial exposure are paleosol horizons, duricrust, and continental deposits. Submarine unconformities may be recognized by mass-movement deposits, glauconitic minerals, and manganese nodules.

3. Recognition of unconformities is critical to various facets of basin analysis such as subdividing stratigraphic units, determining the timing of tectonic activity, interpreting lateral facies relationships, constructing burial and uplift curves, correlating certain stratigraphic boundaries,

interpreting sea-level changes, and reconstructing paleogeography.

4. In exploration, erosional unconformities may be useful in predicting turbidite reservoir facies and in forecasting porosity development. Furthermore, erosional unconformities may act as avenues of hydrocarbon migration, may generate hydrocarbon traps, and they may be favorable sites for economic mineral deposits such as uranium, manganese, and gold.

Acknowledgments. I am thankful to Karen Kleinspehn and Chris Paola for inviting me to submit this paper; A.H. Bouma, R.J. Steel, J.F. Sarg, K.R. Walker, J.G. McPherson, D.W. Kirkland, G.L. Benedict, and D.M. Summers for critically reviewing the manuscript; G.G. Lash, K.A. Alhilali, R.B. Koepnick, and J.E. Damuth for valuable discussions; K.F. Keller and J.G. McPherson for photographs; R.J. Moiola for managerial support; and Mobil Research and Development Corporation for granting permission to publish this paper.

References

AL-GAILANI, M.B. (1981) Authigenic mineralizations at unconformities: Implication for reservoir characteristics. Sedimentary Geology 29:89–115.

ALHILALI, K.A. and DAMUTH, J.E. (1987) Slide block (?) of Jurassic sandstone and submarine channels in the basal Upper Cretaceous of the Viking Graben: Norwegian North Sea. Marine Petroleum Geology 4:35–48.

AL-SHAIEB, Z., WARD, W.C., and SHELTON, J.W. (1981) Diagenesis and secondary porosity evolution of Sarir sandstone, southeastern Sirte basin, Libya (Abstract). American Association Petroleum Geologists Bulletin 65:889–890.

BARRELL, J. (1917) Rhythms and measurement of geologic time. Geological Society America Bulletin 28: 745–904.

BATEMAN, A.M. (1942) Economic Mineral Deposits. New York: John Wiley & Sons, Inc., 808 p.

BATES, R.L. and JACKSON, J.A. (eds) (1980) Glossary of Geology. Falls Church, Virginia: American Geological Institute, 749 p.

BERNER, R.A. (1978) Rate control of mineral dissolution under earth surface conditions. American Journal Science 278:1235–1252.

BJØRLYKKE, K. (1983) Diagenetic reactions in sandstones. In: Parker, A. and Sellwood, B.W. (eds) Sediment Diagenesis. Dordrecht: D. Reidel Publishing Company, pp. 169–213.

BJØRLYKKE, K. and BRENDSDAL, A. (1986) Diagenesis of the Brent Sandstone in the Statfjord field, North Sea. In: Gautier, D.L. (ed) Roles of Organic Matter in Sediment Diagenesis. Society Economic Paleontologists Mineralogists Special Publication 38, pp. 157–167.

BODARD, J.M., WALL, V.J., and CAS, R.A.F. (1984) Diagenesis and the evolution of Gippsland basin reservoirs. Australian Petroleum Exploration Association Journal 24, pt. 1:314–335.

BOWIE, S.H.U. (1979) The mode of occurrence and distribution of uranium deposits. Philosophical Transactions Royal Society London A291:289–300.

BUSHNELL, H. (1981) Unconformities—key to N. Slope oil. Oil Gas Journal 79(2):114–118.

CHENWORTH, P.A. (1967) Unconformity analysis. American Association Petroleum Geologists Bulletin 51:4–27.

CHENWORTH, P.A. (1972) Unconformity traps. In: King, R.E. (ed) Stratigraphic Oil and Gas Fields—Classification, Exploration Methods, and Case Histories. American Association Petroleum Geologists Memoir 16, pp. 42–46.

CLIFFORD, H.J., GRUND, R., and MUSRATI, H. (1980) Geology of a stratigraphic giant: Messla oil field, Libya. In: Halbouty, M.T. (ed) Giant Oil and Gas Fields of the Decade 1968–1978. American Association Petroleum Geologists Memoir 30, pp. 507–524.

CLOETINGH, S. (1986) Intraplate stresses: A new tectonic mechanism for fluctuations of relative sea level. Geology 14:617–620.

CREASE, J. (1965) The flow of Norwegian sea water through the Faroe bank channel. Deep-Sea Research 12:143–150.

CRONAN, D.S. (1977) Deep-sea nodules: Distribution and geochemistry. In: Glasby, G.P. (ed) Marine Manganese Deposits. Amsterdam: Elsevier Scientific Publishing Company, pp. 11–44.

DAMUTH, J.E., JACOBI, R.D., and HAYES, D.E. (1983) Sedimentation processes in the Northwest Pacific Basin revealed by echo-character mapping studies. Geological Society America Bulletin 94:381–395.

DARWIN, C. (1859) On the origin of species by means of natural selection, or the preservation of favoured races in the struggle for life. London: Murray, 502 p.

DAVID, M. (1946) Devonian (?) producing zone, TXL pool, Ector county, Texas. American Association Petroleum Geologists Bulletin 30:118–119.

DICKEY, P.A. and HUNT, J.M. (1972) Geochemical and hydrogeologic methods of prospecting for stratigraphic traps. In: King, R.E. (ed) Stratigraphic Oil and Gas Fields—Classification, Exploration Methods, and Case Histories. American Association Petroleum Geologists Memoir 16, pp. 136–167.

DOTT, R.H. JR. and BATTEN, R.L. (1971) Evolution of the Earth. New York: McGraw-Hill Book Company, 649 p.

DUNBAR, C.O. and RODGERS, J. (1957) Principles of Stratigraphy. New York: Wiley, 356 p.

ESTEBAN, M. and KLAPPA, C.F. (1983) Subaerial exposure environment. In: Scholle, P.A., Bebout, D.G., and Moore, C.H. (eds) Carbonate Depositional Environments. American Association Petroleum Geologists Memoir 33, pp. 1–54.

FAERSETH, R.B., OPPEBOEN, K.A., and SAEBOE, A. (1986) Trapping styles and associated hydrocarbon potential in the Norwegian North Sea. In: Halbouty, M.T. (ed) Future Petroleum Provinces of the World. American Association Petroleum Geologists Memoir 40, pp. 585–597.

FREEZE, R.A. and CHERRY, J.A. (1979) Groundwater. Englewood Cliffs, NJ: Prentice-Hall, Inc., 604 p.

GALLOWAY, W.E. and HOBDAY, D.K. (1983) Terrigenous Clastic Depositional Systems. New York: Springer-Verlag, 423 p.

GDULA, J.E. (1983) Reservoir geology, structural framework, and petrophysical aspects of the DeWijk gas field. Geologie Mijnbouw 62:191–202.

GILES, M.R. and MARSHALL, J.D. (1986) Constraints on the development of secondary porosity in the subsurface: Re-evaluation of processes. Marine Petroleum Geology 3:243–255.

GILREATH, J.A. and MARICELLI, J.J. (1964) Detailed stratigraphic control through dip computations. American Association Petroleum Geologists Bulletin 48: 1902–1910.

GOUDIE, A. (1973) Duricrust in Tropical and Subtropical Landscapes. Oxford: Clarendon Press, 174 p.

GRANT, W.D., WILLIAMS, A.J. III, and GROSS, T.F. (1985) A description of the bottom boundary layer at the HEBBLE site: Low-frequency forcing, bottom stress and temperature structure. In: Nowell, A.R.M. and Hollister, C.D. (eds) Deep Ocean Sediment Transport. Amsterdam: Elsevier, pp. 219–241.

GUANGMING, Z. and QUANHENG, Z. (1982) Buried-hill oil and gas pools in the North China basin. In: Halbouty, M.T. (ed) The Deliberate Search for the Subtle Trap. American Association Petroleum Geologists Memoir 32, pp. 317–335.

HALBOUTY, M.T. (ed) (1982) The Deliberate Search for the Subtle Trap. American Association Petroleum Geologists Memoir 32, 351 p.

HANCOCK, N.J. and TAYLOR, A.M. (1978) Clay mineral diagenesis and oil migration in the Middle Jurassic Brent Sand Formation. Journal Geological Society London 135:69–72.

HANSON, B.M. (1985) Truncated Devonian and Fusselman fields and their relationship to the Permian basin reserve. Transactions Southwest Section American Association Petroleum Geologists 1985 Convention, Fort Worth, Texas, p. 132.

HEA, J.P. (1971) Petrography of the Paleozoic-Mesozoic sandstones of the southern Sirte basin, Libya. In: Gray, C. (ed) Symposium on the Geology of Libya. University of Libya, pp. 100–125.

HOEFS, J. (1980) Stable Isotope Geochemistry. Berlin: Springer-Verlag, 208 p.

HOLLISTER, C.D. and HEEZEN, B.C. (1972) Geologic effects of ocean bottom currents: Western North Atlantic. In: Gordon, A.L. (ed) Studies in Physical Oceanography, Volume 2. New York: Gordon and Breach Science Publishers, pp. 37–66.

HUANG, W.H. and KELLER, W.D. (1970) Dissolution of rock-forming silicate minerals in organic acids: Simulated first-stage weathering of fresh mineral surfaces. American Mineralogist 55:2076–2094.

HUTTON, J. (1788) Theory of the Earth, or an investigation of the laws observable in the composition, dissolution and restoration of land upon the globe. Royal Society Edinburgh Transactions 1:109–304.

JACOBI, R.D. (1981) Peripheral bulge – a causal mechanism for the Lower/Middle Ordovician unconformity along the western margin of the northern Appalachians. Earth Planetary Science Letters 56:245–251.

JENSEN, M.L. and BATEMAN, A.M. (1981) Economic Mineral Deposits, 3rd edition. New York: John Wiley & Sons, 593 p.

JONES, H.P. and SPEERS, R.G. (1976) Permo-Triassic reservoirs of Prudhoe Bay field, North Slope, Alaska. In: Braunstein, J. (ed) North American Oil and Gas Fields. American Association Petroleum Geologists Memoir 24, pp. 23–50.

KELLER, G., HERBERT, T., DORSEY, R., D'HONDT, S., JOHNSSON, M., and CHI, W.R. (1987) Global distribution of late Paleogene hiatuses. Geology 18:199–203.

KING, R.E. (ed) (1972) Stratigraphic Oil and Gas Fields – Classification, Exploration Methods, and Case Histories. American Association Petroleum Geologists Memoir 16, 687 p.

KIRK, R.H. (1980) Statfjord field: A North Sea giant. In: Halbouty, M.T. (ed) Giant Oil and Gas Fields of the Decade 1968–1978. American Association Petroleum Geologists Memoir 30, pp. 95–116.

KRUMBEIN, W.C. (1942) Criteria for subsurface recognition of unconformities. American Association Petroleum Geologists Bulletin 26:36–62.

KRUMBEIN, W.C. and SLOSS, L.L. (1963) Stratigraphy and Sedimentation. San Francisco: W.H. Freeman and Company, 660 p.

KYLE, J.R. (1983) Economic aspects of subaerial carbonates. In: Scholle, P.A., Bebout, D.G., and Moore, C.H. (eds) Carbonate Depositional Environments. American Association Petroleum Geologists Memoir 33, pp. 73–92.

LEVORSEN, A.I. (1934) Relation of oil and gas pools to unconformities in the mid-continent region. In: Wrather, W.E. and Lahee, F.H. (eds) Problems of Petroleum Geology. American Association Petroleum

Geologists Sidney Powers Memorial Volume, pp. 761–784.

LEVORSEN, A.I. (1936) Stratigraphic versus structural accumulation. American Association Petroleum Geologists Bulletin 20:521–530.

LEVORSEN, A.I. (1954) Geology of Petroleum. San Francisco: W.H. Freeman and Company, 703 p.

LEWIS, K.B. (1971) Slumping on a continental slope inclined at 1°–4°. Sedimentology 16:97–110.

LONGSTAFFE, F.J. (1984) The role of meteoric water in diagenesis of shallow sandstones: Stable isotope studies of the Milk River aquifer and gas pool, southern Alberta. In: McDonald, D.A. and Surdam, R.C. (eds) Clastic Diagenesis. American Association Petroleum Geologists Memoir 37, pp. 81–98.

McPHERSON, J.G. (1979) Calcrete (caliche) paleosols in fluvial redbeds of the Aztec Siltstone (Upper Devonian), Southern Victoria Land, Antarctica. Sedimentary Geology 22:267–285.

MIALL, A.D. (1984) Principles of Sedimentary Basin Analysis. New York: Springer-Verlag, 490 p.

MIALL, A.D. (1986) Eustatic sea level changes interpreted from seismic stratigraphy: A critique of the methodology with particular reference to the North Sea Jurassic record. American Association Petroleum Geologists Bulletin 70:131–137.

MIDDLETON, G.V. (1973) Johannes Walther's law of correlation of facies. Geological Society America Bulletin 84:979–988.

MILLER, E.G. (1972) Parkman field, Williston basin, Saskatchewan. In: King, R.E. (ed) Stratigraphic Oil and Gas Fields—Classification, Exploration Methods, and Case Histories. American Association Petroleum Geologists Memoir 16, pp. 502–510.

MORGRIDGE, D.L. and SMITH, W.B. (1972) Geology and discovery of Prudhoe Bay field, eastern Arctic Slope. In: King, R.E. (ed) Stratigraphic Oil and Gas Fields—Classification, Exploration Methods, and Case Histories. American Association Petroleum Geologists Memoir 16, pp. 489–501.

ODIN, G.S. (1985) Significance of green particles (glaucony, berthierine, chlorite) in arenites. In: Zuffa, G.G. (ed) Provenance of Arenites. Dordrecht: D. Reidel Publishing Company, pp. 279–307.

PETTIJOHN, F.J. (1926) Intraformational phosphate pebbles of the Twin-City Ordovician. Journal Geology 34:361–375.

PITMAN, W.C. (1978) Relationship between eustacy and stratigraphic sequences of passive margins. Geological Society America Bulletin 89:1389–1403.

READ, J.F. and GROVER, JR. G. (1977) Scalloped and planar erosion surfaces, Middle Ordovician limestones, Virginia: Analogues of Holocene exposed karst or tidal rock platforms. Journal Sedimentary Petrology 47:956–972.

REEVES, C.C. (1976) Caliche. Lubbock, TX: Estacado Books, 233 p.

ROSE, A.W., HAWKES, H.E., and WEBB, J.S. (1979) Geochemistry in mineral exploration. London: Academic Press, 657 p.

SANFORD, R.M. (1970) Sarir oil field—Libya desert surprise. American Association Petroleum Geologists Memoir 14, pp. 449–476.

SCHLAGER, W. and CAMBER, O. (1986) Submarine slope angles, drowning unconformities, and self-erosion of limestone escarpments. Geology 14:762–765.

SCHLEE, J.S. (ed) (1984) Interregional Unconformities and Hydrocarbon Accumulation. American Association Petroleum Geologists Memoir 36, 184 p.

SCHWAN, W. (1980) Geodynamic peaks in Alpinotype orogenies and changes in ocean-floor spreading during Late Jurassic–Late Tertiary time. American Association Petroleum Geologists Bulletin 64:359–373.

SELLEY, R.C. (1984) Porosity evolution of truncation traps: Diagenetic models and log responses. Stavanger, Norway: Norwegian Petroleum Society Offshore North Seas Conference Paper No. G3, 18 p.

SELLEY, R.C. (1985) Elements of Petroleum Geology. New York: W.H. Freeman and Company, 449 p.

SHANMUGAM, G. (1978) The Stratigraphy, Sedimentology, and Tectonics of the Middle Ordovician Sevier Shale Basin in East Tennessee. Unpublished Ph.D. Dissertation. Knoxville: The University of Tennessee, 222 p.

SHANMUGAM, G. (1985a) Significance of secondary porosity in interpreting sandstone composition. American Association Petroleum Geologists Bulletin 69:378–384.

SHANMUGAM, G. (1985b) Types of porosity in sandstones and their significance in interpreting provenance. In: Zuffa, G.G. (ed) Provenance of Arenites. Dordrecht: D. Reidel Publishing Company, pp. 115–137.

SHANMUGAM, G. (1985c) Significance of coniferous rain forests and related organic matter in generating commercial quantities of oil, Gippsland basin, Australia. American Association Petroleum Geologists Bulletin 69:1241–1254.

SHANMUGAM, G. and HIGGINS, J.B. (1987) Porosity development from chert dissolution beneath the Neocomian Unconformity: Ivishak Formation, North Slope, Alaska: SEPM Mid-Year Meeting Abstracts (Austin, Texas) 4:76.

SHANMUGAM, G. and LASH, G.G. (1982) Analogous tectonic evolution of the Orodovician foredeeps, southern and central Appalachians. Geology 10:562–566.

SHANMUGAM, G. and MOIOLA, R.J. (1982a) Eustatic control of turbidites and winnowed turbidites. Geology 10:231–235.

SHANMUGAM, G. and MOIOLA, R.J. (1982b) Prediction of deep-sea reservoir facies. Transactions-Gulf Coast Association Geological Societies 32:275–281.

SHANMUGAM, G., DAMUTH, J.E., and MOIOLA, R.J. (1985a) Is the turbidite facies association scheme valid for interpreting ancient submarine fan environments? Geology 13:234–237.

SHANMUGAM, G., MOIOLA, R.J., and DAMUTH, J.E. (1985b) Eustatic control of submarine fan development. In: Bouma, A.H., Normark, W.R., and Barnes, N.E. (eds) Submarine Fans and Related Turbidite Systems. New York: Springer-Verlag, pp. 23–28.

SHELTON, J.W. (1967) Stratigraphic models and general criteria for recognition of alluvial, barrier bar and turbidity current sand deposits. American Association Petroleum Geologists Bulletin 51:2441–2460.

SHINN, E.A. (1983) Tidal flat environment. In: Scholle, P.A., Bebout, D.G., and Moore, C.H. (eds) Carbonate Depositional Environments. American Association Petroleum Geologists Memoir 33, pp. 171–210.

SIEVER, R. (1951) The Mississippian-Pennsylvanian unconformity in southern Illinois. American Association Petroleum Geologists Bulletin 35:542–581.

SLOSS, L.L. (1984) Comparative anatomy of cratonic unconformities. In: Schlee, J.S. (ed) Interregional Unconformities and Hydrocarbon Accumulation. American Association Petroleum Geologists Memoir 36, pp. 1–6.

SOMMER, F. (1978) Diagenesis of Jurassic sandstones in the Viking Graben. Journal Geological Society London 135:63–67.

STOW, D.A.V. and LOVELL, J.P.B. (1979) Contourites: Their recognition in modern and ancient sediments. Earth-Science Reviews 14:251–291.

THIRY, M. and MILLOT, G. (1987) Mineralogical forms of silica and their sequence of formation in silcretes. Journal Sedimentary Petrology 57:343–352.

TUCHOLKE, B.E. and EMBLEY, R.W. (1984) Cenozoic regional erosion of the abyssal sea floor off South Africa. In: Schlee, J.S. (ed) Interregional Unconformities and Hydrocarbon Accumulation. American Association Petroleum Geologists Memoir 36, pp. 145–164.

VAIL, P.R., MITCHUM, R.M., TODD, R.G., WIDMIER, J.M., THOMPSON, S., SANGREE, J.B., BUBB, J.N., and HATLELID, W.G. (1977) Seismic stratigraphy and global changes of sea level. In: Payton, C.E. (ed) Seismic Stratigraphy—Application to Hydrocarbon Exploration. American Association Petroleum Geologists Memoir 26, pp. 49–212.

VAIL, P.R. and TODD, R.G. (1981) Northern North Sea Jurassic unconformities, chronostratigraphy and sea-level changes from seismic stratigraphy. In: Illing, L.V. and Hodson, G.D. (eds) Petroleum Geology of the Continental Shelf of Northwest Europe. London: Heyden and Son, Ltd., pp. 216–235.

VAIL, P.R., HARDENBOL, J., and TODD, R.G. (1984) Jurassic unconformities, chronostratigraphy, and sea-level changes from seismic stratigraphy and biostratigraphy. In: Schlee, J.S. (ed) Interregional Unconformities and Hydrocarbon Accumulation. American Association Petroleum Geologists Memoir 36, pp. 129–144.

VISSER, M.J. (1980) Neap-spring cycles reflected in Holocene subtidal large-scale bedform deposits: A preliminary note. Geology 8:543–546.

WALKER, K.R., SHANMUGAM, G., and RUPPEL, S.C. (1983) A model for carbonate to terrigenous clastic sequences. Geological Society America Bulletin 94: 700–712.

WATKINS, N.D. and KENNETT, J.P. (1972) Regional sedimentary disconformities and upper Cenozoic changes in bottom water velocities between Australasia and Antarctica. Antarctic Research Series 19:273–293.

WRIGHT, V.P. (ed) (1986) Paleosols: Their Recognition and Interpretation. Princeton, NJ: Princeton University Press, 315 p.

YOUNG, R.W. (1987) Sandstone landforms of the tropical east Kimberley region, northwestern Australia. Journal Geology 95:205–218.

ZIEGLER, W.H. (1975) Outline of the geological history of the North Sea. In: Woodland, A.W. (ed) Petroleum and the Continental Shelf of Northwest Europe. London: Institute Petroleum, pp. 165–190.

6

Analysis of Eustatic, Tectonic, and Sedimentologic Influences on Transgressive and Regressive Cycles in the Upper Cenozoic Merced Formation, San Francisco, California

H. Edward Clifton, Ralph E. Hunter, and James V. Gardner

Abstract

The Merced Formation consists of approximately 2,000 m of shallow marine and coastal nonmarine sediment of late Cenozoic age that accumulated in a structural trough south of the city of San Francisco. About 1,750 m of these deposits crop out in a well exposed tilted sequence in sea cliffs on the north side of the San Andreas fault. The part of the section north of the fault appears to be of Pleistocene age.

Depositional facies within the Merced Formation can be identified with a high degree of confidence based on biota, sedimentary structures, and textural character. In vertical succession, the facies define alternating episodes of transgression and regression. The individual transgressive/regressive cycles almost certainly reflect the eustatic sea-level fluctuations that occurred during Pleistocene time. The general pattern of transgression and regression can be fairly well matched in much of the section to a Pleistocene sea-level curve determined from oxygen-isotope data. If eustatic fluctuations can be subtracted from the stratigraphic record, tectonic and sedimentologic influences become evident. The accumulation of such a thick succession of shallow-marine and coastal deposits implies long-term subsidence at an average rate of 1 m/10^3 yr. Evidence suggests that the rate of subsidence may have been greater in the section deposited prior to emplacement of the Rockland ash (at 0.4 Ma) than in the section deposited subsequent to the ash. A pronounced shift to predominantly nonmarine deposits about 290 m below the top of the section may be due in part to diminished subsidence rates. However, the fact that this change coincides with a change in provenance from local Franciscan sources to the Sacramento-San Joaquin River systems indicates that the shift to nonmarine facies more likely resulted from increased rates of sedimentation.

Introduction

Sedimentary basins that contain thick intervals of shallow-marine or shoreline deposits character-istically display a succession of marine transgressions and regressions (Ryer 1977; Clifton 1981). Shifts in the position of a shoreline depend on a combination of three factors: eustatic sea-level change, vertical motion of the land (by tectonism, compaction or isostatic adjustment), and rate of sediment accretion at the shoreline. Eustatic sea-level change is not known to exceed 100–200 m over an interval of tens of millions of years (Pitman 1978). A succession of shallow-marine deposits thicker than this deposited in such an interval of time or less can be attributed only to vertical motion of the depositional surface, due to either tectonic subsidence or compaction.

The transgressions and regressions may reflect various combinations of eustatic, tectonic, and sedimentologic processes (Pitman 1978). They may be due solely to eustatic fluctuations. Alternatively, sea level and the rate of sediment supply may be constant; alternations between transgression and regression may be either due to alternating uplift or subsidence or, for prograding shorelines, simply to fluctuations in the rate of subsidence. Conversely, transgression and regression may be caused by variations in the rate of sedimentation in a constantly subsiding basin, under conditions of unchanging sea level. Alternating transgressions and regressions in a shallow-marine succession imply at least episodic progradation.

It would be useful in any study of a sedimentary basin to be able to separate the effects of eustatic change, vertical movement of the depositional surface, and sedimentation rate. Identifying eustatic effects, for example, can provide a basis for inter-basin correlation and for recognition of tectonic influences. Because of the complicated interplay of these processes, however, the individual influences on establishing shoreline position may be difficult to isolate.

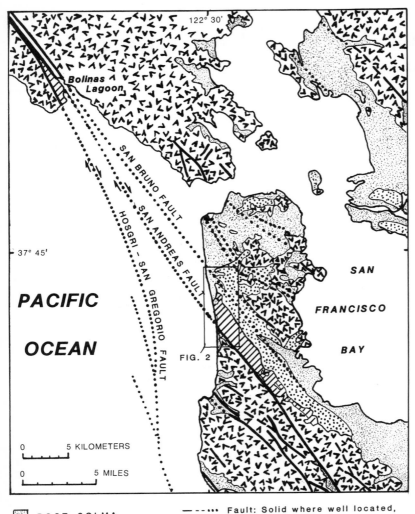

Fig. 6.1. Map showing distribution of the Merced Formation, Colma Formation, pre-Merced rocks, and post-Colma deposits and major faults on the San Francisco Peninsula. Geology generalized from Bortugno *et al.* (in press).

POST-COLMA

COLMA FORMATION

MERCED FORMATION

PRE-MERCED

— — --- ••• Fault: Solid where well located, dashed where approximately located or inferred, dotted where concealed by younger rocks or water

In this contribution we examine the possible origin of transgressions and regressions in a very young sedimentary basin on the Pacific Coast. The Pleistocene basin fill allows comparison of the pattern of deposition with the well documented pattern of Pleistocene eustatic sea-level change. Major strike-slip faults that now bound the basin are likely to have influenced its development and evolution. A major change in provenance, which almost certainly was accompanied by a significant change in sedimentation rate, is known to occur within the sequence. Our goal is to isolate as much as possible the influences of each of these factors and to evaluate the effectiveness of such an analysis.

The Merced Formation
The Depositional Basin

The strata under consideration are mostly part of the Merced Formation, a unit of very late Cenozoic age. The Merced Formation lies in a northwest-trending basin that crosses the shoreline near the northern end of the San Francisco Peninsula (Fig. 6.1).

Fig. 6.2. Map showing geologic features in the vicinity of the type section of the Merced Formation. The cross section (Fig. 6.4) and stratigraphic column (Fig. 6.5) of beds studied for this report extend along the beach from point A to point C, and the complete type section of the Merced extends along the beach from point B to point D.

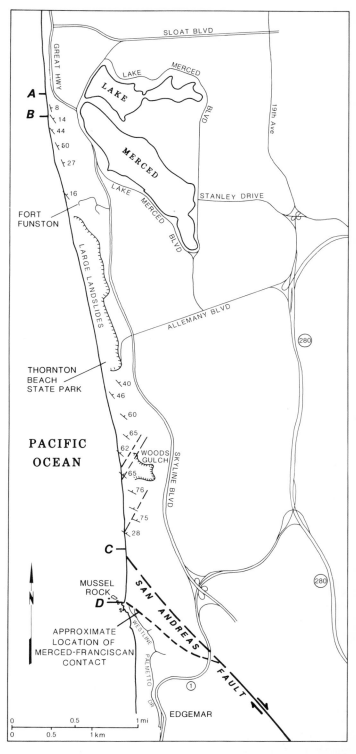

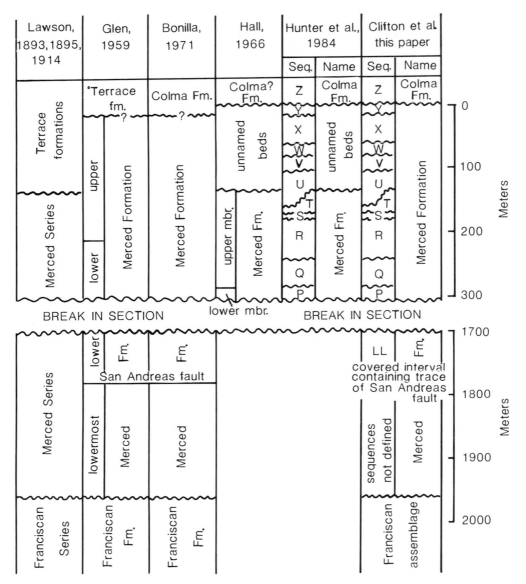

Fig. 6.3. Comparison of stratigraphic nomenclature for the Merced Formation and adjacent units by previous workers and ourselves. The Colma Formation (3–12 m thick) is not drawn to scale. The vertical scale of thickness applies to only our own section; other sections are correlated with ours by key beds.

►

Fig. 6.4. Generalized cross section of the part of the Merced Formation and younger strata exposed in sea cliffs between Lake Merced and the San Andreas fault, northwestern San Francisco Peninsula. Southern end of section lies about 500 m north of Mussel Rock, the base of the type section of the Merced; northern end lies 80 m north of the north boundary of Fort Funston as shown on the 1956 (photorevised 1968) edition of the San Francisco South 7.5-minute quadrangle, U.S. Geological Survey. Letters and numerical subscripts refer to sequences and lithologic units shown in stratigraphic column (Fig. 6.5). Sections tie together end-to-end from northernmost section (top) to southernmost section (bottom). No vertical exaggeration. Beach level is usually about 5 m above sea level. Significant horizons indicated by arrows: 1) Base of section measured here, underlying interval to 1906 trace of San

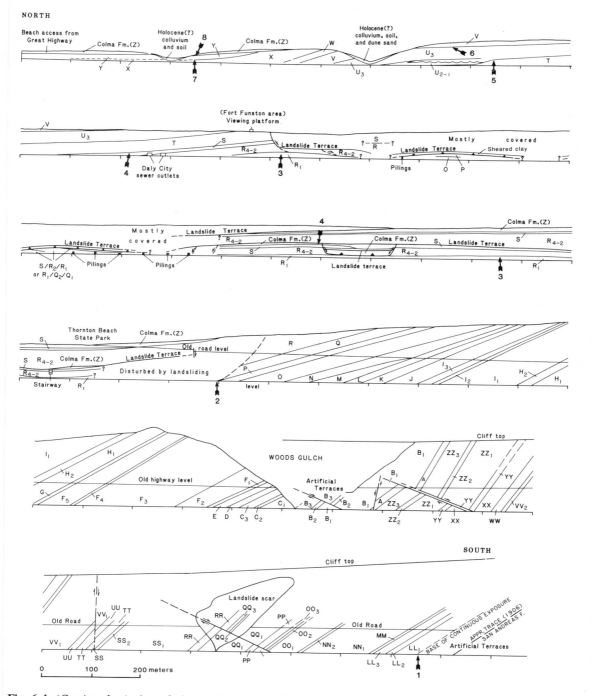

Fig. 6.4. (*Continued*). Andreas fault mostly covered; 2) Location of mineralogic change indicating a change in provenance from Franciscan rocks to those in the drainage basin of the Sacramento and San Joaquin Rivers (Hall 1965, 1966), boundary between Hall's (1966) upper and lower members of the Merced Formation; 3) "Upper Gastropod Bed" of Ashley (1895), boundary of Glen's (1959) upper Merced and lower Merced; 4) Rockland ash bed; 5) Erosional surface with about 30 m of relief, Lawson's (1893, 1895) boundary between the Merced and overlying "Terrace formations," Hall's (1966) boundary between the Merced and overlying unnamed beds; 6) Stratigraphic top of Glen's (1959) measured section of the Merced Formation; 7) Mapped position of the top of the Merced by Glen (1959) and Bonilla (1971); 8) Stratigraphic top of the Merced as defined here.

Inland, the basin is bounded on the southwest by the San Andreas Fault, but at the coast, near Pacifica, California, the basin overlaps the fault. At its northern limit, at Bolinas Lagoon, the basin is bounded on the west by the Hosgri-San Gregorio Fault (Fig. 6.1), and seismic-reflection profiles suggest that the Hosgri-San Gregorio Fault truncates the basin on the shelf between Bolinas Lagoon and Pacifica (D.M. McCulloch, oral communication, 1987). On its northeastern side, the basin appears to be truncated by the San Bruno fault, a fault proposed by Lawson (1895) and subsequently delineated by geophysical data and drill-hole information (Bonilla 1971) and observed in seismic-reflection profiles offshore (Mustart and McCulloch 1984).

The best exposure of the Merced occurs in sea cliffs just north of the San Andreas fault (Fig. 6.2). There, about 1,750 m of uppermost Cenozoic strata are exposed in a northeast-dipping succession. Based on facies analysis, all of these strata were deposited within a few kilometres landward or seaward of a shoreline. South of the San Andreas fault the Merced consists of about 200 m of sandstone and siltstone that depositionally overlie the Franciscan assemblage (Glen 1959). These deposits are largely covered by landfill and were not examined for this report.

Stratigraphic Definition

The name "Merced" was proposed by Lawson (1893) for a stratigraphic unit, now called the Merced Formation, exposed in sea cliffs that extend from Lake Merced southward to Mussel Rock on the west side of the northern San Francisco Peninsula (Fig. 6.2). This exposure was first referred to as the type section of the Merced by Martin (1916). Glen (1959) described the type section in greater detail.

Although it is well agreed that the Merced Formation rests uncomformably on the Franciscan assemblage, here of Jurassic and Cretaceous age, the upper boundary of the formation has been variously placed by previous workers (Fig. 6.3). All workers have agreed that the proper stratigraphic level to designate as the top of the Merced is an angular unconformity that separates the tilted beds of the Merced from the uplifted but only slightly tilted overlying beds. The disagreements on the position of this unconformity have arisen because the degree of tilt-

ing decreases toward the northern, stratigraphically highest end of the section (Fig. 6.4), making the angular discordance between the stratigraphically highest beds of the Merced and the overlying beds difficult to recognize.

As defined by Lawson (1893, 1895, 1914), the Merced was regarded as being overlain unconformably by "Terrace formations," at the base of which was a peaty bed that has been recognized by later geologists and is equivalent to the basal part of our sequence U (Figs. 6.3, 6.5). We consider the contact between sequence U and the underlying sequence T to be a disconformity surface of high paleotopographic relief (about 30 m) but not a demonstrably angular unconformity (Hunter et al. 1984).

Glen (1959) revised the upper boundary of the Merced, placing it along the beach at a point 2,200 feet (670 m) south of the northern boundary of Fort Funston. (The Fort Funston boundary is now differently located than it was on Glen's map.) This point corresponds approximately to the point where our sequence Y/sequence X contact intersects the beach surface (Fig. 6.4). Glen considered the revised top of the Merced to lie 140 feet (43 m) stratigraphically above the level defined as the top of the Merced by Lawson, but our measurements indicate 110–140 m of section between the Y/X contact and the U/T contact (Figs. 6.3, 6.5). We consider the contact between sequence Y and sequence X to be a disconformity, not a demonstrably angular unconformity. The uppermost part of sequence U is a crossbedded eolian sand in which the major bedding planes cannot safely be assumed to be nearly horizontal; under those conditions, the small amount of angular discrepancy that may exist between the bedding planes and the upper contact of sequence U does not prove that the contact is an angular unconformity.

At about the same time that Glen (1959) was studying the Merced, Schlocker et al. (1958) proposed the name "Colma Formation" for beds approximately equivalent to Lawson's (1914) "Terrace formations," but they did not discuss whether the Colma overlies the Merced at the type section of the Merced. Bonilla (1959) mapped the Colma as overlying the Merced with angular unconformity in a small area around Thornton Beach State Park (Fig. 6.2), but his mapped area did not include the stratigraphically highest beds in the Merced. Bonilla (1971) later mapped a larger area and placed the Colma-Merced contact at about the same position as

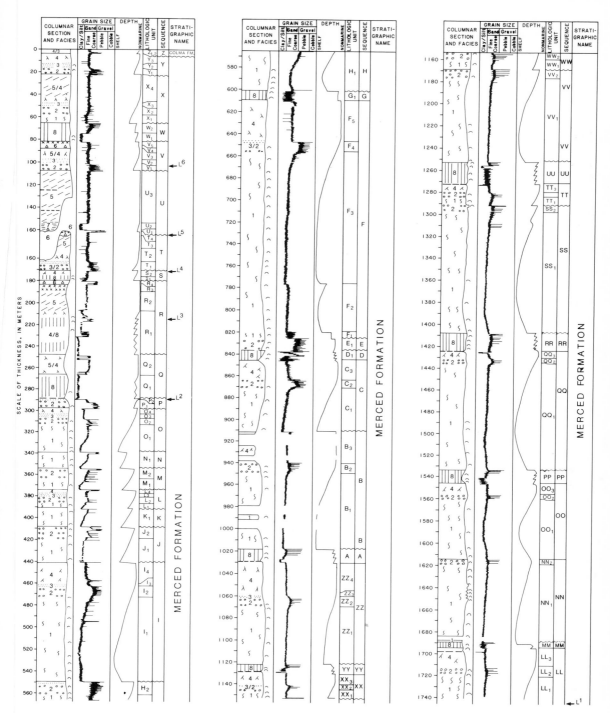

Fig. 6.5. Part of the type section of the Merced Formation and younger strata exposed in sea cliffs north of the San Andreas fault, San Francisco Peninsula. Sequences and lithologic units as introduced in Hunter *et al.* (1984). Facies identified by numbers in the stratigraphic column: 1) shelf (including outer-, mid-, and inner-shelf facies); 2) nearshore (may include foreshore facies in some sequences); 3) foreshore; 4) backshore; 5) eolian dune; 6) alluvial; 7) fresh-water pond/swamp/marsh; 8) embayment. Combinations of facies shown with short diagonal lines. Heavy horizontal dashed lines in the stratigraphic column indicate location of lignitic layers; x's mark well-developed paleosols. Occurrence of mollusks shown schematically by symbols to right of column. Predominant grain size was estimated visually in the field. Wavy lines between sequences indicate erosional surfaces. Gaps in section reflect covered intervals. Thickness in metres. Significant horizons indicated by arrows, numbered as in Figure 6.4.

on Glen's (1959) map; that is, the contact intersected the beach surface at the approximate position of the contact between our sequence Y and sequence X (Fig. 6.3).

In the most detailed study of the upper part of the Merced Formation and overlying beds preceding our own study, Hall (1966) identified the beds that make up our sequence Z (Fig. 6.3) as Colma (?) Formation. Hall referred to the beds between the top of the Merced as defined by Lawson (1893, 1895, 1914) and the base of his Colma (?) Formation as the "unnamed beds" (Fig. 6.3). The angular nature of the unconformity at the base of Hall's (1966) Colma (?) Formation is evident on his cross section showing relationships from Fleishhacker Zoo to Thornton Beach; the unconformity at the base of Hall's Colma (?) Formation (our sequence Z) truncates the contact between our sequence Y and sequence X (Fig. 6.4). We believe that sequence Z is unquestionably Colma (Bonilla 1971; Hunter *et al.* 1984) and agree with Hall's interpretation of the lower contact of the Colma as an angular unconformity. We here propose that the top of the Merced be clarified to correspond to the base of the Colma as used in this report; this contact is the contact between our sequence Z and sequence Y. This contact (point 8, Fig. 6.4) is only about 6 m above beach level at the northern end of Glen's section (point 7, Fig. 6.4). The contact becomes nearly horizontal at this point and does not intersect beach level at any outcrop now visible. By this definition, the unnamed beds of Hall (1966), corresponding to our sequences U through Y (described in detail by Hunter *et al.* 1984), are part of the Merced Formation. Most of these beds were included in the Merced as defined by Glen (1959) and as mapped by Bonilla (1971), and the inclusion of these beds in the Merced was advocated by Hall (1965) in an unpublished thesis.

Age

The Merced Formation was regarded until recently as of Pliocene and Pleistocene age, with the boundary between the two epochs placed high in the section (Glen 1959). However, advances in correlation based on radiometric dating, micropaleontology, tephrochronology, and magnetostratigraphy have forced a downward shift of the Pleistocene-Pliocene boundary in other upper Cenozoic sections of

California, and these changes necessitate a reexamination of the age of the Merced.

The only well dated bed in the Merced Formation is the Rockland ash bed (at the 170 m level in Fig. 6.5), now regarded as having an age of 0.4 Ma (million years) on the basis of fission-track dating (Sarna-Wojcicki *et al.* 1985). A less well dated horizon is the mineralogic change from detritus that was locally derived to detritus from the Sacramento-San Joaquin drainage system (at the 290 m level in Fig. 6.5). This change in drainage is dated roughly at 0.6 Ma on the basis of a complex chain of evidence from deposits in the San Joaquin Valley (Sarna-Wojcicki *et al.* 1985).

Another speculatively dated horizon is the top of the formation (as here defined), which must be older than the overlying Colma Formation. The Colma Formation was considered by Hall (1965, 1966) to be probably no younger than the last interglacial stage, the Sangamon. Considering the complex series of events, including five transgressions and regressions plus an episode of major tilting and erosion represented in the part of the section between the Rockland ash bed and the Colma Formation, a last-interglacial age for the Colma is more likely than an older age. The last major interglacial stage is now regarded to correspond to oxygen-isotopic stage 5, dated at 0.07–0.13 Ma, or to one of the substages of stage 5 (Shackleton and Opdyke 1973, 1976; Pisias and Moore 1981). The top of the Merced, therefore, is probably older, but not much older, than 0.07–0.13 Ma.

For the bulk of the formation beneath the horizon of mineralogic change, the best evidence for age assignment is the invertebrate macrofauna. Comparison with a recent study in the Eel River Basin (also called Humboldt Basin) of northwestern California by Roth (1979) suggests a Pleistocene age for this fauna. The strongest evidence is the occurrence of the echinoid *Scutellaster major* (Kew) in the lower Merced (Fig. 6.6; Faustman 1964; Roth 1979). This echinoid is a defining element of Roth's (1979) *Scutellaster major* zone in the Eel River Basin. At the Scotia Bluffs section in the Eel River Basin, the *Scutellaster major* zone occurs in the upper Rio Dell Formation and lower Scotia Bluffs Sandstone above a zone that Roth named for *Scutellaster* new species "A" Durham (Fig. 6.6; Roth 1979). In that section Roth located the Pleistocene-Pliocene contact near the base of the *Scutellaster* n. sp. A zone on the basis of the uppermost occurrence of the mollusk

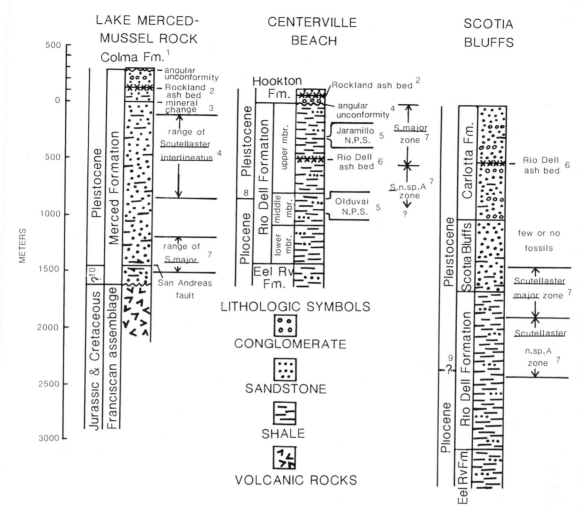

Fig. 6.6. Columnar sections of upper Cenozoic deposits at Scotia Bluffs and Centerville Beach in the Eel River Basin, northwestern California, and at the type section of the Merced Formation at San Francisco-Daly City, showing data pertinent to the age of the Merced Formation. Pertinent data are: 1) age of Colma Formation, about 0.07–0.13 Ma (Hunter *et al.* 1984); 2) age of Rockland ash bed, 0.4 Ma (Sarna-Wojcicki *et al.* 1985; the ash bed shown in the Centerville Beach section is actually at an inland exposure); 3) age of mineralogical change in upper Merced Formation, about 0.6 Ma (Sarna-Wojcicki *et al.* 1985); 4) age of top of Rio Dell Formation at Centerville Beach, no younger than 0.7 Ma according to paleomagnetic data (Dodd *et al.* 1977; Kodama 1979); 5) Jaramillo Normal Polarity Subchron (age 0.90–0.97 Ma) and Olduvai Normal Polarity Subchron (age 1.67–1.87 Ma) identified by paleomagnetic data of Dodd *et al.* (1977) and Kodama (1979); 6) age of Rio Dell ash bed, 1.2–1.5 Ma (Izett 1981; Sarna-Wojcicki *et al.* 1987); 7) macrofaunal zonation in Eel River Basin and approximate ranges of *Scutellaster major* and *S. interlineatus* in the Merced according to Roth (1979, p. 165)—in the Centerville Beach section, the lower boundary of the *S. major* zone is tentatively placed about 248 m above the base of the upper member of the Rio Dell, but not on the basis of *Scutellaster* occurrences; 8) Pleistocene-Pliocene boundary at Centerville Beach identified by foraminifers (Keller and Ingle 1981) and diatoms (Burckle *et al.* 1980); 9) Pleistocene-Pliocene boundary at Scotia Bluffs suggested by uppermost occurrence of *Patinopecten dilleri* (Dall), near base of *Scutellaster* n. sp. A zone (Roth 1979); 10) the part of the Merced Formation south of the San Andreas fault, here considered Pliocene? and Pleistocene.

Patinopecten dilleri (Dall), which is overlapped slightly and succeeded upward in the section by *Patinopecten caurinus* (Gould).

At the Centerville Beach section in the Eel River Basin, neither *S. major* nor *S.* n. sp. A is identified with certainty. On the basis of mollusk species in that section, however, Roth (1979) tentatively identified the *Scutellaster major* zone as extending from the top of the Rio Dell Formation, which is no younger than 0.7 Ma according to paleomagnetic interpretations (Dodd *et al.* 1977; Kodama 1979), to a little below the Rio Dell ash bed, dated at 1.2–1.5 Ma (Fig. 6.6; Izett 1981; Sarna-Wojcicki *et al.* 1987). On the basis of mollusk species, Roth also tentatively identified the *Scutellaster* n. sp. A zone at Centerville Beach as extending downward to near the base of the upper member of the Rio Dell Formation, below which level macrofossils are too rare to provide a basis for zonation of the formation. The Pleistocene-Pliocene contact in the Centerville Beach section has been identified as occurring in the middle member of the Rio Dell Formation on the basis of foraminifers (Keller and Ingle 1981), diatoms (Burckle *et al.* 1980), and a zone of normal paleomagnetic polarity identified as the Olduvai Normal Polarity Subchron (Fig. 6.6; Dodd *et al.* 1977; Kodama 1979). There is little doubt that in both the Scotia Bluffs and Centerville Beach sections of the Eel River Basin the *Scutellaster major* zone is well above the Pleistocene-Pliocene contact.

In the type section of the Merced Formation, the echinoid species that Glen (1959) identified as *Scutellaster oregonensis* (Clark), together with some of the stratigraphically lower occurrences of what Glen (1959) identified as *S. interlineatus* (Blake), are now regarded to be *S. major* (Kew) (Faustman 1964; Roth 1979). According to Roth (1979), *Scutellaster major* occurs from about 120 m to 430 m above the base of the type section of the Merced, or from about 60 m below to 250 m above the trace of the San Andreas fault. If the range of *S. major* in the Merced is similar to its range in the Eel River Basin, the part of the type section of the Merced north of the trace of the San Andreas fault must be entirely of Pleistocene age. Unfortunately, *S. major* has not been identified elsewhere than in the Eel River Basin and in the Merced Formation, and therefore its stratigraphic significance in the Merced Formation must be regarded as questionable.

Components of the Merced fauna other than the echinoids have been cited as suggesting a Pliocene age for the lower part of the Merced. For example, *Crepidula princeps* Conrad was cited by Glen (1959), and *Spisula (Mactromeris) coosensis* Howe in the Merced near Twelvemile Creek, 4 km southeast of its type section, was cited by Roth (1979) as suggesting a Pliocene age. However, these fossils are now known to occur in Pleistocene strata elsewhere in California (Roth 1979; E.J. Moore, written communications 1985, 1986). Given the absence of exclusively Pliocene or older fossils and the presence of *Scutellaster major* near the base of the section, the part of the type section of the Merced Formation north of the San Andreas fault is here regarded as probably entirely Pleistocene in age.

Sedimentary Facies and Sequences

The biota, sedimentary structures, and textural character of the Merced sediment north of the San Andreas fault provide a basis for interpreting depositional facies with a high level of confidence (Hunter *et al.* 1984). Ten facies types can be delineated: outer-shelf, mid-shelf, and inner-shelf, nearshore, foreshore, backshore, eolian dune, alluvial, marsh/pond/swamp, and coastal-embayment facies. Table 6.1 summarizes the characteristic features of each facies.

The facies recur stratigraphically in consistent patterns (Fig. 6.5). The most common pattern is one of upward shallowing. In a complete upward-shallowing sequence, the basal facies consists of bioturbated fossiliferous siltstone devoid of any evidence of physical structures. A random concave-up/concave-down orientation of bivalve shells and absence of shell lags indicates deposition below storm wave base. This sediment is interpreted to be an outer-shelf deposit, very roughly estimated to have formed at water depths on the order of 100 m.

In an upward-shallowing sequence, the completely bioturbated outer-shelf siltstone grades up section into siltstone that contains sporadic lag deposits of shells and sharp-based intervals 0.05–0.2 m thick of laminated very fine sand or silt. This sediment is inferred to be a mid-shelf deposit, formed at water depths shallower than storm wave base. Hummocky cross-stratification is present, as is ripple lamination. Typically the laminated intervals merge

Table 6.1. Characteristics and inferred origin of facies in sea-cliff exposures of the Merced Formation, San Francisco.

Inferred origin	Texture	Biota	Physical structures	Biogenic structures
Outer shelf	Sandy silt	Mollusks (*in situ* and scattered shells)	None	Intense bioturbation
Mid shelf	Sandy silt	Mollusks	Shell lags, parallel lamination, hummocky cross-stratification, ripple lamination	Sharp-topped bioturbated intervals, between sets of laminae
Inner shelf	Very fine sand, scattered small pebbles near top	Mollusks, echinoids	Shell lags, parallel lamination, hummocky cross-stratification	Locally intense bioturbation
Nearshore	Fine to coarse sand gravel	Mollusks, echinoids	Lenticular gravel and sand beds, high- and low-angle crossbedding, parallel lamination	Vertical burrows, *Macaronichnus* (near top)
Foreshore	Fine to coarse sand, gravel fining-upward trend	Mollusks, echinoids	Parallel lamination, heavy mineral layers near top	*Macaronichnus* (near base)
Embayment	Mud, sand, gravel, some fining-upward cycles	Mollusks, ostracods	Fine sand-silt-clay lamination, ripples and ripple bedding (sand), cross-bedding (sand and gravel), shell lags	Bioturbation, root-rhizome structures
Backshore	Fine sand	None	Parallel lamination, low-angle crossbedding, climbing adhesion ripple bedding, ripple lamination	Root-rhizome structures, vertebrate footprints, vertical tubes
Eolian dune	Fine sand	None	Medium- and large-scale and medium- and high-angle crossbedding	Local mottling, vertical tubes
Alluvial	Gravel and pebbly sand	None	Indistinct stratification, lenticular bedding, medium- and small-scale trough crossbedding	None
Pond/swamp/marsh	Mud, peat or lignite	Freshwater diatoms, insect wings, terrestrial vertebrate bones	Flat-bedded (mud)	Root structures, burrows and intrastratal trails

in the upper part of the facies into an amalgamated succession. A pronounced textural gradation (typically within a few metres of section) to well sorted fine-grained sandstone marks the top of the mid-shelf facies. On the modern shelf of central California, a similar transition occurs at a water depth of approximately 60 m (Drake and Cacchione 1985).

Well sorted fine-grained sand containing scattered shells and shell fragments characterizes the inner-shelf facies. Stratification is obscure, owing in part to bioturbation and in part to the homogeneity of the sediment; where stratification is discernible it consists of planar or gently wavy parallel lamination. In shallowing-upward sequences, the upper part of this facies includes scattered granules and small pebbles and lag deposits of shelly material.

The nearshore facies lies gradationally above the shelf facies and is characterized by 2–5 m of intercalated beds of gravel and crossbedded sand. Mollusk shell fragments and echinoid tests are common in some of the gravel beds. Vertically filled burrows, 10–20 mm across and several centimetres deep, commonly extend downward from the base of the coarser beds. The trace fossil *Macaronichnus* (Clifton and Thompson 1978) is common in finer sand layers near the top of the facies.

The foreshore facies lies gradationally above the nearshore facies and is typified by 2–3 m of planar-parallel laminated sand and fine gravel. The sediment closely resembles that of modern beach foreshores. The facies generally displays upward fining, and placers of dark heavy minerals occur in the upper part of some of the foreshore facies units. Scattered shell fragments may be present in the coarser layers and *Macaronichnus* occurs in some fine sand layers near the base of the facies.

The backshore facies gradationally overlies foreshore or embayment facies and consists of metres to tens of metres of well sorted fine-grained sand that contains sporadic layers 1–30 cm thick of lignitic or clayey sediment assigned to the fresh-water pond/swamp/marsh facies. The sand generally displays parallel or gently inclined planar lamination. Root or rhizome structures, climbing adhesion-ripple stratification, paleosols and, in a few beds, vertebrate footprints attest to deposition under conditions of subaerial exposure. Thin intervals of current-ripple-laminated sand indicate the presence of small streams sporadically within the succession.

The embayment facies is several metres to a few tens of metres thick and is typified by the presence of muddy sediment and molluscan fauna. In most parts of the section this facies overlies backshore sediments; a thin lignitic bed commonly separates the two facies. Many of the embayment deposits, particularly in the lower half of the section, consist of a succession of upward-fining cycles 1–3 m thick. The base of a cycle is typically marked by a lag deposit of shells or pebbles, and bivalves in living position are commonly concentrated in the sediment just beneath the lag. The cycles become progressively coarser from mud at the base of an embayment unit to gravelly deposits at its top. The upper parts of a few cycles contain rhizome structures that imply subaerial (or nearly subaerial) exposure. Much of the sediment in the embayment facies is intensively burrowed. Low-angle, large-scale foresets that overlap and interfinger with the lag deposits indicate the presence of migrating tidal channels, and a distinctive rhythmic alternation of laminae in one unit suggests neap-spring tidal fluctuations.

The eolian dune facies consists of several tens of metres of well sorted, fine-grained sand that typically displays high-angle crossbedding in units one to several metres thick. Large-scale deformational structures are present in a few of the crossbedded units, and a few thin layers of mud and pebbly sand occur locally in the flat-bedded intervals between crossbedded units. Paleosols are present at several horizons within many of the dune units. A few parts of the dune facies contain simple trace fossils—a mottled pattern or small vertical tubes. The dune facies typically overlies the backshore facies.

The alluvial facies is uncommon in the succession. Where present, alluvial units consist of up to a few metres of angular to subrounded gravel of local derivation, and pebbly sand. Stratification is indistinct or in the form of lenticular bedding and decimetre-scale trough cross-stratification. The alluvial facies typically occurs as broadly lenticular deposits at the top of the backshore or dune facies.

The fresh-water pond/swamp/marsh facies occurs sporadically within intervals otherwise composed of the backshore or dune facies. The facies is characterized by broadly lenticular muddy, peaty, or lignitic beds typically a few decimetres thick (an exception occurs just above the base of sequence U in Fig. 6.5 where lignitic mud several metres thick overlies an alluvial deposit). These fresh-water deposits commonly contain simple burrows or intrastratal trails, and root structures or burrows extend from the bases of many muddy or carbonaceous beds into the underlying sand. Fossils in this facies include fresh-water diatoms (Glen 1959) and insect and terrestrial mammalian remains (Hall 1965, 1966).

Most of the facies in the Merced Formation (Fig. 6.5) lie in successions that define an upward-shallowing trend: shelf—nearshore—foreshore—backshore. Dune, alluvial, and fresh-water pond/swamp/marsh deposits are restricted to the interval that includes or overlies the backshore facies. Many of the shallowing-upward sequences are bounded at their base by an erosional surface overlain by lags of pebbles, cobbles, and, commonly, shell fragments. Others begin with the finest part of a shelf deposit. Upward-deepening intervals occur mostly in the lower two-thirds of the section (Fig. 6.5). They are defined by backshore-to-embayment successions and by upward-fining inner- to mid- to outer-shelf sequences. An erosional surface and pebble-shell lag mark the contact between embayment deposits and superjacent shelf sediment in an upward-deepening succession.

The pattern of facies development in this section records at least 30 alternations of transgressive and regressive episodes (as derived from the "depth" curve alongside the columnar section in Fig. 6.5). The inferred depth-of-deposition curve mostly follows a saw-toothed pattern; intervals of gradual upward shallowing alternate with abrupt increases in depth of deposition. This pattern probably does not reflect differences in rates of transgression and regression, but rather results from transgressive erosion of the shoreface along a ravinement surface as a shoreline retreats (Swift 1968).

Origin of the Transgressive and Regressive Episodes

As noted in the introduction, the cycles of transgression and regression may be due to eustatic sea-level changes, vertical tectonic movements, fluctuations in the rate of sedimentation, or some combination of the three. The Merced Formation accumulated during a time of pronounced Pleistocene glacio-eustatic fluctuations of sea level; it also was deposited in a tectonically active area, influenced by an array of strike-slip faults, including the presently active San Andreas fault. The climatic fluctuations characteristic of Pleistocene time create a potential for

episodic variation in sedimentation rate. In the discussion that follows we will attempt to isolate the influence of each of these factors on the Merced succession. Our approach is to begin by comparing the pattern of marine transgressions and regressions with a eustatic sea-level curve developed for the Pleistocene from oxygen isotope data. If the impact of eustatic sea-level change on the section can be identified, the influence of tectonism and changes in sedimentation rate can be explored.

Sea-Level History

We have generated a model for Pleistocene global sea levels based on an oxygen-isotope record. We will describe how we generated this sea-level record and point out the assumptions used. We want to stress that this record undoubtedly is not accurate in its fine detail but we argue that it is a good first approximation of Pleistocene global sea level and is far better than using a Vail-type curve (i.e., Haq *et al.* 1987) for investigating Quaternary sequences. In addition, we believe that this is a valid and fruitful approach that has not been fully exploited to date.

Accurate histories of eustatic sea levels can be difficult to reconstruct because of the interaction of several global events. Pitman (1978) estimated that the effects on eustatic sea level can be: 1) about 1×10^{-2} m/10^3 yr caused by changes in the volume of the Mid-Ocean Ridge system; 2) about 2×10^{-4} m/10^3 yr produced by crustal shortening; 3) about 2×10^{-4} m/10^3 yr through lithospheric degassing; 4) 10^{-4} m/10^3 yr by sediment infill; and 5) as much as 16 m/10^3 yr by the buildup and decay of global ice. Because the section of the Merced and Colma Formations that we are investigating represents only about 1.7 million years, only the effect of global glaciations on eustatic sea levels has sufficient magnitude to cause the observed changes in the lithostratigraphy.

Quaternary time was marked by glaciations in which large quantities of seawater were extracted from the oceans and stored on land in ice sheets and alpine glaciers. The extraction of seawater through evaporation preferentially extracts isotopically light oxygen (^{16}O) relative to isotopically heavy oxygen (^{18}O). This fractionation leaves the seawater isotopically heavier in oxygen, and this heavy oxygen is eventually incorporated into the tests of marine organisms. When the glaciers melt, the isotopically light oxygen is returned to the oceans. A 10 m drop in sea level, other factors (temperature and salinity) being unchanged, would produce a 0.11 per mil increase in $^{18}O/^{16}O$ (Fairbanks and Matthews 1978). However, during global glaciations, the temperature of both the surface waters and bottom waters of some areas of the ocean cooled, and a 1°C decrease in temperature, the other factors (salinity and sea level) being unchanged, can produce a 0.23 per mil increase in $^{18}O/^{16}O$ (Epstein et al. 1953). In addition, ocean salinity can change because of global glaciations (the extraction of fresh water by evaporation and ice formation) and a change of 4.8 per mil in salinity can cause a 1.4 per mil change in $^{18}O/^{16}O$ (Craig 1966).

If there were significant amounts of floating ice during global glaciation, then there could have been a change in the isotopic composition of the ocean without a corresponding sea-level signal. Arguments have been made for an influence in the $^{18}O/^{16}O$ record by the presence of a floating ice cover over the Arctic Ocean (Williams et al. 1981; Fillon and Williams 1983). They suggest that the 0.3 per mil discrepancy during glacial intervals between sea-level records generated from dated uplifted terraces and those generated from the benthonic Foraminifera of deep-sea records can be accounted for by a significant ice cover over the Arctic Ocean during all of Pleistocene time. The work of Zahn et al. (1985) and Markussen et al. (1985), however, shows conclusively that there was no significant floating ice cap over the Arctic Ocean at least during the last glacial maximum and that there was "modest" biological productivity there during this time. They also present oxygen-isotope records from the Arctic Ocean that can be correlated with the Norwegian-Greenland Sea records. These newer data invalidate the model of Williams et al. (1981). The Arctic Ocean is the only area where enough floating ice could have accumulated to have a significant effect on the global oxygen-isotope signal.

A recent discussion by Chappell and Shackleton (1986) suggests that the differences in Pacific sea levels determined from dated uplifted terraces and those derived from oxygen isotopes of benthonic Foraminifera, especially in the time interval from about 110 Ka to 18 Ka, resulted from a cooling of the bottom waters by about 1.5°C. It is interesting that the 1.5°C cooling suggested by Chappell and Shackleton (1986) converts to a 0.33 per mil increase in $^{18}O/^{16}O$, a value almost identical to that used by Williams et al. (1981).

Both of the above arguments used benthonic Foraminifera to determine the oxygen-isotope com-

position of the ocean and hence the corresponding sea-level record. A short discussion by Matthews (1986) argues that one should use tropical or low-latitude planktonic rather than benthonic Foraminifera to generate the best continuous record of global ice volume. Matthews (1986) reasons that changes of bottom-water temperatures could be expected (see Chappell and Shackleton 1986) but sea-surface temperatures in the tropics and low latitudes did not change from glacial to interglacial conditions (Matthews and Poore 1980; CLIMAP 1981). Consequently, the surface-dwelling planktonic Foraminifera are the preferred organism for analysis.

The now-classic record of marine oxygen-isotope changes for Quaternary time is that of V28-239 from the western equatorial Pacific Ocean (Shackleton and Opdyke 1976). This record was generated from planktonic Foraminifera and it has been argued (Shackleton and Opdyke 1973) that this record is mainly one of ice volume (hence sea level). The changes in $^{18}O/^{16}O$ between glacial and interglacial portions of the isotope record are about 1.4 per mil. A change in $^{18}O/^{16}O$ of 1.4 per mil would require a 6°C cooling in sea-surface temperature (not supported by CLIMAP 1981), a 3.7 per mil increase in salinity (highly improbable), a 130 m drop in global sea level, or some combination of all these changes. It seems clear that sea level is the primary signal.

Consequently, we have assumed for our approximation that no temperature and salinity changes occurred and that all of the $^{18}O/^{16}O$ changes were the result of eustatic sea-level fluctuations. This assumption yields a very detailed chronology of sealevel changes with reasonably well constrained ages. The sea-level curve was constructed from the published oxygen-isotope values from V28-239 (Shackleton and Opdyke 1976) using the following equation:

$$SL(i) = 10 \left[\frac{ISO(0) - ISO(i)}{0.11} \right] \quad (1)$$

where SL(i) is the ith value of sea level, ISO(0) is the zero-depth oxygen-isotope value, and ISO(i) is the ith oxygen-isotope value.

This procedure yielded a sea level-versus-depth curve. The depth curve was converted to age using the oxygen-isotope stage/paleomagnetic boundary age model of Pisias and Moore (1981), and linearly interpolating the age of each sample depth from the surrounding age datums. The 23 age datums are

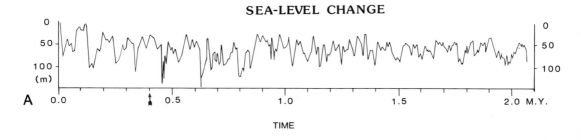

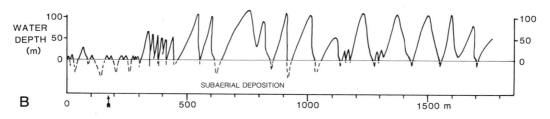

STRATIGRAPHIC DISTANCE BELOW TOP OF SECTION

Fig. 6.7. Comparison of (A) sea-level change in the Pleistocene relative to present sea level with (B) the paleobathymetry of Merced deposition. See text for explanation of the generation of the sea-level curve. Paleobathymetric curve in B based on facies and sequences shown in Figure 6.5. Arrows indicate the age and the stratigraphic position of the Rockland ash bed.

fairly evenly distributed between 11 Ka and 1,820 Ka. This process produced the sea level-versus-age curve shown in Figure 6.7A.

Minor variations of the calculated sea level (Fig. 6.7A) may be due to changes of salinity and sea-surface temperatures. Tectonic factors may have influenced the general slope of the glacio-eustatic curve shown in Figure 6.7A. For instance, Haq *et al.* (1987) present a sea-level curve for the Cenozoic that has a long-term lowering of sea level by several tens of metres during Pleistocene time. Such a trend could partly offset the general increase in sea level seen in the past 0.8 million years suggested in Figure 6.7A. We feel that the curve shown in Figure 6.7A is a good approximation of at least the high frequency (10^5–10^6 yr) signal of Quaternary eustatic sea level.

Relation of the Merced Transgressive/ Regressive Cycles and the Pleistocene Sea-Level Curve

Figure 6.7 allows for a direct comparison between the sea-level curve as derived in the foregoing section and the inferred paleobathymetry as developed in Figure 6.5. Both curves are drafted at the same vertical scale, and the horizontal scale of thickness in the lower curve (Fig. 6.7B) is drafted to correspond approximately with the horizontal scale of time for the past 2 million years used in the upper (sea-level) curve (Fig. 6.7A).

A key element in the correlation of Figures 6.7A and B is age control. Unfortunately, this element is largely lacking. The 400 Ka ash bed at 170 m and the approximately 100 Ka age of the Colma Formation (Sequence Z on Fig. 6.5) provide the best basis for correlation. The only other lithologic event that has a specific age context is the change at 290 m in provenance of the coastal sediment to the Sacramento-San Joaquin River systems, which, on the basis of tephrochronology associated with drainage changes in California's Central Valley, is thought to have occurred between 400 Ka and 620 Ka (Sarna-Wojcicki *et al.* 1985). The underlying 1,470 m of section contains no presently definitive bases for age correlation. Attempts to use magnetostratigraphic methods have proven unsuccessful, apparently because the sediment has been remagnetized (Verosub, K., oral communication 1985).

With so few age data, the curves can be correlated only by visual inspection. Cursory examination of Figure 6.7 demonstrates the difficulty in attempting a direct comparison of the curves. The inability to match the curves in detail may result from: 1) the possibility that the sea-level curve is inaccurate; 2) unrecognized loss of part of the section by erosion (units in this part of the section are known to be locally missing beneath erosional surfaces); 3) unrecognized changes in water depth within the shelf successions that relate to smaller scale fluctuations of sea level; or 4) tectonic or sedimentologic effects in either of the two curves. It seems clear, however, that the scale of relative sea-level change implied by the inferred depth of deposition of the Merced resembles that which would be imposed by the Pleistocene eustatic fluctuations that occurred during the same interval. The individual transgressions and regressions during deposition of the Merced Formation are therefore inferred to result primarily from glacio-eustatic fluctuations.

The inferred paleobathymetry curve for the upper 300 m of the Merced Formation differs from that below in that most of the uppermost section was deposited under nonmarine conditions. In contrast, relatively little of the section below this level is represented by nonmarine deposits. Instead, most of the lower section is dominated by shelf deposits, which occur in only one sequence in the upper 300 m of section. Figure 6.7B suggests that in the section below 300 m shoreline deposits accumulated here during the low stands of the sea, whereas in the section above 300 m shoreline deposits formed at highstands. This change cannot be reconciled with any aspect of the sea-level curve shown in Figure 6.7A. Accordingly, this lithologic variation would seem to reflect tectonic and/or sedimentologic influences.

Tectonic and Sedimentologic Effects

As stated in the introductory section, a significant amount of subsidence was required to accommodate the more than 1,740 m of coastal and shallow marine sediment deposited within the Merced Formation. Assuming a maximum age of 1.6 Ma for the base of the section shown in Figure 6.5, the Merced Formation north of the San Andreas fault accumulated at an overall average rate that was not less than $1 \text{ m}/10^3$ yr. The average rate of accumulation reflects a combination of the average rate of subsidence and the rate of sediment supply. In a marginal marine setting

such as that of the Merced, sedimentation is more likely to result in progradation than vertical accretion unless it is accompanied by subsidence. The fact that the Merced section below about 300 m shows no progressive shift in facies toward deeper or shallower deposits (Figs. 6.5 and 6.7B) implies a balance between sedimentation and subsidence, and also implies that the average accumulation rate actually reflects the average rate of subsidence.

Some evidence exists that the rate of subsidence decreased during Merced deposition. The Merced section above the 0.4 Ma Rockland ash is about 165 m thick. The age of sequence Y, at the top of this section, is not known, but it predates the deposition of the Colma Formation at 0.07–0.13 Ma. Figures 6.5 and 6.7B suggest that sequence Y accumulated during a high stand of the sea. If the sea-level curve shown in Figure 6.7A is reasonably accurate, it is likely that sequence Y was deposited no earlier than the high stand that prevailed in the interval 0.19–0.23 Ma. Otherwise, sequence Y would have accumulated during the high stand at approximately 0.3 Ma, which would leave very little time (about 0.1 million years) to accumulate 150 m of section that includes two well-defined transgressions and regressions. If the age of sequence Y is approximately 0.2 Ma, the rate of subsidence for the upper 165 m is in the range of $0.8–0.9 \text{ m}/10^3$ yr.

If the Merced section below the Rockland ash and above the trace of the San Andreas fault is entirely Pleistocene (with a maximum age of 1.6 Ma), the maximum time available for the accumulation of this section is 1.2 million years. Accordingly, the minimum average subsidence rate for the 1,570 m of section is about $1.3 \text{ m}/10^3$ yr. The nature and timing of changes in the rate of subsidence cannot be more accurately established without better age control.

Some physical evidence exists within the section of more rapid subsidence in its lower part. The individual transgressive-regressive cycles in the lower part of the section are thicker than those in the upper part. Paleosols, which are well developed in the backshore facies in the upper 200 m of the section, are poorly developed in the same facies in the lower part of the succession. This could reflect either more rapid subsidence or shorter periods of subaerial exposure in the lower section. Transgressive embayment deposits, which may require rapid subsidence, are common in the lower half of the section, but are generally absent from its upper part.

The inferred rapid rates of subsidence potentially could contribute to the changes in water depth evident in the Merced deposits. In the absence of other factors, subsidence alone could cause a depth increase on the order of 100 m over a time interval of 100,000 years. The lack of a trend toward deposition in progressively deeper water through time indicates that, over the long term, the rate of sedimentation balanced the rate of subsidence to the extent that subsidence did not influence the paleobathymetry.

Conceivably, the balance between sedimentation and subsidence might change depending on whether sea level was rising or falling. During rapid transgression, sediment is likely to be trapped in coastal river valleys and embayments, thereby reducing the rate of sediment input to the shelf areas (Wanless 1976). In a rapidly subsiding basin, subsidence might not be fully offset by sedimentation during periods of rising sea level, thereby contributing to the increase in water depth at any particular location.

Isostatic readjustments should also be taken into account. The amount of vertical displacement due to isostatic effects has been estimated to be in the range of one third of the depth of the water that was emplaced or removed (Bloom 1967). The isostatic response lags behind the actual addition or removal of water and is a potentially complicating factor in analyzing the paleobathymetric curve.

A major tectonic change occurred between the deposition of the Merced and the Colma Formations. The Colma lies in angular unconformity upon even the youngest sequence of the Merced (Y), indicating rotational uplift and erosion following the deposition of sequence Y. As noted previously, sequence Y probably accumulated during the high stand of sea level that occurred around 0.2 Ma. The deformation therefore appears to have occurred in the interval 0.1–0.2 Ma. Uplift and tilting of the Colma Formation suggest that this deformation may continue actively today.

Sedimentologic influences on the transgressive-regressive record in the Merced Formation may derive from several factors: Changes in the size or relief of the source area, climatic variations, and changes in the locus of deposition. Mineralogic analysis indicates that a major change in provenance occurred near the top of sequence P. The section below this contact contains heavy minerals that reflect local Franciscan-assemblage sources; the section above contains a heavy-mineral suite similar to that carried by the modern Sacramento-San Joaquin River systems (Hall 1965, 1966). These rivers presently drain an area equivalent to one fourth of the state of California and carry a sediment load measured in billions (10^9) of kilograms per year

(Krone 1979). The sudden appearance of this load in the area of Merced deposition undoubtedly effected a major change in the rate of sedimentation along the Pleistocene coast. After this event, the shoreline reached the location of the present sea cliffs only during the highest stands of sea level. It seems probable that the average position of the shoreline shifted some kilometres seaward as a consequence of the influx of Sacramento-San Joaquin River sediment.

Other variations in source area or relief are difficult to delineate in the Merced section. The presence of gravel in nearshore deposits throughout the succession suggests that high gradients were maintained over the entire period of deposition. The abundance and grain size of gravel in the nearshore facies declines in the section below sequence VV, possibly reflecting lower stream gradients and/or diminished discharge. The pattern of transgressions and regressions, however, seems uninfluenced by any such change.

The local climate almost certainly fluctuated with the Pleistocene glacial-interglacial alternations. The influence of these fluctuations on sediment discharge along this part of the coast cannot readily be determined. Conceivably, accelerated sediment discharge during glacial episodes generated progradation that contributed to the regression of the sea as sea level fell. Such a situation, combined with the aforementioned possibility of sediment trapping in river valleys and bays during sea-level rise, may contribute to the dominance of upward-shallowing sequences over upward-deepening sequences within the Merced succession.

Conclusions

The Merced Formation is a remarkably thick, young succession that records multiple episodes of transgression and regression. Deposition occurred in a tectonically active area during a time of rapid eustatic sea-level change and climatic fluctuations. Eustatic, tectonic, and sedimentologic influences are imprinted on the depositional record. The individual transgressive and regressive deposits almost certainly result from eustatic fluctuations of sea level. The extent and duration of the inferred depth changes in the Merced Formation correspond in a general way to the glacio-eustatic excursions of Pleistocene sea level, and it is inconceivable that these sea-level fluctuations would not profoundly influence this coastal succession.

In the absence of better age control, it is difficult to correlate individual transgressive-regressive cycles with specific derived sea-level oscillations. The only well established tie occurs near the top of sequence S where an ash bed is dated at 0.4 Ma. Following the deposition of the ash, this location was inundated to the extent that nearshore deposits accumulated. This transgression may correspond to the small rise in sea level that began at approximately 0.39 Ma.

The influence of tectonic forces is evident in the accumulation of the entire section. More than 1,750 m of subsidence are required to accommodate these shallow-water deposits. At some point probably between 0.1 and 0.2 Ma, the subsidence ceased and the section was tilted and raised. The elevation and tilting of the Colma Formation indicate rotational uplift that probably continues at present.

The preservation of transgressive embayment deposits probably requires rapid subsidence. The overall average rate for accumulation of the Merced Formation seems to have been about 1.0 m/10³ yr. The increase in the average thickness of the transgressive-regressive cycles in the lower part of the section may reflect a greater rate of subsidence in the early phase of basin development.

The average rate of sedimentation was probably substantially increased at about 0.5–0.6 Ma owing to the delivery of sediment from the Sacramento-San Joaquin River systems. As a consequence, the average position of the shoreline shifted seaward and the section, which earlier had been dominated by shelf deposits, became dominated by nonmarine facies. The rate of sedimentation likely fluctuated in response to glacial-interglacial alternations. A higher rate of sediment delivery to the open coast may partly contribute to the dominance of shallowing-upward sequences in the section.

Acknowledgments. We thank Earl Brabb, J. Wyatt Durham, Mike Field, Bill Glenn, Ellen Moore, and Barry Roth for their reviews of the manuscript.

References

ASHLEY, G.H. (1895) The Neocene of the Santa Cruz Mountains. I-Stratigraphy. California Academy Sciences Proceedings (Series 2) 5:273–367.

BLOOM, A.L. (1967) Pleistocene shorelines: A new test of isostacy. Geological Society America Bulletin 78:1477–1494.

BONILLA, M.G. (1959) Geologic observations in the epicentral area of the San Francisco earthquake of March 22, 1957. California Division Mines Special Report 57, pp. 25–37.

BONILLA, M.G. (1971) Preliminary geologic map of the San Francisco South Quadrangle, California. United States Geological Survey Miscellaneous Field Studies Map MF-311, scale 1:24,000.

BORTUGNO, E.J., MCJUNKIN, R.D., and WAGNER, D.L. (in press) Geologic map of the San Francisco-San Jose Quadrangle, California. California Division Mines Geology, scale 1:250,000.

BURCKLE, L.H., DODD, J.R., and STANTON, R.J., Jr. (1980) Diatom biostratigraphy and its relationship to paleomagnetic stratigraphy and molluscan distribution in the Neogene Centerville Beach section, California. Journal Paleontology 54:664–674.

CHAPPELL, J. and SHACKLETON, N.J. (1986) Oxygen isotopes and sea level. Nature 324:137–140.

CLIFTON, H.E. (1981) Progradational sequences in Miocene shoreline deposits, southeastern Caliente Range, California. Journal Sedimentary Petrology 51:165–184.

CLIFTON, H.E. and THOMPSON, J.K. (1978) *Macaronichnus segregatus*: A feeding trace of shallow marine polychaetes. Journal Sedimentary Petrology 48:1293–1302.

CLIMAP Project Members (1981) Seasonal reconstructions of the Earth's surface at the last glacial maximum. Geological Society America Map Chart Series MC-36.

CRAIG, H. (1966) Isotopic composition and origin of the Red Sea and Salton Sea geothermal brines. Science 154:1544–1548.

DODD, J.R., MEAD, J., and STANTON, R.J., Jr. (1977) Paleomagnetic stratigraphy of Pliocene Centerville Beach section, northern California. Earth Planetary Science Letters 34:381–386.

DRAKE, D.E. and CACCHIONE, D.A. (1985) Seasonal variation in sediment transport on the Russian River shelf. Continental Shelf Research 4:495–514.

EPSTEIN, S., BUCHSBAUM, H.A., LOWENSTAM, H.A., and UREY, H.C. (1953) Revised carbonate-water isotopic temperature scale. Geological Society America Bulletin 64:1315–1325.

FAIRBANKS, R.G. and MATTHEWS, R.K. (1978) The marine oxygen isotope record in Pleistocene coral, Barbados, West Indies. Quaternary Research 10:181–196.

FAUSTMAN, W.F. (1964) Paleontology of the Wildcat Group at Scotia and Centerville Beach, California. University California Publications Geological Science 41:97–160.

FILLON, R.H. and WILLIAMS, D.F. (1983) Glacial evolution of the Plio-Pleistocene: Role of continental and Arctic Ocean ice sheets. Palaeogeography, Palaeoclimatology, Palaeoecology 42:7–33.

GLEN, W. (1959) Pliocene and lower Pleistocene of the western part of the San Francisco Peninsula. University California Publications Geological Science 36:147–198.

HALL, N.T. (1965) Petrology of the Type Merced Group, San Francisco Peninsula, California. Unpublished MA thesis, University California, Berkeley, 126 pp.

HALL, N.T. (1966) Fleishacker Zoo to Mussel Rock (Merced Formation)—A Plio-Pleistocene nature walk. California Division Mines Geology Mineral Information Service 19:S22–S25.

HAQ, B.U., HARDENBOL, J., and VAIL, P.R. (1987) Chronology of fluctuating sea levels since the Triassic. Science 235:1156–1166.

HUNTER, R.E., CLIFTON, H.E., HALL, N.T., CSASZAR, G., RICHMOND, B.M., and CHIN, J.L. (1984) Pliocene and Pleistocene coastal and shelf deposits of the Merced Formation and associated beds, northwest San Francisco Peninsula, California. Society Economic Paleontologists Mineralogists Field Trip Guidebook 3, 1984 Midyear Meeting San Jose, California, pp. 1–29.

IZETT, G.A. (1981) Volcanic ash beds: Recorders of upper Cenozoic silicic pyroclastic volcanism in the western United States. Journal Geophysical Research 86:10200–10222.

KELLER, G., and INGLE, J.C., Jr. (1981) Planktonic foraminiferal biostratigraphy, paleoceanographic implications, and deep-sea correlation of the Pliocene-Pleistocene Centerville Beach section, northern California. In Armentrout, J.M. (ed) Pacific Northwest Cenozoic Biostratigraphy. Geological Society America Special Paper 184, pp. 127–135.

KODAMA, K.P. (1979) New paleomagnetic results from the Rio Dell Formation, California. Geophysical Research Letters 6:253–256.

KRONE, R.B. (1979) Sedimentation in the San Francisco Bay system. In: Conomos, T.J. (ed) San Francisco Bay: The Urbanized Estuary. San Francisco: Pacific Section American Association Advancement Science, pp. 85–96.

LAWSON, A.C. (1893) The post-Pliocene diastrophism of the coast of southern California. University California Publications Department Geology Bulletin 1:115–160.

LAWSON, A.C. (1895) Sketch of the geology of the San Francisco Peninsula. United States Geological Survey Yearbook 15, 1893–94, pp. 405–476.

LAWSON, A.C. (1914) San Francisco folio. United States Geological Survey Geologic Atlas of the United States Folio 193, 24 p.

MARKUSSEN, B., ZAHN, R., and THIEDE, J. (1985) Late Quaternary sedimentation in the eastern Arctic basin: Stratigraphy and depositional environment. Palaeogeography, Palaeoclimatology, Palaeoecology 50:271–284.

MARTIN, B. (1916) The Pliocene of middle and northern California. University California Publications Department Geology 9:215–259.

MATTHEWS, R.K. (1986) The del ^{18}0 signal of deep-sea planktonic foraminifer at low latitudes as an ice-volume indicator. South African Journal Science 82: 521–522.

MATTHEWS, R.K. and POORE, R.Z. (1980) Tertiary ^{18}0 record and glacio-eustatic sea-level fluctuations. Geology 8:501–504.

MUSTART, D.A. and MCCULLOCH, D. (1984) The San Bruno fault—A possible eastern strand of the San Andreas fault in the San Francisco region. Geological Society America Abstracts with Programs 16:605–606.

PISIAS, N.G. and MOORE, T.C. (1981) The evolution of the Pleistocene climate: A time-series approach. Earth Planetary Science Letters 52:450–458.

PITMAN, W.C. (1978) Relationship between sea-level change and stratigraphic sequences. Geological Society America Bulletin 89:1389–1403.

ROTH, B. (1979) Late Cenozoic Marine Invertebrates from Northwest California and Southwest Oregon. University California Berkeley Ph.D. thesis, 803 p.

RYER, T.A. (1977) Patterns of Cretaceous shallow-marine sedimentation, Coalville and Rockport area, Utah. Geological Society American Bulletin 88:177–188.

SARNA-WOJCICKI, A.M., MEYER, C.E., BOWMAN, H.R., HALL, N.T., RUSSELL, P.C., WOODWARD, M.J., and SLATE, J.L. (1985) Correlation of the Rockland ash bed, a 400,000-year-old stratigraphic marker in northern California and western Nevada, and implications for middle Pleistocene paleogeography of central California. Quaternary Geology 23:236–257.

SARNA-WOJCICKI, A.M., MORRISON, S.D., MEYER, C.E., and HILLHOUSE, J.W. (1987) Correlation of upper Cenozoic tephra layers between sediments of the western United States and eastern Pacific Ocean, and comparison with biostratigraphic and magnetostratigraphic age data. Geological Society America Bulletin 98:207–223.

SCHLOCKER, J., BONILLA, M.G., and RADBRUCH, D.M. (1958) Geology of the San Francisco North Quadrangle. United States Geological Survey Miscellaneous Investigations Map I-272, scale 1:24,000.

SHACKLETON, N.J. and OPDYKE, N.D. (1973) Oxygen isotope and palaeomagnetic stratigraphy of equatorial Pacific core V28-238: Oxygen isotope temperatures and ice volumes on a 10^5 and 10^6 year scale. Quaternary Research 3:39–55.

SHACKLETON, N.J. and OPDYKE, N.D. (1976) Oxygen-isotope and paleomagnetic stratigraphy of Pacific core V28-239 Late Pliocene to Latest Pleistocene. In: Cline, R.M. and Hays, J.D. (eds) Investigation of Late Quaternary Paleoceanography and Paleoclimatology. Geological Society America Memoir 145, pp. 449–464.

SWIFT, D.J.P. (1968) Coastal erosion and transgression. Journal Geology 76:444–456.

WANLESS, H.R. (1976) Intracoastal sedimentation. In: Stanley, D.J. and Swift, D.J.P. (eds) Marine Sediment Transport and Environmental Management. New York: John Wiley and Sons, pp. 221–239.

WILLIAMS, D.F., MOORE, W.S., and FILLON, R.H. (1981) Role of glacial Arctic Ocean ice sheets in Pleistocene oxygen isotope and sea level records. Earth Planetary Science Letters 56:157–166.

ZAHN, R., MARKUSSEN, B., and THIEDE, J. (1985) Stable isotope data and depositional environments in the late Quaternary Arctic Ocean. Nature 314:433–435.

7

Cambro-Ordovician Eustasy: Evidence from Geophysical Modelling of Subsidence in Cordilleran and Appalachian Passive Margins

GERARD C. BOND, MICHELLE A. KOMINZ, and JOHN P. GROTZINGER

Abstract

Evolution of early Paleozoic passive margins (miogeo-clines) in the Cordilleran and Appalachian orogens was controlled by a tectonic subsidence mechanism during Cambrian and early Ordovician time that is exponential in form resulting from thermal contraction during cooling of heated lithosphere. The subsidence was initiated by a major episode of continental rifting around the edge of the ancient North American continent (Laurentia) in latest Proterozoic or early Cambrian time.

New evidence suggests that two additional mechanisms may have been operating at the same time as the thermally controlled subsidence. The evidence consists of small, systematic differences between the tectonic subsidence curves from the ancient passive margins and best-fit expo-nential cooling curves. These differences can be inter-preted as evidence of two orders of change in relative sea level, one consisting of a single, long-term rise and fall spanning more than 40 million years and the other consist-ing of multiple, short-term changes, lasting from 2 to 10 million years, that are superimposed on the long-term event. The long-term event coincides temporally with the well documented transgression and regression of the craton, as recorded in strata of the Sauk sequence. The short-term events are broadly similar in form to the third order "sea-level" cycles of the Vail et al. (1977) coastal-onlap curve. In the passive-margin sequences of Utah and Virginia-Tennessee, the short-term events may be related to similar, but less well documented, shale-carbonate cycles.

The long-term event appears to have the same timing in the passive margin strata of the southern Canadian Rocky Mountains, in Utah, and in Virginia and Tennessee. The correlation of this event on a regional scale together with its close temporal relation to the Sauk sequence of the craton suggests a eustatic control.

The origin of the short-term events is less clear. In Mid-dle Cambrian time, they appear to be correlative in the Cordilleran and Appalachian passive margins, within the available biostratigraphic resolution, suggesting a eustatic rather than tectonic (local) origin. On the other hand, the cycles are poorly correlated in Late Cambrian time, although details of lithologies in most of the Upper Cam-brian sections are obscured by extensive dolomitization.

Introduction

The possibility that eustatic sea-level changes have occurred during the geologic past has long been a controversial issue. Vail and his co-workers (Vail *et al*. 1977, 1984) advocate a complex history of eustasy during Phanerozoic time, consisting of short-term sea-level changes superimposed on long-term changes (Fig. 7.1). There has been much debate over the Vail coastal-onlap curve, however, particularly with regard to the reality and origin of the short-term events (Watts 1982; Hallam 1984; Thorne and Watts 1984). The most compelling evi-dence for the complex eustatic histories in the Vail curve comes from seismic stratigraphic studies in Mesozoic and Cenozoic sedimentary basins and passive margins. An important question is whether evidence for complex, multiple orders of sea-level change exists in earlier parts of the geologic record where seismic methods are less applicable. Such evidence would then require identification of caus-ative mechanisms that operate over very long geo-logic time scales. Recently, short-term, presumably eustatic, events, comparable to the short-term events of Vail *et al*., have been suggested for Carboniferous to Permian (Ross and Ross 1985), for Devonian (Johnson *et al*. 1985), and for Ordovician time (Barnes 1984; Fortey 1984).

The Cambrian Period is of particular interest. Limited evidence, mostly from North America,

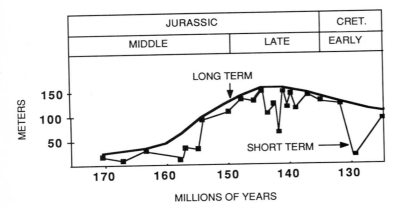

Fig. 7.1. An example of superimposed long-term and short-term cycles of relative sea-level change of Mesozoic age that are thought to have been caused by eustatic changes in sea level. (Adapted from Vail *et al.* 1984).

raises the possibility that two distinct orders of superimposed eustatic events occurred during that interval of time. One of these is a long-term event that occurred over a period of 40 to 60 million years and may have been responsible for the well known Sauk transgression and regression on the craton (Sloss 1963; Vail *et al.* 1977). The other consists of a number of much shorter-term events, each perhaps only a few million years long. These may have produced distinct shoaling upward cycles, termed Grand Cycles, that were first recognized in the southern Canadian Rockies by Aitken (1966, 1978) and now are thought to occur elsewhere in the Cordilleran and Appalachian passive margins (Fritz 1975; Palmer and Halley 1979; Aitken 1981; Mount and Rowland 1981). The possibility of short-term events in Cambrian time is particularly interesting because there is no evidence of glaciation that might be invoked as an explanation. A eustatic mechanism for these events has been difficult to test, however, because biostratigraphic correlations of Cambrian strata are too imprecise to determine if the short-lived shoaling upward cycles are synchronous on an inter-regional or global scale. A test for Cambrian eustasy, then, must rely heavily on physical criteria.

Recent work on subsidence mechanisms in the early Paleozoic passive margins of the Cordillera and Appalachians has suggested that procedures developed recently for the analysis of physical stratigraphy might help resolve the question of eustasy without requiring highly precise biostratigraphic correlations. Armin and Mayer (1983), Bond *et al.* (1983, 1985), and Bond and Kominz (1984) have shown that an approximation of the driving or tectonic component of subsidence can be recovered quantitatively from stratigraphic sections. To first order, subsidence was found to decay exponentially with time in response to slow cooling and thermal

contraction of lithosphere that was heated during rifting and continental breakup during late Proterozoic time. Bond *et al.* (1983) suggested that many of the subsidence curves contain systematic deflections from the smooth exponential subsidence that seem to require a long-term eustatic rise and fall with the same timing as the Sauk transgression and regression. This paper describes a modification of the methods used by Bond *et al.* (1983). The results not only further substantiate the existence of a long-term eustatic signal in the data, but also have produced some unexpected evidence for short-term eustatic changes, some of which appear to be correlative with Grand Cycles in the southern Canadian Rockies, which are superimposed on the long-term signal. The two scales of apparent eustatic cycles appear to be strikingly similar to two scales of cycles recognized by Vail *et al.* (1984) for Jurassic time (Fig. 7.1).

Construction of Subsidence Curves from Early Paleozoic Passive Margins and Identification of Eustatic Sea-Level Changes

We give only a brief discussion of the procedures used to construct and interpret subsidence curves for lower Paleozoic passive margin strata, using examples from our previous work in the Cordilleran and Appalachian passive margins. The steps in construction of the curves have been fully discussed in Bond and Kominz (1984). It is important to emphasize that the subsidence curves are always constructed from strata that were deposited entirely during the post-rift stage in the evolution of a passive margin or ancient passive margin. This minimizes the possibility of confusing complex tectonic events with eustatic signals, since the post-rift tectonic subsidence in a passive margin has a smooth, gradually decaying rate that is controlled mainly by cooling and thermal contraction of heated lithosphere (Sleep 1971; McKenzie 1978; Watts 1981).

Construction of the Subsidence Curves

The subsidence curves are constructed from cumulative stratigraphic-thickness curves by removing the effects of sediment lithification, sediment loading, and water depth or paleobathymetric changes. This reduces the cumulative thickness curves to a form that reflects only the effects of the tectonic or driving component of subsidence and the eustatic change in sea-level. For purposes of discussion, we refer to the subsidence curves produced by this first reduction as the R1 subsidence curves.

The general equation for calculating the tectonic subsidence plus eustatic sea-level or R1 component for a given stratigraphic unit is, according to Steckler and Watts (1978):

$$Y + \Delta SL \frac{\rho_w(\Phi - 1) + \rho_m}{\rho_m - \rho_w} = \Phi S* \frac{\rho_m - \rho_s}{\rho_m - \Phi_w}$$
$$+ Wd = R1 \qquad \text{(1st reduction)}$$

where Y = tectonic subsidence, $S*$ = sediment thickness corrected for lithification, ρ_s = mean bulk density of sediments, ρ_m = mean density of the mantle, ρ_w = mean density of sea water, ΔSL = eustatic change in sea-level relative to its present elevation, Φ = a basement response or weighting function relating sediment and water loads to tectonic subsidence, and Wd = average depth of water in which the unit was deposited.

For fully lithified sediment, as is the case for the early Paleozoic passive margins, we calculate $S*$ from empirical maximum and minimum delithification factors (Bond and Kominz 1984). Briefly, the maximum factors assume that all strata, including the abundant carbonate rocks, were lithified during burial by mechanical compaction. This process includes both the squeezing of grains closer together and the solution of grains at pressure points followed by the precipitation of the dissolved material in adjacent pore space. The minimum factors assume that all strata except the shales were lithified by the addition of a cement that was precipitated early in the

burial history of the sediments and that was derived from an external source, i.e., a source located some distance from the section being analyzed. This procedure was adopted to avoid the problem of assigning specific values to the complex and largely unknown processes of compaction and cementation in the passive-margin strata, especially in the carbonates. The function Φ in equation (1) is set equal to 1; that is, we assume that the subsidence caused by the sedimentary and water loads can be removed using a simple Airy compensation model and that the effects of lateral heat flow can be ignored. All of the sections that we have analyzed in the Cordilleran and Appalachian passive margins have been taken from the inner parts of the margin wedges where the strata are mainly shallow-marine platform carbonate rocks. These rocks most likely were deposited in inner-shelf environments where water depths probably did not exceed 100 m (Lohmann 1976; Aitken 1978; Demicco 1983; Walker et al. 1983). Consequently, the water depth or paleobathymetry changes [Wd in eq. (1)] are small relative to the total amount of subsidence, which is on the scale of kilometres, and changes in water depth can be ignored in interpreting the overall form of the R1 subsidence curves.

The values for R1 subsidence obtained from equation (1) are plotted as a function of age to produce cumulative curves for R1 (Figs. 7.2b, 7.3). The absolute ages for time-stratigraphic boundaries are from the time scale of Harland et al. (1982). Although the subsidence data are routinely calculated for sections extending to or near the base of the Lower Cambrian, the oldest point plotted in the curves is the base of the Middle Cambrian because it is the oldest stratigraphic boundary with a reliable absolute age.

The next step in the analysis is to compare the form of the R1 subsidence curves from the passive margin with curves calculated from models that simulate the subsidence of a passive margin during its post-rift or cooling stage. The post-rift or cooling model curves that we use for comparison are calculated from the one-layer stretching model of McKenzie (1978) in which the post-rift subsidence is modelled in terms of a cooling lithospheric plate. Recent studies have shown that this model is a good approximation of tectonic subsidence in modern passive margins after rifting ends and cooling begins (Watts 1981; Keen 1982). The McKenzie model is calibrated for 5 km of oceanic crust following Steckler (1981) and Cochran (1983) and for an equilibrium thermal thickness of the lithosphere of 125 km.

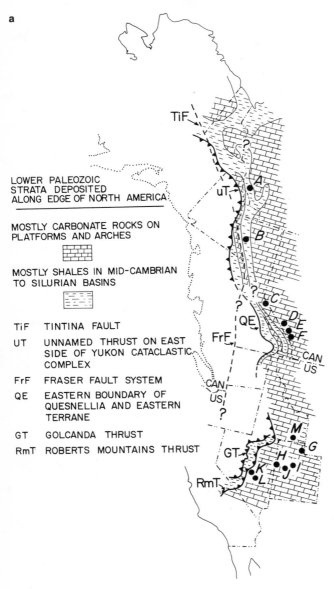

a

LOWER PALEOZOIC
STRATA DEPOSITED
ALONG EDGE OF NORTH AMERICA

MOSTLY CARBONATE ROCKS ON
PLATFORMS AND ARCHES

MOSTLY SHALES IN MID-CAMBRIAN
TO SILURIAN BASINS

TiF   TINTINA FAULT

UT   UNNAMED THRUST ON EAST
        SIDE OF YUKON CATACLASTIC
        COMPLEX

FrF   FRASER FAULT SYSTEM

QE   EASTERN BOUNDARY OF
        QUESNELLIA AND EASTERN
        TERRANE

GT   GOLCANDA THRUST

RmT   ROBERTS MOUNTAINS THRUST

Fig. 7.2a. Distribution of the lower Paleozoic mio-geoclinal facies in North American Cordillera. uT, QE, GT, and RmT are terrane boundaries (Monger and Price 1979; Coney *et al.* 1980; Monger *et al.* 1982). TiF-FrF is the Tintina-Fraser fault (Tempelmen-Kluit 1979). Dotted outline of western North America gives the position of this region after restoration of movement along TiF-FrF (Tempelmen-Kluit 1979). Filled circles give locations of sections for which R1 curves have been calculated. Reprinted by permission from Macmillian Journals Ltd.

These parameters constrain the model ocean-floor curve to be the same as the average age-depth curve for modern ocean floor according to Parsons and Sclater (1977).

As has been shown previously (Armin and Mayer 1983; Bond *et al.* 1983, 1984; Bond and Kominz 1984), the R1 subsidence in the early Paleozoic passive margins in the Cordilleran and Appalachian orogens is relatively smooth and decays steadily through time. The first-order form of the R1 subsidence curves resembles the form of the curves calculated from the cooling plate model (Figs. 7.2b, 7.3).

The fit of the R1 subsidence curves with the model curves is regarded as strong evidence that, after correcting for lithification and sediment and water loading, most of the post-rift subsidence was controlled by cooling and thermal contraction of a heated lithospheric plate.

Recognition of the Eustatic Component of Subsidence

If eustatic sea-level changes occurred during the subsidence of the early Paleozoic margin, they

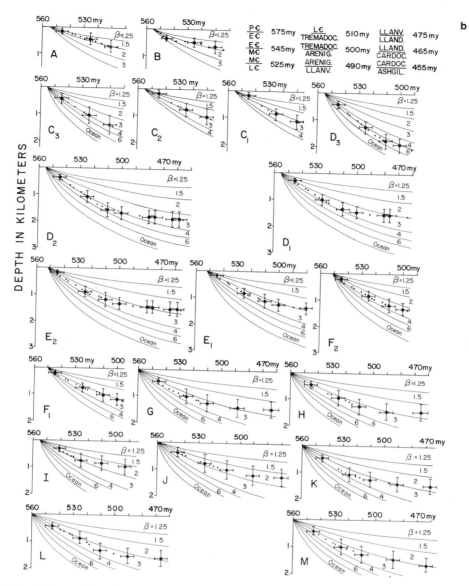

Fig. 7.2b. R1 subsidence (dots) for sections located in 7.2a compared with model post-rift cooling curves for a passive margin (solid curves; from McKenzie 1978). Large dots are stratigraphic boundaries to which absolute ages in the table in the upper right have been assigned. Horizontal bars give uncertainty in stratigraphic correlations, and vertical bars are ranges between values of R1 subsidence for maximum and minimum values of delithification. Letters with subscripts designate different sections lying along east-west lines at locations that are too close to indicate separately in 7.2a; higher subscript numbers indicate more westerly sections. (Figure from Bond *et al.* 1983. Reprinted by permission from Macmillian Journals Ltd.)

should be indicated by a misfit, possibly very small, between the R1 subsidence curves and curves that can be assumed for post-rift tectonic subsidence (Y) of the margins. In principle, then, if the tectonic subsidence can be properly identified and subtracted from R1 curves from several passive margins, the residual curves should be a good approximation of the eustatic sea-level change. For example, there is a slight, but nevertheless persistent, misfit between the R1 curves from the passive margins and the cooling curves (corrected for water loading) that is remarkably similar in both the southern Canadian Rockies and in the western United States (Bond *et al.* 1983). The difference between an R1 curve and

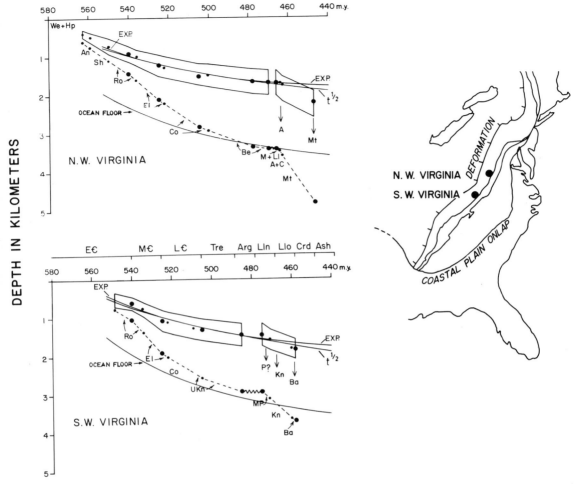

Fig. 7.3. R1 subsidence curves for SW and NW Virginia from Bond *et al.* (1984) and Bond (unpublished field data). Unlabeled light-face lines are maximum (lower) and minimum (upper) R1 curves for corresponding delithification factors. Large dots are absolutely dated horizons; small dots are located by interpolation. Dashed curves are present cumulative stratigraphic thickness curves. Abbreviations on curves are formations: We = Weaverton, Hp = Harpers, An = Antietam, Ro = Rome, Sh = Shady, El = Elbrook, Co = Conococheague, Ukn = Upper Knox, Be = Beekmantown, M + Li = Mosheim and Lenoir, MP = Mosheim + Lenoir + Paperville, A + C = Athens and Chambersburg, Kn = Knobs, Mt = Martinsburg, Ba = Bays. Ocean floor refers to exponential curve for subsidence of model ocean floor. Note that the fit is about the same between the R1 curves and the model cooling curves for both time$^{1/2}$ and the exponential (EXP) with a decay constant of 62.8 million years. Note also the slight but similar misfit between the two R1 curves and model cooling curves. Vertical arrows indicate that the tectonic subsidence curves would move downward (or would have additional tectonic subsidence) if water depths were added for corresponding formations. These formations contain turbidites and graptolitic shales, suggesting much deeper water conditions than for the predominately shelf-type carbonate strata that underlie them. On location map, solid dots indicate locations of analyzed sections.

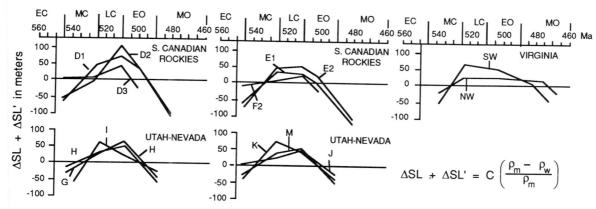

Fig. 7.4. R2 curves calculated from the misfit between the R1 curves and the 62.8 million year exponential for selected sections in Figure 7.2a and for the Appalachian sections in Figure 7.3. Canadian and Utah data are from Bond *et al.* (1983). ΔSL refers to the difference between the zero datum of the curves (the point at which the best-fit exponential crosses the R1 curves) and present sea level. C is the difference between R1 and the best-fit exponen-tial. Other symbols are defined in the text. The vertical scale should not be read as the actual magnitude of sea-level change relative to present sea level, because the present sea-level reference elevation is not known for the Cambrian strata and because the magnitudes of ΔSL are somewhat model-dependent. EC = Early Cambrian, MC = Middle Cambrian, LC = Late Cambrian, EO = Early Ordovician, MO = Middle Ordovician.

the best-fit cooling curve can be regarded as a second reduction of the cumulative subsidence curves and is referred to as an R2 subsidence curve. The form of the R2 curves for the Cordilleran passive margin (Fig. 7.4) was regarded as evidence of an apparent eustatic rise in Middle and Late Cambrian time followed by an apparent eustatic fall from Late Cambrian through early Ordovician time (Bond *et al.* 1983). The pattern of change is remarkably similar to the well documented Sauk transgression and regression on the craton (Sloss 1963). Recent subsidence analyses of strata in the Appalachian passive margin in northwestern and southwestern Virginia (Bond *et al.* 1984 and Bond, unpublished field data) now appear to indicate essentially the same deflection from the model curves (Fig. 7.4), a striking result that further strengthens the argument that the R2 subsidence curves contain evidence of a long-period eustatic cycle.

The results of the subsidence studies in the Cordilleran and Appalachian passive margins suggest that a rigorous application of the procedures to the lower Paleozoic stratigraphic record has the potential of producing new constraints on the occurrence and timing of eustatic sea-level changes in early Paleozoic time. An important question that we have not fully addressed previously, however, is whether the R1 and R2 curves contain unacceptably large and systematic errors resulting from the assumptions regarding time scales, delithification factors, and the use of simple models to calculate the tectonic component of subsidence (Y).

Sources and Magnitudes of Possible Error in the Construction and Interpretation of R1 and R2 Subsidence Curves

Choice of an Early Paleozoic Time Scale

More than one time scale has been proposed recently for early Paleozoic time, and the difficulties in constructing a reliable time scale have been reviewed by Harland *et al.* (1964), Harland and Francis (1971), and Fitch *et al.* (1976). It is important to recognize, however, that the form of the R1 curves is most critical and that the form is affected by the differences in the duration of time-stratigraphic units to which absolute ages have been assigned, not by the actual ages assigned to the time-stratigraphic boundaries.

Recent time scales for Early, Middle, and Late Cambrian time have been proposed by: 1) Cowie and Cribb (1978) who used the decay constants recommended by the IUGS Subcommission on Geochronology (Steiger and Jager 1977; 2) Van

Table 7.1. Radiometric ages assigned to stratigraphic boundaries in different time scales for early Paleozoic time.

Cambrian	Base of:			
	Atd	MC	LC	Trem
Armstrong (1978)	575 30	545 20	525 10	510
Cowie and Cribb (1978)	560 30	530 25	505 20	485
Van Eysinga (1978)	575 35	540 25	515 15	500
Harland et al. (1982)	570 30	540 15	525 20	505

Ordovician	Base of:						
	Trem	Areg	LLan	LLad	Card	Ash	LLand
Armstrong (1978)	510 10	500 10	490 10	475 10	465 10	455 10	445
Harland et al. (1982)	505 17	488 10	478 10	468 10	458 10	448 10	438
McKerrow et al. (1980)	519 15	504 15	489 10	479 12	467 22	445 7	438
Gale et al. (1979,1980)	490 8	482 18	464 11	453 19	434		418
Ross et al. (1978,1982)		494 7	487		474 14	460 25	435

The duration of each stage is listed between age columns. The age listed for each stratigraphic boundary is the mid point of the range of uncertainty for the age determination. Numbers are in millions of years. Atd = Atabanian; MC = Middle Cambrian; LC = Late Cambrian; Trem = Tremodocian; Areg = Arenig; LLan = Llanvirn; LLad = Llandeilo; Card = Caradoc; Ash = Ashgill; LLand = Llandovery.

Eysinga (1978) who does not give sources and decay constants; 3) Armstrong (1978) who used the decay constants proposed by the Subcommission except for K/Ar; and 4) Harland et al. (1982) and Palmer (1983) whose early Paleozoic scales are essentially the same, except for the base of the Cambrian, and are based on the recommended constants for all dating schemes. We do not include the scale of Odin et al. (1983) based on the comments of Harland (1983). The scale we have used to construct the curves in Figure 7.2 is from Armstrong (1978) which, for the early Paleozoic, is similar to the more recent scale of Harland et al. (1982). All curves in subsequent figures and calculations are based on the Harland et al. scale.

The data in Table 7.1 show the differences in duration for Early, Middle, and Late Cambrian time in all of the scales, based on the mid-points of the range of numerical error for each boundary. The differences are not greater than about 5 million years. This is probably of about the same order as the uncertainty in the intercontinental biostratigraphic correlations between North American stratigraphic boundaries and those to which absolute ages have been assigned, mainly in Britain (Cowie et al. 1971; Sweet and Bergstrom 1976; Williams 1976; Robison et al. 1977; Landing et al. 1978). Some of the Ordovician time scales, however, give larger differences in durations for some of the stages, as shown in Table 7.1.

However, we have found that the form of a thermal subsidence curve is retained regardless of which

of the time scales is used. In Figure 7.5a we have assumed a T_o (age for beginning of cooling) of 560 Ma for a post-rift cooling curve from McKenzie (1978) with a stretching factor (β) equal to 2 (solid curve). The curve is divided into segments using the ages for time stratigraphic boundaries given by Armstrong (1978) in Table 7.1. Those segments are then shifted horizontally to fit the McKerrow et al. (1980) and Harland et al. (1982)-Palmer (1983) time scales and a combination of the Cowie and Cribb (1978) and Gale et al. (1979, 1980) time scales. The Cowie-Cribb and Gale et al. scales are joined at 490 Ma. The dashed curves are the solid McKenzie curve but shifted horizontally to a best fit with the points of the other scales. The excellent fit between the dashed curves and the points indicates that the form of the cooling curve is not especially sensitive to the differences between the different Cambrian to Ordovician time scales. In Figure 7.5b, we assume a T_o of 520 Ma and apply the same test with similar results but to only the Ordovician time scales. This is necessary because for a T_o of 560 Ma, the Ordovician points fall on the flat part of the R1 subsidence curve where the sensitivity to the time scale differences is low. We should emphasize that this test does not take into account the ranges of uncertainty for each of the absolute ages in the Harland et al. (1982) time scale. These ranges are mostly between 10 and 20 million years but are as large as 36 million years for the end of Middle Cambrian time, due to the small number of radiometric determinations for this part of the time scale (Harland et al. 1982).

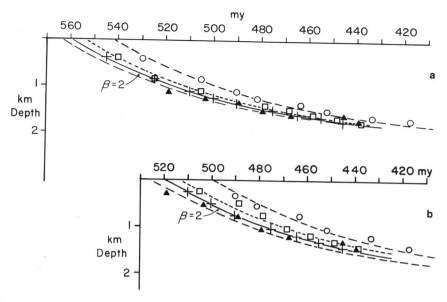

Fig. 7.5. Effect of assuming different early Paleozoic time scales on the thermal form of a McKenzie (1978) curve for post-rift or thermally controlled subsidence in a passive margin. Solid triangles = McKerrow *et al.* (1980) scale; plusses = Armstrong (1978) scale; squares = Harland *et al.* (1982) scale; circles = Cowie and Cribb (1978) and Gale *et al.* (1979, 1980) combined scales joined at 490 million years in the upper diagram. Only the Gale *et al.* scale is shown in the lower diagram. The upper diagram (a) is for Cambrian to Ordovician time; the lower diagram (b) is for Ordovician time only. The ranges of uncertainty for the absolute ages given by Harland *et al.* (1982) are not included; hence, the comparison applies to the mid-points of the ranges of absolute age only.

Delithification of Fully Lithified Strata

In studies of modern passive margins, where tectonic subsidence curves were shown to be thermal in form, the calculation of S* in Equation (1) was based on observed porosity-depth data from drill holes (Keen 1979; Steckler and Watts 1982). Porosity-depth data are not useful for lower Paleozoic strata in the passive margins because the sedimentary rocks are fully lithified. The calculation of S*, therefore, requires use of empirical maximum and minimum delithification factors as was described briefly in the preceding section.

The maximum and minimum delithification factors produce maximum and minimum values for the R1 curves. Although the range between these values is relatively large, that range is not a true indication of the possible variability in the form of R1. If the R1 subsidence data in Figures 7.2 or 7.3 were recalculated using any specific delithification factor from the range of values assumed, the recalculated curves could not jump from one extreme limit to the other (Fig. 7.6). This is the case because any subsidence value depends not only on delithification factors assumed for a bed at a given point in the section, but also upon factors assumed for all of the beds lower in the section. The maximum fluctuation that can occur in the curve between successive points is approximately 20 to 30% of the range between the maximum and minimum R1 subsidence values and over the total distance of the curve approximately 40% of that range (Fig. 7.6). Thus, specific values for delithification could produce R1 subsidence curves that lie along the minimum curve (most beds cemented early), along the maximum curve (most beds mechanically compacted or cemented by pressure solution), or that lie between the maximum and minimum curves and nearly parallel to them (some beds cemented early and some beds compacted).

The delithification procedure produces porosity-depth values for typical sections in the early Paleozoic passive margin of the southern Canadian Rockies that are very similar to those observed in the modern carbonate passive margin off the coast of Florida (Bond and Kominz 1984). We further tested the procedure by using our empirical delithification

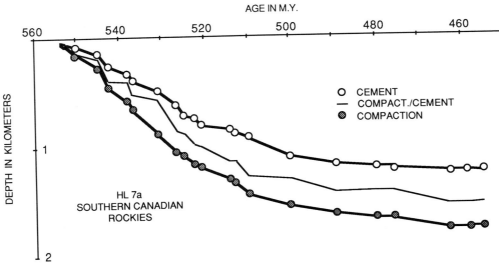

Fig. 7.6. R1 curves for different delithification options. The amount of variation in R1 subsidence that can occur assuming maximum changes in delithification factors is given by the light-face curve without ornamentation (compaction/cementation). See text for discussion.

factors to calculate an R1 subsidence curve for the section in the COST B2 well in the passive margin off the coast of New York (Smith *et al.* 1976; Scholle 1977). The factors were adjusted appropriately for the presence of porosities greater than zero in the strata in the well. Steckler and Watts (1978) have previously calculated the "tectonic subsidence" (the same as our R1 subsidence) for that well, using values for S* that were calculated from the observed down-hole porosity-depth data. Except for the calculation of S*, our procedure and data were the same as those used by Steckler and Watts. Our curve is slightly shallower than theirs but has an identical form (Fig. 7.7). This result is important because Watts and Steckler (1981) showed that the form of R1 subsidence in the COST B2 well was thermal. Moreover, they used the same procedure that we described above in our calculation of the apparent long-term early Paleozoic eustatic cycle, and they obtained apparent eustatic sea-level changes that were consistent with the timing of relatively long-term transgressions and regressions on the continents. We conclude that while our delithification schemes are undoubtedly an oversimplification, they do not appear to generate anomalous porosities relative to those observed in the modern southeastern Atlantic carbonate shelf and platform, and they do not distort the R1 curves for a passive margin that were determined using observed rather than estimated porosity-depth data.

Derivation of the Model Post-Rift Tectonic Subsidence Curves (Y) from Passive Margin Models

We assume that a reasonable approximation of the form of the true tectonic subsidence [Y in Equation (1)] of a passive margin during the post-rift stage can be calculated from simple cooling-plate models. This assumption must be examined in terms of: 1) whether the form could be different for different models of the evolution of passive margins; 2) whether the form depends on the choice of lithospheric parameters such as initial thicknesses of the crust and final equilibrium thermal thickness of the lithosphere, and, especially; 3) whether the form is affected by ignoring the important two-dimensional aspects of passive margins, in particular flexure, lateral heat flow, and different shapes and widths.

Models for the Evolution of Passive Margins

A number of different models have been proposed to account for the rifting and subsidence histories of passive margins. Two of the models, the stress-induced crustal thinning model (Bott 1971) and deep crustal metamorphism model (Falvey 1974; Neugebauer and Spohn 1982), are concerned mainly with syn-rift mechanisms and, although unspecified, the post-rift history in each is essentially the same as

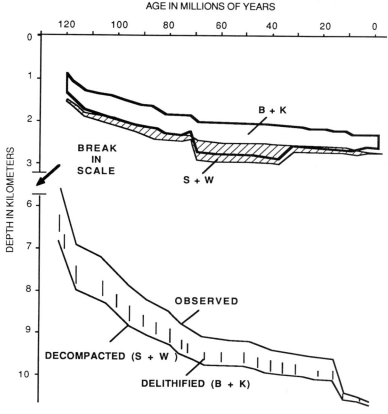

Fig. 7.7. R1 curves for the COST B1 Well offshore of New York calculated by Steckler and Watts (1978) using observed down-hole porosities (referred to as S + W curves) and by using empirical delithification factors as described in the text (referred to as B + K curves). Short bars are ranges of values for decompaction from Bond and Kominz (1984).

that in other models based on thermo-mechanical processes of rifting and subsidence.

Four types of passive-margin model based on thermo-mechanical processes of rifting and crustal heating have been proposed. These models quantitatively predict the form of post-rift subsidence, and were shown to be consistent with observations of thinned continental crust and post-rift thermal subsidence in modern margins (Royden and Keen 1980; Steckler 1981; Keen and Barrett 1981; Watts 1981; Steckler and Watts 1982). Three of the thermomechanical models assume that the rifting process is instantaneous. These are: 1) the uniform stretching mechanism of McKenzie (1978) in which the entire lithosphere is stretched and thinned, and as a result heated; 2) the depth-dependent model of Royden and Keen (1980) in which differential amounts of instantaneous stretching (or thinning) are assumed for the upper (brittle) and lower (ductile) lithosphere; and 3) the dike intrusion (or melt segregation) model of Royden et al. (1980), in which hot ultrabasic dikes are intruded into the lithosphere. The fourth type of model includes the finite extension models of Jarvis

and McKenzie (1980) and Cochran (1983) in which rifting is of finite duration, causing part of the lithospheric cooling to occur during the rift stage. The finite rifting models can be incorporated into any of the first three instantaneous models.

The most important aspect of all of these models with regard to the post-rift subsidence history of a passive margin is that after the first 10 to 15 million years of post-rift cooling, the form of the subsidence curves is identical with that of average ocean floor. The difference in the first 10 to 15 million years of the cooling history is due to the different initial temperature conditions required by each model at the time that rifting ends (Royden et al. 1980; Beaumont et al. 1982). After that, the form of subsidence predicted by all of the models is the same as that calculated for a simple cooling plate (Fig. 7.8).

In the finite-rifting models, the form of subsidence is the same, but the asymptotic value is decreased. This is because as the duration of the rifting period increases, more of the lithospheric cooling occurs during the rifting and thus, relative to the instantaneous models, there is less post-rift thermal

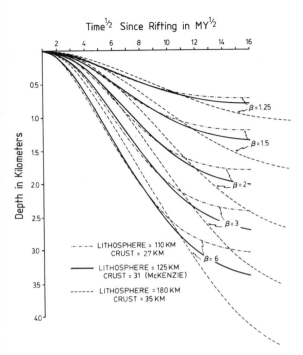

Fig. 7.8. The effect of assuming different values for lithospheric thickness on post-rift thermal subsidence curves calculated from the one-dimensional McKenzie (1978) model. Solid curves are those calculated using the lithospheric and crustal thickness in McKenzie (1978). β refers to stretching factor. For other values of lithospheric thickness, the thickness of the crust has been adjusted so that in all cases the surface is at sea level before stretching.

subsidence for any given amount of stretching (Jarvis and McKenzie 1980; Cochran 1983). The asymptotic value is increased somewhat when stretching results in partial melting of the lower crust and upper mantle, thus decreasing the depth of syn-rift subsidence and increasing the depth of post-rift subsidence.

Choice of Model Parameters

Certain parameters used in the thermo-mechanical models significantly affect the calculated syn-rift and post-rift subsidence. The most important are the initial thickness and density of the crust and the initial and equilibrium thickness and density of the lithosphere. The choice of model parameters, however, does not affect the basic form of post-rift tectonic subsidence. For example, in Figure 7.8, for different lithospheric thicknesses, the subsidence beginning about 10 to 15 million years after rifting is always characterized by a time$^{1/2}$ segment followed

by an exponentially decaying segment. Both the final depth of subsidence and the length of time in which the subsidence maintains a linear relation to the square root of time increase as the lithospheric thickness increases. The rate of tectonic subsidence in the square root of time segments increases only slightly with increasing lithospheric thickness and cannot exceed the subsidence rate of oceanic crust [about 350 m/(million years)$^{1/2}$; Parsons and Sclater 1977]. The initial thickness of the crust affects the amount of stretching required for a given depth of initial and thermal subsidence (McKenzie 1978; Royden and Keen 1980; Steckler 1981). That is, a thicker initial crust will require less stretching to reach the same final sediment thickness and, thus, subsidence. The form of thermal subsidence, however, is unaffected by this factor.

Effect of Ignoring the Two-Dimensional Properties of Flexure, Lateral Heat Flow and Different Margin Geometries

Our procedure for calculating R1 subsidence from the passive margins is purely one-dimensional. It ignores all two-dimensional or laterally varying processes such as flexure and lateral heat flow that not only affect the subsidence of a passive margin but also vary as a function of the shape of the margin and as a function of time as the margin cools (Watts 1981; Beaumont et al. 1982). The model tectonic subsidence curves we obtain from cooling models are also calculated by using one-dimensional procedures and ignoring the same two-dimensional processes.

Simple tests, however, show that ignoring the two-dimensional aspects of passive margins does not introduce error in calculating and analyzing the form of subsidence curves, either from the passive-margin or from the cooling-plate model. In these tests, we have used an instantaneous two-dimensional forward stretching model developed by Steckler (1981) and Steckler and Watts (1982) to simulate the post-rift tectonic subsidence in hypothetical passive margins of different shapes and widths and in a narrow basin where the effects of lateral heat flow are extreme (Fig. 7.9). The parameters used in this model are the same as those in McKenzie (1978). The model assumes that the equilibrium thermal thickness of the lithosphere is 125 km and is adjusted for a continental crustal thickness of 32 km and an oceanic crustal thickness of 5 km. Using the Steckler model, a compacting sediment load was

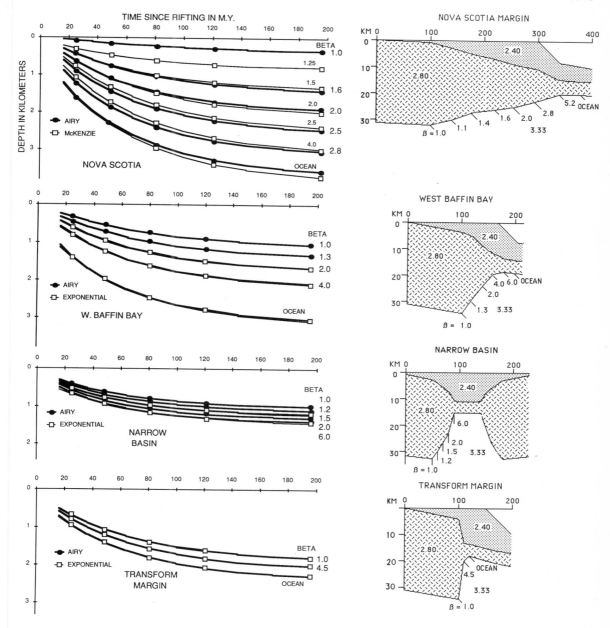

Fig. 7.9. Hypothetical margins on right were constructed using a computer model from Steckler (1981) that simulates the formation of a passive margin in terms of the uniform stretching model of McKenzie (1978) and with the effects of time-dependent flexure and lateral heat flow included. β refers to the amount of stretching; 2.40, 2.80, and 3.33 are sediment, crust, and mantle densities, respectively. Other parameters are the same as those used for model margins in Bond and Kominz (1984). The heavy subsidence curves on the left were constructed by removing the sediment load in the hypothetical margins on the right using an Airy model so that the effects of flexure and lateral heat flow are ignored. The light-face curves for the Nova Scotia margin are the McKenzie post-rift subsidence curves for the stretching factors in the model. These do not include the effects of flexure or lateral heat flow. For a wide margin such as this one, McKenzie post-rift curves adequately describe all of the subsidence. In the three plots below, exponentials with decay constants of 62.8 million years are fitted to the Airy subsidence curves beginning 15 million years after rifting.

added flexurally to the hypothetical margins. Tectonic subsidence curves were then constructed at different points across the sediment-filled margins using an Airy unloading model and no correction for lateral heat flow; that is, using the same one-dimensional procedures we use to calculate tectonic subsidence in ancient passive margins. In all cases, beginning about 10 to 15 million years after rifting, we found that the forms of the one-dimensional tectonic subsidence curves clearly are exponential, even in a narrow basin (Fig. 7.9).

We conclude that as long as only the *form* of subsidence is of interest, simple, one-dimensional procedures of analysis are fully justified. They can be used without introducing significant error to calculate the form of subsidence in the passive margins and to calculate curves from cooling models to obtain a good approximation of the thermally controlled subsidence, Y, in equation (1). We emphasize, however, that the true amount of stretching or crustal thinning cannot be calculated from the

forms of one-dimensional tectonic-subsidence curves. Provided that the thermal equilibrium thickness of the lithosphere is not much greater than 125 km, which is thought to be the case in modern passive margins (Watts 1981), the portion of the curve that is linear with respect to the square root of time closely approximates an exponential with a decay constant of 62.8 million years (McKenzie 1978). Thus, for modern passive margins and presumably ancient passive margins as well, all but the first 10 to 15 million years of post-rift subsidence can be described as exponential in form with a decay constant close to 60 million years. For a given duration of the rifting period the asymptotic depths for the exponential curves vary according to the amount of stretching or crustal thinning (McKenzie 1978; Royden and Keen 1980), or the amount of dike injection (Royden *et al.* 1980). Consequently, a family of subsidence curves can be generated that reflect the characteristic increasing amounts of stretching or crustal thinning that are observed in a crustal section across a typical passive margin. It follows that for passive margins, misfits between R1 and exponential curves, such as in Figure 7.4, are evidence that an event not associated with a simple post-rift cooling process, such as a eustatic change in sea level, must have contributed to the subsidence of the margin.

Results of Modified Analytical Procedures in Selected Sections

In order to test our previous hypotheses for Cambro-Ordovician eustasy further, we analyzed representative sections from each of three areas in the early Paleozoic passive margins using both published data and our field data. Three sections are located in the southern Canadian Rockies, one is in the western United States and one is in southwestern Virginia (Figs. 7.10, 7.11). The stratigraphic data from these sections that were used to construct the R1 curves are shown in Table 7.2.

The procedures applied to the strata in these three areas have been modified in a number of ways from the methods used previously by Bond *et al.* (1983). First, we used an exponential with a decay constant of 62.8 million years for the tectonic subsidence (Y) rather than a time$^{1/2}$ curve (Fig. 7.12). The exponential is easier to use than a time$^{1/2}$ function and, as emphasized in the preceding sections, it

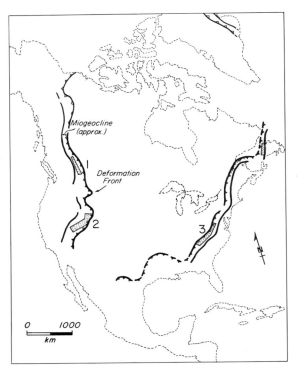

Fig. 7.10. Location of areas within early Paleozoic miogeoclines from which stratigraphic sections in Figure 7.11 were selected. Area 1 is in the southern Canadian Rocky Mountains, area 2 is in Utah, and area 3 is in the Virginia–Tennessee Appalachians.

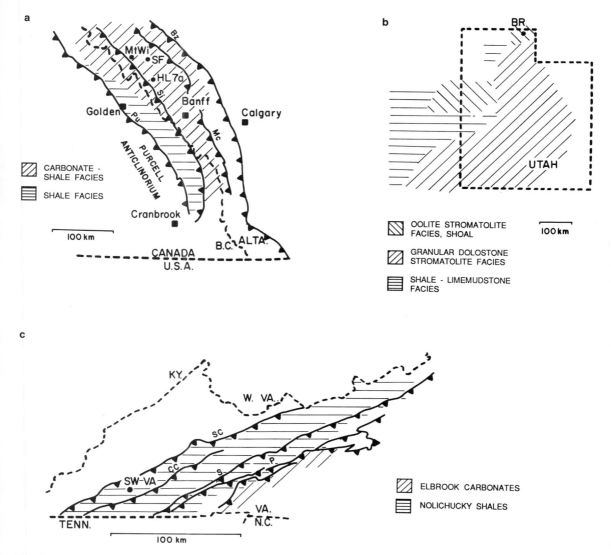

Fig. 7.11. Locations of the five selected sections in Figure 7.12 and the regional facies in which they occur. Data in 7.11a are from Bond and Kominz (1984). In Figure 7.11b, BR is the Bear River Range section. Facies data in Figure 7.11b showing distribution of facies in Middle and early Late Cambrian time are from Kepper (1972). The data in

Figure 7.11c are from Rodgers and Kent (1948) and Markello and Read (1982). The Elbrook carbonate rocks are part of the middle carbonate belt and the Nolichucky shales occur within a widely developed shale "basin" (detrital facies) to the west and southwest of the carbonate belt.

best describes the full subsidence history of modern passive margins. For a typical R1 subsidence curve, however, there is only a slight difference between the exponential and time½ fit (e.g., Fig. 7.3). The R1 curves all show essentially the same slight increase in amount of subsidence in Late Cambrian time with respect to the best-fit exponential (Fig. 7.12).

A second modification involves the procedure for fitting the exponential to the points on the R1 subsi-

dence curves. The modified procedure fits the exponential to both the series or stage boundaries with absolute ages from the Harland *et al.* (1982) time scale (540, 525, 505 Ma, etc.) and to the boundaries of stratigraphic units, mostly formations, that are located by interpolation between the points with absolute ages (Fig. 7.12). The ages of the interpolated points are determined by assuming that the cumulative-thickness (observed) curve is linear

Table 7.2. Stratigraphic and age data used to construct the curves in Figures 7.12 and 7.14.

	Bear River, Utah		NTE–SVA, Appalachians			Hector Lake		Mt. Wilson		Siffluer River	
								Canadian Rocky Mountains			
Early Cambrian	Prospect Mt$_1$	824	Unicoi	152	Early Cambrian	Gog	296	Gog	580	Gog	448
	Prospect Mt$_1$	91	Shady	190		Mt. Whyte$_1$	41	Mt. Whyte$_1$	45	Mt. Whyte$_1$	41
	Pioche$_1$	10	Rome	300							
540					540						
	Pioche$_2$	43									
	Naomi Peak	10									
	Spence	56				Mt. Whyte$_2$	83	Mt. Whyte$_2$	92	Mt. Whyte$_2$	83
	U. Langston	82	Pumpkin Valley	98		Cathedral	247	Cathedral	366	Cathedral	376
Middle Cambrian	Ute$_1$	81	Rutledge	96	Middle Cambrian	Stephen	74	Stephen	122	Stephen	89
	Ute$_2$	140	Rogersville	46		Eldon	336	Eldon	381	Eldon	322
	Blacksmith	148	Maryville	183		Pika	242	Pika	151	Pika	183
	Hodge	165	Nolichucky$_1$	44		Arctomys	109	Arctomys	112	Arctomys	124
	Bloomington	220									
	Calls Fort	55									
525					525						
			Nolichucky$_2$	21							
	Nounan	343	Nolichucky$_3$	49		Waterfowl	109	Waterfowl	184	Waterfowl	149
Late Cambrian	Worm Creek	25	Nolichucky$_4$	72		Sullivan	74	Sullivan	376	Sullivan	99
	St. Charles$_1$	259	Nolichucky$_5$	56	Late Cambrian	Lyell	306	Lyell	304		
			Maynardsville	108		Bison Creek	50	Bison Creek	192	Lynx	505
			Copper Ridge	282		Mistaya	153	Mistaya	102		
505					505						
			Chpultepec$_1$	110							
Tremadocian	St. Charles$_2$	25	Chpultepec$_2$	110	Tremadocian	Survey Peak$_1$	127	Survey Peak$_1$	132	Survey Peak$_1$	127
	Garden City$_1$	61	Longview	75		Survey Peak$_2$	127	Survey Peak$_2$	132	Survey Peak$_2$	127
488					488						
						Survey Peak$_3$	127	Survey Peak$_3$	132	Survey Peak$_3$	127
Arenigian	Garden City$_2$	222	Newala	110	Arenigian	Outram	183	Outram	258	Outram	158
	Garden City$_3$	145	483								
478					478						
Llanvirnian					Llanvirnian	Skoki	173	Skoki	140	Skoki	173
	Swan Peak$_2$	44			471						
						Hiatus		Hiatus		Hiatus	
468					468						
	Swan Peak$_2$	51									
Llandeilian					Llandeilian	Owen Creek	124	Owen Creek	187	Owen Creek	99
458					458						
						Hiatus		Hiatus			
455					456						
					Caradocian	Mt. Wilson	74	Mt. Wilson	171		
					448						

Numbers at the ends of the dashed lines (540, 525, etc.) are absolute ages of stratigraphic boundaries in millions of years from Harland et al. (1982). Numbers in vertical columns are unit thicknesses in metres. Locations of sections are given in Figure 7.11. Data for Hector Lake, Mt. Wilson, and Siffluer River (all from Alberta, Canada) are from Bond and Kominz (1984). Data for the Bear River Range in northern Utah are from Hintze (1973), Landing (1981), Maxey (1958), Rigo (1968), Deiss (1938), and Ross (1977). Data for the southwest Virginia section are from King (1949), Markello and Read (1981, 1982), and Read (in press). The southwestern Virginia section is not the same section as in Figure 7.3c.

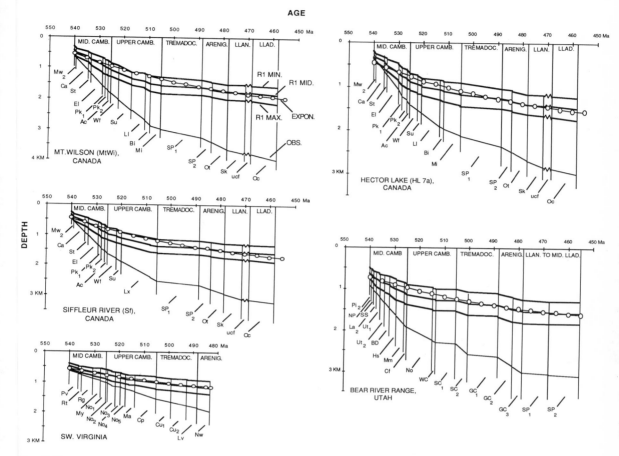

Fig. 7.12. R1 curves constructed for the five selected sections. Locations of sections are given in Figure 7.11. The solid vertical lines extending to the horizontal axis are stratigraphic boundaries with numerical ages from the Harland *et al.* (1982) time scale. Short vertical lines are formation boundaries positioned by assuming a linear relation of thickness and time between the numerical ages.

EXPON. is the best-fit exponential with a decay constant of 62.8 million years to the mid-points of the range of values for R1 (R1 MID). Small abbreviations indicate formations that can be read from Table 7.2. Note the overall decaying rate of subsidence and its similarity with the exponential.

with respect to time between the absolutely dated series boundaries.

The most important modification involves the scheme for delithification. Rather than assume a range between maximum and minimum limits to delithify the strata as we did in the past, we used three different delithification schemes, each of which specifies a single, but generally different, delithification factor for different lithologies. This procedure is discussed in detail in a following section.

Analyses in the Southern Canadian Rockies

Our modified procedures produced unexpected results. As an example, for the Mount Wilson (MtWi)

section (Fig. 7.13a) in the southern Canadian Rockies, we constrained the delithification factors so that all siliciclastic and calcareous shales were delithified assuming maximum factors, that is, assuming they were compacted along the deeper porosity-depth empirical curves. All sand to silt-sized carbonate rocks were delithified assuming the minimum factors; that is, assuming that porosity was eliminated mainly by addition of an early, external cement. While we have little direct evidence that this is a correct delithification scheme for all lithologies, it is well known that many coarser grained carbonate rocks contain evidence of early cementation and that siliciclastic and calcareous shales undergo generally more compaction than calcareous sand-

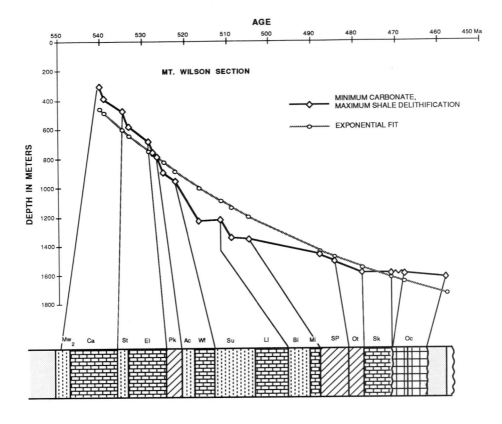

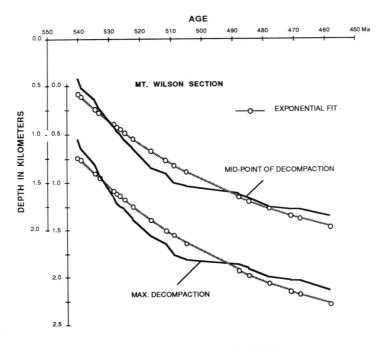

Fig. 7.13.

stones and siltstones (Fuchtbauer 1974; Chilingarian and Wolf 1975, 1976; Magara 1980). These differences in compaction as a function of lithology are consistent with our observations in the passive margins. For example, open packing occurs commonly in oolitic rocks and in many calcarenites, suggesting early cementation in coarse-grained sediments, and fluid escape structures and severely flattened worm tubes in the plane of bedding are common in both siliciclastic and calcareous shales, indicating substantial amounts of compaction relative to early cementation during burial in fine-grained sediments.

The R1 subsidence curve produced by this delithification scheme has a distinct zig-zag pattern with relatively higher rates of subsidence implied for certain units, usually the calcareous and siliciclastic shales, and relatively lower rates implied mainly for the coarser grained carbonate units (Fig. 7.13a). The pattern is essentially the same as in the minimum or upper set of the R1 subsidence curves in Figure 7.12 that were produced with the minimum delithification options in our standard method of delithifying between maximum and minimum ranges. Figure 7.13b shows that a similar zig-zag pattern is obtained for MtWi with two other delithification schemes, one using the maximum delithification factors for all lithologies (no early cementation in the carbonates), and the other using the mid-point of the ranges between the maximum and minimum values. Thus, although the pattern of the MtWi R1 curves is dependent on delithifying the strata, it is not particularly sensitive to the specific procedure for delithification or to the presence or absence of an external cement. In essence, the zig-zag subsidence of MtWi results from the fact that regardless of the specific delithification scheme, more pore water is always added to the delithified calcareous and siliciclastic shales, which are most abundant in the shaly half

cycles, than to the delithified coarse sediments such as siliceous and calcareous sands, which are most abundant in the carbonate half cycles. As a result, relative to the present-day thicknesses, the delithified thicknesses of the shaly half cycles are always increased slightly more than the delithified thicknesses of the carbonate half cycles.

The R2 subsidence of the MtWi section, which is the residual after removing the best-fit exponential with a decay constant of 62.8 million years from the R1 curve in Figure 7.13a, has a distinct and complex form (Fig. 7.14a). The rise through Cambrian and fall in early Ordovician time, which we inferred from our earlier work (Bond et al. 1983) to be a long-term eustatic signal, is clearly evident. In addition, however, the zig-zag pattern of the R1 curve produces several short-term excursions from the exponential that are superimposed on the long-term signal. The other two R2 test curves from the southern Canadian Rockies (HL7a and Sf, Fig. 7.14) were produced using the same delithification scheme as for MtWi and are remarkably similar to the MtWi curve in Middle Cambrian time. In Late Cambrian time, however, some differences exist between the curves. In particular, the two distinct cycles in MtWi and HL7a are not evident in Sf (Fig. 7.14a).

The most significant result of the analyses of the Canadian test sections is that the short-term events have a direct physical relation to the Grand Cycles of Aitken (1966; 1978). Each Grand Cycle consists of a lower predominantly shaly half cycle, containing large amounts of both siliciclastic and calcareous shales, that passes gradationally upward into a predominantly carbonate half cycle with large amounts of coarser grained carbonate sediments (Fig. 7.13a). The cycle is terminated by a relatively sharp contact with the overlying shale of the next cycle. Each rise or acceleration in the rate of rise in

◄

Fig. 7.13a. R1 curve for the Mt. Wilson (MtWi) section constructed assuming the minimum delithification factors for the carbonates and maximum factors for the shales (for location, see Fig. 7.11). This has the effect of creating a markedly irregular subsidence pattern directly related to the Grand Cycles as seen in the MtWi stratigraphic column below (not drawn to scale). The exponential is a best-fit exponential with a decay constant of 62.8 million years. Abbreviations Mw, Ca, etc., are formation names given in Table 7.2. Patterns are as follows: fine dots = mature

quartz sandstone, dashes = shaley, fine-grained half cycle of Grand Cycle, bricks = carbonate, coarser-grained half cycle of Grand Cycle, diagonal lines = undifferentiated lithologies constituting a complete Grand Cycle, grid = mainly dolomitic rocks. Figure 7.13b. R1 curves for MtWi section calculated using the mid-point (R1 MID) for MtWi in Figure 7.12, and using maximum delithification factors. The diagram indicates that the irregular subsidence of the section is evident regardless of the delithification scheme used.

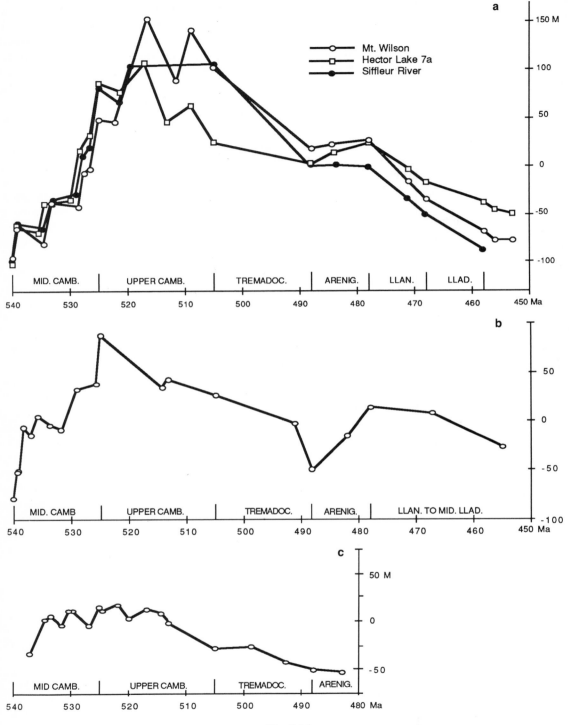

Fig. 7.14.

the R2 curve corresponds to the shaly or lower half cycle and each fall or reduction in rate corresponds to the carbonate or upper half cycle. It is also significant that, except in the Upper Cambrian segment of the Sf curve, the short-term events correlate remarkably well for the three curves. This is striking in view of the fact that the ages of most of the points on the curves were located by interpolation, sedimentation rates were assumed to have been constant, and the curves are from different thrust plates in different parts of the passive margin. The correlations are consistent with Aitken's contention that the cycle boundaries are coeval, at least within the resolution of the faunal data (Aitken 1981). The absence of the two Upper Cambrian cycles in the Sf section is characteristic of most of the northern and eastern parts of the passive margin in the southern Canadian Rockies. Their absence is due to the presence of a monotonous succession of limestones and dolomites through Late Cambrian time in those areas that have been assigned to a single stratigraphic division, the Lynx Group (Aitken and Greggs 1967).

It is tempting to regard the short-term events in the Middle Cambrian segments of the curves as approximations of small-scale eustatic changes that are superimposed on a long-term eustatic change in a manner similar to that proposed by Vail et al. (1984) for mid-Mesozoic time (Fig. 7.1). A eustatic origin for some of the shale-carbonate Grand Cycles in the Cordillera was, in fact, suggested previously (Palmer and Halley 1979; Aitken 1981; Mount and Rowland 1981). If a eustatic mechanism is correct, the magnitudes of the changes in slopes of the small-scale events in the R2 curves are not constrained well enough by the analytical methods to determine the direction of change in the eustatic signal. In other words, the data do not indicate whether the eustatic events could be actual reversals in the sea-level change, as might be implied by the forms of the R2 curves in the Upper Cambrian segments, or could be changes in only the rate of sea-level rise, as is suggested by the shapes of some of the R2 curves in the Middle Cambrian. Palmer and Halley (1979) argued that the absence of an unconformity at

the termination of the early Middle Cambrian Carrara cycle was evidence that a reduction in the rate of relative rise (not necessarily eustatic) caused the shoaling and deposition of the carbonate half of the cycle.

Even though a case can be made for short term eustatic changes based only on the shapes of the curves from the southern Canadian Rockies and the similarity in the timing of deflections, the results alternatively could be only a peculiarity of the local setting or an artifact of ignoring water-depth changes [Wd in Equation (1)]. Important additional tests for a eustatic origin of the cycles include determining whether similar events occur at the same time in other R2 subsidence curves, such as from the passive margin in the Great Basin of the western United States and in the Appalachians, and how the forms of the R2 curves are affected by the addition of reasonable estimates of water depths.

Analyses in the Great Basin and Appalachians

To determine if there are correlatives of the Grand Cycles in R2 curves from other areas, we applied the same procedures used for the Canadian sections to one section in the Bear River Range-Wellsville area in northern Utah (point BR in Figure 7.11b) and to one section from the Appalachian passive margin from the northeastern edge of the Nolichucky Basin in southwestern Virginia-northeastern Tennessee (point SW-VA in Figure 7.11c; the SW Virginia section in Figure 7.3 is not the same as the section in Fig. 7.11c). These three areas are in widely separated parts of the early Paleozoic passive margin and, if the cycles are eustatic in origin, they should be the same age in all three localities. Recently, however, an important question has been raised about this apparently straightforward assumption.

Thorne and Watts (1984) and Parkinson and Summerhayes (1985) argued that if two or more passive margins are subsiding at different rates, the facies patterns that form in response to a given eustatic

◄

Fig. 7.14. R2 curves constructed from the misfit between the R1 curves and the exponential in Figure 7.12. These curves suggest an overall eustatic rise and fall in Cambrian time on which are superimposed a number of smaller, possibly eustatic events. Compare with the short-term and long-term events in the Vail et al. (1984) sea-level curve in Figure 7.1. a = Southern Canadian Rocky Mountains; b = Wellsville-Bear River Range, Utah; c = Virginia-Tennessee Appalachians.

signal may not always develop in each margin at the same time. For example (Parkinson and Summerhayes 1985), suppose that sea level is falling and three passive margins, A, B, and C, are subsiding exponentially such that A is subsiding faster than B and that B is subsiding faster than C and that all three are subsiding more slowly than sea level is falling. At this time all three of the margins will experience a regression. Subsequently, if the rate of sea-level fall slows so that A and B now subside faster than sea level is falling, but margin C continues to subside more slowly than the sea-level fall, C will continue to experience regression, but A and B will undergo a transgression. Because the subsidence of all three margins is decaying exponentially, eventually A and B will subside more slowly than sea level is falling and will experience a regression again, but the regression will occur in B somewhat earlier than in A.

Although this hypothetical scenario is correct in principle, it applies rigorously only to the strandline and does not accurately describe events offshore where sedimentation rates and water depths vary (Neumann and Land 1975; Hine and Neumann 1977; Schlager and Ginsburg 1981; Halley et al. 1983). Even if the exponential subsidence is fast enough that a strandline transgression continues during a rapid eustatic fall or a slow down in rate of rise, the offshore areas will experience a net reduction in the rate at which deepening is occurring. Given the proper rate of sedimentation and appropriate water depths, the offshore areas could begin to fill with sediment and a shoaling-upward vertical sequence (a Grand Cycle) could develop. The shoaling-upward sequence would begin at about the time of the eustatic change and would occur at the same time as the strandline transgression. Other examples of the interplay between rates of subsidence, rates of eustatic change, and rates of sedimentation can also be produced in which the offshore facies patterns record shoaling during eustatic falls or slow-downs and deepening during eustatic rises, while the strandline is recording essentially the opposite history in the sense of transgressions and regressions. It is noteworthy that James (1984), in reference to shallowing-upward cycles in carbonate rocks, recommends against using the term "regressive" for vertical changes from deep-water to shallow-water facies because it implies a shoreline event that, in fact, may not have occurred. We conclude that, although significant differences in timing between eustatic changes and transgressive-regressive strandline histories might occur between or within certain margins, many offshore settings will exist in those same margins where the facies patterns, especially in vertical sequences, will record the sense and timing of the eustatic changes with reasonable accuracy.

Consequently, we limited our selection of test sections to specific paleoenvironments. We selected sections that are located at or near the transition between shallow-water shoal or bank complexes that contain mainly clean carbonate rocks and somewhat deeper-water flanking environments that contain a large proportion of siliciclastic sediments. Palmer (1960, 1971a) was the first to recognize that these facies are regionally developed in the Great Basin. In general, an outer belt is characterized by subtidal environments and the inner belt by shelf and lagoonal environments, while the carbonate shoal complex has been compared to modern carbonate banks such as in the Bahamas (Robison 1964; Palmer 1971a; Brady and Koepnick 1979). Our reasoning was that if small eustatic changes in sea level were taking place, perhaps causing changes in only the relative rates of net subsidence, the transitional environments between the major facies are the most likely place to find interlayered shales and carbonates that accurately record the timing and direction of the eustatic changes. If, for example, the rate of eustatic sea-level rise slowed, but the basin continued to deepen owing to a high rate of exponential subsidence, the reduction in net rate of subsidence might tend to cause carbonate lenses to prograde laterally from the shoal into the nearby deeper water environments. Conversely, if the rate of eustatic sea-level rise increased, the increase in net subsidence rate might be large enough to drown the carbonate facies and cause a rapid shift of the deeper water shaly facies toward the shoal and over the carbonate tongues, or even cause a shift in the position of the shoal itself. In fact, the Grand Cycles in the southern Canadian Rockies are thought to record a periodic westward expansion and drowning of a carbonate bank located on the western edge of a carbonate platform (Aitken 1971). Similarly, the interfingering tongues of deeper water shales and shallower water carbonates in the selected sections in Utah and in the Virginia Appalachians are regarded as evidence of periodic, rapid, lateral shifts of a carbonate shoal facies and a deeper water shale facies (Robison 1964; Palmer 1971b; Koepnick 1976; Brady and Koepnick 1979; Pfeil and Read 1980; Markello and Read 1981, 1982).

Table 7.3. Facies categories identified in the sections measured in the Bear River area, northern Utah, and ranges of water depths assigned to each.

Facies category	Water depths
1) Cryptalgalaminites with prism cracks.	0–1 m
2) Cross-bedded, bioclastic, ooid, and pellet grainstones; stromatolitic and thrombolitic boundstone.	1–10 m
3) Fine to coarse, winnowed bioclastic ribbon limestone and dolomite with burrows; fine to medium, thin-bedded siliciclastic sandstone with glauconite. Both facies contain current and wave-generated ripples, and lack mudcracks.	5–20 m
4) Pervasively bioturbated calci/dolosiltite, with distinctive rod-shaped calcite-filled burrows, thin beds of intraclastic conglomerate (tempestites?), and locally abundant thrombolites; thinly interbedded siliciclastic siltstone and fine to medium sandstone, burrowed, cross-laminated, possible wave ripples.	15–30 m
5) Nodular, burrowed calci/dolosiltite with locally abundant *Girvenella* oncoliths, lacks rod-shaped calcite-filled burrows and contains uncommon whole-body fossils; burrowed, laminated siliciclastic siltstone and fine sandstone with parallel lamination and possible hummocky cross-stratification.	20–40 m
6) Laminated to nodular limestone and minor dolomite with locally abundant spicules and whole-body fossils, minor burrowing; siliciclastic shale with minor limy layers and concretions.	30–>60 m

See text for discussion and sources of water-depth values.

The R2 curves for the Utah and Appalachian sections (Fig. 7.14b,c) contain distinct short-term events that resemble many of those in the Canadian sections (Fig. 7.14). In particular, four short-term events occur during Middle Cambrian time in all three areas. To a first approximation, these four events appear to occur at approximately the same time on the curves in all three areas, suggesting a degree of correlation that is consistent with a eustatic mechanism.

Water Depth Corrections

The effect of including water depth corrections [values for Wd in Equation (1)] on the form of R2 curves was tested using new stratigraphic data gathered from sections measured on a metre scale in Middle Cambrian strata in the Bear River-Wellsville area of Utah (BR in Fig. 7.11b). The various facies in the measured sections were grouped into six categories according to their inferred ranges of water depths (Table 7.3). Although the specific value for a given category of water depth is open to dispute, the relative arrangement of water depths between categories is probably generally correct. The limits on assigned water depths were obtained from a variety of sources, including: 1) summaries of well documented ancient platforms (Halley *et al.* 1983; Wilson and Jordan 1983); 2) detailed analysis

of specific lithofacies and modern analogs (Aitken 1978; Mullins *et al.* 1980; Pratt and James 1982; Demicco 1983); 3) studies of modern sediment/water depth relationships (Neumann and Land 1975; Hine and Neumann 1977; Mullins *et al.* 1980; Schlager and Ginsburg 1981); 4) geometric analysis of facies/depositional slope/water depth relationships of ancient platforms (Lohmann 1976; Grotzinger 1986); and 5) stratigraphically controlled forward modelling of water depths over ancient carbonate platforms (Grotzinger 1986; Read *et al.* 1986). Overlap of water-depth ranges between categories is one measure of the degree of uncertainty inherent in such assignments. In some instances, a specific facies may actually belong in several different categories of water depth (e.g., thrombolites), and must be treated on an individual basis where water depths are probably most accurately derived from information contained in associated facies (Pratt and James 1982). In other cases, the chosen range of water depths for a specific category [e.g., that for "ribbon rocks": rhythmically interbedded thin (few cm) shales and limestones] is subject to arbitration between various sources of information and represents a reasonable compromise.

In order to accommodate the water-depth values, the program for calculating the R1 curves in Figure 7.12 was modified. The modified program includes the water depths assigned to each unit in calculating the R1 subsidence curves. In addition, while the

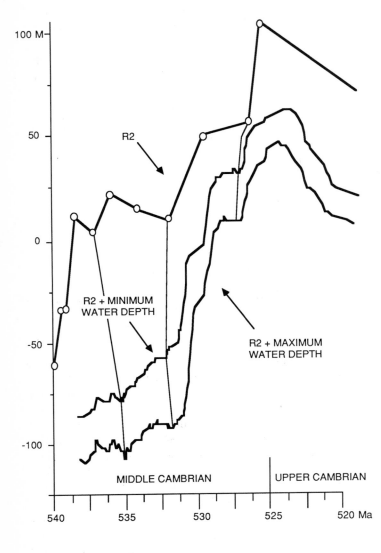

Fig. 7.15. R2 curve for Middle Cambrian strata in Bear River area from Figure 7.14 compared with R2 curves from the same area constructed using detailed facies data from sections measured on a metre scale and water-depth corrections given in Table 7.3. Curve labeled R2 is the R2 curve for the Bear River area in Figure 7.14 without water-depth corrections. Curve labeled R2 + MINIMUM WATER DEPTH incorporates the minimum water-depth estimates for each facies in Table 7.3. Curve labeled R2 + MAXIMUM WATER DEPTH incorporates the maximum water-depth estimates for each facies in Table 7.3. The light-face lines connecting the curves indicate the probable correlation of cycle boundaries in the R2 curve with those in the R2 + minimum and maximum water-depth corrections.

ages of the boundaries of each measured unit are determined by interpolation between the absolutely dated series boundaries as before, the interpolation is based on the assumption that the delithified thicknesses plus the water depths are linear with respect to time between the absolutely dated boundaries. This procedure for determining ages results in a slight shift in time of the units relative to ages based on interpolation using the observed thicknesses.

Although differences are evident between the R2 curves with and without water-depth corrections (Fig. 7.15), the cycles in the R2 curve without water-depth corrections clearly are present in the two R2 curves with the water-depth corrections. Moreover, the curves with water-depth corrections appear to contain within-cycle small-scale events that are not evident in the upper R2 curve (Fig. 7.15). Based on

the results of the analyses of the sections in northern Utah, it appears that reasonable water-depth corrections do not affect the general form that is evident in the uncorrected R2 curves, and may even produce evidence of additional smaller-scale events.

We also found from the detailed sections in northern Utah that there is a general pattern in the vertical changes in facies that appears to correspond to the cycles in the R2 curves. Facies 4) through 6) tend to predominate in the lower halves of the cycles while facies 1) through 3) tend to predominate in the upper halves (Table 7.3). This suggests a broad similarity, at least in terms of grain size, between the lithologies in the Utah cycles and in the Canadian Grand Cycles. In our measured sections, however, true siliciclastic shales are not as common as in the Canadian Grand Cycles.

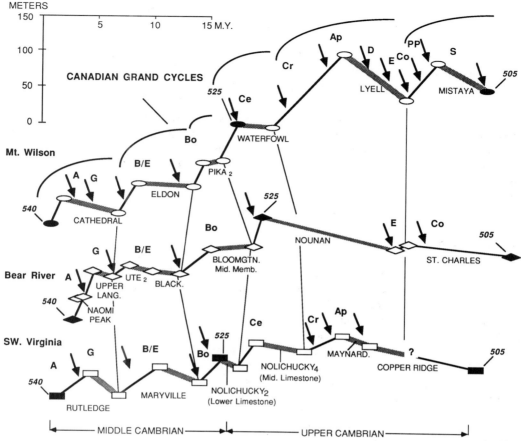

Fig. 7.16. Comparison of the Cambrian segments, on an enlarged scale, of the R2 curves in Figure 7.14 for widely separated localities in the miogeoclines: Mt. Wilson from Canada; Bear River from Utah; SW Virginia from the Appalachians. The thick, patterned lines in the Mt. Wilson curve are the carbonate halves of the Grand Cycles. The thin black lines are the shaly halves. In the two lower curves the thick patterned lines are predominately carbonate units that may be correlative with the carbonate half cycles in the Mt. Wilson R2 curve. The unnamed, thin black lines in the two lower curves are units that are predominately shales. The thin black lines in the two lower curves in the Upper Cambrian segments with formation names added are predominately carbonate units that do not appear to correlate directly with the Upper Cambrian carbonate half cycles in Canada. Fine solid (near-vertical) lines indicate possible correlatives of the Canadian Grand Cycles in Utah and Virginia. The positions of stratigraphic boundaries with absolute ages are indicated by a solid marker and the numerical value for the absolute age, based on the Harland *et al.* (1982) time scale. Bold letters stand for trilobite zones, delimited by short, near-vertical arrows as follows: A = *Albertella*, G = *Glossopleura*, B/E = *Bathyuriscus-Elrathina*, Bo = *Bolaspidella*, Ce = *Cedaria*, Cr = *Crepicephalus*, Ap = *Aphelapsis*, D = *Dunderbergia*, E = *Elvinia*, Co = *Conaspis*, PP = *Prosaukia-Ptychapsis*, S = *Saukia*.

A comparison of the Cambrian segments of R2 curves derived from all three of the localities (Fig. 7.16), suggests that, at least in Middle Cambrian and earliest Late Cambrian time, the Grand Cycles, as defined in the Canadian Rockies, could have equivalents in Utah and in southwestern Virginia. A horizontal (age) shift in the R2 curves of no more than 1.8 million years was required to bring the ages of the Middle Cambrian cycles into reasonably good agreement, and that shift results in a good biostratigraphic correlation of the cycles as well (Fig. 7.16).

On the other hand, the short-term events in the Upper Cambrian sections of the R2 curves do not appear to correlate as well as those in the Middle Cambrian sections (Fig. 7.16). The lower limestone in the Nolichucky seems to form an upper half cycle that is correlative with the Waterfowl upper half of the Arctomys-Waterfowl Grand Cycle in Canada. Possibly, the Maynardville and part of the Copper Ridge carbonate deposits in the Appalachians are an upper half cycle that correlates with the Lyell carbonate half cycle in Canada. Other distinct Upper Cambrian strata, such as in the *Elvinia* and *Dunderbergia* trilobite zones in Utah, seem to have no correlatives in Canada, and the Upper Cambrian cycles in Canada in the MtWi and Sf sections appear to have no counterparts in Utah. Palmer (1981b) noted similar difficulties in attempting to correlate cycle tops around the North American continent. Moreover, Palmer (1981a) suggested that two interregional unconformities are present within the Sauk sequence of North America, each of which may record a eustatic lowering of sea level. One occurs in latest Early Cambrian or earliest Middle Cambrian time and the other at about the position of the *Dunderbergia* faunal zone in Late Cambrian time. Interestingly, the *Dunderbergia* event occurs in a marked fall or slow down in the rate of sea-level rise in the MtWi and HL7a sections associated with the carbonate half of the Sullivan-Lyell cycle (Fig. 7.16). The earlier event is too old to be resolved on the curves.

The fact that the boundaries of the trilobite zones in the R2 curves have nearly the same numerical age in all of the sections, suggests that the assumption of dating units by linear interpolation between the absolute ages at series boundaries, that is, between 540, 525, 505 Ma, etc., is correct to a first approximation. We suggest that this is the case because the exponential subsidence is nearly linear with respect to time between the numerically dated stratigraphic boundaries (Figs. 7.12, 7.13). Because the strata were deposited in shallow marine conditions with relatively small fluctuations in water depths probably not exceeding 100 m (Lohmann 1976; Aitken 1978; Demicco 1983; Walker *et al.* 1983), the average rate of sediment accumulation must have been relatively close to the average rate of net subsidence. Consequently, between the numerically dated series boundaries, the average rate of sediment accumulation must have been nearly linear with respect to time.

Discussion

Based on the results of analyses of sections in the Cordilleran and Appalachian passive margins, we suggest a working hypothesis for eustasy and facies development during Cambrian time that can explain the different results for Middle and Upper Cambrian cycle correlations. For Middle Cambrian time, the shapes of the R2 curves and the presence of equivalents of the Canadian Grand Cycles in Utah and Virginia are permissive evidence for a eustatic origin of the cycles. We suggest that each eustatic event begins with a rise, or an increase in the rate of a long-term rise, which, in combination with the exponential subsidence of the margins, increases the net subsidence sufficiently to reduce the production of carbonate sediment, thereby causing deposition of slightly deeper-water shales on top of preexisting carbonate shoals. This is followed by a eustatic fall, or more likely a reduction in the rate of rise, which allows expansion of the carbonate shoal and gradual progradation of carbonate facies over shales.

A critical fact in both the Cordilleran and Appalachian passive margins is that the exponential subsidence of the margins was relatively high in Middle Cambrian time (Figs. 7.12, 7.13), which, together with the apparent long term sea-level rise, results in relatively high net subsidence rates in the margins. Consequently, even though the rate of sedimentation of carbonates is normally high, an increase in rate of sea-level rise could tend to overwhelm the rate of carbonate production in certain depositional settings, particularly in transitional environments at the edges of carbonate banks or shoals. Thus, shale to carbonate cycles related to eustasy would tend to be found in such transitional environments during the time when rates of thermal subsidence and rates of eustatic sea-level rise both are high.

Within the carbonate shoals, however, where sediment production rates would have been greatest, increases in net subsidence accompanying the eustatic rises might be sufficient to increase water depths only slightly. Consequently, within the carbonate shoal complex, equivalents of the thick shaly halves of the Grand Cycles might be only thin lenses of shales or intervals composed entirely of carbonate sediments, but of slightly deeper-water origin and, probably, finer grain size. An important additional test of eustasy, then, is to examine carefully the vertical succession of microfacies within the carbonate sequences that lie within the middle carbonate belt in the Great Basin and within comparable carbonate shoals in the Appalachians, such as in the Honaker and Elbrook carbonate sequences in southwestern Virginia.

An explanation for the difficulty in identifying and correlating Upper Cambrian shale-carbonate cycles is suggested by the fact that in Late Cambrian time, the exponential subsidence of both the Cordilleran and Appalachian passive margins was significantly reduced, probably by as much as 30% to 35% of that in Middle Cambrian time (Fig. 7.12). The long-term sea-level rise appears to have slowed as well (Figs. 7.4, 7.14). As a result of the large reduction in the rate of net subsidence of the margins, the carbonate shoals would be able to expand laterally into the deeper flanking environments, thereby reducing the extent of the shale-carbonate transitional environments. This is consistent with the fact that in Late Cambrian time throughout the Cordilleran and Appalachian passive margins carbonate shoal and bank complexes were widespread and shale facies were fairly limited, relative to their distribution in Middle Cambrian time (Lochman-Balk 1970; Palmer 1971a, 1971b). Even in the southern Canadian Rockies, where distinct shale to carbonate cycles occur in the Upper Cambrian deposits, extremely shallow peritidal carbonate facies are much more widely developed than in Middle Cambrian time, and the Upper Cambrian Grand Cycles cannot be traced as far to the north and northeast as the Middle Cambrian cycles (Aitken and Greggs 1967). This is reflected in the absence of the two Upper Cambrian cycles in the Sf section from the southern Canadian Rockies (Fig. 7.14a).

We suggest that if short-term eustatic events like those proposed for Middle Cambrian time occurred during Late Cambrian time as well, the evidence of those events will be only sporadically developed as easily recognized shale to carbonate cycles. Rather, the evidence might occur mainly as subtle, gradual changes over tens of metres in grain sizes and microfacies of predominantly carbonate strata. In the Newfoundland segment of the Appalachian early Paleozoic passive margin, for example, Chow and James (1987) recognized only one Grand Cycle in the entire Upper Cambrian succession. These Upper Cambrian strata consist predominantly of shallow-water oolite-laminite assemblages. Although much work has been done on Cambrian cycles in the predominantly carbonate sequences, including Upper Cambrian strata (e.g., Lohmann 1976; Pfeil and Read 1980; Markello and Read 1982), these studies have focused mainly on cycles a few tens of metres or less in thickness, whereas the equivalents of the Canadian Grand Cycles in such rocks should be tens to hundreds of metres thick.

The results of the subsidence analyses and construction of R2 curves, particularly as shown in Figure 7.16, indicate some of the advantages of this procedure relative to direct physical correlations of facies successions in measured sections as a means of testing the rock record for evidence of eustasy:

1. The construction of R2 curves reduces the difficulty in identifying the scale of the cyclic events being compared. For example, Palmer (1981b) noted that the abrupt change from the Naomi Peak Limestone to the overlying shale in the Bear River section of Utah has no correlative in the Grand Cycles in Canada (Fig. 7.16) and that a thin shale member in the Cathedral Formation in Canada (the Ross Lake Shale) is not present in Utah. However, the Naomi Peak Limestone produces an extremely small, short-term event in the R2 curve (Fig. 7.16) and the shale member in the Cathedral is so thin that it cannot be shown on the R2 curves. These thin deposits, whatever their origin, represent distinctly smaller-scale events than the Grand Cycles in Canada and, therefore, should not be compared with them. On the other hand, the next limestone higher in the Bear River Range section, the upper member of the Langston Formation, produces an event in the R2 curve that is comparable in scale to that of the Canadian Grand Cycles, and it correlates well with the Cathedral carbonate half (Cathedral Formation) of the Mt. Whyte-Cathedral cycle.

   The advantage of using R2 curves for correlating cycles is especially evident when comparing sections that were deposited in inner and outer parts of a passive margin. Because thermal subsidence increases substantially from the inner to outer part of a passive margin, correlative cycles thicken significantly in the same direction. The R2 curves correct for this by, in effect, normalizing the time span of the cycles from any position on the margin with respect to the rate of thermal subsidence. Similarly, R2 curves constructed for sections on cratons where thermal subsidence does not occur can be compared directly with R2 curves from passive margins.

2. Comparison of R1 and R2 curves provides insights into the interaction between the development of cycles and the relative rates of components of subsidence that contribute to long-term net subsidence of a basin. For example, in the lower Paleozoic strata that were analyzed in this study, the long-term components are thermal subsidence of the passive margins and a long-term sea-level rise and fall. As we noted previously, changes in the rates of long-term net subsidence resulting from interaction of the long-term

components appear to correspond closely to distinct changes in the lithologic character of the cycles from Middle to Late Cambrian time.

In addition, although the absolute rates of subsidence given by the R1 and R2 curves are model-dependent, the curves give an approximation of relative magnitudes and rates of the different components of subsidence. For example, in Cambrian time the thermal subsidence is by far the dominant long-term component of subsidence and the cycles appear to have much smaller magnitudes of change than the long-term sea-level change.

3. By correcting thicknesses for the effects of compaction and including corrections for water depths, the R2 curves provide a more accurate measure of the relative proportions of lower and upper halves of the cycles and the shapes of the cycles with respect to time than is possible with measured thicknesses alone.

4. Identification and correlation of cycles by means of R2 curves place emphasis on the cumulative effects of gradual changes in grain size and facies. This offers distinct advantages over identification and correlation of cycles solely by means of specific facies such as shales, ribbon rock, or cryptagal laminites in measured sections. For example, in sections we measured in the Bear River-Wellsville area, not all of the cycle tops are marked by cryptalgal laminites (shallowest facies) nor are all cycle bases marked by laminated, nodular limestones and siliciclastic shales (deepest facies). In fact, it is not uncommon to find beds of cryptagal laminites within cycles whose tops are marked by oolitic grainstones.

The analytical procedures described in this paper provide a fresh means of addressing the problem of Cambrian eustasy. The analytical procedures can be applied to other early Paleozoic passive margins such as in the Arctic regions where Palmer (1981b) has noted events in the stratigraphy that do not appear to match those in western North America. Work in such areas, as well as elsewhere on the globe, will be a crucial test of the synchroneity of the cycles. If the cycles are eustatic in origin, they will provide a means of correlation that is as good, if not better, than the available Cambrian zonal correlation, especially on a global scale. This would be particularly worthwhile for sparsely fossiliferous Cambrian sequences, where detailed correlation could be accomplished by matching the shapes of R2

curves, provided that the series boundaries are reasonably well identified stratigraphically. On the other hand, if the evidence suggests that the cycles are only locally correlative, it will underscore the need to identify local mechanisms that were responsible for their origin. The evolutionary history of faunas in shallow epeiric seas has been related, in part, to relative sea-level changes (e.g., Fortey 1984; Hallam 1984). If some or all of the changes in Cambrian sea level are found to be eustatic, then paradigms for evolution of life in Cambrian epeiric seas should be examined in terms of the eustatic sea-level history. Finally, the long-term eustatic rise and fall implied by our R2 curves could be evidence that important changes occurred in the volume of the early Paleozoic spreading ridge system, as seems to be the case for the long-term sea-level fall in Late Cretaceous and Cenozoic time (Pitman 1978; Kominz 1984). Bond *et al.* (1984) have suggested that the long-term sea-level rise might have been caused by an increase in ridge-system volume following breakup of a Late Proterozoic supercontinent. An intriguing possibility for the long-term fall at the end of Cambrian and in early Ordovician time is that parts of the ridge system were consumed as oceans began to close.

Acknowledgments. The authors are indebted to N. Christie-Blick and W. Pitman III for helpful comments on earlier drafts of this manuscript. This work was supported by NSF grants EAR 83-13230, EAR 85-18644 and EAR 84-17439 to G. Bond. Lamont-Doherty Geological Observatory of Columbia University Contribution No. 4201.

References

AITKEN, J.D. (1966) Middle Cambrian to Middle Ordovician cyclic sedimentation, Southern Canadian Rocky Mountains of Alberta. Bulletin Canadian Petroleum Geology 14:405–441.

AITKEN, J.D. (1971) Control of lower Paleozoic sedimentary facies by the Kicking Horse Rim, Southern Canadian Rocky Mountains, Canada. Bulletin Canadian Petroleum Geology 19:557–569.

AITKEN, J.D. (1978) Revised models for depositional grand cycles, Cambrian of the Southern Rocky Mountains, Canada. Bulletin Canadian Petroleum Geology 26:515–542.

AITKEN, J.D. (1981) Generalizations about grand cycles. In: Taylor, M.E. (ed) Short papers for the Second Inter-

National Symposium on the Cambrian System. United States Geological Survey Open-File Report 81–743: 8–14.

AITKEN, J.D. and GREGGS, R.G. (1967) Upper Cambrian Formations, Southern Rocky Mountains of Alberta, an Interim Report. Geological Survey Canada Paper 66–49, 91 p.

ARMIN, R.A. and MAYER, L. (1983) Subsidence analysis of the Cordilleran miogeocline: Implications for timing of late Proterozoic rifting and amount of extension. Geology 11:702–705.

ARMSTRONG, R.L. (1978) Pre-Cenozoic Phanerozoic time scale: Computer file of critical dates and consequences of new and in progress decay constant revisions. In: Cohee, G.V., Glassner, M.F., and Hedberg, H.D. (eds) Contributions to the Geologic Time Scale. American Association Petroleum Geologists Studies Geology 6:73–91.

BARNES, C.R. (1984) Early Ordovician eustatic events in Canada. In: Bruton, D.L. (ed) Aspects of the Ordovician System. Palaeontological Contributions University Oslo, Norway: Universitetsforlaget, pp. 51–63.

BEAUMONT, C.R., KEEN, C.E., and BOUTILLIER, R. (1982) On the evolution of rifted continental margins; comparisons of models and observations for the Nova Scotia Margin. Geophysical Journal Royal Astronomical Society 70:667–715.

BOND, G.C., CHRISTIE-BLICK, N., KOMINZ, M.A., and DEVLIN, W.J. (1985) An early Cambrian rift to post-rift transition in the Cordillera of western North America. Nature 316:742–745.

BOND, G.C. and KOMINZ, M.A. (1984) Construction of tectonic subsidence curves for the early Paleozoic miogeocline, southern Canadian Rocky Mountains: Implications for subsidence mechanisms, age of breakup and crustal thinning. Geological Society America Bulletin 95:155–173.

BOND, G.C., KOMINZ, M.A., and DEVLIN, W.J. (1983) Thermal subsidence and eustasy in the lower Paleozoic miogeocline of western North America. Nature 306:775–779.

BOND, G.C., NICKESON, P.A., and KOMINZ, M.A. (1984) Breakup of a supercontinent between 625 Ma and 555 Ma: New evidence and implications for continental histories. Earth Planetary Science Letters 70:325–345.

BOTT, M.T.H. (1971) Evolution of young continental margins and formation of shelf basins. Tectonophysics 11:319–327.

BRADY, M.J. and KOEPNICK, R.B. (1979) A Middle Cambrian platform-to-basin transition, House Range, west of central Utah. Brigham Young University Geological Studies 26:1–7.

CHILINGARIAN, G.V. and WOLF, K.H. (1975) Compaction of Coarse-Grained Sediments, I. Amsterdam: Elsevier, 808 p.

CHILINGARIAN, G.V. and WOLF, K.H. (1976) Compaction of Coarse-Grained Sediments, II. Amsterdam: Elsevier, 808 p.

CHOW, N. and JAMES, N. (1987) Cambrian Grand Cycles: A northern Appalachian perspective. Geological Society America Bulletin 98:418–429.

COCHRAN, J. (1983) Effects of finite rifting times on the development of sedimentary basins. Earth Planetary Science Letters 66:289–302.

CONEY, P.J., JONES, D.L., and MONGER, J.W.H. (1980) Cordilleran suspect terranes. Nature 288:329–333.

COWIE, J.W. and CRIBB, S.J. (1978) The Cambrian System. In: Cohee, G.V., Glassner, M.F., and Hedberg, H.D. (eds) Contributions to the Geologic Time Scale. American Association Petroleum Geologists Studies Geology 6:355–362.

COWIE, J.W., RUSHTON, A.W.A., and STUBBLEFIELD, C.J. (1971) A Correlation of Cambrian Rocks in the British Isles. Geological Society London Special Report 2, 42 p.

DEISS, C. (1938) Cambrian formations and sections in part of the Cordilleran Trough. Geological Society America Bulletin 49:1067–1168.

DEMICCO, R.V. (1983) Wavy and lenticular-bedded carbonate ribbon rocks of the Upper Cambrian Conococheague limestone, central Appalachians. Journal Sedimentary Petrology 53:1121–1132.

FALVEY, D.A. (1974) The development of continental margins in plate tectonic theory. Australian Petroleum Exploration Association Journal 14:95–106.

FITCH, F.J., FORSTER, S.C., and MILLER, J.A. (1976) The dating of the Ordovician. In: Bassett, M.G. (ed) The Ordovician System. Proceedings Palaeontological Association Symposium. Cardiff, Wales: University Wales Press, pp. 15–27.

FORTEY, R.A. (1984) Global earlier Ordovician transgressions and regressions and their biological implications. In: Bruton, D.L. (ed) Aspects of the Ordovician System. Palaeontological Contributions, University Oslo, Norway: Universitetsforlaget, pp. 37–50.

FRITZ, W.H. (1975) Broad correlations of some lower and middle Cambrian strata in the North American Cordillera. Geological Survey Canada Paper 75–1, Part A, pp. 533–540.

FUCHTBAUER, H. (1974) Sediments and Sedimentary Rocks, Part II. New York: Halsted Press, 464 p.

GALE, N.H., BECKINSALE, R.D., and WADGE, A.J. (1979) A Rb-Sr whole rock isochron for the Stockdale rhyolite of the English Lake District and a revised mid-Palaeozoic time scale. Journal Geophysical Society London 136:235–242.

GALE, N.H., BECKINSALE, R.D., and WADGE, A.J. (1980) Discussion of a paper by McKerrow, Lambert and Chamberlain on the Ordovician, Silurian and Devonian time scales. Earth Planetary Science Letters 57:9–17.

GROTZINGER, J.P. (1986) Cyclicity and paleoenvironmental dynamics, Rocknest platform, northwest Canada. Geological Society America Bulletin 97:1208–1231.

HALLAM, A. (1984) Pre-Quaternary sea-level changes. Annual Reviews Earth Planetary Science 12:205–243.

HALLEY, R.B., HARRIS, P.M., and HINE, A.C. (1983) Bank margin. In: Scholle, P.A., Bebout, D.G., and Moore, C.H. (eds) Carbonate Depositional Environments. American Association Petroleum Geologists Memoir 33:463–506.

HARLAND, W.B. (1983) More time scales. Geological Magazine 120:393–400.

HARLAND, W.B., COX, A.V., LLOWELLYN, P.G., PICHON, C.A.G., SMITH, A.G., and WALTERS, R. (1982) A Geologic Time Scale. Cambridge University Press, 131 p.

HARLAND, W.B. and FRANCIS, E.H. (eds) (1971) The Phanerozoic Time-Scale—A Supplement. Special Publication 5, Geological Society London, 120 p.

HARLAND, W.B., SMITH A.G., and WILCOCK, B. (eds) (1964) The Phanerozoic Time Scale. Quarterly Journal Geological Society London, 1205 p.

HINE, A.C. and NEUMANN, A.C. (1977) Shallow carbonate-bank-margin growth and structure, Little Bahama Bank, Bahamas. American Association Petroleum Geologists Bulletin 61:376–406.

HINTZE, L.F. (1973) Geologic History of Utah. Brigham Young University Geology Studies 20, 181 p.

JAMES, N.P. (1984) Shallowing-upward sequences in carbonates. In: Walker, R.G. (ed) Facies Models, Second Edition. Ontario: Geological Association Canada Reprint Series 1, pp. 213–228.

JARVIS, G.T. and McKENZIE, D.P. (1980) Sedimentary basin formation with finite extension rates. Earth Planetary Science Letters 48:42–52.

JOHNSON, J.G., KLAPPER, G., and SANDBERG, C.A. (1985) Devonian eustatic fluctuations in Euramerica. Geological Society America Bulletin 96:567–587.

KEEN, C.E. (1979) Thermal history and subsidence of rifted continental margins—evidence from wells on the Nova Scotian and Labrador shelves. Canadian Journal Earth Sciences 15:505–522.

KEEN, C.E. (1982) The continental margins of eastern Canada: A review. In: Scrutton, R.A. (ed) Dynamics of Passive Margins. Washington, D.C.: American Geophysical Union, Geodynamics Series 6:45–58.

KEEN, C.E. and BARRETT, D.L. (1981) Thinned and subsided continental crust on the rifted margin of eastern Canada: Crustal structure, thermal evolution and subsidence history. Geophysical Journal Royal Astronomical Society 65:443–465.

KEPPER, J.C. (1972) Paleoenvironmental pattern in middle to lower Upper Cambrian interval in eastern Great Basin. American Association Petroleum Geologists Bulletin 56:503–527.

KING, B.P. (1949) The base of the Cambrian in the southern Appalachians. American Journal Science 247:513–530.

KOEPNICK, R.B. (1976) Depositional history of the upper Dresbachian-lower Franconian (Upper Cambrian) Pterocephaliid biomere from west central Utah. Brigham Young University Geology Studies, 23:123–138.

KOMINZ, M.A. (1984) Oceanic ridge volumes and sea level change—an error analysis. In: Schlee, J. (ed) Interregional Unconformities and Hydrocarbon Accumulation. American Association Petroleum Geologists Memoir 36, pp. 109–127.

LANDING, E. (1981) Conodont Biostratigraphy and Thermal Color Alteration Indices of the Upper St. Charles and Lower Garden City Formations, Bear River Range, Northern Utah and Southeastern Idaho. United States Geological Survey Open File Report 81–740, 25 p.

LANDING, E., TAYLOR, M.E., and ERDTMANN, B.D. (1978) Correlation of the Cambrian-Ordovician boundary between the Acado-Baltic and North American faunal provinces. Geology 6:75–78.

LOCHMAN-BALK, C. (1970) Upper Cambrian faunal patterns on the craton. Geological Society America Bulletin 81:3197–3224.

LOHMANN, K.C. (1976) Lower Dresbachian (Upper Cambrian) platform-to-basin transition in eastern Nevada and western Utah: An evaluation through lithologic cycle correlation. In: Robison, R.A. and Rowell, A.J. (eds), Paleontology and Depositional Environments: Cambrian of Western North America, Brigham Young University Geology Studies 23, Part 2:111–122.

MAGARA, K. (1980) Comparison of porosity-depth relationships of shale and sandstone. Journal Petroleum Geology 3:175–185.

MARKELLO, J.R. and READ, J.F. (1981) Carbonate ramp-to-deeper shale shelf transitions of an upper Cambrian intrashelf basin, Nolichucky Formation, southwest Virginia Appalachians. Sedimentology 28:573–597.

MARKELLO, J.R. and READ, J.F. (1982) Upper Cambrian intrashelf basin, Nolichucky Formation, southwest Virginia Appalachians. American Association Petroleum Geologists Bulletin 66:860–878.

MAXEY, G.B. (1958) Lower and middle Cambrian stratigraphy in northern Utah and southeastern Idaho. Geological Society of America Bulletin 69:647–688.

McKENZIE, D.P. (1978) Some remarks on the development of sedimentary basins. Earth Planetary Science Letters 40:25–32.

McKERROW, W.S., LAMBERT, R.S.T.J., and CHAMBERLAIN, V.E. (1980) The Ordovician, Silurian and Devonian time scales. Earth Planetary Science Letters, 51:1–8.

MONGER, J.W.H. and PRICE, R.A. (1979) Geodynamic evolution of the Canadian Cordillera—progress and problems. Canadian Journal Earth Sciences 16:770–791.

MONGER, J.W.H., PRICE, R.A., and TEMPELMAN-KLUIT, D.J. (1982) Tectonic accretion and the origin of the two major metamorphic and plutonic welts in the Canadian Cordillera. Geology 10:70–76.

MOUNT, J.F. and ROWLAND, S.M. (1981) Grand Cycle A (lower Cambrian) of the southern Great Basin: A product of differential rates of sea-level rise. In: Taylor, M.E. (ed) Short Papers for the Second International Symposium on the Cambrian System. United States Geological Survey Open File Report 81–743, pp. 143–146.

MULLINS, H.T., NEUMANN, A.C., WILBER, R.J., and BOARDMAN, M.R. (1980) Nodular carbonate sediment on Bahamian slopes: Possible precursors to nodular limestones. Journal Sedimentary Petrology 50:117–131.

NEUGEBAUER, H.J. and SPOHN, T. (1982) Metastable phase transitions and progressive decline of gravitational energy: Aspects of Atlantic type margin dynamics. In: Scrutton, R.A. (ed) Dynamics of Passive Margins, American Geophysical Union Geodynamics Series 6:166–183.

NEUMANN, A.C. and LAND, L.S. (1975) Lime mud deposition and calcareous algae in the Bight of Abaco, Bahamas: A budget. Journal Sedimentary Petrology 45:763–786.

ODIN, G.S., GALE, N.H., AUVRAY, B., BIELSKI, M., DORE, F., LANCELOT, J.R., and PASTEELS, P. (1983) Numerical dating of Precambrian-Cambrian boundary. Nature 301:21–23.

PALMER, A.R. (1960) Some aspects of the early Upper Cambrian stratigraphy of White Pine County, Nevada and vicinity. Intermountain Association of Petroleum Geologists, Eleventh Annual Field Conference, pp. 53–58.

PALMER, A.R. (1971a) The Cambrian of the Great Basin and adjacent areas, western United States. In: Holland, C.H. (ed) Cambrian of the New World, London: Wiley-Interscience, pp. 1–78.

PALMER, A.R. (1971b) The Cambrian of the Appalachian and eastern New England regions, eastern United States. In: Holland, C.H. (ed) Cambrian of the New World, London: Wiley-Interscience, pp. 169–217.

PALMER, A.R. (1981a) Subdivision of the Sauk sequence. In: Taylor, M.E. (ed) Short Papers for the Second International Symposium on the Cambrian System, United States Geological Survey Open File Report 81–743, pp. 160–163.

PALMER, A.R. (1981b) On the correlatibility of Grand Cycle tops. In: Taylor, M.E. (ed) Short Papers for the Second International Symposium on the Cambrian System, United States Geological Survey Open File Report 81–743, pp. 156–157.

PALMER, A.R. (1983) The decade of North American geology 1983 geologic time scale. Geology, 11:503–504.

PALMER, A.R. and HALLEY, R.B. (1979) Physical Stratigraphy and Trilobite Biostratigraphy of the Carrara Formation (Lower and Middle Cambrian) in the Southern Great Basin. United States Geological Survey Professional Paper 1047, 131 p.

PARKINSON, N. and SUMMERHAYES, C. (1985) Synchronous global sequence boundaries. American Association Petroleum Geologists Bulletin 69:685–687.

PARSONS, B. and SCLATER, J.G. (1977) An analysis of the variation of ocean floor bathymetry and heat flow with age. Journal Geophysical Research 82:803–828.

PFEIL, R.W. and READ, J.F. (1980) Cambrian carbonate platform margin facies, Shady Dolomite, southwestern Virginia, U.S.A. Journal Sedimentary Petrology 50:91–116.

PITMAN, W.C., III (1978) Relationship between eustasy and stratigraphic sequences of passive margins. Geological Society America Bulletin 89:1389–1403.

PRATT, B.R. and JAMES, N.P. (1982) Cryptalgal-metazoan bioherms of early Ordovician age in the St. George Group, western Newfoundland. Sedimentology 33:543–569.

READ, J.F. (in press) Evolution of Cambro-Ordovician passive margin, U.S. Appalachians: Decade of North American Geology Synthesis, volume. Appalachian-Ouichitas, Geological Society of America.

READ, J.F., GROTZINGER, J.P., BOVA, J.A., and KOERSCHNER, W.F. (1986) Models for generation of carbonate cycles. Geology 14:107–110.

RIGO, R.J. (1968) Middle and Upper Cambrian stratigraphy in the autochthon and allochthon of northern Utah: Brigham Young University Geology Studies, 15:31–66.

ROBISON, R.A. (1964) Upper Middle Cambrian stratigraphy of western Utah. Geological Society American Bulletin 75:995–1010.

ROBISON, R.A., ROSOVA, A.V., ROWELL, A.J., and FLETCHER, T.P. (1977) Cambrian boundaries and divisions. Lethaia 10:257–262.

RODGERS, J. and KENT, D.F. (1948) Stratigraphic section at Lee Valley, Hawkins County, Tennessee. State Tennessee Department Conservation, Division Geology 55, 19 p.

ROSS, R.J., JR. (1977) Ordovician paleogeography of the western United States. In: Stewart, J.H., Stevens, C.H., and Fritsche, A.E. (eds) Paleozoic Paleogeography of the Western United States. Pacific Coast Paleogeography Symposium 1, Society Economic Paleontologists Mineralogists, pp. 19–38.

ROSS, C.A. and ROSS, J.R.P. (1985). Late Paleozoic depositional sequences and synchronous and worldwide. Geology 13:194–197.

ROSS, R.J., JR., NAESER, C.W., and LAMBERT, R.S. (1978) Ordovician Geochronology. In: Cohee, G.V., Glassner, M.F., and Hedberg, H.D. (eds) Contributions to the Geologic Time Scale. American Association Petroleum Geologists Studies Geology 6:347–354.

ROSS, R.J., JR., NAESER, C.W., IZETT, G.A., OBRADOVICH, J.D., BASSETT, M.G., HUGHES,

C.P., COCKS, L.R.M., DEAN, W.T., INGHAM, J.K., JENKINS, C.J., RICKARDS, R.B., SHELDON, P.R., TOGHILL, P., WHITTINGTON, H.B., and ZALASIE-WICZ, J. (1982) Fission-track dating of British Ordovician and Silurian stratotypes. Geological Magazine 119:135–153.

ROYDEN, L. and KEEN, C.E. (1980) Rifting process and thermal evolution of the continental margin of eastern Canada determined from subsidence curves. Earth Planetary Science Letters 51:343–361.

ROYDEN, L., SCLATER, J.G., and VON HERZEN, R.P. (1980) Continental margin subsidence and heat flow: Important parameters in formation of petroleum hydrocarbons. American Association Petroleum Geologists Bulletin 64:173–187.

SCHLAGER, W. and GINSBURG, R.N. (1981) Bahama carbonate platforms—the deep and the past. Marine Geology 44:1–24.

SCHOLLE, P.A. (1977) Geological studies on the COST No. B-2 well, United States Mid-Atlantic outer shelf area. In: Scholle, P.A. (ed) Geological Studies on the COST B-2 Well United States Mid-Atlantic Outer Continental Shelf Area. United States Geological Survey Circular 750.

SLEEP, N.H. (1971) Thermal effects of the formation of Atlantic continental margins by continental breakup. Geophysical Journal Royal Astronomical Society 24:325–350.

SLOSS, L.L. (1963) Sequences in the cratonic interior of North America. Geologic Society America Bulletin 74:93–113.

SMITH, M.A., AMATO, R.V., FURBISH, M.A., PERT, D.M., NELSON, M.E., HENDRIX, J.S., TAMM, L.C., WOOD, G., JR., and SHAW, D.R. (1976) Geological and Operational Summary, COST No. B-2 Well, Baltimore Canyon Trough Area, Mid-Atlantic Ocean. United States Geological Survey Open file Report 76–774.

STECKLER, M.S. (1981) Thermal and Mechanical Evolution of Atlantic-Type Margins (unpublished Ph.D. thesis) New York, Columbia University, 261 p.

STECKLER, M.S. and WATTS, A.B. (1978) Subsidence of the Atlantic-type continental margins off New York. Earth Planetary Science Letters 41:1–13.

STECKLER, M.S. and WATTS, A.B. (1982) Subsidence history and tectonic evolution of Atlantic-type continental margins. In: Scrutton, R.A. (ed) Dynamics of Passive Margins. American Geophysical Union Geodynamics Series 6:184–196.

STEIGER, R.H. and JAGER, E. (1977) Subcommission on geochronology: Convention on the use of decay constants in geo- and cosmochemistry. Earth Planetary Science Letters 36:359–362.

SWEET, W.C. and BERGSTROM, S.M. (1976) Conodont biostratigraphy of the Middle and Upper Ordovician of the United States midcontinent. In: Bassett, M.G. (ed) The Ordovician System: Proceedings Paleontological Association Symposium. Cardiff: University of Wales Press, pp. 121–151.

TEMPELMEN-KLUIT, D.J. (1979) Transported Cataclasite, Ophiolite and Granodiorite in Yukon: Evidence of Arc-Continent Collision. Geological Survey Canada Paper 79–14, 27 p.

THORNE, J. and WATTS, A.B. (1984) Seismic reflectors and unconformities at passive continental margins. Nature 311:365–368.

VAIL, P.R., HARDENBOL, J., and TODD, R.G. (1984) Jurassic unconformities, chronostratigraphy, and sea-level changes from seismic stratigraphy and biostratigraphy. In: Schlee, J.S. (ed) Interregional Unconformities and Hydrocarbon Accumulation. American Association Petroleum Geologists Memoir 36:129–144.

VAIL, P.R., MITCHUM, R.M., and THOMPSON, S. III. (1977) Seismic stratigraphy and global changes of sea level, Part 4: Global cycles of relative changes of sea level. In: Payton, C.E. (ed) Seismic Stratigraphy—Applications to Hydrocarbon Exploration. American Association Petroleum Geologists Memoir 26:83–97.

VAN EYSINGA, F.W.B. (1978) A geological time table, 3rd edition. Amsterdam: Elsevier.

WALKER, K.R., SHANMUGAM, G., and RUPPEL, S.C. (1983) A model for carbonate to terrigenous clastic sequences. Geological Society America Bulletin 94:700–712.

WATTS, A.B. (1981) The U.S. Atlantic Continental Margin: Subsidence history, crustal structure and thermal evolution. In: Bally, A.W., Watts, A.B., Grow, J.A., Manspeizer, W., Bernoulli, D., Schreiber, C., and Hunt J.M. (eds) Geology of Passive Continental Margins: History, Structure and Sedimentologic Record. American Association Petroleum Geologists Education Course Note Series 19, chapter 2, 75 p.

WATTS, A.B. (1982) Tectonic subsidence, flexure and global changes of sea level. Nature 297:469–474.

WATTS, A.B. and STECKLER, M.S. (1981) Subsidence and tectonics of Atlantic-type continental margins. In: Colloque C3 Geologie Morges Continentales, Oceanologica Acta, pp. 143–153.

WILLIAMS, A. (1976) Plate tectonics and biofacies evolution as factors in Ordovician correlation. In: Bassett, M.G. (ed) The Ordovician System: Proceedings of a Palaeontological Association Symposium. Cardiff: University of Wales Press, pp. 29–66.

WILSON, J.E. and JORDAN, C. (1983) Middle Shelf. In: Scholle, P.A., Bebout, D.G., and Moore, C.H. (eds) Carbonate Depositional Environments. American Association Petroleum Geologists Memoir 33:297–344.

8
Coal Correlations and Intrabasinal Subsidence: A New Analytical Perspective

W. NEMEC

Abstract

The evolving patterns of differential subsidence in coal-basin interiors appear to exert an important control on the stratigraphic patterns of peat (coal) development. This study demonstrates how these resultant stratigraphic patterns, when deciphered from the rock record on the basis of correlatable coal beds (and possibly other marker horizons), can then be used to recognize the underlying patterns of differential subsidence. Conventional stratigraphic methods (correlation graphs) are adapted for this purpose, and small-scale, spatial and temporal variations in basin-floor subsidence are identified, quantified, and mapped. Examples from different coal basins are presented, with emphasis on the tectonic settings. The analysis indicates that the local-scale differential subsidence, accommodating differential sediment aggradation in a basin, tends to be constant for appreciable geological time periods ($>10^5$ yr) and is subject to sudden, discrete changes (or "disturbances"). A direct relationship also appears to exist between the magnitude and spatial/temporal frequency of such disturbances and the actual tectonic regime of a basin. The methodological approach outlined in this study has several important ramifications from the practical point of view of basin analysis. The correlation graphs can be used to correlate coeval sedimentary facies, formed simultaneously under different environmental conditions in the basin, and to evaluate the origin of particular types of intrabasinal facies variations.

Introduction

Coal basins are sites of pronounced sediment accumulation, where clastic sedimentation, typically deltaic or alluvial-lacustrine, combines with peat-forming processes to fill an evolving depression in the Earth's crust. Economically important coal measures occur in various intracratonic and craton-margin basins of late Paleozoic, Mesozoic, and Cenozoic age.

Coal-bearing basins are known to form in a variety of tectonic settings, ranging from continental rifts/pull-apart features and passive-margin basins, where extensional processes predominate, to foreland basins associated with compressional regimes. Many other coal basins owe their development to broad thermal sagging of the continental crust, with only subsidiary faulting.

In general, the basins that originate in any of these settings may be coal-bearing or virtually coal-free, even when apparently filled by similar fluvio-deltaic processes and containing a similar range of clastic facies. Suitable biological and physico-chemical conditions for peat formation (see recent reviews by Teichmüller and Teichmüller 1982; McCabe 1984) are not obviously specific to any particular tectonic setting. Basin tectonism, however, affects the sedimentary environment and the local base level/shoreline position, whose control on the actual development and extent of the peat-forming mires is usually profound (Williams and Bragonier 1974; McCabe 1984, and other discussions in same volume). Tectonic subsidence then controls also, in part, the length of time that the peat-forming processes operate at the depositional surface before burial.

In this study, the subsidence of coal-bearing basins is considered, and the main focus is on how correlatable coals can be used to analyze differential *intrabasinal* subsidence. Conventional stratigraphic methods are adapted for this purpose, and small-scale, spatial and temporal variations in subsidence are identified, quantified, and mapped. Other important ramifications of this analytical approach are discussed from the practical point of view of basin analysis. Examples from different coal basins are presented, with emphasis on their tectonic settings.

Coal basins were chosen because they are an important, if not almost unique, source of data. They develop in a broad range of tectonic settings, and most importantly, they contain correlative, often almost basin-wide, stratigraphic markers — the coal beds.

Subsidence and Coal

The methodological part of the study is preceded by a discussion of relevant aspects of basin subsidence and the stratigraphic significance of coal beds. The discussion provides a background for the aspects of basin evolution that are subsequently analyzed, and for the methodological reasoning and analytical approach presented.

Basin Subsidence

Sedimentary basins are the main source of information on the styles and mechanisms of crustal subsidence in cratonic areas (e.g., Bott 1976, and other papers in same volume; McKenzie 1978). The subsidence of the continental crust in any area is a response to thermal and/or mechanical stresses, and the evolving depression fills with sediment through a complex interplay of surface and near-surface processes. Sediment loading eventually becomes significant, although in cratonic settings its main role is that of amplifying subsidence caused by the primary mechanisms (Bott 1976). It is then the sedimentary fill of a basin that provides the ultimate record of subsidence. In a coal basin, this sedimentary record is further punctuated by the development of extensive peat (coal) horizons, making such basins particularly significant from a stratigraphic point of view (as further discussed below).

A sedimentary basin may thus be regarded as the cumulative product of mechanical and thermal stresses operating within the sub-basinal crust, the processes of sedimentation and diagenesis that operate within the basin, and the factors that operate externally upon the basin, such as climate and eustatic sea-level changes. A major focus of modern research on sedimentary basins has been to identify and isolate these different formative mechanisms, so

that the specific nature of each and its actual control on the basin history can be understood.

Crustal subsidence has always been of major interest in basin research, and the paramount role that subsidence plays, or might potentially play, in basin evolution has been particularly well revealed by the recent progress in two-dimensional synthetic stratigraphic modelling and related thermo-mechanical simulations (e.g., Jarvis and McKenzie 1980; Beaumont 1981; Beaumont et al. 1982; Turcotte and Schubert 1982; Watts et al. 1982). This new type of approach, by integrating geological and geophysical insight, provides an excellent methodology for understanding of the long-term evolution and broad geometry of sedimentary basins, and of some large-scale stratigraphic patterns of the basin-fill successions.

Given this emerging breakthrough in large-scale basin modelling, investigations of secondary, intra-basinal mechanisms capable of modifying a basin's overall subsidence pattern have been generally ignored. Consequently, relatively little is known about *intrabasinal* differential subsidence, occurring on a short-term and more local scale.

A broad view on basin subsidence is obviously crucial, and is used as a major, fundamental framework in basin analysis. However, this approach alone fails to explain most of the small-scale stratigraphic complexities observed in basin interiors (except, perhaps, in some very small basins or narrow rifts). At least some of these stratigraphic variations, as observed on a formation scale, are likely to be related to intrabasinal differential subsidence, whose actual impact on sedimentation patterns may be more critical than commonly inferred. For example, it has been demonstrated from a variety of coal basins that there is a close statistical relationship between the number (and thickness) of coal-measure "cycles" and net local subsidence (Duff and Walton 1962; Duff 1967; Read and Dean 1967, 1968, 1976; Read et al. 1971; Johnson and Cook 1973; Nemec and Ćmiel 1979). It is, therefore, rather surprising that so little analytical consideration has been given to the small-scale, spatial and temporal variations in subsidence *within* an evolving basin, even though any such variation is potentially recorded as a stratigraphic variation in the basin's sedimentary fill (e.g., Williams and Bragonier 1974).

The shallower, sub-basinal crust is thought to deform in a brittle fashion when responding to the more ductile deformation of the deeper crust. The anisotropic continental crust is then most likely to subside under the control of pre-existing and newly created fractures, no matter whether the general subsidence is driven by mechanical or thermal stresses. Sub-basinal deformation would almost certainly affect the basin-fill strata, in terms of synsedimentary faulting or local warping, and be propagated directly to the depositional surface. Such local-scale deformation may thus play an important role by adding a differential component, or "noise," to the overall, smoothly varying subsidence of a basin (e.g., Adams 1965; Shelton 1968; Alexander 1969; Williams and Bragonier 1974; Busch 1975; Nemec *et al.* 1984). Differential subsidence in basin interiors is the main topic of the present study.

Significance of Coal Horizons

The presence of thick, extensive coal seams implies certain physical conditions in the sedimentary basin, as recently discussed and emphasized by McLean and Jerzykiewicz (1978). A bed of bituminous coal with a thickness of 2 m can be considered as representing roughly 12 m of original peat (Teichmüller and Teichmüller 1982). Such thickness might require approximately 4,000–12,000 years of peat accumulation, assuming accumulation rates of 1–3 $m/10^3$ yr as representative of a broad range of reasonably favorable settings (Teichmüller and Teichmüller 1982; McCabe 1984). Otherwise, even more time might be required, and the estimates by Teichmüller and Teichmüller (1982) indicate as much as 12,000–18,000 years. Although seams thinner than 2 m are much more common, there are coal measures that contain seams up to 10 m thick or even much thicker. At a rate of 1 $m/10^3$ yr, 60,000 years would be required to accumulate 60 m of peat (10 m of bituminous coal), or 20,000 years at 3 $m/10^3$ yr rate. Therefore, large portions of the depositional surface must have been free of fluvial or marine sedimentation for periods of thousands of years, or even for a few tens of thousand years. These are relatively long periods. Accordingly, McLean and Jerzykiewicz (1978) compared an average depositional rate (stratigraphic thickness divided by geological time interval), estimated for the Cretaceous coal-bearing alluvial succession they studied in Alberta, with the short-term accumulation rates of clastics and the temporal frequencies of channel avulsion and major floods reported from alluvial environments. They concluded that there must have been long periods of nondeposition of

clastics, and that the processes of peat accumulation had the tendency to fill those hiatuses. The non-deposition periods would simply favor peat accumulation, if the biological and groundwater conditions were favorable.

The same conclusion is probably valid for any coal basin (Nemec 1984), and for sedimentary basins in general (see relevant discussions of the episodic aspect of sedimentation by Ager 1973 and Sadler 1981). Hiatuses in cratonic basins may be represented by various condensed sequences, such as paleosol or calcrete horizons, or may remain unfilled and be hidden in the sedimentary record.

A basin, however, does not cease to subside during periods of extensive plant growth and peat accumulation (McLean and Jerzykiewicz 1978). The accumulation and preservation of peat requires that the water table be sufficiently high to cover the decaying vegetation and prevent its oxidation, but also sufficiently low that living vegetation is not drowned. Such a balance can be maintained if the depositional surface is gradually subsiding. Given the long time periods involved (see above), the subsidence of the depositional surface ensures that: 1) the groundwater table is gradually rising (i.e., stays roughly constant in the area of aggrading peat mire); 2) the aggrading peat is brought below an erosional base; and 3) active fluvial/distributary channels are not encouraged to shift their courses, and hence stay away from the mires.

Therefore, the rate of peat aggradation might be roughly equal to the current rate of subsidence (including compaction). Following the reasoning of McLean and Jerzykiewicz (1978), the 12 m of peat required to form 2 m of bituminous coal (Teichmüller and Teichmüller 1982) would require 4.5 m of subsidence if the water table remains approximately constant, and if there is a continuous reduction in the volume of peat during deposition from 1:1 at the surface to 3:1 at the lowest layer (Teichmüller and Teichmüller 1982). With 4.5 m of subsidence over 4,000–12,000 years, the accumulation rates assumed above (1–3 $m/10^3$ yr) imply subsidence rates of 0.4–1.1 $m/10^3$ yr. These are moderate rates, but still several orders greater than the average subsidence rates reported from cratonic areas, where average rates of 0.005–0.03 $m/10^3$ yr prevail and the highest average rate scarcely exceeds 0.05 $m/10^3$ yr (Fischer 1969).

The development of peats, their thicknesses, location, and areal extent may thus all be controlled by intrabasinal differential subsidence. Although the rates of peat growth are probably compatible with any rates of geological subsidence, the peat-forming process cannot withstand invasions of fluvial processes or marine incursions. When an area of peat accumulation begins to subside more rapidly than the adjoining areas occupied by other environments, this differential subsidence necessarily attracts river channels, or causes inundation by marine or lacustrine water. On the other hand, intrabasinal differential subsidence may be crucial in keeping active fluvial tracts in particular areas for appreciable time periods, so that the accumulation of thick, extensive peat elsewhere in the basin is made possible. However, if an area adjacent to peat-forming mire begins to subside too rapidly, the water table in the mire may be critically lowered. The peat growth then ceases, and degradation of an upper part of the peat may begin; this may even lead to a complete destruction of the peat deposit. If peat is to be preserved in the sedimentary record, the peat-forming mire must be drowned or be rapidly buried by clastic sediment (McCabe 1984).

The evolving patterns of differential subsidence in a basin interior are, therefore, of primary importance in controlling spatial, and thus also stratigraphic, patterns of peat (coal) development. The present study demonstrates how the resultant stratigraphic patterns, when deciphered from the rock record, can be used to recognize and quantify the underlying patterns of differential subsidence.

Another important stratigraphic aspect of coal (peat) horizons is their lateral development (or extent) versus time interval required. From a variety of field evidence, it has long been realized that coal seams, as well as the associated marine layers present in many coal measures, are essentially chronostratigraphic units (e.g., Trueman 1947, p. 54; Cross 1952, p. 36; Weller 1956, p. 34). Of course, "it does not follow . . . that the beginning or ending of deposition of any [such] member was exactly synchronous throughout its extent" (Weller 1956, p. 34). The onset of peat-forming conditions would be local, and the marsh/swamp vegetation would gradually expand with time over a larger area; a similar tendency pertains to the termination of peat growth (see McCabe 1984).

Given the long time periods involved in the development of peat layers, it is however almost certain that the horizontal rates of peat accumulation (lateral expansion of mire vegetation) are substantially higher than the actual vertical accumulation rates. It does not take an expedition to a tropical forest to notice that the plants are able to expand their territory at rates may orders of magnitude higher than the $1-3$ m/10^3 yr estimated above for vertical peat aggradation. From an analysis of clastic sedimentation rates, Schwarzacher (1975) concluded that a lithological boundary may migrate 10 km horizontally in a rapid-sedimentation setting, or 100 km in a cratonic, shelf-type setting, while only 1 cm is deposited in the vertical succession; this means lateral rates are possibly 10^6-10^7 times higher than vertical rates. A somewhat similar ratio does not seem unrealistic, as an extreme, also for peat accumulation, though ratios as low as 10^3 or 10^2 can be anticipated in some settings (e.g., Bloom 1964).

The boundaries of a coal (peat) bed most probably are diachronous to some extent, but the bulk of the bed is not necessarily so. This study demonstrates that coal seams can generally be considered as chronostratigraphic units, as their inherent degree of diachroneity appears to be geologically insignificant.

Graphic Correlation Method in Basin Analysis

The quantitative, graphic technique of stratigraphic correlations used in this study is based on the concept of geochronological dating that was originally proposed and applied in the field of biostratigraphy in a benchmark book by Shaw (1964). The concept involves correlation of a particular type of biological time event (appearance and disappearance of fossil taxa) recorded in fossiliferous sedimentary successions. In its present application to coal-bearing profiles, the methodological approach of Shaw becomes much more straightforward and technically simpler than originally designed, hence it is necessary first to review briefly the original concept and its previous use.

Shaw's Method

Shaw (1964) introduced the use of a conventional two-axis graph, with the axes representing a pair of fossiliferous stratigraphic profiles, as a means of determining time equivalence between the profiles. The observed lowest and highest stratigraphic occurrences of fossil taxa (i.e., the taxon-range zones) were used as indicators of geological time. When projected on a graph, the corresponding points of two compared profiles form what Shaw has termed the Line of Correlation (LOC) and recommended as an important tool for chronostratigraphic correlation ("point-to-point" matching) of fossiliferous profiles. With the use of an established LOC, any point in one profile can be matched, at least approximately, with the correct (coeval) point in another profile. The LOC has been further interpreted by Shaw in terms of a relative "rock accumulation" rate for a given pair of profiles. (Shaw correctly considers "rock" rather than "sediment" accumulation, as the stratigraphic record naturally bears the effect of compaction; this difference, however, tends to be somewhat ignored in later discussions by other authors.)

Shaw's graphic method was initially accorded with little practical interest, most probably because it required fossil material of a quality that was neither commonly available nor easy to obtain. Only more recently have other biostratigraphers applied the method to fossil data to achieve reasonable and informative results (Miller 1977a,b; Murphy and Edwards 1977; Hay and Southam 1978; Sweet 1979a,b; Hazel et al. 1980; Murphy and Berry 1983). The reliability and accuracy of Shaw's method have been tested by Edwards (1984, 1985), using a computer simulation on hypothetical data sets, and by Rubel and Pak (1984) in terms of the formal logic and stochastic theory of biostratigraphic correlations. Edwards raised no formal objections to the method, but emphasized (1984) that "the method is only as good as the available [fossil] data." Rubel and Pak (1984) concluded that "the concept of a time line interpreted in terms of relative sedimentation rates, as developed by Shaw (1964), is one of the best tools for dating [because it reveals] the time relations between geologic sections."

In the field of sedimentary basin analysis, Miall (1984, pp. 102–105) recently reviewed Shaw's method and recommended it as a useful biostratigraphic technique for basinal studies. Miall states (1984, p. 105): "The value of the graphic method for correlating sections with highly variable lithofacies and no marker beds is obvious, and it is perhaps surprising that the method is not more widely used. An important difference between this method and conventional [fossil] zoning schemes is that zoning methods provide little more than an ordinal level of correlation, whereas the graphic method provides interval data."

The graphic technique in the present approach is rather straightforward, and no "composite standards" (Shaw 1964) will need to be constructed. Hence, the reader should refer to Shaw's book, or to the reviews by Miller (1977b) and Miall (1984), for complete details of the original biostratigraphic method.

Present Approach

The development of a peat (coal) layer is considered here as a biological "event," where the appearance and extinction of peat-forming vegetation punctuate physical time. The boundaries of a coal layer are then natural datum planes. These planes, however, will tend to coalesce in coal-bearing stratigraphic successions, simply because most coal seams are very thin compared to the intervening clastics. A coal layer then normally needs to be considered as one "thick" datum plane for the purpose of stratigraphic correlations (but with true thickness shown wherever graphically distinguishable). In other words, the appearance, growth, and extinction of peat-forming vegetation punctuate time essentially as a single event.

An observed stratigraphic sequence of coal layers (datum planes), taken independently of the temporal increments between them, forms an ordinal time scale of biological events. Such a sequence of events may be disturbed to some extent by sedimentary processes, but will never be fully lost in the rock record (Edwards 1979).

The clastic sediments between the coal layers may thus be considered as representing time-controlled, aggradational increments of the rock succession that are bounded by the datum planes. When two profiles are compared and their time-equivalent increments matched, the relative rock-accumulation rates (interval mean ratios) can readily be estimated. The scales are ordinal, hence only relative rates can be inferred. Inasmuch as more sediment normally accumulates and is preserved in faster subsiding areas, the relative rate of accumulation may be taken essentially as the relative rate of subsidence. Subsidence is considered as the process accommodating sediment vertical accumulation, but it does not follow, of course, that the space created by subsidence at a particular rate is necessarily filled by the sediment instantaneously, at strictly the same rate, or at a constant rate. It is the interval (interseam) mean rate that is considered.

The term "subsidence" should formally be understood here as the subsidence of a depositional surface (i.e., "bulk" subsidence, including both basement subsidence and compactional effects). This is simply because the rock accumulation rate (Shaw 1964) also includes sediment compaction. Such definition, however, should not raise any serious problem to the geologist. Firstly, the thickness of the rocks is what geologists usually measure and refer to, and is what the existing, quantitative sedimentation-stratigraphic concepts have been founded upon (e.g., Krumbein and Graybill 1965; Vistelius 1967; Merriam 1972; Schwarzacher 1975). Secondly, and more importantly, it will be shown later in this paper that the effect of compaction in coal measures is usually negligible as far as the relative rates are concerned, and hence the entire presented analysis pertains essentially to the relative rates of sediment accumulation and relative rates of sub-basinal (or basement-induced) subsidence.

The Technique and Results

The analytical technique requires that, for a given pair of profiles, a number of coal beds have already been correlated with confidence. This requirement is not a problem in most coal measures that have been mined extensively and/or where sufficiently detailed research has been carried out. Other correlatable "marker beds," such as tonstein/tuff layers or marine horizons, can also be used. The method relies on correct correlations, hence its result should be considered only as reliable as the available correlation data. However, the method itself may aid correlation, particularly where the miscorrelated or doubtfully correlated seams are minor and accompanied closely by correctly matched horizons.

The correlated beds are plotted on a two-axis graph (Fig. 8.1, left-hand part), according to their heights above an assumed zero datum in the compared profiles. This datum may either correspond to a lithostratigraphic boundary (e.g., base of a formation) or be taken arbitrarily in the profiles.

The projection points on a graph tend to be distributed along a straight line, and this often appears to be variously segmented. The line, or a particular segment, can usually be directly drawn in the graph (Fig. 8.1, left-hand part); a minimum of three points is used in this study. If the projection points are somewhat scattered, the line can possibly be fitted statistically by means of least-squares linear regression (Shaw 1964; Hay and Southam 1978). This latter solution, however, may be less appropriate

for paleontological data (see Miall 1984, pp. 103–104), although the scatter of fossil projection points is normally much greater than in the present case of coal beds.

An established line on a correlation graph, the LOC, can then be used to make a number of inferences that appear to be of prime importance to basin analysis. The main application of LOCs and the stratigraphic information they provide can be summarized as follows (Fig. 8.1):

1. An established LOC (Fig. 8.1, bottom left) provides approximate point-to-point correlation of the compared profiles. Any point in one profile can be projected, via the LOC, onto the other profile, and hence correlated with an approximate time-equivalent point (Shaw 1964). This offers the attractive possibility of a detailed reconstruction of the facies distribution, in a basin, along "time-lines" (in terms of cross sections) or "time-surfaces" (in terms of maps). This particular application of LOCs, though not exploited further in the present study, was also suggested by Miall (1984) in the context of Shaw's biostratigraphic method.

2. A straight LOC can be considered, for convenience, in terms of its linear equation (see Fig. 8.1A), $Y = \beta X + \alpha$, where β is the line's gradient and α is the ordinate of the line's intersection point with the Y axis. The α value, in units of thickness, is the difference between the stratigraphic position of the zero datum in Profiles X and Y, respectively. For example, $\alpha = 5$ means that the zero datum taken in Profile X is equivalent to a point 5 m above the zero datum in Profile Y; similarly, a negative α means below. In the case of a lithostratigraphic boundary (e.g., zero datum taken at a formation base), the α-value may serve as a quantitative estimate of the boundary's diachroneity.

3. A straight LOC (or segment) implies a constant ratio between the accumulation rates, or subsidence rates, in the respective intervals of the two profiles. This ratio, or the *relative rate*, is given by the LOC gradient (β). For example (Fig. 8.1D), $\beta = 1.4$ means that 1 m accumulation in Profile X corresponds to 1.4 m accumulation in Profile Y, in terms of interval mean rates. Graphs B and C in Figure 8.1 show equal rates and a lower rate in Profile Y, respectively.

4. A projection point that is located clearly off the LOC may suggest coal-seam diachroneity. For

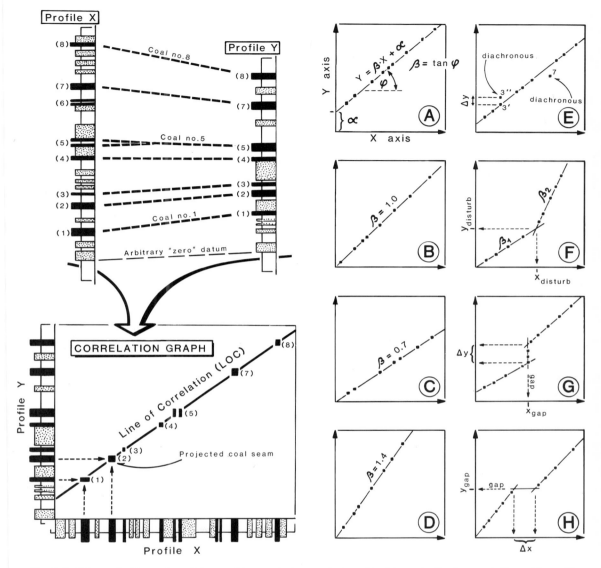

Fig. 8.1. Schematic illustration of the graphic correlation method used in this study. Rock types indicated in profiles X and Y are: coal (black), sandstone (dotted pattern), and argillaceous strata (blank). Further explanation in text.

example, graph E (Fig. 8.1) contains two seams (no. 3″ and no. 7) that are diachronous. The single seam no. 3 in Profile X splits into two seams, labelled 3′ and 3″, in Profile Y, and this second, "rider" seam (3″) apparently is younger than the corresponding seam in Profile X. The amount of its diachroneity is represented by the interval Δy indicated in Profile Y (this would be, again, a quantitative estimate). Similarly, seam no. 7 is clearly "too high" (= significantly younger) in Profile X, or "too low" (= significantly older) in

Profile Y. In principle, however, profiles should be decompacted if a detailed study of coal diachroneity is to be carried out. One must be aware that the projection points may be offset to some extent due to differential compaction, not to mention coal-seam miscorrelation.

5. A segmented LOC (Fig. 8.1F) implies a respective change in the ratio of accumulation rates; in this example, from a lower (β_1) to a higher relative rate (β_2). Such changes in the relative rates will be referred to as *disturbances*. The stratigraphic

point of a disturbance can be readily identified in the profiles (as shown in Fig. 8.1F).

6. A vertical or horizontal segment of a truncated LOC reveals a stratigraphic gap in one of the profiles (Fig. 8.1G–H). In graph G, the thick interval Δy in Profile Y, comprising clastics and three coals, is stratigraphically equivalent to a single coal layer in Profile X; the gap is readily located in Profile X, and its size is given by Δy. It should be noted, however, that the gaps displayed by LOCs are only relative for a given pair of profiles, and no gap will be revealed if exactly the same stratigraphic interval is missing in both profiles.

The reader must also be aware that an apparent gap may be due to tectonic offset within a profile (i.e., when a fault is present within the profile and has significantly changed the stratigraphic thickness of this profile). In order to discriminate between a true stratigraphic gap, as due to nondeposition or erosion, and a secondary, fault-induced gap, it is usually necessary to re-examine a particular interval of the rock sequence. Cross-correlation, in terms of LOCs, with other adjacent profiles may be helpful as well.

It is also worth noting that a LOC segment in some cases is subvertical or subhorizontal, implying a great difference between the accumulation rates. This apparently outlines the scope for a more precise, quantitative definition of the terms "stratigraphic gap" and "condensed sequence" with respect to nonfossiliferous successions.

Application and Appraisal

Preliminary applications of this graphic method have been done to test and evaluate it (see also Nemec 1984). Examples are reviewed below. The selected coal-bearing successions differ in their stratigraphic ages, depositional environments, and geographic/tectonic settings. Despite these differences, the correlation graphs reveal well defined linear trends, and the LOCs, when fitted to the data, are easy to interpret in terms of the guide given in Figure 8.1. The linear patterns apparently are much more distinct than those normally displayed by fossil taxa (cf. Shaw 1964; and other applications reviewed earlier). Very few coal seams show geologically significant diachroneity, and the same pertains

to correlative marine horizons and some other marker units.

Review of Examples

Westphalian Measures of South Wales Basin

The Westphalian coal measures of the South Wales Basin, Great Britain, were deposited near the southern margin of the so-called Midland microcraton (Welsh Massif), in the evolving foreland basin created north of a rising Variscan fold belt (Owen 1964; Anderson and Owen 1968; Kelling 1986). The Silesian (Upper Carboniferous) succession in the basin was deposited mainly in deltaic to fluvial environments, and may be considered as a late syn-orogenic to post-orogenic, Variscan "external molasse."

The tectono-sedimentary evolution of the South Wales Basin is considered by Kelling (1986) in terms of a peripheral foreland basin. The Namurian and early Westphalian sedimentation (Ammanian, or pre-Pennant succession) is thought to have been controlled by an initial thrust-loading of the basin's southern flank, whereby there was peripheral upwarping and enhanced erosion of the foreland margin to the north (Welsh Massif). The late Westphalian sedimentation (Morganian, or Pennant succession) resulted from further loading by northward-advancing thrust nappes, with an increase in both basin subsidence and sediment supply from the nearby southerly source. This was accompanied by a concurrent northward retreat of the foreland's northern margin and led to a reversal of the basinal paleoslope and to a gross upward coarsening of the entire succession (see also Kelling 1964, 1968). The development of this foreland basin is thought to have been further modified by syn-depositional and post-depositional strike-slip faulting, whereby numerous pre-existing faults were subject to reactivation (Owen 1964; Anderson and Owen 1968; Kelling 1986; see also Leeder 1976).

The South Wales Coal Measures were thus deposited in a very active, compressional/transpressional tectonic setting, and the basin itself actually came into existence during Westphalian time (Anderson and Owen 1968, p. 93). The coal-measure succession contains numerous correlative coal seams and a number of marine horizons (or "bands"), all of which form important markers.

The graphic correlation method has been applied separately to three consecutive stratigraphic inter-

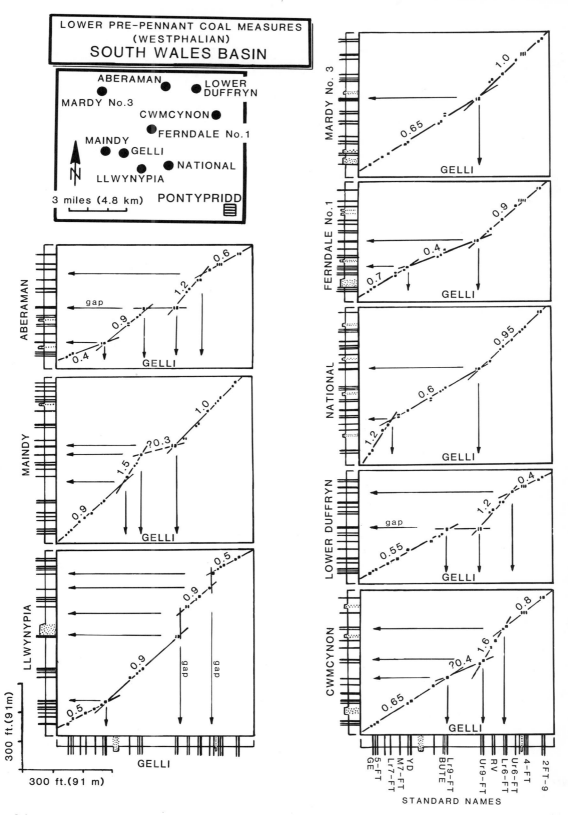

Fig. 8.2. Correlation graphs for selected profiles of the Lower Pre-Pennant Measures, South Wales Basin. The values attached to LOCs are their gradients (β-values). Rock types are indicated by same pattern as in Figure 8.1. (Data are graphically modified from Woodland and Evans 1964, Plate III.)

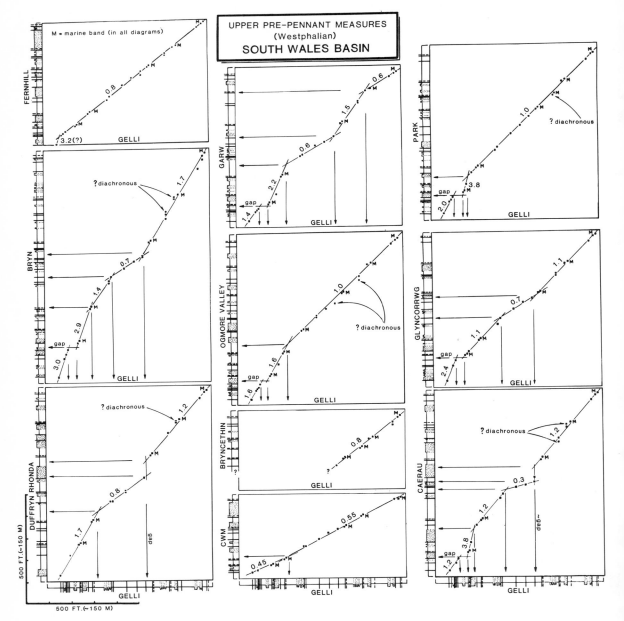

Fig. 8.3. Correlation graphs here and on the facing page for profiles of the Upper Pre-Pennant Measures, South Wales Basin. Explanation as for Figure 8.2; letter M indicates marine bands or similar, correlative fossiliferous horizons of brackish/lacustrine origin. (Data are graphically modified from Woodland and Evans 1964, Plate IV.)

vals of the coal measures. The coal-mine profiles, including correlated markers, are from Woodland and Evans (1964). The data pertain to the Ponty-pridd-Maesteg area of the South Wales Basin (see later index map in Fig. 8.10). In ascending order, the three coal-bearing series are as follows (informal names are used for simplicity):

1. *Lower Pre-Pennant Measures* (interval between the Gellideg and Two-Feet-Nine coal seams), which are 100–200 m thick and consist of predominantly argillaceous deposits and minor sandstones. Some representative examples of correlation graphs are given in Figure 8.2, which shows eight selected profiles correlated with an

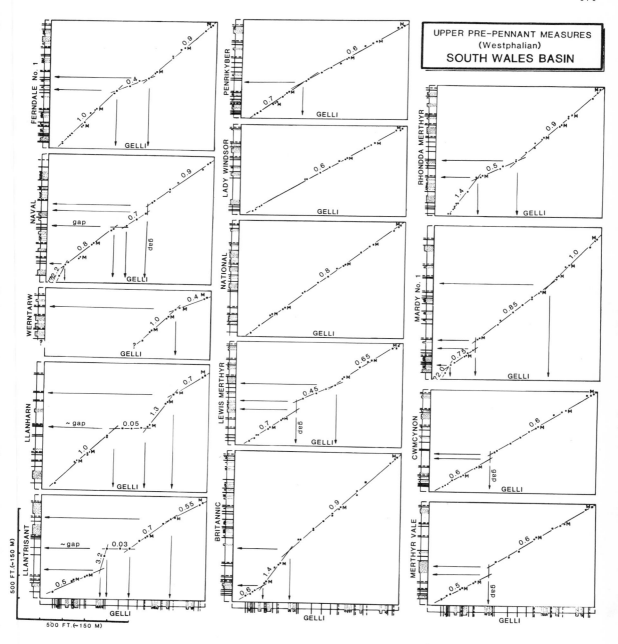

Fig. 8.3. (*continued*)

arbitrary reference profile (Gelli).

2. *Upper Pre-Pennant Measures* (interval between the Two-Feet-Nine coal seam and the Upper Cwmgorse Marine Band), which are 100–300 m thick and consist of argillaceous deposits and more abundant sandstones. Figure 8.3 shows twenty-four individual profiles correlated with a

reference profile (Gelli).

3. *Pennant Measures* (interval between the Upper Cwmgorse Marine Band and the Graigola coal seam), which are 70–100 m thick and consist predominantly of sandstones and subordinate argillaceous deposits. Representative examples of correlation graphs are given in Figure 8.4.

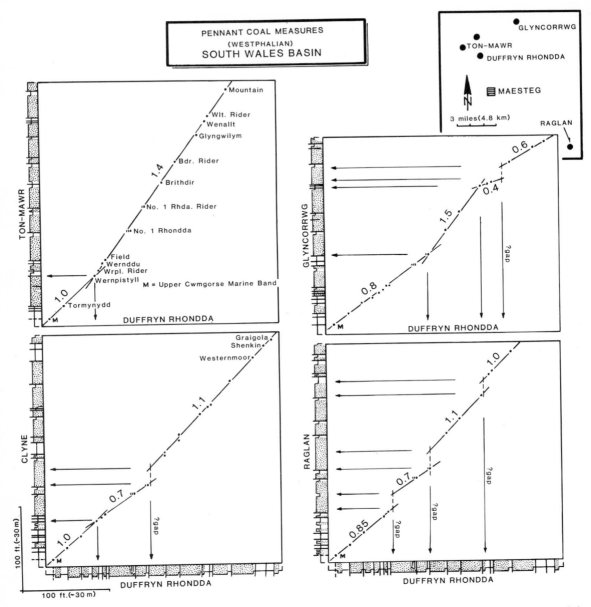

Fig. 8.4. Correlation graphs for profiles of the Pennant Measures, South Wales Basin. Explanation as for Figure 8.2. (Data are graphically modified from Woodland and Evans 1964, Plate V.)

The correlation graphs and the patterns of their LOCs in Figures 8.2–8.4 provide stratigraphic information that is readily interpretable in terms of the explanations given in Figure 8.1 and those included in the graphs. Therefore, only general comments are given below. The marine bands in the graphs (Fig. 8.3) display the same synchroneity as the majority of coal seams, which they often overlie.

All three coal-measure series reveal numerous disturbances and gaps. Most LOCs are strongly segmented and relatively few lack disturbances; the latter case pertains mainly to some closely spaced profiles. Moreover, the LOC segments, even adjacent ones, display a great range of gradients. These features indicate that the subsidence (accumulation) rates in the basin were highly variable, in both space and time. The subsidence was differential not only in terms of its rate differences and the local scale of its spatial variation, but also in terms of its remarkable unsteadiness (time-variability). The degree of relative (recorded) unsteadiness of the subsidence rates can be quantified roughly as the number of disturbances per stratigraphic thickness of a reference profile (or vice versa). Given the moderate stratigraphic thicknesses involved, the three coal-bearing series appear to have recorded a subsidence pattern that is unparalleled by any of the other examples discussed below. Accordingly, this style of intrabasinal differential subsidence is inferred to characterize some particularly active (compressional or transpressional?) tectonic regimes in cratonic settings.

The individual profiles have been correlated deliberately against a single reference profile, as shown in Figures 8.2–8.4. (The reference profile has been selected arbitrarily, though in principle the analysis is most informative when a most complete profile, with minimum relative gaps, is chosen as a reference; this is not the case in Fig. 8.2, for example, where Llwynypia would be more appropriate for a regional study.) All the disturbances and gaps are projected onto one profile, so that their relative stratigraphic order and ordinal time-distribution become more conspicuous. For example, the diagrams for Upper Pre-Pennant Measures (Fig. 8.3), where the Gelli profile (means "Gelli-time") is used as a reference, reveal that some of the disturbances occurred at the same time in several profiles, whereas others were more local or merely affected a couple of profiles (see also Figs. 8.2 and 8.4). The ordinal time-distribution of these disturbances, relative to the Gelli record, will be used later and is shown in Figure 8.10 (see Gelli profile; the positions of 12 recorded disturbances are indicated on the profile's right-hand margin). Data such as those in Figure 8.3 can be used further to map differential subsidence, as discussed in a later section.

Namurian Measures of the Intra-Sudetic Basin

The Intra-Sudetic Basin of Lower Silesia region, southwest Poland, is an extensional, intramontane trough filled with Lower Carboniferous–Lower Triassic, Variscan "internal molasse" (Augustyniak and Grocholski 1968; Nemec *et al.* 1982). The basin was initiated by Variscan, wrench-related extensional tectonism, whereby a small (ca 12 km × 17 km) pull-apart feature, the Walbrzych Coal Basin,

evolved and was eventually filled with Silesian (Upper Carboniferous) coal measures. It later became part of a broader extensional trough. The pull-apart basin subsided rapidly (Silesian thicknesses up to 2,000 m), and sedimentation was accompanied by strong volcanism and associated phreatic eruptions.

The Silesian succession is nonmarine, though underlain by a deltaic/fan-deltaic unit, and contains three coal-bearing alluvial series separated by barren alluvium. The oldest, lower Namurian series, which is known as the Walbrzych Formation (or Beds), has been selected for this study. The succession is 150–250 m thick, and its depositional environment evolved from an upper, nonmarine deltaic plain to a meandering alluvial plain (Nemec 1984). It contains some 30 coal seams, the thickest of which are fairly extensive and correlatable as revealed by mining activity over the past six decades. The data used in this study, including correlated coals, were provided by the geological laboratory of the "Thorez" Mine.

The correlation graphs (Fig. 8.5) are based on eight, relatively closely spaced profiles from near the basin center, which have been cross-correlated in various combinations. Again, the LOCs are readily interpretable in terms of the guide given in Figure 8.1, and hence require little comment. Volcanogenic tonstein beds, which are accepted generally as chronostratigraphic units, match the coal LOCs very well (Fig. 8.5). This further supports the contention made previously that most coal seams can also be regarded as chronostratigraphic units.

The stratigraphic record from the Walbrzych Basin reveals clearly that the intrabasinal subsidence was differential and varied on a local scale. Compared with the Welsh example, the relative rates appear to have been fairly similar, but much steadier. The stratigraphic thicknesses in the two examples are roughly comparable, but the disturbances in the present case are fewer and their effects less drastic (cf. Fig. 8.5 with Fig. 8.3). These features are consistent with the style of general subsidence in the Walbrzych Basin. The subsidence was rapid and due to mechanical stresses (fault-induced subsidence), but the basin was small and most of the vertical displacements and volcanic activity took place along its marginal faults. The basin floor subsided differentially but with relatively little disturbance. The excess stresses potentially arising from mismatches between active basement fault blocks most probably

174   W. Nemec

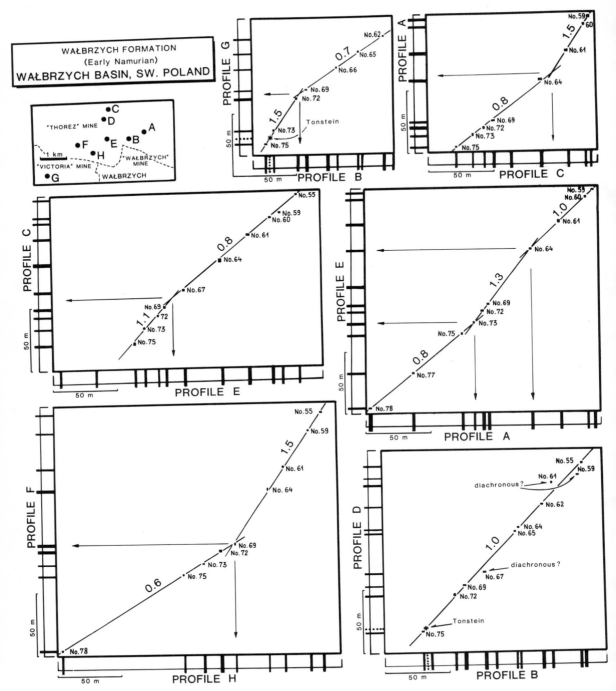

Fig. 8.5. Correlation graphs for selected profiles of the Walbrzych Formation, southwest Poland. The values attached to LOCs are their gradients (β-values). Rock types between coal seams (black stripes) not shown. (Data are from the geological archives of "Thorez" Mine, Poland.)

were accommodated largely by movements at the basin margins, rather than by creation of complementary sub-basinal fractures (Nemec 1984).

Mesozoic Hitra Formation of Haltenbank

The Hitra Formation (late Triassic-earliest Jurassic age) of the Haltenbank area, mid-Norway Continental Shelf, is the lowest subdivision of the Halten Group (Hollander 1984). The succession is known only from offshore wells drilled during recent petroleum exploration. The Haltenbank represents part of a larger Mesozoic basinal complex, and is one of the world's most prolific hydrocarbon provinces (Heum *et al.* 1986). The Hitra Formation, which is coal-bearing and 250–400 m thick (locally up to 500 m), has been recognized as an important source rock for both liquid and gaseous hydrocarbons. Its abundant coals and shaly coals have been correlated tentatively, from some closely spaced wells, on the basis of petrographic/geochemical data, interseam clastic facies, and wireline logs. The thicker Mesozoic coals in offshore Norway tend to display characteristic vertical zonings, in terms of microlithotype composition, so that the correlation of at least these thicker seams is fairly reliable (A. Ryseth, personal communication 1987).

The depositional environment of the Hitra Formation was fluvio-deltaic, although marine intercalations are present only in the uppermost part of the sequence. Sedimentation was accompanied by sub-basinal faulting, mainly listric, related to the extensional tectonism of the Norwegian-Greenland Sea region (Hollander 1984; Heum *et al.* 1986). Thus, subsidence was driven at least partly by mechanical stresses (rift-induced subsidence).

The correlation graph presented (Fig. 8.6) is based on two selected profiles, whose locations in the basin are about 5 km apart in a west-east direction (at a high angle to the strike of syndepositionally active faults). The segmented LOC indicates that the intrabasinal subsidence was highly differential and disturbances (some at least) were quite strong, though not particularly frequent. A similar pattern is displayed by a few other LOCs that have been constructed (A. Dalland, personal communication 1987). Given these limited data, the Haltenbank style of differential subsidence during late Triassic-early Jurassic time seems somewhat intermediate between the two previous examples. Importantly, the Haltenbank LOCs and their disturbances correspond quite well with independently inferred,

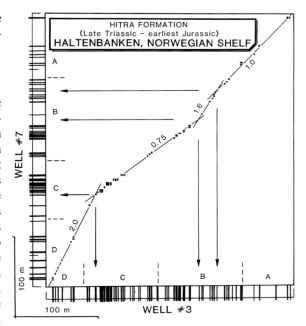

Fig. 8.6. Correlation graph for two profiles of the Hitra Formation, offshore mid-Norway. Explanation as for Figure 8.5. The four units (labelled A–D) are informal lithostratigraphic subdivisions distinguished by regional petroleum geologists. Location of profiles and interseam lithology not shown for the sake of confidentiality. (Data are from Statoil's exploration wells, Norway.)

stratigraphically dated events of synsedimentary intrabasinal faulting (A. Dalland, personal communication 1987).

Pennsylvanian Measures of Illinois Basin

The Illinois Basin, U.S.A. eastern interior, is a large southward-plunging trough (ca. 450 km long and 350 km wide), with the Pennsylvanian Coal Measures generally thinner than 600 m. The subsidence history of the basin (McKee *et al.* 1975; Sloss 1979) began in Early Paleozoic time, when fault-controlled subsidence initiated a cratonic marine embayment open to the south. Subsidence was initially rapid, and the Cambrian-Lower Ordovician strata locally attain 2,700 m in thickness. During Late Visean-Early Namurian time, the basin started to subside much less differentially and its subsidence became controlled largely by very broad flexure, probably driven by thermal stresses. In Pennsylvanian time, the basin continued to subside as a broad depression, but some high-angle faults became reactivated in the basement and spasmodically affected the deposi-

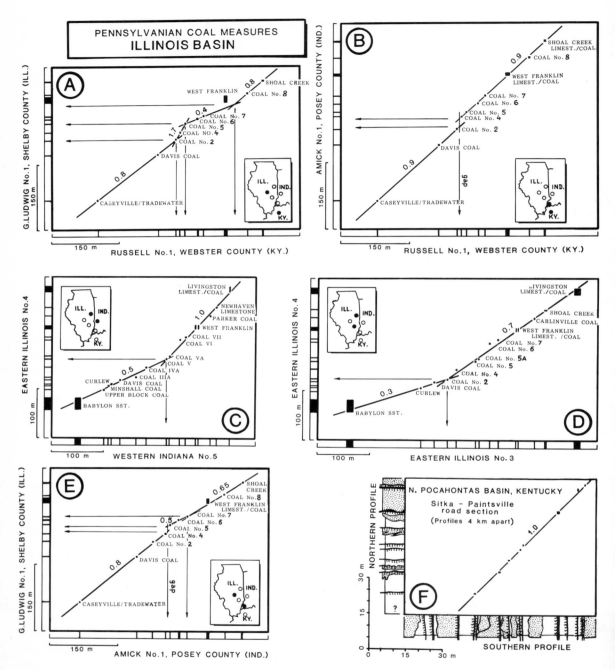

Fig. 8.7. Correlation graphs for selected profiles of the Pennsylvanian Measures, Illinois Basin. Explanation as for Figure 8.5. Coal seams and other basinal markers are plotted in graphs A–E; inter-marker lithology not shown.

(Graphs A–E are based on data from Figures 9 and 10 of Wanless 1955; graph F is based on two arbitrarily chosen profiles from a cross section on Figure 8 of Horne and Ferm 1978.)

tional surface. The subsidence then apparently ceased in Stephanian/Early Permian time.

The Pennsylvanian depositional environment was predominantly deltaic, and the fluvial systems feeding the deltas were fairly conformable with the basin's broad geometry. Details of the Pennsylvanian stratigraphy and sedimentation are beyond the scope of this review, but it is important to note that the basin provides a stratigraphic record of numerous oscillations from subaerial to subaqueous conditions, typically involving an abrupt change from subaerial sedimentation (commonly terminated in peat) to marine inundation. This aspect of the well known Pennsylvanian "cyclothems" (e.g., Weller 1956) closely resembles that documented in the South Wales Coal Measures (see Fig. 8.3), where coal seams are commonly followed by thin but extensive marine horizons, or by similar horizons marking rapid inundation by brackish or fresh water (Woodland and Evans 1964).

Both correlative coal seams and other marker horizons are used in the correlation graphs (Fig. 8.7). The data pertain to the southeastern part of the Illinois Basin, which is commonly considered as the most tectonically active part of the basin during Pennsylvanian time (Wanless 1955; McKee *et al.* 1975; Sloss 1979). The selected profiles, except graph F, are widely spaced (approximately 70–200 km apart); such a data set is probably required for large intracratonic troughs like the Illinois Basin.

The LOCs in Figure 8.7 reveal a pattern of differential subsidence that is quite different from the previous examples. Given the great stratigraphic thicknesses considered (Fig. 8.7A–E), the relative rates of subsidence (accumulation) tend to be remarkably constant and disturbances or gaps are infrequent, with only minor exceptions (Fig. 8.7A). Also, the relative rates themselves tend to be lower (i.e., closer to $\beta = 1.0$) than in the previous examples. These results correspond quite well with what has been summarized above on the Illinois Basin and its Pennsylvanian development.

Appraisal of Compaction Factor

The graphic method used in this study, as well as Shaw's original (1964) concept, both necessarily deal with rock (i.e., compacted sediment) accumulation rates, or with the rates of "bulk" subsidence as defined above. If the analyzed relative rates of rock accumulation are to be translated into actual relative

rates of sediment accumulation, then it becomes necessary also to evaluate the influence of sediment compaction on the LOC patterns.

Shaw (1964, p. 139) contended intuitively that: "If the types of rock in two sections differ fundamentally in their percentage of compaction we may find that the slope of our Line of Correlation is significantly changed." Whatever potential magnitudes of such an angular change Shaw anticipated (1964, p. 140, Fig. 20-7), his contention may still seem rather optimistic. Differential compaction may not merely affect the LOC slope, but also the relative positions of the projection points. How much "chaos" in the relative stratigraphic record might then be introduced or concealed by compaction, or might the linear trends and their disturbances be merely a product of this last factor?

To appraise the role of compaction, several profiles were decompacted numerically and new correlation graphs were produced. In the decompaction procedure, all sediment layers were brought consecutively to the surface, so that an original, compaction-free stratigraphic record was revealed. Examples of such uncompacted correlation graphs, along with their present equivalents, are shown in Figures 8.8 and 8.9. The profiles are from the Upper Pre-Pennant Measures of the South Wales Basin, which are among the most heterolithic and hence potentially most affected by differential compaction. Deliberately, two very different pairs of profiles are shown. The two profiles in Figure 8.8A have very similar sandstone/shale ratios (15% difference), whereas those in Figure 8.9A have sandstone/shale ratios that differ by 61%; also, their LOCs differ in degree of complication. The profiles were decompacted under the following assumptions:

1. Their "depth of burial" (for the top of each profile) was assumed to be 3,000 m. Although the South Wales Basin apparently ceased to subside in post-Carboniferous time, the 3,000 m burial was taken arbitrarily as a moderate equivalent of the thrust-loading to which the coal measures were subjected (their coals now are high-rank, anthracitic).

2. The compaction ratio for all coals was taken as 7:1. This is the degree of compaction of Upper Carboniferous Pteridosperm peats, up to a coking-coal stage, estimated by Teichmüller and Teichmüller (1982).

3. The sandstones and argillaceous sediments were assumed as texturally homogeneous, and their

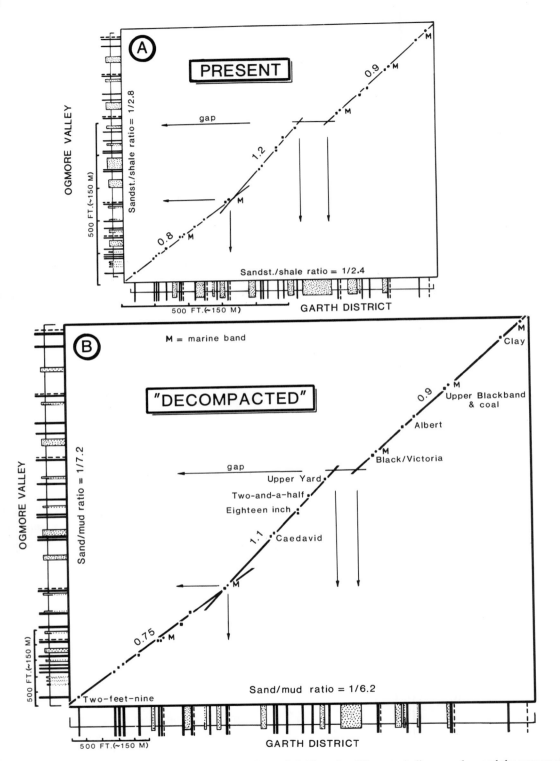

Fig. 8.8. Correlation graph (A) for two selected profiles (with a small difference in sandstone/shale ratio) of the Upper Pre-Pennant Measures, South Wales Basin, and its "decompacted" equivalent (B). Explanation as in Figure 8.2. Note the difference in linear scales, and the nature of the gap revealed by LOC (gap in Ogmore profile corresponds to thick sandstone in Garth profile). (Graph A is based on data from Woodland and Evans 1964, Plate IV.)

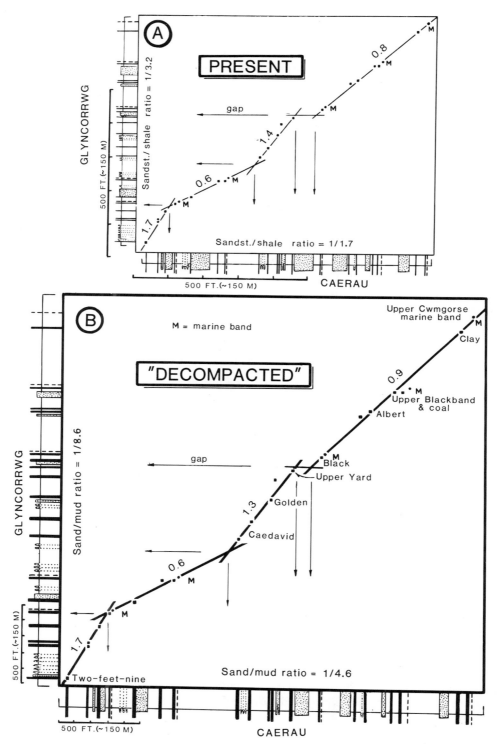

Fig. 8.9. Correlation graph (A) for two selected profiles (with a large difference in sandstone/shale ratio) of the Upper Pre-Pennant Measures, South Wales Basin, and its "decompacted" equivalent (B). Explanation as in Figure 8.2. Note the difference in linear scales, and the nature of the gap revealed by LOC (same as in Fig. 8.8). (Graph A is based on data from Woodland and Evans 1964, Plate IV.)

degree of compaction was calculated according to the equations and curves postulated by Baldwin and Butler (1985). The initial porosities of sand and mud were taken as 0.48 and 0.78, respectively (Baldwin and Butler 1985, their figs. 4, 5).

The decompacted profiles (Figs. 8.8B and 8.9B) display stratigraphic relationships that, in terms of the LOCs, are very similar to those revealed by the rock profiles. From Figures 8.8 and 8.9, it is also clear that neither the disturbances nor gaps are possibly related to compaction. Therefore, the effects of compaction may simply be neglected for the practical purpose of this study as far as the relative rates of accumulation are concerned. In other words, the analysis presented pertains directly to the sediment accumulation rates, and hence to sub-basinal subsidence rates, and this inherent simplicity thus makes the postulated method particularly attractive.

It does not follow, of course, that the potential effects of differential compaction should be ignored in all instances, especially where the relative heterogeneity of profiles is more pronounced. Coal-seam projection points may be offset and certain gaps induced by differential compaction, as the latter also tends to affect facies distribution (e.g., Williams and Bragonier 1974; Brown 1975). However, one must be aware that decompaction itself may still be, in reality, a somewhat doubtful exercise as far as detailed stratigraphic relationships are concerned. Any decompaction procedure is necessarily based on a number of artificial assumptions and parameters, and the amount of actual distortion they introduce to profiles may be difficult to assess.

Mapping Differential Subsidence Rates

Data like those presented in Figure 8.3, where all available profiles have been correlated individually with one reference profile, can be used directly to produce maps of the relative rates of subsidence (or sediment accumulation). Such a map can be made for any selected point of the stratigraphic time interval involved. For example (Fig. 8.3), the selected point in the reference (Gelli) profile is projected successively on the LOCs in the individual graphs, and the gradients of the LOCs (β-values) in the indicated places are then assigned to the respective profiles on their locality map; Gelli, as the reference data point, will always have assigned $\beta = 1.0$. Isograms of equal relative subsidence can be drawn on the map, by

means of interpolation. The resulting map shows the distribution of the relative subsidence rates (i.e., relative to the Gelli rate) at the selected point of the stratigraphic time interval.

Temporal variation in the relative subsidence rates can be analyzed by producing a series of such maps for some closely spaced points in the reference profile. However, successive maps derived from the same straight-line segments of the LOCs will appear identical, simply because such segments represent intervals with constant β-values. It is more efficient, therefore, to project first all the disturbances from individual LOCs onto the reference profile, and then to make just one map for each undisturbed interval of this reference time.

For illustration, this procedure has been applied to the Upper Pre-Pennant Measures of South Wales Basin (Fig. 8.10). The profiles in Figure 8.3 reveal a total of twelve disturbances in their subsidence rates relative to the Gelli rate. The disturbances have been projected on the Gelli "reference-time" profile (Fig. 8.10), and these twelve points divide Gelli-time into thirteen undisturbed intervals, during which the subsidence rates in the individual profiles (localities) were constant relative to the Gelli locality. Figure 8.10 shows the thirteen corresponding maps of the relative subsidence rates, for the time periods separated by disturbances. The maps were produced by interpolation, essentially by distributing the differences in β-values arithmetically between any two data points. The interpreted result was then compared with a structural map of the study area (Fig. 8.11) and slightly adjusted to it.

It can be seen from the maps that some of the disturbances were mild and very local, whereas others caused major rearrangements in the intrabasinal patterns of differential subsidence. It is reasonable to link the evolving subsidence patterns (maps in Fig. 8.10) with some of the faults, or faulted "terraces," presently manifested at the surface of the coal basin (Fig. 8.11).

Shaw (1964) used somewhat similar technique to produce an artificial map of rock accumulation rates, and the reader is referred to his book (pp. 205–207) for some useful comments.

Graphic Perspective Correlation: A Test

Other graphic techniques for stratigraphic correlation, such as "perspective" graphs (Haites 1963) or

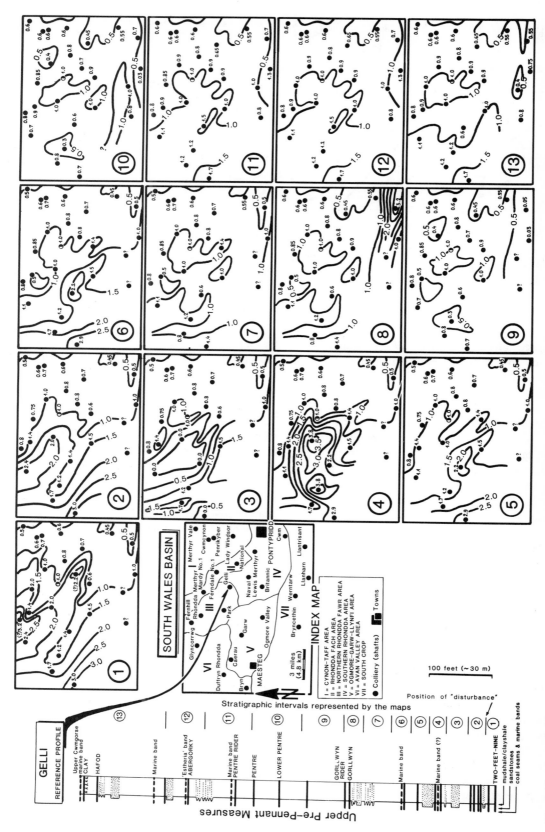

Fig. 8.10. Maps of differential subsidence (subsidence rate relative to the Gelli locality) for the Upper Pre-Pennant Measures, South Wales Basin. Gelli profile used as reference (graphically modified data from Woodland and Evans 1964, Plate IV). Further explanation in text.

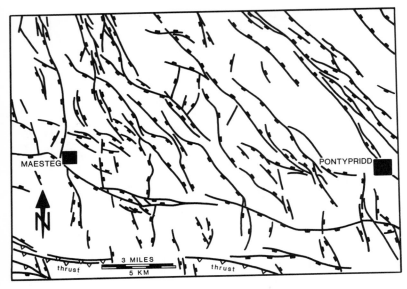

Fig. 8.11. Structural map of Pontypridd-Maesteg area, South Wales Basin, showing faults observed at the surface in outcrop of coal measures. Cross-bars indicate downthrown side of fault. This map corresponds to the index map in Figure 8.10. Data are from a "one-inch" geological map (Pontypridd sheet) published by the Geological Survey of Great Britain (1963).

"no-space" graphs (Edwards 1979), can be applied similarly to coal measures in order to gain insight into the differential aspect of basin-floor aggradation/subsidence. The application of *perspective-correlation graphs* is illustrated (Fig. 8.12), with profiles of the Upper Pre-Pennant Measures (South Wales Basin) used again as an example.

The data requirement for the perspective-correlation method is the same as before: For a given pair of profiles there must be a number of coal beds, or other approximately chronostratigraphic marker horizons that have already been correlated with confidence. In terms of these markers, one profile is projected onto the other profile, simply by linking their correlative points with straight lines. These lines are then extended beyond the profiles, to the left and/or right, so that such *projective rays* intersect in a single center of perspective, called a *focus*. There may be two or more foci, as the projective rays will normally tend to form *clusters*, and the foci may occur on either side of the graph (Fig. 8.12). The graphs refer to dimensionless space, where absolute vertical and horizontal scales are unimportant ("no-space" graphs).

The three diagrams (Fig. 8.12) reveal stratigraphic relationships between six selected profiles, also with respect to an inferred depositional paleoslope (see isopach map in Fig. 8.12, top). For any pair of profiles, the depositional surface appears to have aggraded in stages, and during each stage there

was a constant ratio between the local aggradation rates (as implied by the projective clusters and their foci); the aggradation arrows (Fig. 8.12), when inverted, can be considered in terms of differential subsidence. Foci close to the profiles imply highly differential subsidence, and those farther away indicate subsidence that was probably broader and hence less differential. The see-saw-like reversals (Fig. 8.12), where a focus switches from one side of the graph to the other, indicate that the upper parts of the depositional slope periodically subsided faster than its lower parts.

The perspective graphs provide, in fact, information similar to the LOCs. The "projective clusters" (with their "foci") correspond to the straight-line segments of the LOCs. The ratio of cluster thicknesses in two correlated profiles, if in same scale, is the β-value of particular LOC segment. The change from one cluster to another successive one is a disturbance, as earlier defined for the LOCs. The Upper Pre-Pennant Measures in Figure 8.12 display a total of twelve disturbances (i.e., up to thirteen undisturbed clusters), the same number as were revealed by the LOCs (see Fig. 8.3 and profile in Fig. 8.10).

The perspective-correlation method has been applied here primarily as a test. This application demonstrates clearly that the stratigraphic relationships revealed by the LOCs, notably the segmented linear patterns, cannot be attributed by any means to

Fig. 8.12. Perspective-correlation graphs for six selected profiles of the Upper Pre-Pennant Measures, South Wales Basin. For detailed explanation see text. The thick arrows mark aggradational "clusters" corresponding to individual "foci" (F). The inset map (top) shows thickness distribution of the Upper Pre-Pennant Measures in the study area. (Data as in Fig. 8.3, simplified.)

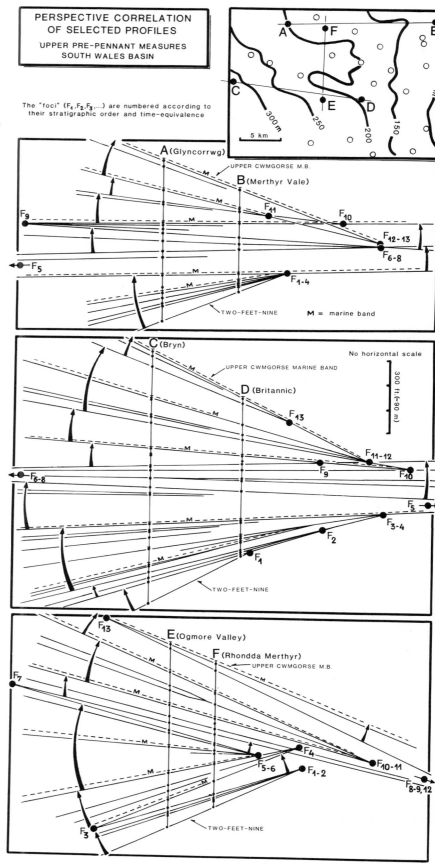

the graphic method itself. The patterns are inherent in the stratigraphic record, and are recognizable by more than one method.

Discussion and Conclusions

The methodological framework for a new, quantitative approach to the analysis of the stratigraphic record in coal basins has been established and provides a promising basis for future research. The proposed method is useful for simple quantitative modelling, and its operational procedures can be fully computerized. This latter approach may be particularly useful in selecting "reference" profiles, which should be those with the fewest relative gaps. It has been demonstrated that the correlation graphs can be used as an effective tool in the analysis of coal-bearing basins, although not all of the potential applications have been exploited by the present study. For example, the graphs can be used to correlate time-equivalent sedimentary facies, formed simultaneously under different environmental conditions in the basin, and hence also to analyze the evolving facies distributions along time lines or time surfaces. The analysis presented pertains directly to coal-bearing basins, but the results and their implications are thought to be of a general significance to cratonic basins.

The present study focuses on the relative rates of sediment accumulation, interpreted in terms of intrabasinal differential subsidence and ascribed to sub-basinal tectonism. It has become widely accepted, at least since Bubnoff's (1950) classical thesis, that the thicknesses of sediment developed in a particular basin, and in its depositional loci, are the best measure of tectonic activity. This general principle, however, may not seem sufficient as an assumption when detailed stratigraphic relationships rather than maximum thicknesses are considered. The interpretational reasoning assumed in the present study, notably its reference to tectonic subsidence, requires additional discussion.

Given the number and spectrum of basin examples considered, the evidence from the correlation graphs is sufficiently compelling to accept basin-floor subsidence as the prime cause for the observed stratigraphic relationships (Figs. 8.2–8.7 and 8.12). Other possible factors, such as compaction, autocyclic sedimentation mechanisms, or sediment loading (isostatic adjustments), appear to be of minor importance and can be ruled out essentially on the basis of the correlation graphs.

Differential compaction and autocyclic processes, such as delta-switching or meander-belt avulsion, can be ruled out because of the evidence that the LOC linear trends and their disturbances occur irrespective of lithological (facies) changes in the profiles (Figs. 8.2–8.4). Constant β is maintained during deposition of clay/mud as well as sand, and the recognized disturbances occur regardless of facies. Consequently, it was neither the areal density variations of the sediments at the time of deposition, nor the depositional framework that caused the differential subsidence and triggered the disturbances.

Moreover, compaction has already been eliminated earlier by removing its influence from the profiles (Figs. 8.8–8.9). The effects of compaction, as well as those of autocyclic processes, appear to be "built-in" elements of the stratigraphic record, at least as far as the interseam accumulation increments are concerned. Maximum compaction probably takes place fairly soon after the sediments have been deposited (e.g., Wanless 1952, p. 155; Bloom 1964; see also review by Ferm and Staub 1984, pp. 283–284), so that its effects are compensated for rather rapidly by subsequent sedimentation. The effects of the remaining, slower compaction become averaged throughout the stratigraphic succession (i.e., the effect of gradual compaction is probably taken up by the successive aggradational increments, more or less proportionally to the time spans of their accumulation and hence also to their thicknesses).

Differential subsidence of the basin floor is, therefore, the most likely causal factor. Local isostatic adjustments due to sediment loading can be precluded. It is simply inconceivable that the differential loading associated with a single interseam clastic unit (see thicknesses involved and their profile-to-profile variations, in Figs. 8.2–8.7) could have propagated to depths, to cause differential, accelerated subsidence of the depositional surface. Moreover, such a mechanism apparently is contradicted by the observed LOC patterns (see behavior of β-values, in Figs. 8.2–8.7 and 8.13). Isostasy probably influenced broad crustal warping during the infilling of the coal basins, but there is little to indicate that it controlled the local-scale differential subsidence revealed by the present study. Accordingly, it is concluded that the differential subsidence

must have been controlled tectonically, most probably by active deformation of the subsiding floor and
basin fill. Faulting could be important, as it seems to
be the only conceivable mechanism to explain the
observed sudden disturbances (Fig. 8.13); faults
move suddenly, and are known to be subject to sudden switching.

The LOCs and their segmented patterns (Fig.
8.13) indicate that the differential subsidence
accommodating local, differential sediment
accumulation in a basin tends to be constant for
appreciable geological time periods, and is subject
to sudden, discrete changes. Since the examples
show that this basic pattern (Fig. 8.13) is essentially
independent of tectonic and depositional settings, its
causes in cratonic basins must be quite universal.
Moreover, there seems to be a close relationship
between the detailed appearance of segmented
LOCs, in terms of the magnitude and spa-
tial/temporal frequency of disturbances, and the
actual tectonic regime of a basin.

Further research in carefully selected areas is
necessary to explain the LOC patterns in terms of
basin tectonism, and the implications of such studies
may be of crucial importance. It is to be noted, for
example, that the extensive marine horizons in coal
measures are often associated with the disturbances
revealed by LOCs (see Fig. 8.3 and cluster boundaries in Fig. 8.12); the relationship may not seem
clear for a particular pair of profiles, but the marine
incursion or incursions often appear to have accompanied basinal disturbances recorded by other pairs
of profiles. It may appear, therefore, that not only
the large-scale (8–10 million years) transgressive-
regressive sequences of Pennsylvanian-type cyclothems (Busch and Rollins 1984) can be ascribed to
crustal tectonism (see Cloetingh *et al.* 1985; Karner
1986), but also that a resurrection of long-forgotten
"diastrophism" (Trueman 1947; Weller 1956) will
prove necessary to explain the cyclothems themselves (see discussion by Sloss 1979). It is more
important, however, to realize that the differential
"noise" that accompanies general, smoothly varying
subsidence of a sedimentary basin may have a considerable impact on the depositional environment
(e.g., Williams and Bragonier 1974; Ouchi 1985),
and hence also on the sedimentary facies development and relationships within the basin. This study
offers an attractive, detailed approach to such a
topic. It may be possible, for example, to evaluate
the origin of particular types of intrabasinal facies

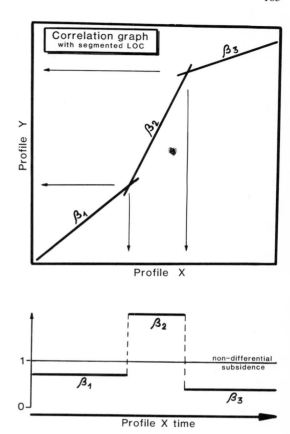

Fig. 8.13. Schematic representation of the main features
of coal-measure LOCs (correlation graphs) revealed by
this study, and their interpretation. The subsidence rates
in this model represent interval mean rates (as discussed in
text).

variations, which have been ascribed so far, rather
uncritically, to eustatic sea-level changes or autocyclic mechanisms of a sedimentary environment. The
main aim of this study is to stimulate further
research on the diverse aspects of the intrabasinal
differential subsidence.

Acknowledgments. The author thanks Karen Kleinspehn and Chris Paola for their invitation to contribute to this book. An earlier version of the manuscript
was reviewed by Arne Dalland, John Ferm, Julie
Jones, Krzysztof Mastalerz, Andrew Miall, Johan
Michelsen, Fred Rich, Signe-Line Røe, Ron Steel,

Michael Talbot, one anonymous reviewer, and the editors. The author appreciates their help, and also acknowledges personal discussions with John Crowell and Alan Gibbs. The Geological Institute, University of Bergen, provided technical assistance, and Aslaug Pedersen typed the manuscript.

References

ADAMS, J.E. (1965) Stratigraphic-tectonic development of Delaware Basin. American Association Petroleum Geologists Bulletin 49:2140–2148.

AGER, D.V. (1973) The Nature of the Stratigraphical Record. London: Macmillan, 371 p.

ALEXANDER, K. (1969) Stratigraphie und Tektonik der Erdöl-Felder Voigtei und Siedenburg (Niedersachsen): Ausgewählte Beispiele zur synsedimentären Tektonik in der tiefen Unter-Kreide. Neues Jahrbuch Geologie Paläontologie Abh. 132:239–256.

ANDERSON, J.G.C. and OWEN, T.R. (1968) The Structure of the British Isles. New York: Pergamon Press, 162 p.

AUGUSTYNIAK, K. and GROCHOLSKI, A. (1968) Geological structure and outline of the geological development of the Intra-Sudetic Depression. Instytut Geologiczny (Warszawa) Biuletyn 227:87–114.

BALDWIN, B. and BUTLER, C.O. (1985) Compaction curves. American Association Petroleum Geologists Bulletin 69:622–626.

BEAUMONT, C. (1981) Foreland basins. Geophysical Journal Royal Astronomic Society 65:291–329.

BEAUMONT, C., KEEN, C.E., and BOUTILIER, R. (1982) A comparison of foreland and rift margin sedimentary basins. Philosophical Transactions Royal Society London (Series A) 305:295–317.

BLOOM, A.L. (1964) Peat accumulation and compaction in a Connecticut coastal marsh. Journal Sedimentary Petrology 34:599–603.

BOTT, M.H.P. (1976) Mechanisms of basin subsidence — an introductory review. Tectonophysics 36:1–4.

BROWN, L.F. (1975) Role of sedimentary compaction in determining geometry and distribution of fluvial and deltaic sandstones. In: Chilingarian, G.V. and Wolf, K.H. (eds) Compaction of Coarse-Grained Sediments I. Developments in Sedimentology 18A. Amsterdam: Elsevier, pp. 247–292.

BUBNOFF, S. (1950) Die Geschwindigkeit der Sedimentbildung und ihr endogener Antrieb. Miscellenia Academia Berolinensis (Berlin), pp. 1–32.

BUSCH, D.A. (1975) Influence of growth faulting on sedimentation and prospect evaluation. American Association Petroleum Geologists Bulletin 59:217–230.

BUSCH, R.M. and ROLLINS, H.B. (1984) Correlation of Carboniferous strata using a hierarchy of transgressive-regressive units. Geology 12:471–474.

CLOETINGH, S., McQUEEN, H., and LAMBECK, K. (1985) On a tectonic mechanism for regional sea level variations. Earth Planetary Science Letters 75:157–166.

CROSS, A.T. (1952) The geology of the Pittsburgh Coal. In: Origin and Constitution of Coal, 2nd Conference Proceedings. Crystal Cliffs: Nova Scotia Department of Mines, pp. 32–111.

DUFF, P.McL.D. (1967) Cyclic sedimentation in the Permian Coal Measures of New South Wales. Journal Geological Society Australia 14:293–308.

DUFF, P.McL.D. and WALTON, E.K. (1962) Statistical basis for cyclothems: A quantitative study of the sedimentary succession in the East Pennine Coalfield. Sedimentology 1:235–255.

EDWARDS, L.E. (1979) Range charts and no-space graphs. Computers Geosciences 4:247–255.

EDWARDS, L.E. (1984) Insights on why graphic correlation (Shaw's method) works. Journal Geology 92:583–597.

EDWARDS, L.E. (1985) Insights on why graphic correlation (Shaw's method) works: A reply. Journal Geology 93:507–509.

FERM, J.C. and STAUB, J.R. (1984) Depositional controls of mineable coal bodies. In: Rahmani, R.A. and Flores, R.M. (eds) Sedimentology of Coal and Coal-Bearing Sequences. International Association Sedimentologists Special Publication 7, pp. 275–289.

FISCHER, A.G. (1969) Geological time-distance rates: The Bubnoff unit. Geological Society America Bulletin 80:549–552.

HAITES, T.B. (1963) Perspective correlation. American Association Petroleum Geologists Bulletin 47:553–574.

HAY, W.W. and SOUTHAM, J.R. (1978) Quantifying biostratigraphic correlation. Annual Review Earth Planetary Sciences 6:353–375.

HAZEL, J.E., MUMMA, M.D., and DUFF, W.J. (1980) Ostracode biostratigraphy of the Lower Oligocene (Vicksburgian) of Mississippi and Alabama. Gulf Coast Association Geological Society Transactions 30:361–401.

HEUM, O.R., DALLAND, A., and MEISINGSET, K.K. (1986) Habitat of hydrocarbons at Haltenbanken (PVT-modelling as a predictive tool in hydrocarbon exploration). In: Spencer, A.M. (ed) Habitat of Hydrocarbons on the Norwegian Continental Shelf. London: Graham and Trotman Ltd., pp. 259–274.

HOLLANDER, N.B. (1984) Geohistory and hydrocarbon evaluation of the Haltenbank area. In: Spencer, A.M. (ed) Petroleum Geology of the North European Margin. London: Graham and Trotman Ltd., pp. 383–388.

HORNE, J.C. and FERM, J.C. (1978) Carboniferous Depositional Environments: Eastern Kentucky and Southern West Virginia. Field Guide, University of South Carolina, Columbia, 151 p.

JARVIS, G.T. and McKENZIE, D.P. (1980) Sedimentary basin formation with finite extension rates. Earth Planetary Science Letters 48:42–52.

JOHNSON, K.R. and COOK, A.C. (1973) Cyclic characteristics of sediments in the Moon Island Beach Subgroup, Newcastle Coal Measures, New South Wales. Mathematical Geology 5:91–110.

KARNER, G.D. (1986) Effects of lithospheric in-plane stress on sedimentary basin stratigraphy. Tectonics 5:573–588.

KELLING, G. (1964) Sediment transport in part of the Lower Pennant Measures of South Wales. In: van Straaten, L.M.J.U. (ed) Deltaic and Shallow-Marine Deposits. Developments in Sedimentology 1. Amsterdam: Elsevier, pp. 177–184.

KELLING, G. (1968) Patterns of sedimentation in Rhondda Beds of South Wales. American Association Petroleum Geologists Bulletin 52:2369–2386.

KELLING, G. (1986) Upper Carboniferous sedimentation in the South Wales: Review and appraisal. In: Controls of Upper Carboniferous Sedimentation in North-West Europe. British Sedimentology Research Group Meeting Abstracts, Keele, pp. 40–42.

KRUMBEIN, W.C. and GRAYBILL, F.A. (1965) An Introduction to Statistical Models in Geology. New York: McGraw-Hill, 475 p.

LEEDER, M.R. (1976) Sedimentary facies and the origins of basin subsidence along the northern margin of the supposed Hercynian ocean. Tectonophysics 36:167–179.

McCABE, P.J. (1984) Depositional environments of coal and coal-bearing strata. In: Rahmani, R.A. and Flores, R.M. (eds) Sedimentology of Coal and Coal-Bearing Sequences. International Association Sedimentologists Special Publication 7, pp. 13–42.

McKEE, E.D., CROSBY, E.J., FERM, J.C., KELLER, W.D., SCHOPF, J.M., WALKER, TH.R., and WANLESS, H.R. (1975) Paleotectonic investigations of the Pennsylvanian System in the United States. United States Geological Survey Professional Paper 853, Part II, 192 p.

McKENZIE, D.P. (1978) Some remarks on the development of sedimentary basins. Earth Planetary Science Letters 40:25–32.

McLEAN, J.R. and JERZYKIEWICZ, T. (1978) Cyclicity, tectonics and coal: Some aspects of fluvial sedimentology in Brazeau-Paskapoo Formations, Coal Valley Area, Alberta, Canada. In: Miall, A.D. (ed) Fluvial Sedimentology. Canadian Society Petroleum Geologists Memoir 5, pp. 441–486.

MERRIAM, D.F. (1972) Mathematical Models of Sedimentary Processes. New York: Plenum Press, 348 p.

MIALL, A.D. (1984) Principles of Sedimentary Basin Analysis. New York-Berlin: Springer-Verlag, 490 p.

MILLER, F.X. (1977a) Biostratigraphic correlation of the Messaverde Group in southwestern Wyoming and northwestern Colorado. Rocky Mountains Association Geologists 1977 Symposium Proceedings, pp. 117–137.

MILLER, F.X. (1977b) The graphic correlation method in biostratigraphy. In: Kauffman, F.G. and Hazel, J.F. (eds) Concepts and Methods of Biostratigraphy. Stroudsburg, PA: Dowden, Hutchinson and Ross, pp. 165–186.

MURPHY, M.A. and BERRY, W.B.N. (1983) Early Devonian conodont-graptolite collection and correlations with brachiopod and coral zones, central Nevada. American Association Petroleum Geologists Bulletin 67:371–379.

MURPHY, M.A. and EDWARDS, L.E. (1977) The Silurian-Devonian boundary in central Nevada. In: Murphy, M.A., Berry, W.B.N., and Sandberg, C.A. (eds) Western North America: Devonian. University California (Riverside) Campus Museum Contributions 4:183–189.

NEMEC, W. (1984) Walbrzych Beds (Lower Namurian, Walbrzych Coal Measures): Analysis of alluvial sedimentation in a coal basin. Geologia Sudetica 19:7–73.

NEMEC, W. and ĆMIEL, S. (1979) An application of Markov chain analysis to the Žacleř Beds succession (Upper Carboniferous), Walbrzych Coal Basin, SW. Poland. Acta Universitatis Wratislaviensis, Prace Geologiczno-Mineralogiczne 7:69–105.

NEMEC, W., PORĘBSKI, S.J., and TEISSEYRE, A.K. (1982) Explanatory notes to the lithotectonic molasse profile of the Intra-Sudetic Basin, Polish Part (Sudety Mts., Carboniferous-Permian). In: Schwab, G. (ed) Tectonic Regimes of Molasse Epochs. Potsdam: Veröffnungen Zentralinstitut Physik der Erde Akademie Wissenschaft. DDR, pp. 267–278.

NEMEC, W., STEEL, R.J., PORĘBSKI, S.J., and SPINNANGR, Å. (1984) Domba Conglomerate, Devonian, Norway: Process and lateral variability in a mass flow-dominated, lacustrine fan-delta. In: Koster, E.H. and Steel, R.J. (eds) Sedimentology of Gravels and Conglomerates. Canadian Society Petroleum Geologists Memoir 10, pp. 295–320.

OUCHI, S. (1985) Response of alluvial rivers to slow active tectonic movements. Geological Society America Bulletin 96:504–515.

OWEN, T.R. (1964) The tectonic framework of Carboniferous sedimentation in South Wales. In: van Straaten, L.M.J.U. (ed) Deltaic and Shallow-Marine Deposits. Developments in Sedimentology 1. Amsterdam: Elsevier, pp. 301–307.

READ, W.A. and DEAN, J.M. (1967) A quantitative study of a sequence of coal-bearing cycles in the Namurian of Central Scotland, 1. Sedimentology 9:137–156.

READ, W.A. and DEAN, J.M. (1968) A quantitative study of a sequence of coal-bearing cycles in the Namurian of Central Scotland, 2. Sedimentology 10:121–136.

READ, W.A. and DEAN, J.M. (1976) Cycles and subsidence: Their relationship in different sedimentary and tectonic environments in the Scottish Carboniferous. Sedimentology 23:107–120.

READ, W.A., DEAN, J.M., and COLE, A.J. (1971) Some Namurian (E2) paralic sediments in Central Scotland: An investigation of depositional environment and facies changes using iterative-fit trend-surface analysis. Journal Geological Society London 127:137–176.

RUBEL, M. and PAK, D.N. (1984) Theory of stratigraphic correlation by means of ordinal scales. Computers Geosciences 10:43–57.

SADLER, P.M. (1981) Sediment accumulation rates and the completeness of stratigraphic sections. Journal Geology 89:569–584.

SCHWARZACHER, W. (1975) Sedimentation Models and Quantitative Stratigraphy. Developments in Sedimentology 19. Amsterdam: Elsevier, 382 p.

SHAW, A.B. (1964) Time in Stratigraphy. New York: McGraw-Hill, 365 p.

SHELTON, J.W. (1968) Role of contemporaneous faulting during basinal subsidence. American Association Petroleum Geologists Bulletin 52:399–413.

SLOSS, L.L. (1979) Plate-tectonic implications of the Pennsylvanian System in the Illinois Basin. In: Palmer, J.E. and Dutcher, R.R. (eds) Depositional and Structural History of the Pennsylvanian System of the Illinois Basin, Part 2: Invited Papers. 9th International Congress of Carboniferous Stratigraphy and Geology, Field Trip 9, Illinois State Geological Survey Guidebook Series 15A, pp. 107–112.

SWEET, W.C. (1979a) Late Ordovician conodonts and biostratigraphy of the western Midcontinent province. Brigham Young University Geological Studies 26:45–85.

SWEET, W.C. (1979b) Graphic correlation of Permo-Triassic rocks in Kashmir, Pakistan and Iran. Geologica Palaeontologica 13:239–249.

TEICHMÜLLER, M. and TEICHMÜLLER, R. (1982) The geological basis of coal formation. In: Stach, E., Mackowsky, M.-TH., Teichmüller, M., Taylor, G.H., Chandra, D., and Teichmüller, R. (eds) Stach's Textbook of Coal Petrography. Berlin: Gebruder Borntraeger, pp. 5–86.

TRUEMAN, A.E. (1947) Stratigraphical problems in the coal measures of Europe and North America. Quarterly Journal Geological Society London 102:49–93.

TURCOTTE, D.L. and SCHUBERT, G. (1982) Geodynamics: Applications of Continuum Physics to Geological Problems. New York: John Wiley, 450 p.

VISTELIUS, A.B. (1967) Studies in Mathematical Geology. New York: Consultants Bureau, 294 p.

WANLESS, H.R. (1952) Studies of field relations of coal beds. In: Origin and Constitution of Coal, 2nd Conference Proceedings. Crystal Cliffs: Nova Scotia Department of Mines, pp. 148–180.

WANLESS, H.R. (1955) Pennsylvanian rocks of Eastern Interior Basin. American Association Petroleum Geologists Bulletin 39:1753–1820.

WATTS, A.R., KARNER, G.D., and STECKLER, M.S. (1982) Lithospheric flexure and the evolution of sedimentary basins. Philosophical Transactions Royal Society London (Series A) 305:249–281.

WELLER, J.M. (1956) Argument for diastrophic control of late Paleozoic cyclothems. American Association Petroleum Geologists Bulletin 40:17–50.

WILLIAMS, E.G. and BRAGONIER, W.A. (1974) Controls of Early Pennsylvanian sedimentation in western Pennsylvania. In: Briggs, G. (ed) Carboniferous of the Southeastern United States. Geological Society America Special Paper 148, pp. 135–152.

WOODLAND, A.W. and EVANS, W.B. (1964) The Geology of South Wales Coalfield, Part IV: The Country around Pontypridd and Maesteg. Geological Survey Great Britain Memoir, 3rd Edition, 391 p.

9

The Use of Magnetic-Reversal Time Lines in Stratigraphic Analysis: A Case Study in Measuring Variability in Sedimentation Rates

N.M. Johnson,† Khalid A. Sheikh, Elizabeth Dawson-Saunders, and Lee E. McRae

Abstract

Magnetic-reversal time lines of known age have been traced over a 15 km front of the Middle Miocene Chinji Formation of the Potwar Plateau in northern Pakistan, providing an example by which magnetic-reversal time lines can be used to measure stratigraphic variability in a fluvial sequence. The intrinsic lateral fluctuation, or nonuniformity, of sedimentation during early Chinji time averages 28%; vertical fluctuation, or unsteadiness of sedimentation, averages 41%. These fluctuations in sediment accumulation from place to place and from time to time characterize the fluvial system operating over the Potwar Plateau during Miocene time.

Introduction

In recent years paleomagnetism has found increasing use in stratigraphy as an analytical tool. Principal among these uses has been to establish the age and previous orientation of sedimentary basins. However, there is also a rather new application for paleomagnetic data in stratigraphy, one that utilizes the concept and properties of reversals in the Earth's magnetic field as a stratigraphic tool.

The Earth's magnetic field has reversed itself on numerous occasions in the geologic past. Even though the cause of these reversals is a matter of some dispute, and their specific behavior and duration generally unknown, the evidence for their existence in the geologic past is compelling. The time interval over which a magnetic reversal takes place seems to be less than 10,000 years (Opdyke *et al.* 1973). Therefore, in a long-term geologic context, a magnetic-reversal event can be considered essentially as an instantaneous event. If a magnetic-reversal transition can be recognized within a stratigraphic sequence, it can serve as a unique time marker for that sequence. If the same magnetic transition can be located in two adjacent stratigraphic sections, then a correlation can be made between the two sections, and a time line can be defined for the intervening space.

In a pioneering study by Barndt *et al.* (1978) such time lines and time surfaces were traced laterally over a wide belt of fluvial outcrops in Pakistan to correlate stratigraphic sections. In a comprehensive sequel to Barndt *et al.* (1978), Behrensmeyer and Tauxe (1982) used these time lines to demonstrate quantitatively the temporal and spatial interplay between two fluvial systems over the Indo-Gangetic plain during Miocene time. Following the techniques of Behrensmeyer and Tauxe (1982), Sheikh (1984) traced five superposed paleomagnetic time lines across a 15 km front of the Chinji Formation in Pakistan. Kappleman (1986) also traced magnetic reversals in the Chinji Formation to correlate widely separated homonid fossil localities. In this paper we will use the paleomagnetic data of Sheikh (1984) to measure quantitatively the variability in sedimentation rate of the Chinji Formation.

Geologic Setting and Methods of Investigation

The magnetic-polarity stratigraphy and age of the Chinji Formation were previously established at several localities, including its type section (Johnson *et al.* 1985). The Chinji Formation of the Potwar Plateau, Pakistan (Fig. 9.1) is a Middle Miocene flood-plain deposit of the ancestral Indus River. Its

---

† Deceased.

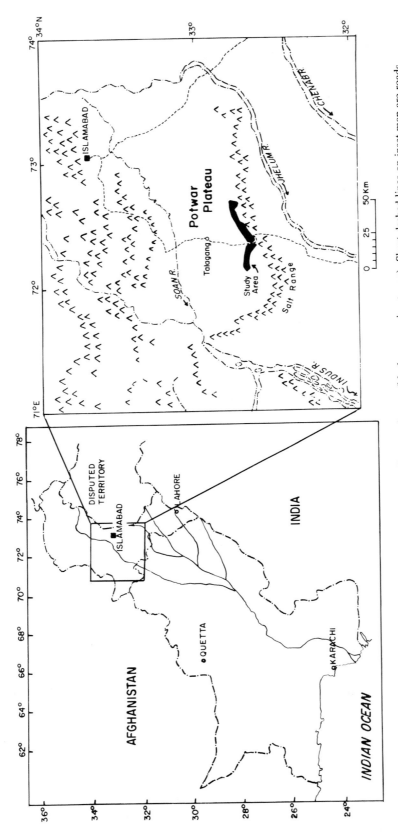

Fig. 9.1. Index map of Pakistan showing location of the Chinji Village study area (black area on inset map). Short dashed lines on inset map are roads.

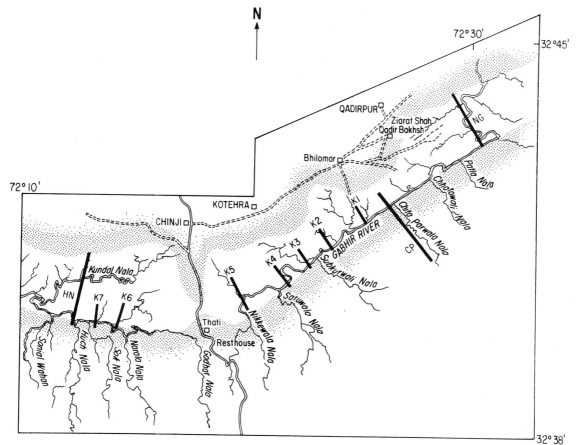

Fig. 9.2. Detailed geography of the Chinji Village study area. Black bars show general location of paleomagnetic sampling sections within the Chinji Formation (white area). Section abbreviations refer to the Chitta Parwala (CP) and Gabhir Kas (NG) sections of Johnson *et al.* (1985); Huch Nala (HN) and K1 through K7 are from this study.

stratigraphy is essentially a sequence of fluvial cycles (Allen 1965) whose return period ranges from 1×10^5 yr to 4×10^5 yr (Johnson *et al.* 1985).

The type area of the Chinji Formation presents a broad, exposed front of gently dipping Chinji strata (Fig. 9.2), a situation that lends itself to a study of chronostratigraphy. Our strategy has been to determine the magnetic polarity of a series of sections at convenient intervals along strike. In this effort we have collected samples at 232 paleomagnetic sites spread over a 15 km distance. Most of the sites are from the lower Chinji Formation where conspicuous units of sand could be traced easily along strike (Fig. 9.3). Seven partial stratigraphic sections were established and correlated to these tracer beds (Figs. 9.2, 9.3). An eighth section was established also at Huch Nala (Fig. 9.2), the western end of our study area, where the entire Chinji Formation was sampled.

To measure stratigraphic thicknesses we used an optical alidade to fix the position of key stratigraphic horizons, and then interpolated positions between these horizons using an Abney level and Jacob's staff. Paleomagnetic samples were collected exclusively from unweathered silts and clays and occasionally sandy mudstones; triplicate block samples were taken from each stratum sampled. The stratigraphic spacing between our paleomagnetic sites averaged 8.0 ± 1.0 m.

The magnetic mineralogy and magnetic cleaning techniques used for the Chinji Formation are described by Tauxe *et al.* (1980), Sheikh (1984), and Johnson *et al.* (1985). The primary remanence in these rocks is carried essentially by hematite. Each sample was thermally demagnetized starting at 500°C and carried to higher temperatures if required. The details of the laboratory methods and

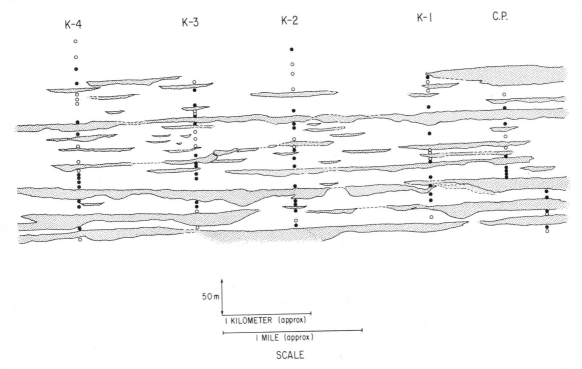

Fig. 9.3. Vertical and lateral distribution of sand in part of the lower Chinji Formation. See Figure 9.2 for location. Filled circles represent sites of normally magnetized rock; open circles represent sites of reversely magnetized rock. Sampling array and paleomagnetic data taken from Sheikh (1984).

results are given in Sheikh (1984). The set of magnetic directions obtained in this study yields a positive reversal test (Sheikh 1984), so that our demagnetization techniques can be considered adequate. Also, because of the positive reversal test the Chinji Formation rocks can be considered as faithful recorders of the Earth's magnetic field.

Analysis and Results

Figure 9.4 shows the superposition of paleomagnetic directions found at Huch Nala, which defines its magnetic-polarity stratigraphy. Also shown in Figure 9.4 is an abstract or log of the polarity data in the form of a black-and-white bar graph. In the Huch Nala section the Chinji Formation, as mapped (Raza 1983; Sheikh and Shah 1984), extends from the bottom of polarity zone N_0 downward to the middle of N_9.

Figure 9.5 shows the number and distribution of magnetic polarity zones in the seven stratigraphic sections between Chitta Parwala on the east and

Huch Nala on the west (Fig. 9.2). The black-and-white bars in Figure 9.5, like those in Figure 9.4, are actually an interpretation of the data in terms of normal and reversed magnetization and not the paleomagnetic data themselves. The complete paleomagnetic data set is described by Sheikh (1984). In Figures 9.4 and 9.5 we position the magnetic reversal midway between two paleomagnetic sites of unlike polarity, following the assumption that in the stratigraphic interval between two sites of unlike polarity, the reversal takes place exactly halfway between. Of course, the actual reversal could take place anywhere within the interval. We will discuss the implications of this assumption later.

Our data show that four polarity zones are continuously present through the entire 15 km distance sampled (Fig. 9.5). Five magnetic reversals were used as time lines for purposes of lateral tracing and correlation. These have been correlated to the Chitta Parwala section (Johnson *et al.* 1985), which we have in turn correlated to the magnetic-polarity time scale. Figure 9.6 shows the results of this identification and correlation between the Huch Nala and the Chitta Parwala sections. Visual inspection of the

Fig. 9.4. Local magnetic-polarity stratigraphy of the Huch Nala section (see Fig. 9.2 for location). Stratigraphy shown in the column on the left is the distribution of sand (white) and overbank silts and muds (black). Superposition of virtual geomagnetic latitudes is given in center panel. Filled circles are statistically significant means, open circles are data not statistically significant, but whose polarity is not in doubt. Right panel is an interpretation of the center panel, showing zones of normal magnetic polarity in black and zones of reversed magnetic polarity in white. Magnetic-polarity transitions are located halfway between adjacent sites of opposite polarity. Adjacent sites with like magnetic polarity are assumed to have no reversals in the intervening space.

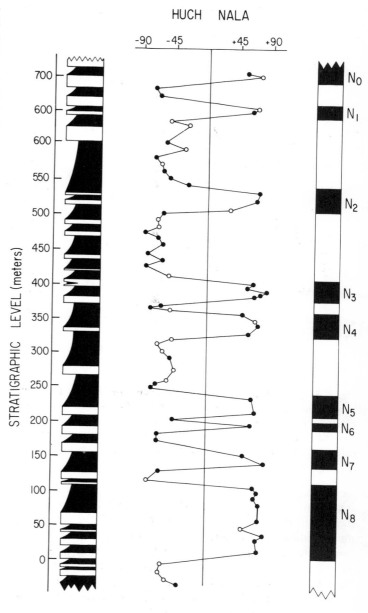

number and the relative thicknesses of the polarity zones in both the Chitta Parwala and Huch Nala sections shows them to be similar, allowing for a reasonable correlation of the sections with the magnetic-polarity time scale (Fig. 9.6).

The Chinji Formation as calibrated at Huch Nala extends from 14.5 Ma to 10.0 Ma (Fig. 9.6). This compares with the 14.5 Ma to 10.8 Ma age span for the same interval in the Chitta Parwala section (Johnson *et al.* 1985). As it is presently mapped, the upper boundary of the Chinji Formation is distinctly time transgressive along strike, varying by some 800,000 years over a 15 km distance. However, the base of the Chinji Formation over the same 15 km distance is basically synchronous.

Closer scrutiny of the two magnetic-polarity stratigraphies (Fig. 9.6) reveals some subtle differences underlying their basic similarities. Although the number of magnetic-polarity zones is the same for both sections (Fig. 9.6), the relative thicknesses of these zones are not quite the same. These differences imply that sedimentation was variable between the

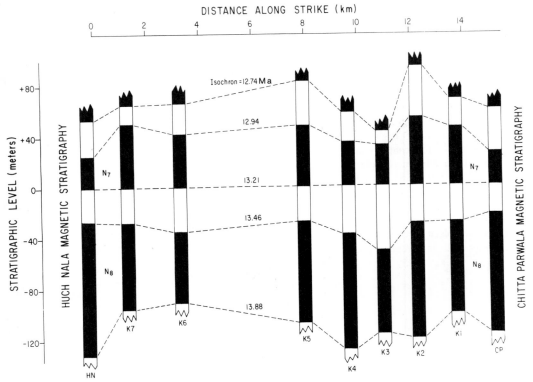

Fig. 9.5. Distribution of magnetic-reversal time lines (isochrons) between Huch Nala and Chitta Parwala (see Fig. 9.2 for location). Polarity time scale is that of Mankinen and Dalrymple (1979). Complete descriptions of the paleomagnetic sampling array and the paleomagnetic data are given in Sheikh (1984). In each stratigraphic sequence the position of each magnetic reversal is assumed to be midway between adjacent sites of opposite magnetic polarity.

two areas. Sediment accumulation curves for both these sections (Fig. 9.7) suggest that although the mean sedimentation rates for both sections are essentially equal, instantaneous sedimentation rates vary considerably. Sediment accumulation at Chitta Parwala follows a smoothly accelerating trajectory, while that at Huch Nala follows a sigmoidal course, showing an inflection point (Fig. 9.7).

This lateral variability in sediment-accumulation rate is also observed in the seven intermediate sections between Huch Nala and Chitta Parwala (Fig. 9.5). Note that we have made nine separate attempts at measuring the magnetic-polarity zonation in the 12.74 to 13.88 Ma time interval. Significantly, no two of these results appear the same (Fig. 9.5). Further inspection of these data also shows that longer chronostratigraphic intervals tend to have relatively less variability than shorter chronostratigraphic intervals (Fig. 9.5). For example, compare the rela-

tive variation for the entire interval (12.74–13.88 Ma) against a part of the interval, say the N_7 interval (12.94–13.21 Ma). This visual impression is reinforced by a statistical assessment of the data (Tables 9.1 and 9.2). There is clearly a systematic decrease in variation as larger chronostratigraphic increments are measured (Table 9.2). In other words when we attempted to measure a short chronostratigraphic interval, our precision (or reproducibility) is rather poor (Table 9.2). However, when we attempted to measure a longer chronostratigraphic interval, our precision became much better (Table 9.2).

Some of this apparent variability must be attributed to the sample spacing that was used to locate our magnetic-reversal time lines (Table 9.2). We should expect then that some fraction of the observed variability in our magnetic zonation is merely an artifact of our sample spacing and not a characteristic of the stratigraphy itself. The question

Fig. 9.6. Magnetic-polarity stratigraphy of the Chinji Formation stratotype, Potwar Plateau, Pakistan. Data taken from Johnson *et al.* (1985) and Figure 9.4 of this chapter. In the Huch Nala area, the Chinji Formation spans the interval from the top of polarity zone N_8 to the base of N_0. In the Chitta Parwala section, the Chinji Formation spans the interval from the top of polarity zone N_8 to N_2. The magnetic-polarity time scale is that of Mankinen and Dalrymple (1979).

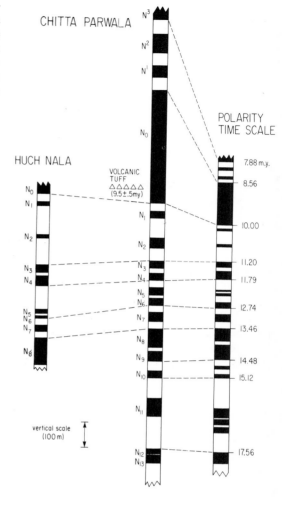

Fig. 9.7. Comparison of sediment accumulation history in the Chinji Formation at two locations, Huch Nala and Chitta Parwala. See Figure 9.2 for locations. As plotted, slope at each point along the curve is equivalent to mean sediment-accumulation rate.

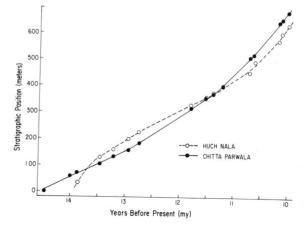

Table 9.1. Lateral variation in sediment accumulation in Lower Chinji Formation.

Time interval	HN	K7	K6	K5	K4	K3	K2	K1	CP	Mean ± 1σ
12.94–12.74 Ma	28 m	15 m	24 m	35 m	23 m	11 m	40 m	22 m	34 m	25.7 ± 9.5 m
13.46–13.21	27	28	35	28	38	51	30	29	23	32.1 ± 8.3
13.21–12.94	25	50	42	48	35	32	54	46	26	39.8 ± 10.7
13.88–13.46	105	68	56	80	91	66	91	72	94	80.3 ± 16.0
13.21–12.74	53	65	66	83	58	42	94	68	60	65.6 ± 15.3
13.46–12.94	52	78	77	76	73	83	84	75	49	71.9 ± 12.7
13.88–13.21	132	96	91	108	129	117	121	101	117	112.4 ± 14.4
13.46–12.74	80	93	101	111	96	94	124	97	83	97.7 ± 13.5
13.88–12.94	157	146	133	156	164	149	175	147	143	152.2 ± 12.4
13.88–12.74	185	161	157	191	187	160	215	169	177	178.0 ± 18.7

Refer to Figures 9.2 and 9.5 for the location and chronostratigraphic data, respectively.

that needs to be addressed is: How much of the observed variability depicted in Table 9.1 is due to sampling and how much is inherent in the stratigraphy itself? In order to demonstrate visually these two sources of fluctuation, the pertinent data from Table 9.1 are presented in graphical form (Fig. 9.8). In this plot the fluctuation due to sampling is depicted as a line whose slope is zero, and whose intercept is a function of sample spacing. On the other hand a linear regression of the observed data (Table 9.1) shows a statistically significant positive slope. This systematic slope can be attributed to the natural chronostratigraphic variability, that is, the fluctuations in the sedimentation rate of the Chinji Formation itself (Fig. 9.5). Had we been able to employ a hypothetically "perfect" sampling program, i.e., an infinite number of paleomagnetic sites, our sampling

fluctuation would approach zero, but we presumably would still have recognized the chronostratigraphic variability as an integral part of the Chinji sedimentary system. Note that in Figure 9.8 the slope estimate is based on a one-standard-deviation basis (1-sigma), or the 68% confidence level. Perhaps a more realistic and conservative measure in this case would be the 95% confidence level, or a two-standard-deviation basis (2-sigma). Table 9.2 presents these data in a 2-sigma form (95% confidence level). In our analysis of these data (Table 9.2) the mean chronostratigraphic fluctuation of the Chinji fluvial system averages some 28% expressed as a 2-sigma dispersion about the mean.

Sadler (1981) introduced the term "stratigraphic unsteadiness" for the variation in the sedimentation rate through time at a fixed point. Using our

Table 9.2. Analysis of lateral variation, Chinji Formation.

Time interval	Δt	Mean thickness	Observed fluctuation (2σ)	Sampling fluctuation*	Intrinsic fluctuation (2σ)
12.94–12.74 Ma	.20 m.y.	25.7 m	74.0%	21.8%	52.2%
13.46–13.21	.25	32.1	51.8	17.4	34.4
13.21–12.94	.27	39.8	53.8	14.1	39.7
13.88–13.46	.42	80.3	39.8	7.0	32.8
13.21–12.74	.47	65.6	26.6	8.5	18.1
13.46–12.94	.52	71.9	35.4	7.8	27.6
13.88–13.21	.65	112.4	25.6	5.0	20.6
13.46–12.74	.72	97.7	27.6	5.7	27.5
13.88–12.94	.94	152.2	16.2	3.7	12.5
13.88–12.74	1.14	178.0	21.0	3.1	17.9
Mean and standard deviation[†]					28 ± 12%

Data taken from Table 9.1.
*On the average there is ±4 metres of uncertainty in locating each time line. So, the uncertainty in the thickness of each time interval is $(4^2 + 4^2)^{.5} = \pm 5.6$ metres.
[†]Note the inverse relationship between Δt and stratigraphic fluctuation, a trend previously noted by Sadler (1981).

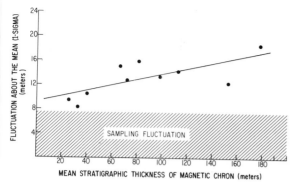

Fig. 9.8. Chronostratigraphic fluctuation of the Chinji Formation as a function of chronostratigraphic interval. Data taken from Table 9.1. Chronostratigraphic interval is taken as the mean vertical separation between two magnetic-reversal time lines (i.e., magnetic chron). Fluctuation or dispersion about the mean is expressed as 1 standard deviation. The regression line has a slope of .045, and an intercept of 9.3; its correlation coefficient is .727, which makes the slope significant at $p < .01$.

paleomagnetic data from the Chinji Formation (Table 9.1), we can estimate the variability in the rate of sedimentation for the Chinji fluvial system. By evaluating the mean sedimentation rate for each of our stratigraphic sections and establishing the dispersion about the mean, we can quantify the vertical variability or unsteadiness of sediment accumulation for each of our nine stratigraphic sections (Table 9.3). Using these estimates of variability (Table 9.3), we then averaged the set of values to characterize the Chinji sedimentary system as a whole (Table 9.4). This analysis yields a variability of 41% for the lower Chinji Formation in its type area (Table 9.4). Note that sampling fluctuation also enters as a factor in the analysis of vertical variability just as it did for lateral variability. As reported in Table 9.4 the values for unsteadiness in sedimentation rate are the composite of sampling fluctuation (column 4) and intrinsic variability (column 5). The reason for the small sampling fluctuation is that an overestimate of the stratigraphic thickness in one magnetic polarity zone causes a compensating underestimate in its adjacent zone.

At this point it may be apparent that our analysis of stratigraphic fluctuations, that is lateral and vertical variability, is verging on the matter of stratigraphic completeness (Sadler 1981). For example, we already have shown that the Chinji Formation is stratigraphically complete at a temporal scale of

10^5 yr. The magnetic-polarity time scale has a time resolution of 10^4–10^5 yr, and as we have observed it, the Chinji Formation reflects this time scale rather well (Fig. 9.6). Note, however, that there are magnetic-polarity zones missing in the Chinji Formation that are present in the time scale (Fig. 9.6). For example, the fine structure of the magnetic-polarity time scale between 11.79–12.74 Ma is not recorded accurately in the Chinji Formation given our present sampling program (Fig. 9.6). This suggests that somewhere at the 10^4 yr level of time resolution the Chinji Formation starts becoming incomplete, i.e., it contains sedimentary hiatuses long enough to miss recording entire magnetic-polarity zones. With closer paleomagnetic sampling and better time-line resolution, perhaps quantitative answers to the question of stratigraphic completeness may be possible.

In summary, these estimates of lateral fluctuations in sediment accumulation (28%), and vertical fluctuations in sediment accumulation (41%), serve to characterize the fluvial system that was responsible for depositing the Chinji sequence. The variation in sedimentation from place to place and through time directly reflects the dynamic processes in operation on the Potwar Plateau during Lower Chinji time.

Overview

In this paper we illustrate a new use for paleomagnetism in sedimentary rocks. We show that magnetic-reversal time lines are useful vehicles for lateral correlation of strata and estimating fluctuations in sedimentation rate. The basis of this technique is in the ability to find and identify magnetic reversals in the sedimentary sequence concerned. The starting point in this effort is the determination of the local magnetic-polarity stratigraphy. The correlation of magnetic-reversal time lines is, then, an extension of the magnetic-polarity stratigraphy method.

The goal of establishing a local magnetic-polarity stratigraphy is to determine the history of the Earth's magnetic field as recorded at the location concerned. The usual procedure is to collect a finite number of paleomagnetic samples at discrete intervals through the stratigraphic sequence. With this kind of sampling procedure there is inevitably a great deal of information missed in the stratigraphic interval between sampling sites. In fact, in most situations, there is likely to be much more chrono-

Table 9.3. Variation in sediment accumulation rate in the lower Chinji Formation.

Time interval	Δt	HN	K7	K6	K5	K4	K3	K2	K1	CP
12.94–12.74 Ma	.20 m.y.	.140 m/10³yr	.075 m/10³yr	.120 m/10³yr	.175 m/10³yr	.115 m/10³yr	.055 m/10³yr	.200 m/10³yr	.110 m/10³yr	.170 m/10³yr
13.46–13.21	.25	.109	.112	.140	.112	.152	.204	.120	.116	.092
13.21–12.94	.27	.093	.185	.156	.178	.130	.119	.200	.170	.096
13.88–13.46	.42	.250	.162	.133	.190	.217	.157	.217	.171	.224
13.21–12.74	.47	.113	.138	.140	.177	.123	.091	.200	.145	.128
13.46–12.94	.52	.100	.150	.148	.146	.140	.160	.162	.144	.094
13.88–13.21	.67	.197	.143	.136	.161	.193	.175	.181	.151	.175
13.46–12.74	.72	.111	.129	.140	.154	.133	.131	.172	.135	.115
13.88–12.94	.94	.167	.155	.141	.166	.174	.159	.186	.156	.152
13.88–12.74	1.14	.162	.141	.138	.168	.164	.140	.189	.148	.155
Mean ± 1σ		.129 ± .050	.139 ± .030	.139 ± .009	.163 ± .022	.154 ± .033	.139 ± .043	.183 ± .027	.145 ± .020	.140 ± .043

Based on chronostratigraphic thicknesses of Table 9.1. Refer to Figure 9.2 for location of stratigraphic sections.

Table 9.4. Analysis of fluctuation in sedimentation rate, Chinji Formation.

Sequence	Mean sediment accumulation rate	Observed fluctuation (2σ)	Sampling fluctuation*	Intrinsic fluctuation (2σ)
HN	.129 m/10³yr	78%	2.1%	76%
K7	.139	44	2.2	42
K6	.139	14	2.3	12
K5	.163	28	2.3	26
K4	.154	42	1.9	40
K3	.139	62	2.0	60
K2	.183	30	2.2	28
K1	.145	28	1.7	26
CP	.140	62	2.0	60
Mean and standard deviation of the mean				41 ± 7%

Data taken from Table 9.3.
*An estimate of sampling fluctuation is based on the fact that there are 10 separate thickness intervals, each of which has an uncertainty of ±5.6 m (see Table 9.2). The error propagated by combining these 10 thickness intervals is $(n\sigma^2)^{1/2}$, which in this specific case is ±17.7 m. Assuming no uncertainty in time, the sampling fluctuation for mean sedimentation rate is thus ±17.7 m/$\Sigma\Delta$m.

stratigraphic information bypassed than there is recorded. Under these circumstances it is necessary to make crucial assumptions about what has happened in the void between sampling sites (Johnson and McGee 1983). A local magnetic-polarity stratigraphy must necessarily be viewed as a thin network of data points filled in with large amounts of inference and assumptions. The portrayal of a magnetic-polarity stratigraphy as a black-and-white bar graph exemplifies this problem. An inspection of Figures 9.3, 9.4, and 9.5 illustrates the role of inference and assumption in establishing a magnetic-polarity stratigraphy.

Sampling strategy is then a critical issue in the conduct of magnetic-polarity stratigraphy and related studies (Johnson and McGee 1983). More paleomagnetic samples are presumably better, but just when are diminishing returns reached? As a general rule, if the ratio between number of magnetic reversals found and number of paleomagnetic data points used to find them is 1 to 7 or greater, then the sampling program probably has not missed any magnetic reversals in the sequence concerned (Johnson and McGee 1983). In a related context, when locating the exact position of a magnetic reversal within a stratigraphic sequence, a mean sample spacing of $< 10^5$ yr is required. Otherwise, the utility of the magnetic-reversal event as a point in time is diminished by sampling uncertainty.

As shown above, a magnetic-reversal time line can be used in a relative or absolute sense. A magnetic-reversal time line whose absolute age is not known is still a valid instrument for purposes of stratigraphic correlation. On the other hand, a magnetic-reversal event that has been uniquely identified in the magnetic-polarity time scale has a designated age and is intrinsically more useful. However, the assigned age for a reversal event is only as accurate as the calibration of the time scale itself. The age assignments in the magnetic-polarity time scale are in a state of flux at the present time (Berggren *et al.* 1985), especially for the older segments of the time scale. As it is presently constructed, the magnetic-polarity time scale is largely a convention, being constrained by only a few radiometric dates and replete with assumptions about what has happened between these isotopic anchor points. Undoubtedly, the magnetic-polarity time scale will become more accurate in the future as more radiometric ages with greater precision become available. In the meantime, however, it is important when using the time scale to specify which of its many versions has been used.

Despite the many complications and pitfalls associated with paleomagnetism in sedimentary rocks, its demonstrable power as a research tool will guarantee its increasing use in stratigraphy in the years ahead. Paleomagnetism has always been present in sedimentary rocks, but until recent years has been inaccessible and largely ignored. New instruments and new conceptual frameworks now make paleomagnetism a practical research tool, and open up new vistas in stratigraphic analysis. In this paper we present a case study illustrating one such use in stratigraphy.

Acknowledgments. We thank R. Tahirkheli, M. Raza, L. Tauxe, A. Johnson, and J. Stix for their active participation in this work. We are especially grateful to I. Khan whose assistance in the field was crucial. Financial support was provided by NSF grants INT-8308069, EAR-8206183, and EAR-8616767.

References

ALLEN, J.R.L. (1965) Fining upward cycles in alluvial successions. Geological Journal 4:229–246.

BARNDT, J., JOHNSON, N.M., JOHNSON, G.D., OPDYKE, N.D., LINDSAY, E.H., PILBEAM, D., and TAHIRKHELI, R.A.K. (1978) The magnetic polarity stratigraphy and age of the Siwalik Group near Dhok Pathan Village, Potwar Plateau, Pakistan. Earth Planetary Science Letters 41:355–364.

BEHRENSMEYER, A.K. and L. TAUXE (1982) Isochronous fluvial systems in Miocene deposits of Northern Pakistan. Sedimentology 29:331–352.

BERGGREN, W.A., KENT, D.V., FLYNN, J.J., and VAN COUVERING, J.A. (1985) Cenozoic geochronology. Bulletin Geological Society America 96:1407–1418.

JOHNSON, N.M. and McGEE, V.E. (1983) Magnetic polarity stratigraphy: Stochastic properties of data, sampling problems, and the evaluation of interpretations. Journal Geophysical Research 88:B2:1213–1221.

JOHNSON, N.M., STIX, J., TAUXE, L., CERVENY, P.F., and TAHIRKHELI, R.A.K. (1985) Paleomagnetic chronology, fluvial process and implications of the Siwalik deposits near Chinji Village, Pakistan. Journal Geology 93:27–40.

KAPPLEMAN, J. (1986) Paleontology and Magnetic Polarity Stratigraphy of the Chinji Formation near Kanatti, Potwar Plateau, Pakistan. Ph.D. Dissertation, Harvard University, Cambridge, Massachusetts, 316 p.

MANKINEN, E.A. and DALRYMPLE, G.B. (1979) Revised K-Ar polarity time scale. Journal Geophysical Research 84:615–626.

OPDYKE, N.D., KENT, D.V., and LOWRIE, W. (1973) Details of magnetic polarity transitions recorded in a high deposition rate deep-sea core. Earth Planetary Science Letters 20:315–324.

RAZA, S.M. (1983) Taphonomy and Paleoecology of Middle Miocene Vertebrate Assemblage Southern Potwar Plateau, Pakistan. Ph.D. Thesis, Yale University, New Haven, Connecticut, 212 p.

SADLER, P.M. (1981) Sediment accumulation rates and the completeness of stratigraphic sections. Journal Geology 89:569–584.

SHEIKH, K.A. (1984) Use of Magnetic Reversal Timelines to Reconstruct the Miocene Landscape near Chinji Village, Pakistan. M.S. Thesis, Dartmouth College, Hanover, New Hampshire, 58 p.

SHEIKH, K.A. and SHAH, S.M.I. (1984) Paleocurrent directions of Chinji Formation. Pakistan Geological Survey Memoir 11, pp. 75–80.

TAUXE, L., KENT, D.V., and OPDYKE, N.D. (1980) Magnetic components contributing to the NRM of Middle Siwalik red beds. Earth Planetary Science Letters 47:279–284.

Part III
Tectonics and Sedimentation: Introduction

HAROLD G. READING

The younger generation of sedimentologists considers basin analysis as a new and exciting field that has suddenly emerged in the 1980s because sedimentologists, stratigraphers, geophysicists, and structural geologists have finally discovered what they can learn by working with each other. Yet to those who remember the 1950s and early 1960s, sedimentation and tectonics, as well as basin analysis, were part of any course on sedimentation whether called "Sedimentary Environments" or "Sedimentary Structures."

The term "basin analysis" was introduced by Potter and Pettijohn (1963) in their influential textbook. At that time the object of basin analysis was to discern the geography of the past and the approach, therefore, was essentially paleogeographical (Potter and Pettijohn 1963, p. 224) with emphasis on the mode of basin fill, by axial or lateral supply. Basins were analyzed by a combination of paleocurrent measurements and facies distributions.

Basin analysis, that is the geographical reconstruction of a sedimentary basin, was simply a tool, along with lithofacies and isopach maps, to understand the relationships between sedimentation and tectonics. As Pettijohn (1957, p. 638) put it "... although sedimentation is affected by many factors, the most fundamental is tectonic." During the 1930s and 1940s this belief had been strongly emphasized in the United States by P.D. Krynine, W.C. Krumbein, F.J. Pettijohn, and L.L. Sloss, and in Europe by E.B. Bailey, O.T. Jones, and J. Tercier. Although there were slight differences in emphasis and terminology, nearly all these authors subscribed to the notion that there were consanguineous associations of sedimentary facies indicating tectonic regimes, such as the graywacke suite (flysch) and the subgraywacke suite (molasse) representing phases of mountain building; the orthoquartzite-carbonate suite sedimentation on cratons, and the arkosic suite sedimentation in grabens.

These concepts were dispersed by the widely read textbooks of Pettijohn (1957), Krumbein and Sloss (1963), and Potter and Pettijohn (1963). Nevertheless in a few years these concepts were largely forgotten. The younger generation of sedimentologists grew up in ignorance of them because of the rapid rediscovery of sedimentary processes. Sedimentologists concentrated on the hydrodynamics of clastic sediments, on the diagenesis of carbonates, and on the monitoring of modern processes and environments.

The new generation of process-oriented sedimentologists of the 1960s was highly critical of the direct matching of sedimentary suites with tectonic regimes. They saw that too little account had been taken of the complexity of sedimentary processes that gave rise to measured data. Too little was known about the effect of tidal currents, storms, or oceanic currents on directional data, or how the composition of sediments could be modified by weathering, transport, and particularly by diagenesis.

A new dogma arose. Sedimentary facies, sequences and cycles, had to be interpreted as responses to observable sedimentary processes, rather than nebulous tectonic movements. Meanwhile structural geologists were besotted with the geometries of folds, and even those concerned with broader scale tectonic patterns were indifferent to the sedimentary consequences of structural deformation.

Following the popularization of the plate-tectonic models at the end of the 1960s there were three lines of development. One line, followed by the majority of sedimentologists, was to continue to delve deeper into processes without much concern for the wider geological implications. The other two lines were two schools of tectonics and sedimentation. One, the W.R. Dickinson/K.A.W. Crook school, was really a continuation of the Krynine tradition. It emphasized the dominant role of the tectonic realm, now expressed in the language of plate tectonics, in controlling sandstone composition. The other school (J.C. Crowell, R.H. Dott, Jr., A.H.G. Mitchell, H.G. Reading, and many others) focussed more on sedimentary facies than on composition. They attempted to understand sedimentation in terms of tectonics established by consideration of the structural pattern, and the magmatic and metamorphic background.

It was only towards the end of the 1970s that tectonics, sedimentation, and basin analysis as we now conceive them, became established. It required improved quality of seismic data, basin modelling by theoretical geophysicists, and the realization of structural geologists and sedimentologists that each required the other. Of particular importance in this integration was the research of petroleum geologists in their search for means to establish the hydrocarbon potential of petroleum source rocks.

The papers in this part of the book indicate where we have reached today and demonstrate that the study of tectonics and sedimentation can be approached from three overlapping angles, by theoretical modelling, geophysical or sedimentological, by classical global tectonic modelling, and by detailed field studies of the relationship between sedimentation and tectonic evolution.

A major issue in basin analysis today concerns the causes of regionally wide unconformities and sea-level changes, which are not due to glaciations or local tectonics but yet are difficult to prove to be world-wide. Of particular importance in this debate is the contribution by Cloetingh. Showing first that the present-day stress patterns within plates are consistently oriented over wide regions, Cloetingh goes on to argue that large stress provinces can be recognized in the interior of plates and that these have changed over periods of a few million years. In doing so they influence vertical motions both in sedimentary basins and at basin margins. These movements may be up to 100 m at rates of $0.01–0.1$ m/10^3 yr. Thus they may well be the cause of sea-level changes, with sea-level rises associated with tensional intraplate stress and sea-level falls with a reduction in tension and increase in compressive stress. If true, apparent sea-level fluctuations can be used to monitor paleo-stress field changes.

To the process-orientated sedimentologist the term "pebble counter" has been a form of abuse over the past two decades; Paola, however, in his contribution, shows how that much neglected subject could be used to better understand the filling of basins. He has erected models that show why, in small, rapidly and/or asymmetrically subsiding basins, penetration of gravels is short and the clasts are sorted mainly by size. In large, slowly and/or uniformly subsiding basins, gravel penetration is large and the sorting of clasts is mainly by durability. Paola then develops a model that predicts the effects of variation in subsidence rate on grain-size variations, but as yet the data to test the model are rarely available in sedimentary basins because so few studies have been made that embrace grain-size, gravel content, and sediment thickness measurements, as well as dispersal patterns. Such studies are badly needed.

Four contributions in this volume address sedimentary basins in terms of models derived from plate tectonics, that is by comparing ancient basins with modern ones. Hsü takes the huge basins of Northwest China that are still little known and even less understood. They are cratonic blocks that have not risen like Tibet but subsided at rates of about 0.1 m/10^3 yr during early Mesozoic time. Hsü compares them with the present-day Black Sea and Caspian Basin and concludes that they are underlain by oceanic crust and were originally back-arc basins.

A well documented and more detailed story emerges from Taiwan where Lundberg and Dorsey reconstruct an oblique arc-continent collision over the last 5 million years in an area of southward migrating zipper closure where the Plio-Pleistocene rocks of eastern Taiwan show the results of present-day processes in the South China Sea to the south. Information is available not only on plate motions but on provenance and on rates of subsidence, uplift and erosion. The importance of their study is to show how a forearc basin in eastern Taiwan changed into a collisional basin with many similarities to the classic foredeep of western Taiwan. Largely due to back thrusting, subsidence rates reached an incredible 5 m/10^3 yr for a short period of time. Innumerable such complex and short-lived basins must have developed around the Pacific on the oceanward side of tectonically accreted fold-thrust belts as intraoceanic arcs collided with the continental margin.

While the reconstruction of basins filled mainly by unmetamorphosed sediments is not easy, unravelling the tectonic evolution of heavily deformed and metamorphosed basins is a lot more difficult. Yet if we are to fully understand the tectonic history of a region we need to delve back into its pre-metamorphic history. Walker describes some of the general methods, with their limitations, using the Precambrian to Triassic rocks of the Mojave desert region as an example. He evaluates not only the paleogeography and source regions, but also the metamorphic and tectonic regimes, and their relationships to each other.

Although allochthonous terranes have become fashionable only in the last decade, the concept is not a new one. Lost microcontinents have been postulated for a century and far-travelled terranes have been implicit in many global reconstructions at least since the advent of plate-tectonic theory. Even now, while it may not be too difficult to identify terranes as allochthonous, their place of origin and line of transport are not so easy to reconstruct. Kleinspehn demonstrates how important it is in a complete basin analysis to separate both the pre-amalgamation and accretion history from the post-amalgamation and accretion history. To do this all available pieces of paleontological, lithostratigraphic, paleomagnetic, and geomorphic data should be used.

The last three contributions are concerned with fold-thrust tectonics and foreland basins and they all demonstrate by detailed mapping that the relationships between thrusting and sedimentation are not simple. As Jordan, Flemings, and Beer show, fault motion should be dated by cross-cutting relations of the faults, by subsidence history, unroofing petrol-

ogy, and by facies migration. However, an understanding of source rock lithology, of climate in the source area, and of sea-level changes is required if the correct deductions are to be made. In addition, many foreland basins are broken by uplifted blocks. The structural complexity of foreland basins is demonstrated even more dramatically by Burbank and Raynolds. Their use of magnetostratigraphy has made the northwestern Himalayan foredeep the most precisely documented foredeep that we have. They show not only the complex sedimentary pattern, and the rotation of some blocks as much as 60°, but, of most importance, their data indicate that thrust-front migration is not smooth and systematic but has complex sequences of deformation unevenly distributed in time and space. Steidtmann and Schmitt also stress the necessity to understand the fault sequence before dating thrusts from the sediments. In particular they show that active sources should be distinguished from passive sources since inactive terranes may be transported over a ramp, also to provide source rocks. In addition, sediment dispersion is not always synthetic to fold-thrust propagation but may be antithetic. Finally, complex clast compositions may be produced by tectonic inversion, by blending, or by cannibalization.

References

KRUMBEIN, W.C. and SLOSS, L.L. (1963) Stratigraphy and Sedimentation. San Francisco: Freeman, 660 p.

PETTIJOHN, F.J. (1957) Sedimentary Rocks. New York: Harper, 718 p.

POTTER, P.E. and PETTIJOHN, F.J. (1963) Paleocurrents and Basin Analysis. Berlin: Springer-Verlag, 296 p.

10
Intraplate Stresses: A New Element in Basin Analysis

SIERD CLOETINGH

Abstract

Evidence reviewed in this paper indicates that intraplate stresses in the lithosphere are of substantial magnitude. Numerical modelling and observation of modern and paleo-stress fields demonstrate the existence of stress provinces of great areal extent in the interiors of the plates. The interaction of intraplate stresses with basin subsidence provides a new element in basin analysis. Fluctuations in intraplate stress fields influence basin stratigraphy and provide a tectonic explanation for short-term, relative sea-level variations inferred from the sedimentary record. Modelling shows that the incorporation of intraplate stresses in models of basin evolution can predict a succession of onlap and offlap patterns similar to those observed at basin flanks. Such a stratigraphy can be interpreted as the natural consequence of short-term changes in basin shape by moderate fluctuations in intraplate stresses, superimposed on long-term broadening of the basin with cooling since its formation. Basin stratigraphy could provide a new source of information for paleo-stress fields.

Introduction

During the last decade considerable progress has been made in the study of the stress field within lithospheric plates. Detailed analysis of earthquake focal mechanisms (e.g., Ahorner 1975; Bergman 1986), in-situ stress measurements (e.g., McGarr and Gay 1978; Illies et al. 1981), and analysis of break-out wells drilled for commercial purposes (Bell and Gough 1979; Blumling et al. 1983; Zoback et al. 1985) have demonstrated the existence of consistently oriented stress patterns in the lithosphere.

Simultaneously, numerical modelling (Richardson et al. 1979; Wortel and Cloetingh 1981, 1983; Cloetingh and Wortel 1985, 1986) has resulted in better understanding of the causes of the observed variations in stress level and stress directions in the various lithospheric plates. Such studies have shown a causal relationship between the processes at plate boundaries and the deformation in the plates' interiors (e.g., Johnson and Bally 1986).

In models of the evolution of sedimentary basins located in the interiors of the plates (e.g., Beaumont 1978; Watts et al. 1982), however, the role of intraplate stresses has been largely ignored. These current geophysical models are not yet able to explain much of the observed evolution of the tectonic component of basin subsidence. Recently, the first steps have been taken towards exploring the consequences of the existence of intraplate stress fields for models of the formation and evolution of intracratonic basins (Lambeck 1983; de Rito et al. 1983). Intraplate stresses have also been demonstrated to be an important element in basin stratigraphy. Work by Cloetingh et al. (1985) and Cloetingh (1986) has shown that temporal fluctuations of the intraplate stress could provide a tectonic explanation for short-term apparent sea-level changes inferred from seismic stratigraphic analysis of the sedimentary record (Vail et al. 1977, 1984; Haq et al. 1987).

The present paper reviews evidence for the existence of intraplate stress fields in the lithosphere. This is followed by a discussion on some implications of intraplate stress for quantitative modelling of the evolution of sedimentary basins. Finally, the potential use of the sedimentary record to extract information on paleo-stress fields in the plates is explored.

Intraplate Stress Fields

The present stress field in the various plates has been studied in great detail by the application of a wide range of observational techniques. The results of a recent compilation of stress-direction data for the northwestern European Platform by Klein and Barr (1986) are displayed in Figure 10.1. The observed modern stress orientations show a remarkably consistent pattern, especially considering the heterogeneity in lithospheric structure in this area. These stress-orientation data indicate a propagation of stresses away from the Alpine collision front over large distances in the platform region. Observations of stress orientation in different continental and oceanic regions have demonstrated the existence of such large stress provinces with preferred stress directions to be a general characteristic of lithospheric plates (e.g., Bell and Gough 1979; Zoback and Zoback 1980; Lambeck et al. 1984; Bergman and Solomon 1985). These observations indicate that regional stress fields are dominated by the effect of plate-tectonic forces acting on the lithosphere.

For the analysis of paleo-stress fields, however, methods used to study the modern stress field cannot be applied. In this case, the information is derived from analysis of the geological record, such as stylolites, microstructures, or fault-orientation data (e.g., Letouzey 1986). As such, the inferred information on paleo-stress fields is less precise than the results of studies of modern stress indicators. The study of paleo-stress fields, however, adds geological time as a parameter crucial to understanding the temporal fluctuations of stress fields in the plates.

In this paper I concentrate on the regional stresses in the lithosphere induced by plate-tectonic forces. Other sources of stress, however, might dominate on a more local scale. Examples are stresses associated with topographic anomalies and crustal-thickness inhomogeneities at passive margins. Further, temperature variations may induce thermal stresses in cooling lithosphere, while membrane stresses are possibly induced by plate motions on a nonspherical earth. A number of these stress mechanisms have been reviewed by Turcotte and Oxburgh (1976). Finally, flexural stresses are generated due to vertical loads on the lithosphere, in particular by sedimentary sequences at passive margins (Cloetingh et al. 1982). Of the various locally induced stress sources the flexural stresses and thermal stresses stand out in magnitude (up to order of kbars), while most of the other mechanisms produce stresses with a characteristic level of the order of a few hundred bars (Turcotte and Oxburgh 1976; Cloetingh et al. 1982).

A complementary approach to collecting stress-indicator data is the study of the intraplate stress field using numerical modelling techniques. In the first phase of modelling intraplate stress fields resulting from plate-tectonic forces, models were tested against stress-orientation data inferred from earthquake focal-mechanism studies to quantify the relative and absolute importance of various possible driving and resistive forces (Solomon et al. 1975; Richardson et al. 1979). These and several other studies (Forsyth and Uyeda 1975; Chapple and Tullis 1977) resulted in the overall understanding that ridge push, which results from the elevation of the spreading ridge above the adjacent ocean floor and the thickening of the lithosphere with cooling, and slab pull, which acts on the downgoing slab in a subduction zone, are the two main driving forces. Since then, deeper understanding has been obtained of the age dependence of the forces acting on the lithosphere (Lister 1975; England and Wortel 1980). This development has benefitted from advances in the analysis of the subduction process (Vlaar and Wortel 1976; Wortel 1984). By implementing these new features and insights in stress modelling, Wortel and Cloetingh (1981, 1983) and Cloetingh and Wortel (1985, 1986) showed that the dynamic basis of their numerical modelling procedure enabled the resulting intraplate stress field to be used to analyze, explain, and even predict various deformational processes within the lithospheric plates.

Because of better constraints on the thermomechanical and tectonic evolution of the oceanic lithosphere, which is relatively well understood, these models have concentrated on the oceanic plates or lithospheric plates with major oceanic parts. However, the comparison between calculated stress fields in various primarily oceanic plates sheds new light on observations of modern and ancient stress fields in oceanic and continental lithosphere. Modelling of the stress field in the Indo-Australian Plate (Cloetingh and Wortel 1985, 1986) has shown that the joint occurrence in this single plate of an exceptionally high level of compressive deformation in the plate interior (McAdoo and Sandwell 1985; Bergman and Solomon 1985; Lambeck et al. 1984) and near-ridge parallel extensional deformation (Wiens and Stein 1984; Stein et al. 1987) is a transient feature unique to the present

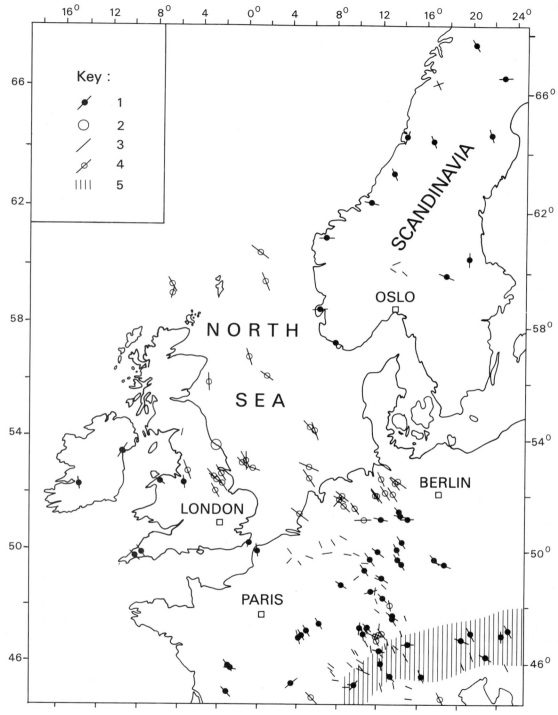

Fig. 10.1. Compilation of observed maximum horizontal present stress directions in the northwestern European Platform. 1 = the direction of maximum horizontal stress from *in-situ* measurements, 2 = a horizontal stress equal in all directions as found from *in-situ* stress measurements, 3 = the direction of maximum horizontal stress inferred from earthquake focal-mechanism studies, 4 = the direction of maximum horizontal stress inferred from break-out analysis, 5 = Alpine fold belt. The data indicate stress propagation away from the Alpine fold belt in the platform region (after Klein and Barr 1986).

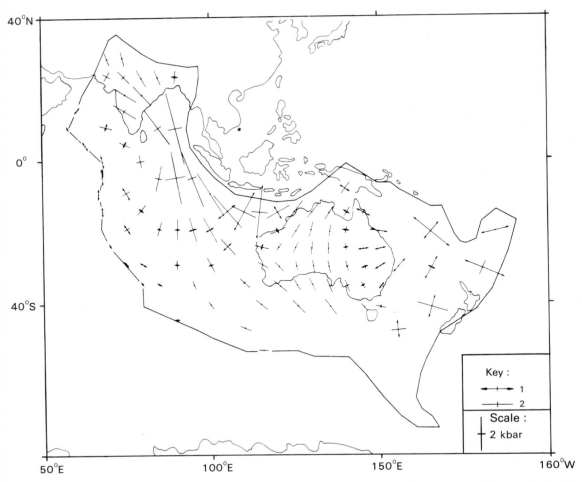

Fig. 10.2. Calculated stress field in the Indo-Australian Plate. Plotted are the horizontal nonhydrostatic stresses averaged over a uniform elastic plate with a reference thickness of 100 km. 1 = tension, 2 = compression. The length of the arrows is a measure of the magnitude of the stresses. The plate includes different stress provinces characterized by high levels of compressional and tensional stresses (after Cloetingh and Wortel 1986).

dynamic situation of the Indo-Australian Plate. The stresses in the Indo-Australian Plate (Fig. 10.2) are an order of magnitude greater than those we have calculated using the same numerical techniques for the Nazca Plate, which is characterized by a low level of intraplate seismicity and deformation. In the Nazca Plate stresses are on the order of 500 bars (Wortel and Cloetingh 1983), a stress level more characteristic of plates not involved in continental collision or rifting processes. The high level of intraplate deformation in the Indo-Australian Plate is manifested as an exceptionally high level of intraplate seismicity. This is particularly the case in the northeastern Indian Ocean sector of the plate, which is at present the most seismically active oceanic intraplate region on Earth. Here stress level in the plate reaches a maximum due to focusing of compressional resistance associated with Himalayan collision and subduction of relatively young oceanic lithosphere in the northern part of the Sunda Arc. The high seismicity level in this area makes it especially suitable for the determination of the intraplate stress-field orientation from earthquake focal-mechanism data (Bergman and Solomon 1985). The stress-orientation data from Bergman and Solomon (1985) given in Figure 10.3b demonstrate a rotation of the observed stress from N-S oriented compression in the north to a more NW-SE directed compression in the southeastern part of the Bay of Bengal region, in agreement with the calcu-

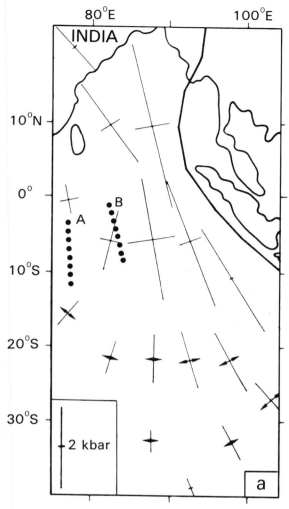
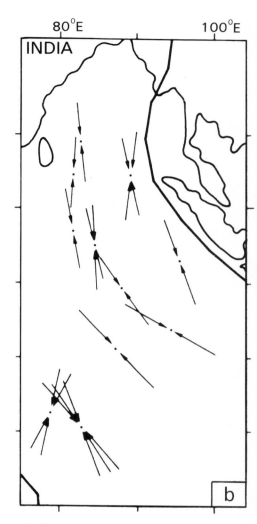

Fig. 10.3. Intraplate stress field in the northeastern Indian Ocean. (a) Calculated stress field after Cloetingh and Wortel (1986). Dotted lines A and B give the location of two long seismic-reflection profiles given in Figure 10.14 that show significant deformation of the basement in the northeastern Indian Ocean (Geller *et al.* 1983). (b) The orientation of maximum horizontal compressive stress inferred from an earthquake focal-mechanism study by Bergman and Solomon (1985).

lated stress field. Furthermore, the intraplate stress field as calculated (Fig. 10.3a) provides a consistent explanation for the observed significant compressional deformation in the oceanic crust in this area (Geller *et al.* 1983; McAdoo and Sandwell 1985).

Analysis of earthquake focal-mechanism data shows that large parts of the Australian continent are also in a state of significant horizontal compression (Lambeck *et al.* 1984). On the basis of observational evidence and modelling of gravity and topography, Lambeck *et al.* (1984) and Stephenson and Lambeck (1985) have suggested a magnitude of the order of 1–2 kbar for the intraplate stress field in large parts of the Australian continent (Fig. 10.4b), a stress level consistent with model predictions (Fig. 10.4a). As displayed in Figure 10.4, the modelling shows that the observed rotation of the compressional intraplate stress field in the western and central parts of the Australian continent is mainly the consequence of its geographic position relative to surrounding plate boundaries. Eastern Australia, an area characterized by recent volcanic activity, probably forms a separate stress province (Fig. 10.4a) with an intraplate tensional stress regime (Duncan

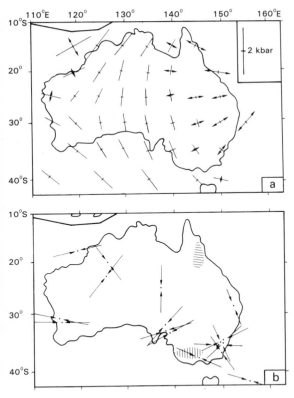

Fig. 10.4. Intraplate stress field in the Australian continent and peripheral areas. (a) Calculated stress field, after Cloetingh and Wortel (1986). (b) Stress-orientation data from earthquake focal-mechanism studies (Lambeck *et al.* 1984). Hatched areas mark the regions with Late Pliocene to present basaltic volcanism in northern Queensland and Victoria (Duncan and McDougall in press).

that stress provinces can vary in size from an entire lithospheric plate to only portions of plates (e.g., Zoback and Zoback, 1980; Gough *et al.* 1983). Modelling results such as displayed in Figure 10.2 corroborate the observed stress patterns (Fig. 10.1) by indicating far-field propagation of stress changes induced at convergent margins and collision zones through the interiors of the plates, to affect passive margins and intracratonic basins. Concentration of slab-pull forces dominates the plate-tectonic stress field (Wortel and Cloetingh 1983; Patriat and Achache 1984; Cloetingh and Wortel 1985). Numerical modelling has demonstrated that temporal changes in stress are not limited to plate collision but also occur through rifting and fragmentation of lithospheric plates (Wortel and Cloetingh 1983). These are especially important for passive margins of plates not involved in collision or subduction processes. As pointed out by Engebretson *et al.* (1985) major plate reorganizations generally occur with characteristic intervals of a few tens of millions of years. The duration of the individual events, and hence the associated stress changes, is however much shorter and on the order of at most a few million years (Engebretson *et al.* 1985). Geological evidence also points to episodic tectonic events on time scales of a few million years (Megard *et al.* 1984; Meulenkamp and Hilgen 1986), possibly caused by the response of individual plates to larger global readjustments of the plate-motion patterns.

Intraplate Stress and Basin Evolution

The evolution of sedimentary basins is in large part controlled by the response of the underlying lithosphere to the various tectonic loads. Lithospheric flexure forms an important element in determining this response (Beaumont 1981; Watts *et al.* 1982). We therefore begin this section with a brief discussion of the flexural response of the lithosphere to intraplate stresses.

The Effect of Intraplate Stress on the Deflection of the Lithosphere

In classical studies, Smoluchowski (1909), Vening Meinesz (see Heiskanen and Vening Meinesz 1958), and Gunn (1944) have investigated the flexural response of the lithosphere to applied horizontal

and McDougall in press), associated with transmission of tensional stresses from the adjacent plate boundaries eastward from the continent. Similarly, the stress province adjacent to the Southeast and Central India Ridges, which exhibits an exceptionally high level of intraplate tensional seismicity (Fig. 10.2), is causally related to far-field stress propagation of tensional stresses induced by the subduction of old oceanic lithosphere at the Java segment of the Sunda Arc (Stein *et al.* 1987).

These numerical models have shown that high-magnitude stresses can be concentrated in the plates' interiors. They have also shown, in agreement with observations (Fig. 10.1), that long-wavelength spatial variations in the stress field may occur, although such features do not necessarily exist in all plates. These models and observations make clear

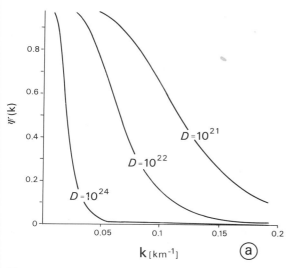

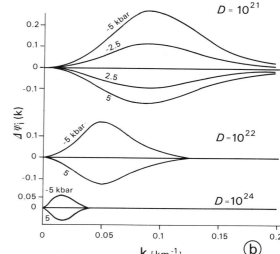

Fig. 10.5. (a) Flexural response functions $\Psi(k)$, in the absence of an intraplate stress field, for elastic thin plates with various flexural rigidities D in units of Newton metres, plotted as a function of wave number k. (b) The effect of intraplate stresses σ_N (tension is positive) on the response function. $\Delta\Psi_i(k) = \Psi_i(k) - \Psi(k)$, with $\Psi_i(k)$ the response function for an applied intraplate stress field. Results are given for the same flexural rigidities for various intraplate stresses σ_N in units of kbars. (After Stephenson and Lambeck 1985.)

forces. The flexural response of a uniform elastic lithosphere at a position x to an applied horizontal force N and a vertical load q(x) is given by:

$$D\frac{d^4w}{dx^4} - N\frac{d^2w}{dx^2} + (\rho_m - \rho_i)gw = q(x)$$

where w is the displacement of the lithosphere, and D the flexural rigidity (D = $ET^3/12(1-v^2)$), with E the Young's modulus, T the plate thickness, and v the Poisson's ratio.

The axial load N is equivalent to the product of the intraplate stress σ_N and the plate thickness T. ρ_m and ρ_i are, respectively, the densities of mantle material and the infill of the lithospheric depression, usually water or sediment, and g is the gravitational acceleration. The solution to this classical equation is easily obtained for some simple loading cases (e.g., Turcotte and Schubert 1982). Assuming zero vertical load on the lithosphere, the early studies made a convincing case for neglecting horizontal forces in modelling the vertical motions of the lithosphere. They showed that for compressional forces below the buckling limit the induced vertical displacements of the lithosphere are negligible. This result, combined with lack of evidence for the existence of such horizontal forces, led for a long time to the withdrawal of attention from this topic. As has

been discussed earlier, significant progress has been made recently in the study of horizontal stress fields in the lithosphere. At the same time, it is now realized that vertical motions of the lithosphere at sedimentary basins are primarily the result of a variety of other processes such as thermally induced cooling of the lithosphere amplified by the loading of sediments that accumulate in these basins (Sleep 1971), isostatic response to crustal thinning (McKenzie 1978), and flexural bending in response to vertical loading (Beaumont 1981). Hence, as pointed out by Cloetingh *et al.* (1985), it is essential to account for the presence of already existing vertical loads on the lithosphere when solving for the response of the lithosphere to applied horizontal loads.

In analytical solutions of the equation describing the flexural behavior of thin elastic plates, the loading response of the plate is traditionally decomposed into its harmonic components by transforming the equation to the Fourier domain (Stephenson and Lambeck 1985). The flexural response function $\Psi_i(k)$ in the presence of a horizontal load N can be written as:

$$\Psi_i(k) = [1 + \frac{D(2\pi k)^4 - N(2\pi k)^2}{\rho_m g}]^{-1}$$

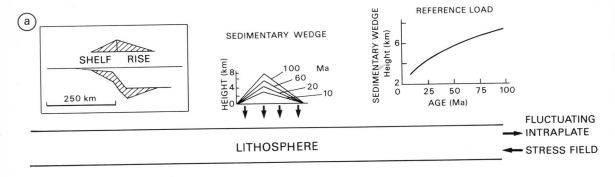

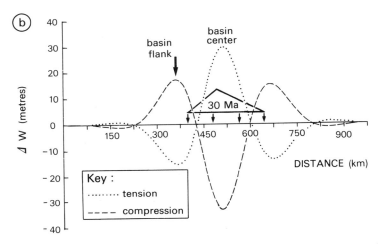

Fig. 10.6. (a) Model for apparent sea-level fluctuations resulting from variations in the intraplate stress field proposed by Cloetingh *et al.* (1985). The vertical displacement of the lithosphere at a passive margin evolves through time because of the thermal contraction and strengthening of the lithosphere and the loading with a wedge of sediments. Inset on the left shows position of this wedge on the outer shelf, slope, and rise. Sedimentation is assumed to be sufficiently rapid to equal approximately subsidence rate (Turcotte and Ahern 1977). The height and the width of the sedimentary wedge (see inset on the right for reference model of sediment loading) are given in kilometres at some selected time steps (specified in million years). The lithosphere is modelled as a uniform elastic layer. Differences between continental and oceanic lithosphere are neglected. (b) Effect of variations in intraplate stress fields on the deflection of a plate with a uniform elastic thickness (22 km) corresponding to a lithospheric age of 30 million years. Differential subsidence or uplift ΔW (metres) from the deflection in the absence of an intraplate stress field is given for intraplate stress fields of 1 kbar compression (dashed curve) and 1 kbar tension (dotted curve). Note the opposite effects at the flanks of the basin and at the basin center.

If N = 0, then $\Psi_i(k) = \Psi(k)$, the flexural response function of the plate in the absence of intraplate stress.

The wave number k at which the intraplate stresses most affect the flexural response of the lithosphere is almost completely determined by the plate's flexural rigidity (Stephenson and Lambeck 1985). These authors showed that the presence of intraplate stresses has a small but perceptible effect on this wave number, but exerts a controlling influence on

the amplitude of the response for a given flexural rigidity. These features are illustrated in Figure 10.5, which shows the effect of intraplate stress fields of a magnitude of a few kbars on the flexural response of an elastic lithosphere.

The analytical formulation of specific simplified problems shows explicitly how the solution depends on various parameters. Numerical modelling techniques have the advantage of allowing more realistic geometries and variation in parameters to be handled, adding flexibility to the analysis. Using numerical models, Cloetingh *et al.* (1985) considered the case of a passive margin evolving through time in response to changing thermal conditions and loading with a wedge of sediments (Turcotte and Ahern 1977). They showed that vertical deflections of the lithosphere of up to a hundred metres may be induced by the action of intraplate stress fields with a magnitude of up to a few bars (Fig. 10.6). To model passive-margin processes we considered for convenience an elastic oceanic lithosphere that cools with time according to the boundary-layer model of thermal conduction. Analysis of the flexural response of oceanic lithosphere to tectonic processes (Bodine *et al.* 1981) as well as seismotectonic studies (Stein and Wiens 1986) show an increase in the elastic thickness of the oceanic lithosphere with age. Thus the response of the oceanic lithosphere to sediment load (Watts *et al.* 1982), and the intraplatehd stress field, is time dependent not only because the sediment load accumulates with time but also because of the changing mechanical properties of the lithosphere.

The vertical motions become more complex when the effective elastic thickness of the lithosphere is laterally variable (Artyushkov 1974; Cloetingh *et al.* 1985). For passive margins, this may result in an additional tilting of the lithosphere at the basin edge. This tilting is a consequence of the change in thickness of the layer that carries the intraplate stress, possibly due to changes in lithospheric thinning for rifted basins, and occurs even in the absence of sediment loading. Vertical motions at the basin edge due to intraplate stresses (Fig. 10.6) are amplified by this mechanism when the effective elastic thickness of continental lithosphere is less than that of oceanic lithosphere. As this effect is highly dependent on the rheological contrast between oceanic and continental lithosphere (e.g., Vink *et al.* 1984), the degree of induced amplification or reduction will vary for different basins.

Consequences for Basin Stratigraphy

In the previous section I pointed out that temporal fluctuations in stress are a natural consequence of the horizontal motions of the lithospheric plates. These findings have important consequences for short-term temporal variations in the basin shape through geological time (Cloetingh *et al.* 1985). This is illustrated by Figure 10.7, which displays the

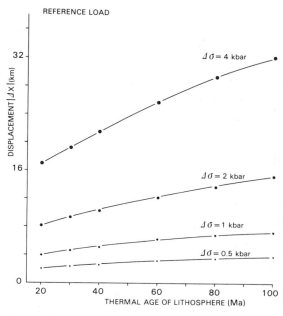

Fig. 10.7. Migration of the flexural node inferred from model calculations shown in Figure 10.6b. Apparent horizontal displacements $|\Delta(x)|$ (km) are plotted as a function of the age of the underlying lithosphere and sediment loading according to the reference model (after Turcotte and Ahern 1977) of Figure 10.6a. Curves give results for stress changes $\Delta\sigma$ of .5, 1, 2, and 4 kbar, respectively. Note that the horizontal expression of stress-induced short-term phases of coastal onlap and offlap increases with age.

horizontal migration of the flexural node under the influence of intraplate stresses. Curves show the horizontal displacement as a function of the age of the lithosphere for changes in stress of 0.5 kbar, 1 kbar, 2 kbar, and 4 kbar. For an intraplate compressional pulse the deflection of the lithosphere is abruptly narrowed and the flexural node migrates inward to the basin center. For an intraplate tensional pulse, rapid widening of the basin occurs and the flexural node migrates outward away from the center of the basin. These short-term widening and narrowing events are equivalent to rapid phases of coastal onlap and offlap, respectively. Inspection of Figure 10.7 shows that for a given stress change the horizontal extent of the basin area involved in a short-term phase of coastal onlap or offlap increases with lithospheric age. This feature is primarily the consequence of the increase in flexural stiffness of the lithosphere with age.

Sedimentation is assumed here to be sufficiently rapid to equal approximately the subsidence rate

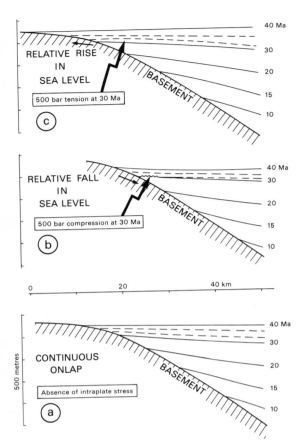

Fig. 10.8. Idealized stratigraphy at the edge of a basin underlain by 40 Ma lithosphere predicted on the basis of the calculations shown in Figure 10.6b. (a) Continuous onlap associated with long-term cooling of the lithosphere in the absence of an intraplate stress field. (b) A transition to 500-bar compression at 30 Ma induces a short-time phase of offlap at that time (indicated by position of short bold arrow) and an apparent fall in sea level superimposed on the thermal cooling subsidence curve. (c) A transition to 500-bar tension produces an additional short-term phase of onlap at 30 Ma (see position of short bold arrow) and an apparent rise in sea level.

(Turcotte and Ahern 1977). Although this reference model is certainly an oversimplification (see Pitman 1978 and Pitman and Golovchenko 1983 for a discussion), the assumption of sedimentation rate equalling subsidence rate allows one to equate onlap and offlap with rises and falls in sea level. The vertical motion of the lithosphere at the flank of the basin is of particular stratigraphic interest. Figure 10.8 schematically illustrates the relative movement between sea level and the lithosphere at the flank of

the basin immediately landward of the principal sediment load as predicted on the basis of numerical calculations. When horizontal compression occurs, the peripheral bulge is magnified while simultaneously migrating in a seaward direction, uplift of the basement takes place, an offlap develops, and an apparent fall in sea level results, possibly exposing the sediments to produce an erosional or weathering horizon. Simultaneously, the basin center undergoes deepening (Fig. 10.6b), resulting in a steeper basin slope. For a horizontal tensional intraplate stress field, the flanks of the basin subside with its landward migration producing an apparent rise in sea level so that renewed deposition, with a corresponding facies change, is possible. In this case the center of the basin shallows (Fig. 10.6b), and the basin slope is reduced. The synthetic stratigraphy at the basin edge is schematically shown for the following three situations: long-term widening of the basin with cooling in the absence of an intraplate stress field (Fig. 10.8a), the same case with a superimposed transition to 500 bar compression at 30 Ma (Fig. 10.8b), and the case of a stress change to 500 bar tension at 30 Ma (Fig. 10.8c).

Beaumont (1978) and Watts et al. (1982) pointed out the differences in stratigraphy predicted on the basin edge for elastic and viscoelastic models in the absence of intraplate stress fields. These authors showed that for elastic models the width of the basin increases with time, inducing progressive onlap of sediments caused by the increase of the thickness of the elastic lithosphere since basin formation. They demonstrated that viscoelastic models for the lithosphere predict narrowing of the infill of the basin with younger sediments restricted to the basin center. However, care should be taken in rigidly classifying basins into elastic type or viscoelastic type strictly on the basis of this criterion. As pointed out by Allen et al. (1986), "from a sedimentologist's or stratigrapher's point of view there has been a largely futile but understandable search for basin geometries that support an elastic lithosphere at one extreme or a viscoelastic lithosphere at the other." The incorporation of intraplate stresses in elastic models of basin evolution can in principle predict a succession of onlaps and offlaps such as observed along the flanks of the Central North Sea Basin during Tertiary time (Fig. 10.9). Such a stratigraphy can be interpreted as a natural consequence of the mechanical widening and narrowing of basins by fluctuations in intraplate stress fields superimposed on the long-term broadening of the basin with cool-

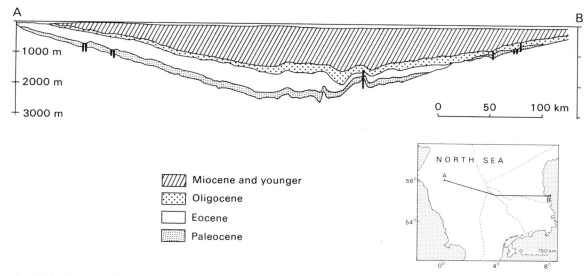

Fig. 10.9. Stratigraphic cross section through Cenozoic Central North Sea Basin (for location see inset). This type of stratigraphy with narrowing of the infilling of the basin with younger sediments restricted to the basin center often has been attributed to the response of a viscoelastic lithosphere to surface loading alone, but can be equally well explained in terms on an elastic plate under the influence of intraplate stresses (after Ziegler 1982).

ing since its formation. In particular, the modelling has shown that the stratigraphy predicted for a basin located on elastic lithosphere and under the influence of a compressional stress regime during deposition of younger strata will be very similar to the stratigraphy characteristically attributed (Watts *et al.* 1982) to the response of viscoelastic lithosphere to surface loading alone (see also Figs. 10.8, 10.9).

Although we have concentrated on the relationship between tectonics and stratigraphy at passive margins and intracratonic basins, the effect of intraplate stress fields is of importance to a wider range of sedimentary environments. Other settings where lithosphere is flexed downward under the influence of sedimentary loads occur in foreland basins (Beaumont 1981; Quinlan and Beaumont 1984). Despite its height of at most a few hundred meters the peripheral bulge flanking forelands basins is of particular stratigraphic interest (Quinlan and Beaumont 1984). These authors interpreted the development of unconformities in the Appalachian foreland basin in terms of uplift of the peripheral bulge caused by viscoelastic relaxation of the lithosphere (Fig. 10.10). However, the presence of intraplate stress of tensional or compressional type, of which the latter is more natural in this tectonic setting, can reduce or amplify the height of the peripheral bulge by an equivalent amount and thus

greatly influence the stratigraphic record in foreland basins. In addition to lithospheric density variations (Stockmal *et al.* 1986), intraplate stresses might remove the need for the existence of hidden loads (Royden and Karner 1984) postulated to improve the fit between foreland basin deflection and the observed topographic load. As pointed out by Allen *et al.* (1986), syntectonic unconformities in basin-margin sequences of foreland basins demonstrate contemporaneous tectonic activity and sedimentation. It is interesting to note that the magnitude of the vertical motions caused by intraplate stresses is of the same order as those attributed to thermal blanketing effects superimposed on long-term foreland basin subsidence (Kominz and Bond 1986).

The dependence of the vertical motions at the edge of the basin on the age of the underlying lithosphere is demonstrated in Figure 10.11. The differential uplift, defined as the difference in deflection for a change in intraplate stress, is calculated for the flank of the basin as a function of the variation in the intraplate stress field. Curves illustrate the deflection for changes in stress of 0.5 kbar, 1 kbar, 2 kbar, and 4 kbar, with the same reference sediment load in each case. As pointed out by Cloetingh *et al.* (1985), stress changes of a few kbar occurring on time scales of a few million years and longer may furnish a tectonic explanation for short-term apparent sea-level

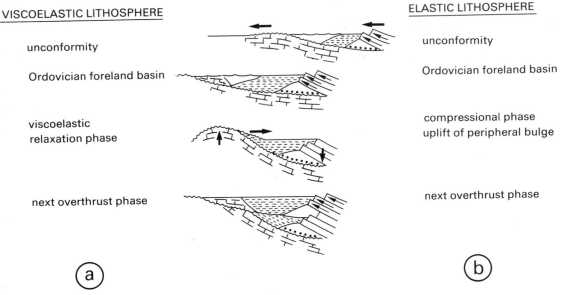

VISCOELASTIC LITHOSPHERE

unconformity

Ordovician foreland basin

viscoelastic
relaxation phase

next overthrust phase

(a)

ELASTIC LITHOSPHERE

unconformity

Ordovician foreland basin

compressional phase
uplift of peripheral bulge

next overthrust phase

(b)

Fig. 10.10a. Diagram illustrating interpretation by Quinlan and Beaumont (1984) for the development of regional unconformities in a multistage foreland basin on viscoelastic lithosphere (column on the left). Bold arrows show overthrust and peripheral bulge migration. Fine arrows illustrate active overthrusting. The column on the right (b) shows interpretation of foreland-basin stratigraphy in terms of the action of intraplate stresses superimposed on flexure of an elastic lithosphere (modified from Quinlan and Beaumont 1984).

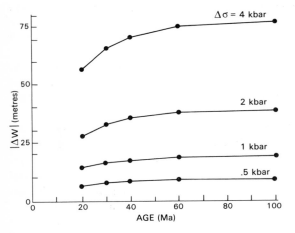

Fig. 10.11. Apparent sea-level fluctuations $|\Delta w|$ (*metres*) at basin edge (position marked by vertical arrow in Fig. 10.6b) due to superposition of variations in intraplate stress field on flexure caused by sediment loading. The data for $|\Delta w|$ are plotted as a function of the age of the underlying lithosphere and sediment loading according to the reference model (after Turcotte and Ahern 1977) of Figure 10.6a. Curves give results for stress changes ($\Delta\sigma$) of .5, 1, 2, and 4 kbar, respectively.

fluctuations with a magnitude of up to a hundred metres and a rate of $0.01-0.1$ m/10^3 yr inferred from the seismic stratigraphic record. Note, however, that the modified Vail *et al.* curves (Vail *et al.* 1984; Haq *et al.* 1987) show a general reduction of the magnitude of the apparent sea-level fluctuations. Similarly, independent recent studies of the magnitude of the mid-Oligocene sea-level lowering point to a value much lower than previously thought. The magnitude of this fall in sea level, which is by far the largest shown in the Vail *et al.* curves, is now estimated to be between 50 metres (Miller and Fairbanks 1985; Watts and Thorne 1984) and 100 metres (Schlanger and Premoli-Silva 1986). Hence, a significant part of the short-term sea-level fluctuations inferred from seismic stratigraphy might have a characteristic magnitude of only a few tens of metres with a time interval of a few million years (Aubry 1985), which can be explained by relatively modest stress variations of a few hundred bars.

The short-term cycles in sea level have been traditionally interpreted in terms of glacio-eustatic controls. This situation has been caused partly by the inferred global character of the changes and partly by the absence of a tectonic mechanism to explain the inferred rapid lowerings in sea level with a magnitude up to a few hundred metres (Pitman and Golovchenko 1983). However, with the exception of the Oligocene events, there is no evidence in the geological and geochemical records for significant Mesozoic and Cenozoic glacial events prior to middle Miocene time (Frakes 1979). Glacio-eustasy, therefore, cannot explain those major parts of the record of short-term sea-level changes where glacial events are thought to have been insignificant. Note that mechanisms for long-term changes in sea level (e.g., Kominz 1984; Heller and Angevine 1985) fall beyond the scope of the present discussion.

Although previous authors (e.g., Bally 1982; Watts 1982) have argued for a tectonic cause for apparent sea-level changes, they were unable to identify a tectonic mechanism operating on a time-scale appropriate to explain the observed short-term lowerings of the sea level. Changes in intraplate stress fields associated with tectonic reorganizations also explain the existence of a strong correlation (Bally 1982) in timing of plate reorganizations and rapid lowerings in sea level shown in the Vail *et al.* (1977) curves. Given current active research in sea-level fluctuation, the possibility that the effect of intraplate stresses is significant has been explored for various regions. Schlanger (1986) suggested that intraplate stress variations provide a possible explanation for the enigmatic high frequency (order of a few million years) variations in Cretaceous apparent sea levels. Hallam (in press) discussed the implications of such variations for the Jurassic sea levels of northwestern Europe. Meulenkamp and Hilgen (1986) explored a causal relationship between Neogene stratigraphy in the eastern Mediterranean and intraplate stress variations. Klein (1987) examined the stratigraphic record of the Paleozoic Appalachian foreland basin and concluded that local variation and local tectonic styles and associated changes in stress field might have masked the possible expression of global fluctuations in sea level in the basin.

Discrimination of Tectonics and Eustasy

It is generally agreed that unconformity-bound units caused by short-term changes in apparent sea level form the natural components of the sedimentary record (Van Hinte 1983). However, the observed boundaries could well reflect local or regional events and their supposed global nature remains to be proven (Parkinson and Summerhayes 1985). Since the publication of the onlap/offlap curves by Vail *et al.* (1977) many deviations have been observed from the inferred global pattern in widely different areas on the globe (Hallam 1984, in press; Harris *et al.* 1984; Jenkin 1984). The regional character of intraplate stresses sheds new light on these observed deviations from "global" sea-level cycles. Whereas such deviations of a global pattern are a natural feature of this tectonic model, the occurrence of short-term deviations does not preclude the presence of global events elsewhere in the stratigraphic record. These are to be expected when major reorganizations in intraplate stress fields occur simultaneously in more than one plate or when glacio-eustasy dominates. Examples of major plate reorganizations are those during mid-Oligocene time (Engebretson *et al.* 1985) and during Early Cenozoic time (Rona and Richardson 1978). Furthermore, differences in rheological structure of the lithosphere, which influence its response to applied intraplate stresses, might also explain differences in magnitude of the inferred sea levels such as observed between the North Sea region and the Gippsland Basin off southeastern Australia (Vail *et al.* 1977).

The discrimination of regional events from eustatic signals in the sea-level record of individual basins is usually a subtle matter, especially if biostratigraphic correlation is imprecise (Hallam 1984, in press). It is interesting to note that the vertical motions induced by the action of intraplate stresses in the center of the basin, although of the same magnitude as the displacements at the flanks, are small compared to the total subsidence in the basin center, which is on the order of several kilometres. The sign and magnitude of the apparent sea-level change will be a function of the sampling point (Figs. 10.6b, 10.8), which may provide a means of testing the effect of intraplate stresses or distinguishing this mechanism from eustatic contributions. In fact, Hallam (in press) has shown that a significant number of Jurassic unconformities are confined to the flanks of North Sea basins, consistent with the predictions of the tectonic mechanism of Figure 10.6.

The effect of intraplate stresses could in principle cause abrupt phases of rapid uplift and subsidence visible in many published subsidence curves of passive margins, intracratonic basins, and foreland basins. An example of such a curve for the Williston Basin of western North America is shown in Figure 10.12. As noted by Fowler and Nisbet (1985) and Sloss (in press), previously proposed subsidence mechanisms fail to explain the observed irregularities in the subsidence pattern. Similarly, changes in

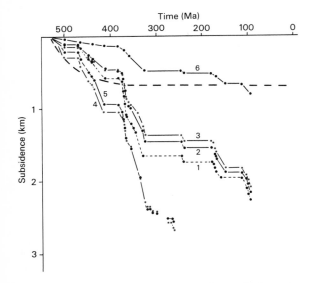

Fig. 10.12. Subsidence-time plots for five wells from the Williston Basin (after Fowler and Nisbet 1985). 1,2,3 = Saskatchewan wells; 4,5 = North Dakota wells; 6 = subsidence in well 2 after backstripping to show subsidence when basin is filled with water alone. Unconformities are shown as horizontal lines. Note that the curves have not been corrected for the erosion of sediments in the unconformities, which gives a distorted picture of the actual magnitude of the irregularities. Heavy dashed line is the subsidence curve calculated for a thermal time constant of 50 Ma. In curve 6, abrupt phases of rapid subsidence are visible, which cannot be explained by thermal processes, but which could possibly result from intraplate stresses.

stress field provide a simple explanation for the observed tilting (Ahern and Mrkvicka 1984) and associated temporal rotation of the source area of the sediments deposited in the Williston Basin. Flexural bulges have probably greatly influenced the extent of erosion that has occurred in the Williston Basin. Episodic subsidence patterns have been observed for many other intracratonic basins, such as the Illinois Basin (Heidlauf *et al.* 1986), passive margins including the United States Atlantic margin (Heller *et al.* 1982), and foreland basins such as the Swiss Molasse basins (Allen *et al.* 1986) and the Appalachian Basin (Klein 1987).

As noted earlier, the action of intraplate stresses causes different and opposite effects on the subsidence at the flanks and in the basin center. These spatial variations have a time-dependent expression due to the shift in the position of the flexural node by

the long-term widening of the basin during its thermo-mechanical evolution. This results in varying expressions of applied stresses with time. The laterally varying effect of stresses applied on various time intervals on subsidence curves of the basin is schematically illustrated in Figure 10.13.

In the present paper I have adopted a reference model in which sedimentation approximately equalled subsidence. As pointed out previously, this assumption allowed me conveniently to equate onlap and offlap with rises and falls in sea level, respectively. Actually, the position of coastal onlap reflects the position where rate of subsidence equals rate of fall in sea level. During the application of stress the rate of subsidence is temporarily changed. Consequently, the equilibrium point of coastal onlap is shifted in position. The thermally induced rate of long-term subsidence strongly decreases with the age of the basin (Turcotte and Ahern 1977). Hence, as pointed out by Thorne and Watts (1984), to produce offlaps during late stages of passive-margin evolution requires much lower values of rate of change of sea level than those needed to produce offlaps during the earlier stages of basin evolution. If these offlaps are caused by fluctuations in intraplate stresses, this implies that the rates of changes of stress needed to create them diminish with age during the flexural evolution of the basin (see also Fig. 10.7). This is of particular relevance for an assessment of the relative contribution of tectonics and eustasy as a cause for Cenozoic unconformities. For example, Cenozoic unconformities developed at old passive margins in association with short-term narrowing of margin basins could be produced by relatively mild changes in intraplate stresses. In this context it is interesting to note that late-stage narrowing of Phanerozoic platform basins and Atlantic margin basins is frequently observed (e.g., Sleep and Snell 1976), without clear evidence for active tectonism. Proterozoic basins, which also show in many cases narrowing during later stages of their evolution, differ, however, by displaying a clear association of the narrowing phase with orogenic events (Grotzinger and Gall 1986; Grotzinger and McCormick 1988: this volume). This possibly reflects closure of the basin and destruction of the basin margins in a regime of strong compression.

Obviously, a need exists to do careful two-dimensional and three-dimensional facies analysis to test the general application of intraplate stresses in sedimentary basin models. It is equally important to obtain precise ages of the sequence boundaries and rate and magnitude of apparent sea-level change to establish their global or regional nature. In this context, critical stratigraphic examination of seismic sections across the inner edge of passive margins to the seaward part is of particular importance (e.g., Thorne and Watts 1984). New criteria will have to be developed to facilitate separation of eustatic effects and effects of vertical motions of the lithosphere on the sedimentary record (e.g., Klein 1982). Of special importance in this respect might be the stratigraphic consequence of spatial and temporal variations of the magnitude of the apparent sea-level change. Similarly, syndepositional deformation (see below) might prove to be a useful criterion to identify intraplate stresses.

Syndepositional Folding of Lithosphere, the Formation of Foreland Arches, and Inversion Tectonics

More data are needed to examine the actual response of the lithosphere under the influence of very large intraplate stresses. The region in the northeastern Indian Ocean south of the Bay of Bengal provides a relatively well understood example of intense deformation caused by intraplate stresses. Seismic reflections profiles (Fig. 10.14; Geller *et al.* 1983) show widespread deformation of originally horizontal sediments. The deformation occurs by broad basement undulations, with wavelengths of roughly 200 km and amplitudes up to 3 km, and numerous high-angle reverse faults. The strike of the undulations and reverse faults is approximately east-west, in agreement with the proposed north-south orientation of the stress field in the area. A prominent Upper Miocene unconformity separates deformed sediments from overlying strata (Geller *et al.* 1983). The absence of syn-sedimentary deformation below the unconformity provides evidence for a Late Miocene age for the timing of the onset of the deformation. The basement undulations coincide with undulations in gravity and geoid anomalies (McAdoo and Sandwell 1985). For elastic plate models, stresses of several tens of kbars are required to induce the observed folding of the basement. McAdoo and Sandwell (1985) pointed out in their study of the folding of the oceanic crust in the Bay of Bengal the importance of incorporating a more realistic rheology in models of the response of the lithosphere to large intraplate stresses. Their work showed that a depth-dependent rheology of the

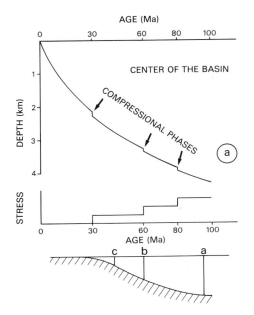

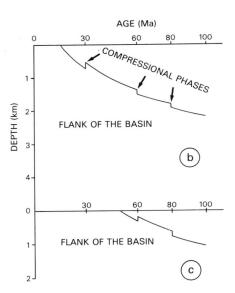

Fig. 10.13. Effect of intraplate stresses on subsidence curves at three different positions (a, b, and c) in the basin. (a) Effect of compression on subsidence predicted for a well in the basin center; (b) and (c) show the effect of compressional stress on subsidence for locations at the flanks of the basin, closer to the position of the flexural node. Note in these cases the different effects of compression at different time intervals, which are caused by shifts in the position of the flexural node caused by widening of the basin during its long-term thermal evolution.

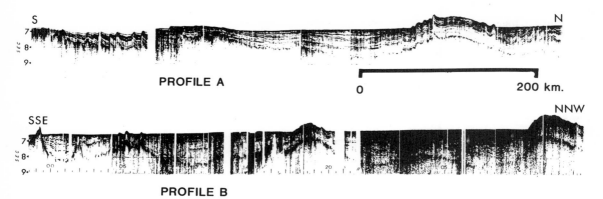

PROFILE A

PROFILE B

Fig. 10.14. Two long seismic reflection profiles from the southern Bengal Fan showing folding of oceanic basement and numerous reverse faults (after Geller *et al.* 1983). Time is given along the vertical scale in terms of seconds of two-way travel time. The location of the profiles is indicated in Figure 10.3a.

lithosphere, inferred from extrapolation of rock-mechanics data on brittle deformation and ductile flow (Goetze and Evans 1979), significantly enhances the effectiveness of intraplate stress in causing folding and syndepositional deformation in a sedimentary basin. Compared to an elastic plate model, the vertical motions of the lithosphere due to a given stress change are magnified, while the horizontal wavelength of the folding is reduced. Taking into account the presence of a thick sedimentary load in the area, a depth-dependent rheology lowers the stress level required to induce the observed folding to approximately 5–6 kbar (Fig. 10.15). Inspection of Figure 10.15 shows that McAdoo and Sandwell's (1985) results are in excellent agreement

with independent estimates inferred from numerical modelling of intraplate stresses in the northeastern Indian Ocean (Cloetingh and Wortel 1985, 1986). Rheological weakening of the lithosphere due to flexural stresses induced by sediment loading (Cloetingh *et al.* 1982) also enhances the effectiveness of the action of intraplate stresses in a more general sense. However, for compressional deformation not reaching the buckling limit, the effect of incorporating more realistic rheologies in the models is less dramatic than in the northeastern Indian Ocean.

Syndepositional folding of oceanic lithosphere might have some interesting analoges in the continents (e.g., Porter *et al.* 1982). Rheological models

Fig. 10.15. Buckling load F (Nm^{-1}) versus age for lithosphere with a depth-dependent rheology inferred from rock-mechanics studies (Goetze and Evans 1979) given by the solid line and for fully elastic oceanic lithosphere given by the dashed line (after McAdoo and Sandwell 1985). Incorporation of depth-dependent rheology magnifies the vertical motions (W) and reduces the horizontal wavelength (λ_c) of stress-induced folding of the lithosphere. The box indicates stress levels calculated for the area in the northeastern Indian Ocean (Cloetingh and Wortel 1985) where folding of oceanic lithosphere under the influence of compressional stresses has been observed (McAdoo and Sandwell 1985).

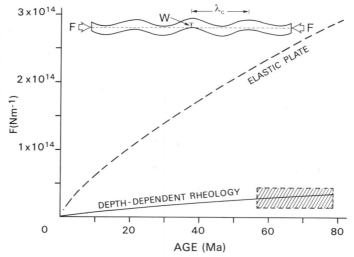

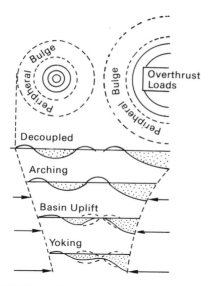

Fig. 10.16. Flexural interaction between foreland basin (right) and an intracratonic basin (left) causes the formation of arches. The upper part of the figure displays a plan view of the basins in the decoupled position, corresponding to the first cross section below. Subsequent cross sections show the nature of the interaction when the basins are closer together. The latter is indicated by an increasing length of horizontal arrows (after Quinlan and Beaumont 1984).

sibly giving rise to vertical motions of several kilometres in the case of large intraplate stresses.

The presence of an anomalously thin elastic lithosphere, such as inferred for the North Sea Basin (Barton and Wood 1984), also enhances the effectiveness of intraplate stresses in causing vertical motions of the basin floor and folding of the basin fill. The same is true for situations, also encountered in the North Sea Basin, where subsidence is primarily fault-controlled (e.g., Ziegler 1987). In fact, vertical motions of several kilometres have occurred during inversion of the North Sea Basins and uplift of intervening highs, associated with compressional phases of the Alpine Orogeny (Ziegler 1982, 1987). These compressional stresses in the European Platform are a mechanical consequence (England and Wortel 1980) of the subduction of young oceanic lithosphere during closure of small oceanic basins during the Alpine Orogeny (Vlaar and Cloetingh 1984). Phases with obduction of ophiolites along the Alpine front coincide with absence of inversion tectonics, which suggests a partial reduction of stress transmission during such phases (Ziegler, P.A. personal communication 1987). The reduced level of stresses, however, may still be of sufficient magnitude to cause the much smaller vertical motions of the North Sea basins reflected in the apparent sea-level record.

indicate that continental lithosphere is weaker than oceanic lithosphere (Vink *et al.* 1984; Cloetingh and Nieuwland 1984). This feature reflects primarily mineralogic differences caused by different stratifications within oceanic and continental lithosphere. Compared to oceanic lithosphere, the strength of which is primarily determined by olivine, the strength of continental lithosphere is reduced by the presence of relatively low strength crustal material, in particular quartz (Stein and Wiens 1986). Therefore, syndepositional folding in continental interiors is probably reached at a lower stress level than required for oceanic plates (Fig. 10.15). In general it is expected that a specific change in intraplate stress will induce larger vertical motions in continental lithosphere than in oceanic lithosphere. The presence of tensional or compressional intraplate stresses can reduce or amplify the height of the peripheral bulge at foreland basins (Fig. 10.10). Similarly, intraplate stresses may affect the postulated interaction (Quinlan and Beaumont 1984) between adjacent foreland arches (Fig. 10.16), pos-

Basin Analysis and Paleo-stresses in the North Sea

In the foregoing sections the implications of the presence of intraplate stress fields in the lithosphere for sedimentary-basin evolution have been discussed. In doing so I have argued for a tectonic cause for short-term apparent sea-level changes. If this tectonic mechanism is accepted, then the apparent sea-level curves can in principle be interpreted in terms of paleo-stress fields. In Cloetingh (1986) (see also Lambeck *et al.* in press for a more detailed discussion) the North Sea region was selected for this purpose. There were several reasons for this choice. Although the Vail *et al.* (1977) "global" coastal onlap/offlap curves are based on data from various basins around the world, they are heavily weighted in favor of North America, the Gulf coast, the northern and central Atlantic margins, and especially the North Sea. Therefore as noted by several authors (e.g., Miall 1986) the global cycles strongly reflect the seismic stratigraphic record of basins in a tectonic setting dominated by rifting events in the northern and central Atlantic. Moreover, in the North Sea area these rifting events have alternated with compressional phases induced by the Alpine Orogeny. Ziegler (1982, 1987) has pointed out the existence of a correlation between tectonic phases in the North Sea and changes in the history of the Atlantic and Tethyan Oceans. Changes in spreading rates in the Atlantic and Tethyan oceans are probably caused by or associated with changes in plate-tectonic forces. Paleostress-field orientation inferred from analysis of microstructures by Letouzey (1986) shows distinctly different stress patterns at different geological periods (Fig. 10.17). Tectonic lineation rosettes determined from spectral analysis of North Sea subsidence show a rotation from a prominent E-W trend in Triassic time to N-S in late Cretaceous time (Thorne 1985), a trend also reflected in the measured paleo-stress directions (Letouzey 1986). The influence of Alpine Orogeny is reflected in differences in orientation between Pliocene and other Cenozoic directions (Thorne 1985).

In the analysis it is assumed that apparent sea level is controlled by regional tectonics. This assumption is certainly an oversimplification, as parts of the sea-level record may be controlled by glacio-eustasy or may be the result of simultaneously acting tectonic or glacio-eustatic processes. However, stratigraphic studies favor an important role for tectonic processes in determining the sea-level record of the North Sea region. Hallam (in press), for instance, recognizes a number of significant regressive events that appear to be caused by regional tectonics rather than global sea level. This applies in particular to the exceptionally high frequency of occurrence of short-term sea-level fluctuations toward the end of Jurassic time, which occur simultaneously with a pronounced increase in fault-controlled tectonic activity (Hallam in press).

Lambeck et al. (in press) constructed a composite sea-level curve for northwestern Europe from regional sea-level curves published by Vail et al. (1977) for Tertiary time, Juignet (1980) and Hancock (1984) for Cretaceous time, and Hallam (1984) for Jurassic time. The composite sea-level curve was calibrated by adopting a magnitude of 50 metres for the mid-Oligocene fall of sea level based on recent estimates inferred from analysis of subsidence in the

North Sea region (Thorne and Bell 1983). On the basis of the model calculations summarized in Figure 10.11, the information on changes in apparent sea level has been used to obtain an estimate of paleo-stress relative to the present stress level (Fig. 10.18). According to the modelling, apparent lowering of sea level is associated with relaxation of tensional intraplate stress or, equivalently, an increase in compressive stress (Figs. 10.6, 10.8). Ziegler's (1982) timing of tectonic events in western Europe is also given in Figure 10.18. The specific model of Figure 10.6 cannot be used to derive quantitative estimates for the stress field in the North Sea region in Mesozoic time. For this purpose, specification of the basin formation mechanism and pertinent thermo-mechanical properties is required. However, the postulated modulation of subsidence by changes in the magnitude of the regional intraplate stress field is relevant for this period. The occurrence of a long-term Mesozoic sea-level rise (and tensional intraplate stress) followed by a long-term Tertiary lowering is in agreement with the observed dominance of wrenching and rifting in Mesozoic time (Ziegler 1982). During Mesozoic time basin subsidence occurred through graben formation and crustal stretching (McKenzie 1978; Sclater and Christie 1980), which requires tensional stresses of the order of a few kbar (Cloetingh and Nieuwland 1984; Houseman and England 1986). At the Mesozoic/Cenozoic transition a reduction in tension or an equivalent change to overall compression could have induced a reversal of this long-term trend. Note, however, that this interpretation is not unique as long-term changes in sea level can equally be explained in terms of changes in long-term spreading rate and area/age distribution of the ocean floor caused by continental break-up (Heller and Angevine 1985).

Of prime interest here are the short-term fluctuations superimposed on these long-term changes in sea level. Figure 10.18 shows that in this case as well, the synthetic paleo-stress curve derived from the seismic stratigraphic record provides a mirror image of the tectonic evolution of northwestern Europe and the North Sea Basin: Rifting episodes correspond to relaxation of tensional paleo-stresses and Alpine orogenic phases correspond to episodes of increased compressive stresses. For reasons mentioned in the previous section, the values of the inferred fluctuations in intraplate stresses are certainly upper limits for the North Sea area.

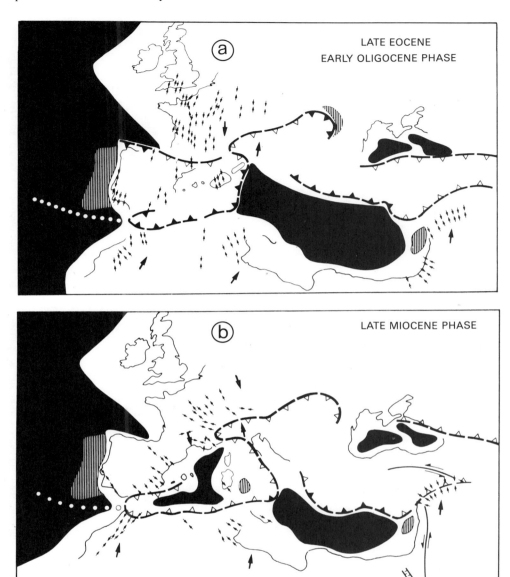

Fig. 10.17. Compressive paleo-stress field directions inferred from microtectonic studies of the northwestern European Platform and surrounding areas. 1 = continental crust and margins, 2 = oceanic crust, 3 = thin continental crust or oceanic crust, 4 = plate boundary, 5 = oceanic subduction, 6 = thrust and collision, 7 = relative Europe-Africa-Arabia azimuth vector of motion, 8 = maximum compressive paleo-stress axis orientation in fossil position relative to continental blocks (represented by present coast line). Position of the main Europe, Arabia, Africa continents is derived from kinematic reconstructions, with Europe arbitrarily fixed (after Letouzey 1986). (a) Late Eocene/Early Oligocene stress field orientations. (b) Late Miocene stress field. Comparison of (a) and (b) shows a rotation of the paleo-stress field in northwestern Europe during Tertiary time.

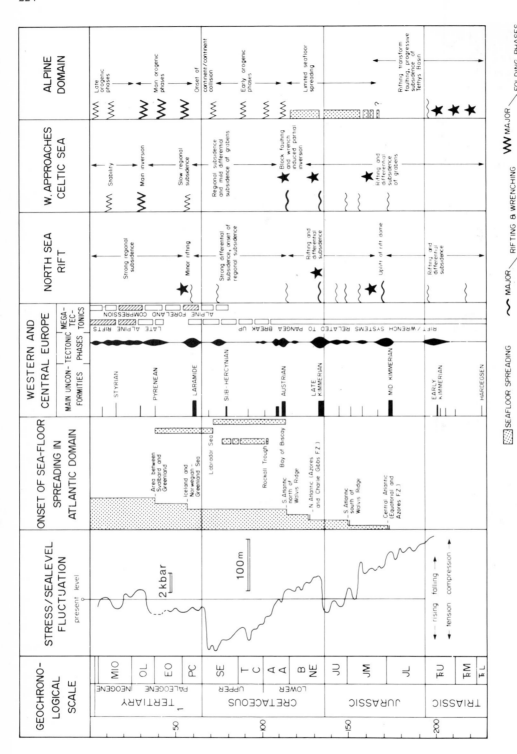

Fig. 10.18. Synthetic paleo-stress curve derived from a composite sea-level curve based on regional sea-level curves for northwestern Europe. Columns on right side show timing of tectonic events (after Ziegler 1982) in western Europe and North Sea Basin, and are given for comparison. Also indicated are timings of onset of sea-floor spreading in the Atlantic domain. Paleo-stresses are plotted relative to present-day stress level. Thin and heavy wavy lines denote minor and major rifting phases; thin and thick zig-zag lines represent minor and major folding phases; stars denote volcanic activity (after Lambeck et al. in press).

Directions for Future Research

The missing link for understanding basin tectonics may turn out to be the interaction of intraplate stresses with basin subsidence. A good understanding of the interaction of intraplate stresses and basin subsidence is thus possibly a crucial element in basin analysis. To incorporate intraplate stresses into basin analysis requires prediction of intraplate stresses from knowledge of plate-tectonic evolution and prediction of the effect of intraplate stresses on tectonic subsidence. Current research is continuing on these topics.

Systematic numerical modelling of paleo-stress fields in the various plates, based on reconstructed geometries and age distributions, is in progress now at the Vening Meinesz Laboratory. The outcome of this research program will be a set of paleo-stress maps for various time slices of the geologic record. The calculated stresses are to be tested by paleo-stress directions inferred from micro-structural analysis. Similarly, the stress patterns will be compared with the timing and the location of phenomena of intraplate tectonic deformation on a larger scale, including for example the inception of rift zones. Independently, paleo-stresses will be derived from the stratigraphic record under the assumption that short-term onlaps and offlaps are caused by fluctuating stress fields. Comparison of the outcome of the two different approaches will in principle allow us to test for the relative importance of intraplate stresses in the stratigraphic record and is expected to be of great value for separating global and primarily eustatic effects from more regional tectonic effects. Furthermore, detailed comparison of paleo-stress calculations with pertinent geological observations might provide better constraints on how fast the individual stress changes actually can occur. Equally important in this respect is the separation of sudden changes in stress (recorded by, for example, sudden depth changes) and stress changes continuous over tens of millions of years.

The motions of a flexural bulge are caused by an interplay of several factors, of which the level of intraplate stresses seems to be an important one. As noted previously, several effects not considered in the present analysis will influence the magnitude of the predicted short-term phases of subsidence or uplift. The effect of a depth-dependent rheology of the lithosphere will enhance the effectiveness of intraplate stresses as a cause of vertical motions, in particular for large values of stress. Note, however, that such a rheology inherently decreases the flexural wavelength of the basement response and hence will reduce the horizontal areal expression of the induced short-term phases of coastal onlap and offlap. For stress levels below a kilobar or so, these consequences of a depth-dependent rheology are probably quite modest, since for small stresses only a relatively thin segment of the plate will reach its limiting stress. In addition to these effects, relaxation of bending stresses by nonlinear ductile flow (Quinlan and Beaumont 1984) is accelerated by intraplate stresses, which causes tectonic subsidence by allowing further isostatic reequilibration of crustal and sedimentary loads.

Reactivation of faults under the influence of intraplate stresses is certainly an important aspect to be incorporated in the modelling. This applies in particular to the earlier phases of basin evolution, in which subsidence is dominantly fault controlled. Although modelling the response of faulted basins to intraplate stresses in terms of flexure of unbroken plates is, strictly speaking, not valid, the results can be used in a qualitative way as long as the faults are formed in an extensional regime. This is true for epicratonic basins and passive margins, which are commonly formed in an extensional regime. In this case, reactivation of extensional faults under compression will promote relative uplift of the segment of the basin close to the major boundary faults that govern the overall basin structure. The following examples might illustrate this. Completion of the rifting phase of basins is associated with relaxation of tensional stresses and a transition to a more compressional regime. The transition from the synrift phase to the postrift phase of basins is commonly reflected in the presence of a break-up unconformity, which agrees with the qualitative prediction of the model discussed in this paper. Break-up faults also play a crucial role in determining the response of the basin to inversion tectonics at later stages of its evolution (e.g., Ziegler 1987). It has been mentioned here that the simple model used by Cloetingh et al. (1985) provides only a very qualitative description which has to be refined in order to explore more quantitatively the dynamics of the actual inversion process. In principle, finite-element techniques such as used in the present analysis allow incorporation of pre-existing faults in the modelling. However, the success of more detailed modelling will be critically dependent on better knowledge of the actual strength of fault zones in the lithosphere under the action of intraplate stresses. Such insight is also crucial for relating sudden, probably episodic, displacements on basin faults to changes in stress of a possibly more continuous character.

Conclusions

The incorporation of intraplate stresses in quantitative models of basin evolution offers new perspectives in the analysis of sedimentary basins. Numerical modelling and observations of modern and ancient stress fields have demonstrated the existence of large stress provinces in the interiors of the plates. These studies have provided strong evidence for far-field propagation of intraplate stresses and established a causal relationship between processes at plate boundaries and deformation in the plate interiors, where stresses affect sedimentary basins both in passive-margin and intracratonic settings. Spatial and temporal fluctuations in stress fields apparently are recorded in the stratigraphic record of sedimentary basins. Numerical modelling has shown that the incorporation of intraplate stresses in models of basin evolution can in principle predict a succession of onlap and offlap patterns frequently observed along the flanks of passive margins and intracratonic basins. Such a stratigraphy can be interpreted as the natural consequence of the short-term mechanical widening and narrowing of basins by moderate fluctuations in intraplate stresses, superimposed on the long-term broadening of the basins with cooling since their formation.

Deformation of the lithosphere under the influence of large stresses generated by plate collision provides an explanation for observed syndepositional folding of oceanic lithosphere, the interaction of foreland peripheral bulges, and the inversion of basins as observed in the Alpine foreland. The seismic-stratigraphic record could provide a new source of information for paleo-stress fields. More detailed examination of the stratigraphic record of individual basins in connection with independent numerical modelling of paleo-stresses and analysis of paleo-stress data is required to exploit fully the role of intraplate stresses in basin evolution.

Acknowledgments. Kurt Lambeck and Herb McQueen made significant contributions to the modelling of the tectonic causes of sea-level changes. Rinus Wortel is thanked for stimulating cooperation in numerical modelling of intraplate stresses. Julian Thorne, Seth Stein, Gordon Shudofsky, Kerry Gallagher, and Peter Ziegler offered valuable suggestions and comments. George de Vries Klein, Anthony Hallam, Larry Sloss, Seymour Schlanger, and Peter Ziegler furnished preprints of papers. Gerard Bond, Julian Thorne, and R.H. Dott Jr. provided thoughtful reviews. The Netherlands Organization for the Advancement of Pure Research (ZWO) is acknowledged for partial support of this research.

References

AHERN, J.L. and MRKVICKA, S.R. (1984) A mechanical and thermal model for the evolution of the Williston Basin. Tectonics 3:79–102.

AHORNER, L. (1975) Present-day stress field and seismotectonic block movements along major fault zones in central Europe. Tectonophysics 29:233–249.

ALLEN, P., HOMEWOOD, P., and WILLIAMS, G.D. (1986) Foreland basins: An introduction. In: Allen, P.A. and Homewood, P. (eds) Foreland Basins. Special Publication International Association Sedimentologists 8, pp. 5–12.

ARTYUSHKOV, E.V. (1974) Can the Earth's crust be in a state of isostasy? Journal Geophysical Research 79:741–750.

AUBRY, M.-P. (1985) Northwestern Europe Paleogene magnetostratigraphy, biostratigraphy and paleogeography: Calcareous nannofossil evidence. Geology 13:198–202.

BALLY, A.W. (1982) Musings over sedimentary basin evolution. Philosophical Transactions Royal Society London A305:325–338.

BARTON, R. and WOOD, P. (1984) Tectonic evolution of the North Sea basin: Crustal stretching and subsidence. Geophysical Journal Royal Astronomical Society 79:471–497.

BEAUMONT, C. (1978) The evolution of sedimentary basins on a viscoelastic lithosphere: Theory and examples. Geophysical Journal Royal Astronomical Society 55:471–498.

BEAUMONT, C. (1981) Foreland Basins. Geophysical Journal Royal Astronomical Society 65:291–329.

BELL, J.S. and GOUGH, D.I. (1979) Northeast-southwest compressive stress in Alberta: Evidence from oil wells. Earth Planetary Science Letters 45:475–482.

BERGMAN, E.A. (1986) Intraplate earthquakes and the state of stress in oceanic lithosphere. Tectonophysics 132:1–35.

BERGMAN, E.A. and SOLOMON, S.C. (1985) Earthquake source mechanisms from body-waveform inversion and intraplate tectonics in the northern Indian Ocean. Physics Earth Planetary Interiors 40:1–23.

BLUMLING, P., FUCHS, K., and SCHNEIDER, T. (1983) Orientation of the stress field from breakouts in a crystalline well in a seismic area. Physics Earth Planetary Interiors 33:250–254.

BODINE, J.H., STECKLER, M.S., and WATTS, A.B. (1981) Observations of flexure and the rheology of the oceanic lithosphere. Journal Geophysical Research 86:3695–3707.

CHAPPLE, W.M. and TULLIS, T.E. (1977) Evaluation of the forces that drive the plates. Journal Geophysical Research 82:1967–1984.

CLOETINGH, S. (1986) Intraplate stresses: A new tectonic mechanism for fluctuations of relative sea level. Geology 14:617–620.

CLOETINGH, S., McQUEEN, H., and LAMBECK, K. (1985) On a tectonic mechanism for regional sea level variations. Earth Planetary Science Letters 75:157–166.

CLOETINGH, S. and NIEUWLAND, F. (1984) On the mechanics of lithospheric stretching and doming: A finite element analysis. Geologie Mijnbouw 63:315–322.

CLOETINGH, S. and WORTEL, R. (1985) Regional stress field of the Indian plate. Geophysical Research Letters 12:77–80.

CLOETINGH, S. and WORTEL, R. (1986) Stress in the Indo-Australian Plate. Tectonophysics 132:49–67.

CLOETINGH, S.A.P.L., WORTEL, M.J.R., and VLAAR, N.J. (1982) Evolution of passive continental margins and initiation of subduction zones. Nature 297:139–142.

DUNCAN, R.A. and McDOUGALL, I. (in press) Time-space relationships for Cainozoic intraplate Volcanism in Eastern Australia, the Tasman Sea, and New Zealand. In: Johnson, R.W. and Taylor, S.R. (eds) Intraplate Volcanism in Eastern Australia and New Zealand. Canberra: Australian Academy of Science.

ENGEBRETSON, D.C., COX, A., and GORDON, R.G. (1985) Relative Motions between Oceanic and Continental Plates in the Pacific Basin. Geological Society America Special Paper 206, 56 p.

ENGLAND, P.C. and WORTEL, R. (1980) Some consequences of the subduction of young slabs. Earth Planetary Science Letters 47:403–415.

FORSYTH, D.L. and UYEDA, S. (1975) On the relative importance of the driving forces of plate motion. Geophysical Journal Royal Astronomical Society 43:163–200.

FOWLER, C.M. and NISBET, E.G. (1985) The subsidence of the Williston Basin. Canadian Journal Earth Sciences 22:408–415.

FRAKES, L.A. (1979) Climates throughout Geologic Time. Amsterdam: Elsevier, 310 p.

GELLER, C.A., WEISSEL, J.K., and ANDERSON, R.N. (1983) Heat transfer and intraplate deformation in the central Indian Ocean. Journal Geophysical Research 88:1018–1032.

GOETZE, C. and EVANS, B. (1979) Stress and temperature in the bending lithosphere as constrained by experimental rock mechanics. Geophysical Journal Royal Astronomical Society 59:463–478.

GOUGH, D.L., FORDJOR, C.K., and BELL, J.S. (1983) A stress province boundary and tractions on the North American plate. Nature 305:619–621.

GROTZINGER, J.P. and GALL, Q. (1986) Preliminary investigations of Early Proterozoic Western River and Burnside River formations: Evidence for foredeep origin of Kilohigok Basin, District of Mackenzie. Current Research, Part A, Geological Survey Canada Paper 86-1A:95–106.

GROTZINGER, J.P. and McCORMICK, D.S. (1988) Flexure of the early Proterozoic lithosphere and the evolution of Kilohigok Basin (1.9 Ga), northwest Canadian shield. In: Kleinspehn, K.L. and Paola, C. (eds) New Perspectives in Basin Analysis. New York: Springer-Verlag, pp. 405–430.

GUNN, R. (1944) A quantitative study of the lithosphere and gravity anomalies along the Atlantic coast. Journal Franklin Institution 237:139–154.

HALLAM, A. (1984) Pre-quaternary sea-level changes. Annual Reviews Earth Planetary Sciences 12:205–243.

HALLAM, A. (in press) A reevaluation of Jurassic eustasy in the light of new data and the revised Exxon curve. In: Wilgus, C.K. (ed) Sea Level Changes—An Integrated Approach. Society Economic Paleontologists Mineralogists Special Publication 42.

HANCOCK, J.M. (1984) Cretaceous. In: Glennie, K.W. (ed) Introduction to the Petroleum Geology of the North Sea. Oxford: Blackwell, pp. 133–150.

HAQ, B., HARDENBOL, J., and VAIL, P.R. (1987) Chronology of fluctuating sea level since the Triassic (250 million years to present). Science 235:1156–1167.

HARRIS, P.M., FROST, S.H., SEIGLIE, G.A., and SCHNEIDERMAN, N. (1984) Regional unconformities and depositional cycles, Cretaceous of the Arabian peninsula. In: Schlee, J.S. (ed) Interregional Unconformities and Hydrocarbon Accumulation. American Association Petroleum Geologists Memoir 36, pp. 67–80.

HEIDLAUF, D.T., HSUI, A.T., and DE VRIES KLEIN, G. (1986) Tectonic subsidence analysis of the Illinois Basin. Journal Geology 94:779–794.

HEISKANEN, W.A. and VENING MEINESZ, F.A. (1958) The Earth and its Gravity Field. New York: McGraw-Hill, 470 p.

HELLER, P.L. and ANGEVINE, C.L. (1985) Sea level cycles during the growth of Atlantic type Oceans. Earth Planetary Science Letters 75:417–426.

HELLER, P.L., WENTWORTH, C.M., and POAG, C.W. (1982) Episodic post-rift subsidence of the United States Atlantic continental margin. Geological Society America Bulletin 93:379–390.

HINTE, J.Van. (1983) Synthetic seismic sections from biostratigraphy. In: Watkins, J.S. and Drake, C.L. (eds) Continental Margin Geology. American Association Petroleum Geologists Memoir 34, pp. 675–685.

HOUSEMAN, G. and ENGLAND, P. (1986) A dynamical model of lithosphere extension and sedimentary basin formation. Journal Geophysical Research 91: 719–729.

ILLIES, J.H., BAUMANN, H., and HOFFERS, B. (1981) Stress pattern and strain release in the Alpine foreland. Tectonophysics 71:157–172.

JENKIN, J.J. (1984) Evolution of the Australian coast and continental margin. In: Thom, B.G. (ed) Coastal Geomorphology in Australia. Sydney: Academic Press, pp. 23–42.

JOHNSON, B. and BALLY, A.W. (eds) (1986) Intraplate deformation: Characteristics, processes and causes. Tectonophysics 132:1–278.

JUIGNET, P. (1980) Transgressions-regressions, variations eustatiques et influences tectoniques de l'Aptien au Maastrichtian dans le Basin de Paris Occidental et sur la bordure du Massif Amoricain. Cretaceous Research 1:341–357.

KLEIN, G. deV. (1982) Probable sequential arrangement of depositional systems on cratons. Geology 10:17–22.

KLEIN, G. deV. (1987) Current aspects of basin analysis. Sedimentary Geology 50:95–118.

KLEIN, R.J. and BARR, M.V. (1986) Regional state of stress in western Europe. In: Stephensson, O. (ed) Rock Stress and Rock Stress Measurements. Lulea, Sweden: Centek Publishers, pp. 33–44.

KOMINZ, M.A. (1984) Oceanic ridge volumes and sea-level change—an error analysis. In: Schlee, J.S. (ed) Interregional Unconformities and Hydrocarbon Accumulation. American Association Petroleum Geologists Memoir 36, pp. 109–127.

KOMINZ, M.A. and BOND, G.C. (1986) Geophysical modelling of the thermal history of foreland basins. Nature 320:252–256.

LAMBECK, K. (1983) Structure and evolution of intracratonic basins of central Australia. Geophysical Journal Royal Astronomical Society 74:843–886.

LAMBECK,K., CLOETINGH, S., and McQUEEN, H. (in press) Intraplate stresses and apparent changes in sea level: The basins of north-western Europe. Canadian Society Petroleum Geologists Memoir 12.

LAMBECK, K., McQUEEN, H.W.S., STEPHENSON, R.A., and DENHAM, D. (1984) The state of stress within the Australian continent. Annale Geophysicae 2:723–741.

LETOUZEY, J. (1986) Cenozoic paleo-stress pattern in the Alpine foreland and structural interpretation in a platform basin. Tectonophysics 132:215–231.

LISTER, C. (1975) Gravitational drive on oceanic plates caused by thermal contraction. Nature 257:663–665.

McADOO, D.C. and SANDWELL, D.T. (1985) Folding of oceanic lithosphere. Journal Geophysical Research 90:8563–8569.

McGARR, A. and GAY, N.C. (1978) State of stress in the Earth's crust. Annual Reviews Earth Planetary Sciences 6:405–436.

McKENZIE, D.P. (1978) Some remarks on the development of sedimentary basins. Earth Planetary Science Letters 40:25–32.

MEGARD, F., NOBLE, D.C., McKEE, E.H., and BELLON, H. (1984) Multiple pulses of Neogene compressive deformation in the Ayacucho intermontane basin, Andes of central Peru. Geological Society America Bulletin 95:1108–1117.

MEULENKAMP, J.E. and HILGEN, F.J. (1986) Event stratigraphy, basin evolution and tectonics of the Hellenic and Calabro-Sicilian arcs. Developments Geotectonics 21:327–350.

MIALL, A.D. (1986) Eustatic sea level changes interpreted from seismic stratigraphy: A critique of the methodology with particular reference to the North Sea Jurassic record. American Association Petroleum Geologists Bulletin 70:131–137.

MILLER, K.G. and FAIRBANKS, R.G. (1985) Oligocene-Miocene global carbon and abyssal circulation changes. American Geophysical Union Geophysical Monograph 32:469–486.

PARKINSON, N. and SUMMERHAYES, C. (1985) Synchronous global sequence boundaries. American Association Petroleum Geologists Bulletin 69:685–687.

PATRIAT, P. and ACHACHE, J. (1984) India-Eurasia collision chronology has implications for crustal shortening and driving mechanism of plates. Nature 311:615–621.

PITMAN, W.C. (1978) Relationship between eustasy and stratigraphic sequences of passive margins. Geological Society America Bulletin 89:1389–1403.

PITMAN, W.C. and GOLOVCHENKO, X. (1983) The effect of sea level change on the shelf edge and slope of passive margins. In: Stanley, D.J. and Moore, G.T. (eds) The Shelfbreak: Critical Interface on Continental Margins. Society Economic Paleontologists Mineralogists Special Publication 33, pp. 41–58.

PORTER, J.W., PRICE, R.A., and McCROSSAN, R.G. (1982) The western Canada sedimentary basin. Philosophical Transactions Royal Society London A305:169–192.

QUINLAN, G.M. and BEAUMONT, C. (1984) Appalachian thrusting, lithospheric flexure and the Paleozoic stratigraphy of the eastern Interior of North America. Canadian Journal Earth Sciences 21:973–996.

RICHARDSON, R.M., SOLOMON, S.C., and SLEEP, N.H. (1979) Tectonic stress in the plates. Reviews Geophysics Space Physics 17:981–1019.

RITO, F. de, COZARELLI, F.A., and HODGE, D.S. (1983) Mechanisms of subsidence of ancient cratonic rift basins. Tectonophysics 94:141–168.

RONA, P.A. and RICHARDSON, E.S. (1978) Early Cenozoic global plate reorganization. Earth Planetary Science Letters 40:1–11.

ROYDEN, L. and KARNER, G.D. (1984) Flexure of the lithosphere beneath the Apennine and Carpathian foredeep basins. Nature 309:142–144.

SCHLANGER, S.O. (1986) High frequency sea-level oscillations in Cretaceous time: An emerging geophysical problem. American Geophysical Union Geodynamics Series 15:61–74.

SCHLANGER, S.O. and PREMOLI-SILVA, I. (1986) Oligocene sea level falls recorded in mid-Pacific atoll and archipelagic apron settings. Geology 14:392–395.

SCLATER, J.G. and CHRISTIE, P.A.F. (1980) Continental stretching: An explanation of the post-mid-Cretaceous subsidence of the central North Sea basin. Journal Geophysical Research 85:3711–3739.

SLEEP, N.H. (1971) Thermal effects of the formation of Atlantic continental margins by continental break up. Geophysical Journal Royal Astronomical Society 24:325–350.

SLEEP, N.H. and SNELL, N.S. (1976) Thermal contraction and flexure of mid-continent and Atlantic marginal basins. Geophysical Journal Royal Astronomical Society 45:125–154.

SLOSS, L.A. (in press) Williston in the family of cratonic basins. American Association Petroleum Geologists Memoir.

SMOLUCHOWSKI, M. (1909) Uber eine gewisse Stabilitatsproblem der Elastizitatslehre und dessem Beziehung zur Entstehung Faltengebirge. Bulletin Interne Academie Scientifique Cracovie 2:3–20.

SOLOMON, S.C., SLEEP, N.H., and RICHARDSON, R.M. (1975) On the forces driving plate tectonics: Inferences from absolute plate velocities and intraplate stress. Geophysical Journal Royal Astronomical Society 42:769–801.

STEIN, S., CLOETINGH, S., WIENS, D., and WORTEL, R. (1987) Why does near ridge extensional seismicity occur primarily in the Indian Ocean? Earth Planetary Science Letters 82:107–113.

STEIN, S. and WIENS, D.A. (1986) Depth determination for shallow teleseismic earthquakes: Methods and results. Reviews Geophysics 24:806–832.

STEPHENSON, R. and LAMBECK, K. (1985) Isostatic response of the lithosphere with in-plane stress: Application to central Australia. Journal Geophysical Research 90:8581–8588.

STOCKMAL, G.S., BEAUMONT, C., and BOUTILIER, R. (1986) Geodynamic models of convergent margin tectonics: Transition from rifted margin to overthrust belt and consequences for foreland basin development. American Association Petroleum Geologists Bulletin 70:181–190.

THORNE, J. (1985) Studies in Stratology: The Physics of Stratigraphy. Unpublished Ph.D. Thesis, Columbia University, 523 p.

THORNE, J. and BELL, R. (1983) A eustatic sea level curve from histograms of North Sea subsidence. EOS Transactions American Geophysical Union 64:858.

THORNE, J. and WATTS, A.B. (1984) Seismic reflectors and unconformities at passive continental margins. Nature 311:365–368.

TURCOTTE, D.L. and AHERN, J.L. (1977) On the thermal and subsidence history of sedimentary basins. Journal Geophysical Research 82:3762–3766.

TURCOTTE, D.L. and OXBURGH, E.R. (1976) Stress accumulation in the lithosphere. Tectonophysics 35:183–199.

TURCOTTE, D.L. and SCHUBERT, G. (1982) Geodynamics. New York: Wiley, 450 p.

VAIL, P.R., HARDENBOL, J., and TODD, R.G. (1984) Jurassic unconformities, chronostratigraphy, and sea level changes from seismic stratigraphy and biostratigraphy. In: Schlee, J.S. (ed) Interregional Unconformities and Hydrocarbon Accumulation. American Association Petroleum Geologists Memoir 36, pp. 129–144.

VAIL, P.R., MITCHUM, R.M., Jr., and THOMPSON III, S. (1977) Seismic stratigraphy and global changes of sea level, part 4: Global cycles of relative changes of sea level. In: Payton, C.E. (ed) Seismic Stratigraphy—Applications to Hydrocarbon Exploration. American Association Petroleum Geologists Memoir 26, pp. 83–97.

VINK, G.E., MORGAN, W.J., and ZHAO, W.L. (1984) Preferential rifting of continents: A source of displaced terranes. Journal Geophysical Research 89:10072–10076.

VLAAR, N.J. and CLOETINGH, S.A.P.L. (1984) Orogeny and ophiolites: Plate tectonics revisited with reference to the Alps. Geologie Mijnbouw 63:159–164.

VLAAR, N.J. and WORTEL, M.J.R. (1976) Lithospheric aging, instability and subduction. Tectonophysics 32:331–351.

WATTS, A.B. (1982) Tectonic subsidence, flexure and global changes of sea level. Nature 297:469–474.

WATTS, A.B., KARNER, G.D., and STECKLER, M.S. (1982) Lithospheric flexure and the evolution of sedimentary basins. Philosophical Transactions Royal Society London A305:249–281.

WATTS, A.B. and THORNE, J. (1984) Tectonics, global changes in sea level and their relationship to stratigraphical sequences at the U.S. Atlantic continental margin. Marine Petroleum Geology 1:319–339.

WIENS, D. and STEIN, S. (1984) Intraplate seismicity and stresses in young oceanic lithosphere. Journal Geophysical Research 89:11442–11464.

WORTEL, M.J.R. (1984) Spatial and temporal variations in the Andean Subduction zone. Journal Geological Society London 141:783–791.

WORTEL, R. and CLOETINGH, S. (1981) On the origin of the Cocos-Nazca spreading center. Geology 9:425–430.

WORTEL, R. and CLOETINGH, S. (1983) A mechanism for fragmentation of oceanic plates. In: Watkins, J.S. and Drake, C.L. (eds) Continental Margin Geology. American Association Petroleum Geologists Memoir 34:793–801.

ZIEGLER, P.A. (1982) Geological Atlas of Western and Central Europe. The Hague: Shell Internationale Petroleum Maatschappij/Elsevier (Amsterdam), 130 p.

ZIEGLER, P.A. (1987) Late Cretaceous and Cenozoic intra-plate compressional deformations in the Alpine foreland—a geodynamic model. Tectonophysics 137: 389–420.

ZOBACK, M.D., MOOS, D., and MASTIN, L. (1985) Well bore breakouts and in situ stress. Journal Geophysical Research 90:5523–5530.

ZOBACK, M.L. and ZOBACK, M.D. (1980) State of stress in the conterminous United States. Journal Geophysical Research 85:6113–6156.

11
Subsidence and Gravel Transport in Alluvial Basins

CHRIS PAOLA

Abstract

In recent years the mechanics of basin subsidence have become much better understood, but exploration of the effects of subsidence on basin filling is just beginning. The primary effect of subsidence is to induce deposition. In alluvial systems transporting coarse, mixed-size sediments, deposition selectively removes the coarsest material from the flow, leading to a fundamental connection between subsidence and downstream fining. In this paper I develop three simple mass-balance models that relate subsidence to sediment distribution. The first model, which considers only selective deposition in bimodal gravel-sand mixtures, shows why slow, uniform subsidence tends to produce thin sheet gravels while rapid, asymmetric subsidence produces thick, areally restricted gravels. This result suggests that areally extensive gravel units may signify periods of reduced subsidence rate rather than source-area tectonism. The second model incorporates clast abrasion as an additional cause of downstream fining and shows that small basins with rapid proximal subsidence tend to sort clasts by size while large, slowly subsiding basins tend to sort clasts by durability. The third model presents a partial solution to the complex problem of detailed quantitative modelling of downstream fining. A simple mathematical transformation is proposed that relates longitudinal grain-size profiles in nonuniformly subsiding basins to a reference profile for uniform subsidence, which is assumed to be exponential. The transformation distorts this uniform-subsidence profile so that most of the fining takes place in areas where subsidence rate is highest. Quantitatively, the third model compares well with data for two available test cases. All three models indicate the need for, and might be useful in planning, further basic studies of grain size, gravel content, dispersal pattern, and subsidence rate in alluvial basins.

Introduction

The transition of basin analysis from a descriptive to an analytical science requires the development of well founded theoretical models for basin development, including both the formation of basins and their infilling with sediments. The last decade has seen vigorous growth in models of basin formation, largely developed in the geophysical community. These models provide methods for calculating the subsidence and thermal history of basins based on simple but powerful models of basin-forming processes, like cooling of heated, thinned lithosphere (McKenzie 1978) or flexure of the lithosphere under the weight of advancing thrust sheets (Beaumont 1981; Jordan 1981). These models are of great theoretical interest, and they have also been enthusiastically applied by field workers in a variety of settings. Examples in this volume include papers by Grotzinger and McCormick (1988) and Jordan *et al.* (1988).

The rapid development of models of basin formation has not, however, been accompanied by a parallel development of models of basin filling. At this point, there have been relatively few attempts to relate quantitatively the distribution of sediments and sedimentary facies to basin subsidence. The major one is a series of papers commencing with those of Leeder (1978) and Allen (1978) that sought to relate subsidence of alluvial basins to the distribution of meander-belt sand bodies through the use of random-avulsion models. Further development of the basic idea was carried out by Allen (1979) and Bridge and Leeder (1979), and attempts to apply it to

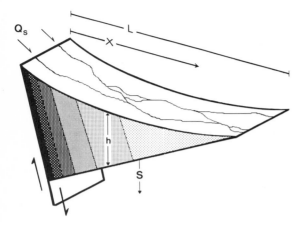

Fig. 11.1. Definition sketch of an alluvial basin. The upstream end of the basin is at x = 0.

field observations have been made by Johnson *et al.* (1979), Read and Dean (1982), Bridge and Diemer (1983), and Gordon and Bridge (1987).

The purpose of this chapter is to present a series of mass-balance models for the filling of an alluvial basin with gravel and sand. The models are quite simple; nevertheless, they provide the beginnings of a theoretical framework that can easily accommodate more complex schemes as our understanding of the sedimentary processes involved grows. The main use of the models is to predict grain size as a function of distance along the direction of sediment transport. Grain size is an important and readily measurable parameter in both the subsurface and in surface exposures. Moreover, since grain-size changes play a major role in governing downstream facies changes in river systems, these models can be viewed as a step toward quantitatively modelling facies variation in alluvial basins.

Basin-Filling Models: General Features

Elements of a Basin-Filling Model

A general definition sketch is shown in Figure 11.1. Sediment is supplied at a specified rate at one end of a basin. The distribution of subsidence rate in the basin is known. It is assumed that the basin is above sea level and that rivers are the main agent of sediment transport. It is also assumed that the transport pattern in the basin is two-dimensional—that is, that the drainage pattern is not strongly tributary or dis-

tributary. The goal of the three models to be developed is to shed light on how the sediment is distributed in the basin.

One expects in general that the grain size will decrease downstream. This is caused by: 1) abrasion and weathering of grains, especially large clasts, a nonconservative process that would take place even in the absence of deposition; and 2) selective deposition of the less transportable grains, generally the largest ones. Selective deposition is a conservative process that is directly driven by deposition in the basin. It can be thought of as a kind of sedimentary fractionation.

There is clear evidence for the importance of both processes of downstream fining in alluvial systems. Clasts of different composition but nearly identical density are observed to become finer at differing rates (Plumley 1948; Bradley 1970; Abbott and Peterson 1978; Shaw and Kellerhals 1982), which implies that the changes result from weathering and/or abrasion. On the other hand, alluvial systems with very high rates of deposition, especially alluvial fans, show downstream fining rates that are often several orders of magnitude higher than those typical of equilibrium or degrading systems, which are commonly one or two orders of magnitude greater than abrasion rates estimated from laboratory experiments (Krumbein 1941; Kuenen 1956; Shaw and Kellerhals 1982; Rust and Koster 1984; Brierly and Hickin 1985). This correlation between rate of deposition and rate of downstream fining implies control by selective deposition.

Selective deposition and physical or chemical breakdown potentially could affect the whole range of grain sizes in alluvial basins. Nevertheless, at present their effects are best understood for gravel-sized and sand-sized sediments, so I will restrict attention to basins or parts of basins where these grain sizes predominate.

To connect abrasion and selective deposition with the long-term evolution of a basin, it is necessary to relate sedimentation to subsidence. In basins containing hundreds of metres or more of alluvial sediments, sedimentation and subsidence must on average remain in balance, especially where facies types are consistent up section relative to the basin margin. This condition, that subsidence rate equals sedimentation rate, is the starting point of the modelling. It means that the results apply strictly over only time scales long enough to average out imbalances between sediment supply and subsidence. It would not be hard to relax this condition, though, to allow for short-term aggradation and degradation relative to the long-term-average surface topography in the basin. In this case results like those calculated here would form the time-averaged profiles on which these short-term fluctuations would be superimposed.

Not only does long-term net deposition in response to subsidence exert a direct control on the rate of fining due to selective deposition; it may also affect fining caused by weathering or abrasion. The nature of this effect depends on the actual mechanism of comminution, which is still poorly understood (Shaw and Kellerhals 1982). If the mechanism is abrasion during transport then it depends only on distance of transport and not on sedimentation rate. Alternatively, if the mechanism depends strongly on weathering of clasts during intervals between transport events (Bradley 1970), or on abrasion in place (Schumm and Stevens 1973), then the rate of burial of clasts should control the rate of fining: High rates of deposition should be associated with low rates of grain-size decrease since in this case the clasts are transported and buried before significant comminution can occur. Since this association has not been observed, I will assume that abrasion during transport is the main effect. The real situation is certainly more complex than this; even if abrasion during transport is dominant, weathering probably plays a role at least in increasing the susceptibility of clasts to abrasion (Bradley 1970). Such effects can easily be incorporated in future models as they become better understood.

Overview: Three Models

Three models are developed in this paper, each meant to examine a different aspect of downstream fining in alluvial basins. The first deals with the overall shape of gravel bodies in cross section as determined by gravel supply and the distribution of subsidence; the second, with the relative effects of selective deposition and abrasion on gravel/sand ratio; the third, with the effects of differential subsidence on grain-size variation within the gravel fraction. The basic idea of all three models is the same, however: Subsidence induces deposition; deposition causes downstream fining by selectively removing the coarsest clasts from the flow; and thus areas of rapid subsidence are areas of rapid downstream fining. In general, basins that subside slowly near their source areas can transport gravels relatively long

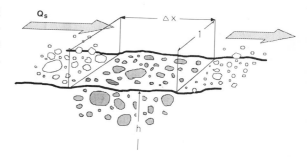

Fig. 11.2. Control volume of unit width for mass-balance modelling.

distances compared with basins that subside rapidly near their source areas.

Derivation of the Governing Equation for Models 1 and 2

Concept

The first two models to be developed are for downstream change in the fractional content of gravel in the sediment. The gravel fraction in the bed is related to the gravel fraction in the flow by a partition coefficient; the equation derived is for gravel fraction in the bed. The equation expresses the idea that gravel is lost from the flow to both selective deposition in the bed and breakdown by abrasion. All the terms in the equation are averages: Transport rate and subsidence rate over time, and gravel fraction over vertical distance in the deposit. As mentioned above, the averaging intervals must be large enough to smooth out imbalances between subsidence and sedimentation.

Analysis

Consider the control volume shown in Figure 11.2. The rate of gravel supply is given by $Q_s f$, where Q_s is the total volumetric sediment-transport rate per unit width (in units of length^2/time) and f is the gravel fraction of the sediment being carried in the flow. Gravel in transport is lost by deposition on the bed and breakdown to sand. Defining a partition coefficient J as f/p where p is the gravel fraction in the bed, the loss to the bed is $\Delta x(f/J)dh/dt$ where h is the height of the bed above basement and x is downstream distance. As mentioned above, it is assumed that over the long-term average, dh/dt is equal to the subsidence rate S. It is also assumed that the porosity of the gravel deposit is small. Selective deposition implies that $J < 1$. If $J = 1$ the loss term is not zero but it does not cause the transported load to become finer downstream.

The loss to breakdown is taken to be $\Delta(fQ_s) = kfQ_s\Delta x$, where k is an abrasion coefficient. As mentioned above it is assumed that the breakdown is related only to distance of transport and not to exposure time as would be the case if weathering were important. The parameter k has units of length^{-1} and must be determined empirically for each clast composition.

Combining the two loss terms and taking the limit as Δx goes to zero, the equation governing downstream grain-size change is:

$$\frac{dp}{dx} = \frac{pS}{Q_s}\left[1 - \frac{1}{J}\right] - kp \qquad (1)$$

The key ingredient needed to solve this equation is a model for J, the partition coefficient. In general the rate at which sediment is selectively removed from the flow and deposited is a complex function of the flow, the range of sizes in the sediment, and the rate of deposition. As defined above, a value $J = 1$ corresponds to no sorting at all. On the other hand, the limit $J \to 0$ implies perfect sorting in which only gravel is deposited until it is exhausted in the flow.

Basin-Filling Models: Three Examples

Perfect Sorting as a Simple Idealization: Model 1

Concept

The limiting behavior of equation (1) as $J \to 0$ (perfect sorting) is useful in providing a general understanding of how selective deposition and subsidence combine to influence the cross-sectional shape of gravel bodies. Although perfect sorting is clearly not realistic in detail, it may be a reasonable first approximation for areas in which centimetre-scale or coarser gravel is abundant enough to dominate the bed load. Bed shear stress is then high enough that sand and mud travel mainly in suspension and the deposit would be mostly gravel.

Under these conditions, the mass balance for gravel is quite simple: In a cross section parallel to transport, the area of basin filled with gravel per unit time must be equal to the supply rate in volume of gravel per unit time per unit width of basin. Thus for a given gravel supply rate there is an inverse relation

between subsidence rate near the source and distance of gravel transport.

Analysis

As $J \to 0$, the first term on the right-hand side of (1), the selective-deposition term, becomes large relative to the second (abrasion) term. The equation cannot be solved directly in the limit, but conservation of gravel requires:

$$f(0) = \frac{R\int_0^{L_g} S(x)dx}{\int_0^L S(x)dx} \qquad (2)$$

where $x = 0$ refers to the upstream margin of the basin, L is the total length of the basin, L_g is the distance to which gravel is transported, and

$$R = (\int_0^L S(x)dx)/Q_s(0)$$

R, the capture ratio, is the ratio of rate of creation of cross-sectional area in the basin (length^2/time) to rate of sediment supply (length^2/time). If $R < 1$ the basin is oversupplied and sediment is transported out the far side; if $R > 1$ the basin is undersupplied and in most cases would fill with sea water. Roehl (1962) estimated that for a number of modern alluvial basins with drainage areas greater than a few hundred square kilometres the short-term capture ratio is about one. This value will be used throughout this paper, although this is largely for computational simplicity and can easily be modified as more detailed models are developed.

In the case $R = 1$ the term on the right-hand side of (2) is the ratio of basin area filled in with gravel per unit time to total basin area created and filled with sediment per unit time (Fig. 11.3).

Equation (2) suggests that the gravel-transport distance L_g is controlled by several factors even in this very simple model. The gravel fraction in the sediment supply, and the overall supply rate, appear to be mainly controlled by climate, lithology, and relief in the source area (Pettijohn 1957; Ahnert 1970; Wilson 1973; Jansen and Painter 1974). For fault-bounded basins, the relief is controlled at least in part by the subsidence rate at the basin margins. (This is one mechanism by which long-term equilibrium between subsidence and sedimentation can be maintained.) The other two controls on L_g are the distribution and, if the assumption $R = 1$ is relaxed, the rate of subsidence. Equation (2) shows that for a given supply gravel fraction $f(0)$, laterally extensive gravels are to be expected where subsi-

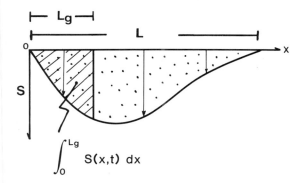

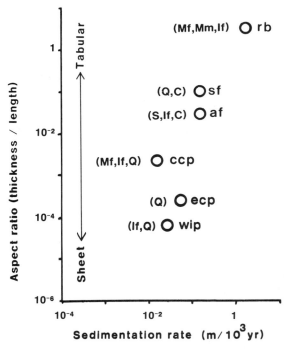

Fig. 11.3. Sketch showing the two integrals appearing in equation (2) for the case R = 1 (sediment supply equals subsidence). The vertical axis is subsidence rate (not total subsidence). The dotted area is the total area created per unit time. The diagonally ruled area is the area filled with gravel per unit time, which is equal to the volume of gravel supplied to the basin per unit width and time. The length of the gravel-filled area is L_g, the gravel-transport distance.

Fig. 11.4. Gravel-body cross-sectional shape as a function of sedimentation rate. Dominant clast types are shown in parentheses: Mf = felsic metamorphic, Mm = mafic metamorphic, If = felsic igneous, C = carbonate, Q = quartzite and chert, S = other sedimentary rocks. Basins and data sources: rb = Ridge Basin, California, Upper Miocene, Violin Breccia and coeval units (Crowell 1982; Ensley and Verosub 1982); sf = Sevier Foreland, western interior U.S.A., Albian-Santonian, Indianola Group (Lawton 1986); af = Alpine Foreland, Switzerland, Chattian-Aquitanian, Lower Freshwater Molasse (Trümpy 1980); ccp = California coastal plain, southern California, Upper Eocene, Ballena gravel (Minch 1979; Steer and Abbott 1984); ecp = central Atlantic coastal plain, U.S.A., Pliocene, Brandywine Formation (Schlee 1957); wip = western interior alluvial plain, U.S.A., Pleistocene, unnamed gravels (Stanley and Wayne 1972).

dence is slow and either uniform or increasing in the direction of transport; thus slow, uniform subsidence tends to produce thin gravel sheets. On the other hand, in cases where subsidence is rapid, especially if it is concentrated near the sediment source (e.g., in foreland basins), L_g is small but the thickness is likely to be large, so that relatively tabular gravels are produced. The association of rapid subsidence with tabular gravels is shown for six alluvial systems in Figure 11.4.

This analysis, although simplified, helps clarify the relation between gravels and tectonism. Because gravels are relatively uncommon in the sedimentary record, the appearance of gravel in a stratigraphic section is often taken as an indication of tectonic activity in the source area (e.g., Armstrong and Oriel 1965; Wiltschko and Dorr 1983; Rust and Koster 1984). Although it is not always stated explicitly, the control that is being invoked is presumably that tectonism in the source area increases relief, thus increasing sediment supply rate and perhaps gravel fraction. Insofar as this is the *only* effect associated with tectonic activity, the above analysis supports this inference. But tectonic activity in the source area might equally well increase subsidence rate in the basin, especially near the source area, which acts to reduce the transport distance of the gravel. Thus the actual effect of tectonic activity on the distance to which

gravel is transported into the basin depends on the outcome of two competing effects: increasing relief, and thus gravel-supply rate, versus increasing subsidence rate.

To take a specific example, in a foreland basin, activity on an associated thrust fault causes coarsening of basin sediments only if the increase in sediment supply caused by the rising thrust is sufficient to offset the rapid asymmetric subsidence in the foreland basin that is associated with thrusting. The two foreland basins shown in Figure 11.4

(Sevier and Alpine) both show relatively tabular gravels restricted to within 100 km of the thrust, so in these two cases rapid proximal subsidence is apparently effective at restricting the transport distance of the gravel. As explained by Jordan *et al.* (1988: this volume), whether an increase in thrust load produces mainly increased subsidence or mainly increased relief depends on the stiffness of the lithosphere. Thus, activity on a thrust could produce either coarsening or fining of alluvial sediments in the central and distal parts of the foreland basin, although it should always produce coarsening near the thrust edge. An apparent example of upward fining in response to Early Cretaceous thrusting in the western interior of North America is discussed by Heller *et al.* (1986). They concluce that except within a few tens of kilometres of the thrust, the onset of thrusting is marked by an upward transition from a thin, laterally extensive gravel to fine-grained lacustrine siliciclastics and carbonates.

Abrasion Versus Selective Deposition: Model 2

Concept

To derive a model for downstream fining in alluvial basins that includes both selective deposition and abrasion, it is necessary to have a value for J, the partition coefficient between gravel in the flow and gravel in the bed. A simple analysis in the Appendix suggests that for rivers with noncohesive banks, J approaches a constant value (0.17) as the gravel fraction in the flow becomes small. With this value and an empirical value for the abrasion coefficient k, Equation (1) can be solved to give an estimate of the rate of downstream fining in both bed and flow due to the combination of selective deposition and abrasion.

Analysis

The variables in (1) may be made nondimensional as follows:

$$S_* = S/\bar{S}; \quad p_* = p/p(0); \quad x_* = x/L$$

where $\bar{S}$ is the average subsidence rate, and S_* and p_* are both functions of x_*. The nondimensional version of (1) is:

$$\frac{dp_*}{dx_*} = -p_* \left[\frac{S_*[(1/J)-1]}{R^{-1} - A_*} + kL \right] \quad (3)$$

in which A_* is defined by $dA_*/dx_* = S_*$.

Three nondimensional parameters appear in (3) that are of fundamental importance in determining the rate of downstream fining in sedimentary basins. These are: J, the partition coefficient, which measures the effectiveness of selective deposition; kL, which measures the characteristic distance for clast breakdown against the size of the basin; $R = \bar{S}L/Q_s(O)$, the capture ratio defined in the previous section.

The two terms on the right-hand side of (2) express the effects of sorting and breakdown, respectively, in causing downstream fining. The balance between the two depends, for the case $R = 1$, on the ratio $[(1/J)-1]/kL$ so that breakdown of grains is of comparable importance to selective deposition if kL ~ 6. Values of basin length L required for this in temperate climates are of the order of tens of kilometres for limestone and schist, and hundreds of kilometres for granite and quartzite, based on a range of laboratory estimates for k from Shaw and Kellerhals (1982). It is clear that mechanical breakdown is important in moderately small basins (or equivalently, relatively near source) for easily weathered clast types like schist and limestone, but becomes important only for large basins and long distances for more durable quartzite and chert clasts.

The spatial distribution of the subsidence rate also plays a major role in determining the rate of downstream fining. In equation (3), the shape of the subsidence profile enters via the parameters A_* and S_*. S_* is the subsidence rate normalized by its average value, so $S_* (x_*)$ expresses the shape of the subsidence distribution along the direction of sediment transport. A_* is the integral of this shape function; for any x_*, $A_* (x_*)$ is the ratio of basin area created upstream of x_* to basin area created over the whole length of the basin, all per unit time. For example, for uniform subsidence $S_* = 1$ and $A_* = x_*$, so that at a point halfway across the basin ($x_* = 0.5$), half of the total area created per unit time lies upstream, and half downstream, of the point in question (thus $A_* = 0.5$).

Several solutions to equation (3) are shown in Figure 11.5, which compares calculated rates of downstream fining for a basin with constant subsidence rate (Fig. 11.5a) to fining rates for a basin in which the subsidence rate varies linearly from zero at the downstream end to a maximum at the upstream end (Fig. 11.5b). As expected, the rate of fining is much larger in the asymmetric case. In addition, in the asymmetric case the fining rate is much less sensitive to variation in kL, our nondimensional

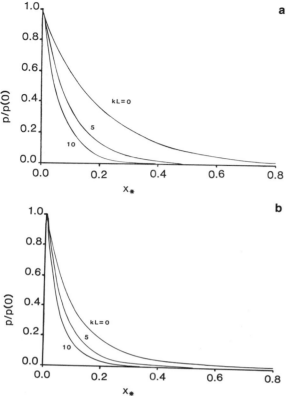

a

Fig. 11.5. Deposited gravel fraction p relative to its initial value p(0), calculated from equation (3) for (a) uniform subsidence and (b) linear asymmetric subsidence [that is, $S_* = 2(1-x_*)$]. The three curves in each graph have different values of kL, a nondimensional measure of the effectiveness of abrasion relative to selective deposition in causing downstream fining.

measure of clast durability. Evidently, in short basins with rapid proximal subsidence, various gravel clast types penetrate into the basin to similar but relatively short distances; such basins are efficient at size fractionation. In large basins with low, uniform subsidence rates gravel clasts travel farther in general but the distance is relatively sensitive to clast composition. These basins are efficient at composition fractionation.

It is difficult to find data to test these results quantitatively. Nonetheless, Figure 11.4 indicates that the general trend is correct. Gravels in small, asymmetric basins like the Ridge Basin include a wide variety of clasts, including a significant number of easily weathered types. In large, slowly subsiding systems like the western Great Plains and eastern coastal plain examples, gravels are domi-

nated by relatively durable types like quartzite and chert.

Effect of Subsidence Distribution on Grain-Size Variation: Model 3

Concept

The foregoing models provide estimates of variation in gravel fraction across alluvial basins, but do not treat grain-size variation within the deposited gravel. However, most of the data available on grain-size variation in alluvial basins are in the form of spatial distribution of pebble size. Thus to look for quantitative effects of subsidence-rate variation on downstream fining, a version of the model that treats grain-size variation in the gravel fraction is required. This is complicated considerably by two circumstances: 1) the processes by which size sorting takes place in coarse bed load are still poorly understood (e.g., Komar 1987; Wilcock 1987); and 2) most of the grain-size data available are in the form of some measure of "average largest clast size," which is unfortunate because variation in extreme sizes is difficult to model physically and is relatively sensitive to details of the coarse tail of the input size distribution.

As a result of these two complications, the strategy in this section will be somewhat different than in the preceding two. I will begin by developing a simple conservation equation for each grain size (analogous to Equation (1) but without abrasion) to show how spatial variation in subsidence rate affects continuous grain-size distributions. The general behavior of this equation is similar to those of the previous sections: Relatively rapid subsidence is expected to produce relatively rapid downstream fining. However, because many aspects of selective deposition are not well understood, I will not try to develop a complete theory using continuous grain-size distributions. Instead the conservation equation in nondimensional form will be used to relate grain-size variation in uniformly subsiding basins to grain-size variation in basins with complex subsidence patterns. The relation takes the form of a simple mathematical transformation in which the non-dimensional area-creation function A_* defined in the previous section serves as a measure of the spatial distribution of subsidence.

Analysis

An equation analogous to (1) but for a continuous grain-size distribution is:

$$\frac{d}{dx}(FQ_s) = -PS \qquad (4)$$

where F and P are continuous probability-density functions for grain size in the flow and the bed, respectively, and Q_s and S are the sediment-transport rate and subsidence rate, respectively. Equation (4) does not include abrasion effects; to do so for continuous grain-size distributions involves complexities that are beyond the scope of this paper. To provide any useful information, equation (4) must be averaged (that is, integrated over all or part of the distribution). For comparison with available maximum pebble-size data we are most interested in the coarse tail of the deposits. I define D_ε as the grain size at the $100(1-\varepsilon)$ percentile; for example the ninetieth percentile size, commonly measured in modern streams, is in this notation $D_{0.1}$. Defining the partition coefficient J in a manner analogous to the previous sections as $J = F/P$ and assuming that for small ε, J is nearly constant, we can integrate (4) to get:

$$\int_{D_\varepsilon}^{\infty} \frac{d}{dx}(PQ_s)\,dD = -\frac{\varepsilon S}{J}$$

where I have used the definition of D_ε:

$$\int_{D_\varepsilon}^{\infty} P\,dD = \varepsilon$$

Using this definition again along with Leibnitz's rule the left-hand side becomes:

$$\int_{D_\varepsilon}^{\infty} \frac{d}{dx}(PQ_s)\,dD = \varepsilon\frac{dQ_s}{dx} + Q_s\,P(D_\varepsilon)\frac{dD_\varepsilon}{dx}$$

and rearranging gives:

$$\frac{dD_\varepsilon}{dx} = \frac{\varepsilon S}{Q_s P(D_\varepsilon)}\left(1 - \frac{1}{J}\right) \qquad (5)$$

The term $P(D_\varepsilon)$ is an unknown function that depends on how the grain-size distribution in the bed changes downstream. Development of a relation for this function would require a lengthy discussion of alluvial sediment-transport mechanics that is beyond the scope of this paper. However, the information in equation (5) that is most directly useful in basin analysis can be obtained without a model for P, because for a given input grain-size distribution P depends *only* on D_ε. The form of this dependence is still unknown. Nondimensionalizing x gives, for uniform subsidence over the basin:

$$\frac{dD_\varepsilon}{dx_*} = \frac{\varepsilon}{1 - x_*}\frac{(1 - 1/J)}{P(D_\varepsilon)} \qquad (6)$$

Suppose the solution to (6) for this uniform case is $D_0(x_*)$. Then by direct substitution it may be veri-

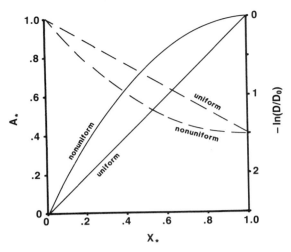

Fig. 11.6. The effect of nonuniform subsidence on grain-size variation. The grain-size profile (dashed line) for the uniform case is assumed to follow Sternberg's exponential form. The nonuniform case is the linear asymmetric subsidence profile shown in Figure 11.5b. A_* (solid line) is a nondimensional function that expresses the distribution of rate of area creation in the basin along lines parallel to sediment transport, and x_* is the distance from the upstream edge of the basin divided by the total length of the basin. Both are further explained in model 2. The effect of redistributing most of the subsidence toward the source is to deform the grain-size profile so that the rate of change is high near the source and diminishes quickly going into the basin.

fied that the general solution to (5) is the composite function $D_0[A_*(x_*)]$. In other words, the effect of variation in subsidence rate is to deform the uniform-subsidence solution by replacing its argument with the nondimensional function A_* that describes the rate of creation of area across the basin. The transformation this entails is shown graphically in Figure 11.6. Where subsidence is relatively rapid, A_* increases faster than x_* and so the sediment fines relatively quickly (proximal section, asymmetric basin, Fig. 11.6). As in the previous two models, rapid subsidence is associated with rapid downstream fining. Note that in the uniform case $A_* = x_*$ so that the uniform solution $D_0(x_*)$ is properly recovered.

Clearly, to make sense of the distribution of grain size in a nonuniformly subsiding basin it is necessary to have some idea of the form the grain-size variation would have if the subsidence were uniform. Since this depends on the input grain-size distribution, in principle it could be different for each basin. However, exponential downstream size decrease (Schultheis and Mountjoy 1978; Shaw and Kellerhals 1982; Rust and Koster 1984) is common enough in both modern rivers and ancient alluvial deposits to suggest that this exponential form is a reasonable first approximation to the general form of the variation.

Substantiation of this hypothesis is difficult because in most studies of downstream fining, the profile of rate of deposition is unknown. Where exponential downstream fining has been observed over short distances (i.e., a few kilometres, as is commonly the case in very proximal alluvial systems; e.g., Eckis 1928; Blissenbach 1964; Bradley *et al.* 1972), large downstream changes in sedimentation rate seem unlikely. Firmer evidence comes from studies of ancient gravels of known, uniform thickness (Schlee 1957; Stanley and Wayne 1972), for which the form of grain-size variation is also exponential.

To look for the effect of nonuniform subsidence rate on downstream fining data are needed on grain size, sediment thickness, and dispersal pattern taken over a well constrained and preferably small time interval. Although none of these are unusual types of data in alluvial gravels, it is surprisingly difficult to assemble data sets containing all three. Two examples for which such data are available will be discussed below. Both data sets fulfill the basic condition required for applying a two-dimensional model: Their paleocurrent patterns indicate that each had a relatively two-dimensional transport system. Furthermore, in both cases the clasts measured were mainly durable types such as vein quartz, so that ignoring abrasion should not produce major errors.

Case 1

Tuscarora Quartzite (Lower Silurian, central Appalachian Mountains, U.S.A.) — The data used for this example all come from a comprehensive study of this unit by Yeakel (1962) that includes isopach, paleocurrent, and maximum grain-size data over a distance of some 200 km in the direction of sediment transport. This work was part of a series of papers on paleocurrents, petrology, and paleogeography carried out by students of Francis J. Pettijohn beginning in the 1950s. They are remarkable for their thoroughness and consistency of observation; it is no coincidence that three of the studies used to provide data for the present analysis (Schlee 1957; Yeakel 1962; Meckel 1967) come from this series.

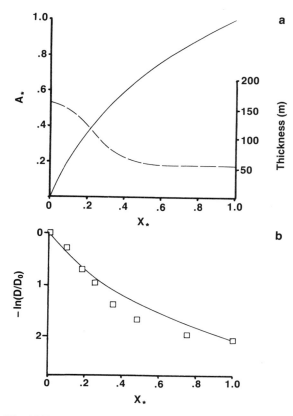

Fig. 11.7. (a) Thickness variation (dashed line) in the Tuscarora Formation, Lower Silurian, central Appalachians, U.S.A. (from Yeakel 1962) and the resulting subsidence-area function A_* (solid line). (b) Comparison of observed grain size (squares) with the profile (solid line) calculated from A_* and the transformation described in model 3.

Figure 11.7 shows a grain-size profile parallel to transport provided by Yeakel (1962; his Fig. 15, line 3), and then corrected for subsequent crustal shortening using values provided for the same area by Meckel (1967). The maximum correction is about 10% of total basin length. Also shown is the nondimensional subsidence-area function A_* defined above, derived from isopach data provided by Yeakel for the entire Tuscarora (his Fig. 4), also corrected for crustal shortening. A_* is found by: 1) digitizing the thickness profile; 2) finding the average thickness; 3) dividing thickness by average thickness to get S_*; and 4) numerically integrating S_* to get A_*.

The Tuscarora deposits show a pronounced increase in rate of change of grain size going toward source, and that this increase is associated with a significant increase in sediment thickness. The

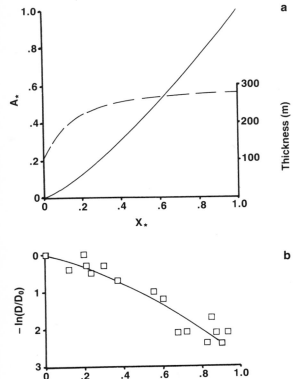

a

b

Fig. 11.8. (a) Thickness variation (dashed line) in the Middle Buntsandstein, Triassic, eastern France and south-western Germany (from Forche 1935) and the resulting subsidence-area function A_* (solid line). (b) Comparison of observed grain size (squares) with the profile (solid line) calculated from A_* and the transformation described in model 3.

grain-size variation predicted from transforming a simple exponential using the observed area function A_* is also shown. The agreement between theory and observation is good, especially considering that the isopachs represent the whole of the Tuscarora while the grain sizes represent a single horizon at each location. Evidently the downstream fining was in this case mainly controlled by selective deposition. This is further corroborated by the fact that Yeakel selected only vein-quartz pebbles for measurement, for which kL (based on abrasion rates cited in Shaw and Kellerhals 1982) is of the order of 0.1.

Case 2

Middle Buntsandstein (Triassic, eastern France and southwestern Germany) – The data for this example

are those of Forche (1935; also summarized in Potter and Pettijohn 1977, Fig. 4-10). In this case, as in the previous one, there are marked changes in thickness of the unit in question but now the thickness increases *away* from source. The selective-deposition model developed here predicts that in this case the rate of change of grain size should also increase going away from source, and Figure 11.8 shows that this is indeed the case. The quantitative agreement between theory and observation is also very good, although, as in the above example, the theory somewhat underestimates the effect of differential subsidence.

Discussion

In this paper I present three models for downstream decrease in grain size in gravel-sand alluvial systems. The first and simplest one, which assumes a bimodal gravel-sand input and perfect sorting, is useful in developing a basic understanding of controls on the cross-sectional shape of gravel bodies in alluvial basins. The model shows that the distance to which gravel penetrates in a basin is largest when gravel supply rate is high and subsidence is slow and either uniform or increasing in the direction of transport. This suggests that widespread alluvial gravels, rather than being indicators of tectonism in their source areas, may record instead periods of reduced subsidence rate, when gravel is carried across the basin rather than being fractionated out near its upstream end.

The second model, which retains the bimodal input but uses a more reasonable sorting coefficient and also allows for the effects of abrasion, can be used to generate approximate estimates of vertically averaged gravel fraction as a function of downstream distance. This model suggests that small basins with rapid proximal subsidence sort clasts mainly by size while large basins with slow, uniform subsidence sort clasts mainly by durability.

The third model provides estimates of downstream variation of maximum grain size in basins with differential subsidence, relative to an assumed exponential form in uniformly subsiding basins. The weakness of having to assume such a reference profile would be partially alleviated by a clear understanding of how selective deposition works, which would presumably tell us under what conditions an exponential profile should occur. However, any model that uses maximum grain size will be sensitive to the very coarse tail of the input grain-size distribution, the details of which are likely to remain difficult to predict. Hence, calculations in terms of gravel fraction are preferable even though gravel fraction is more difficult to measure in the field than the average maximum clast size at each outcrop.

There is potentially much to be gained from the testing and further refinement of basin-filling models. To develop quantitative models for major facies transitions in alluvial basins using the kind of approach developed here would require, among other things, an extension of selective-deposition models to include sand and mud. This would allow the mud fraction in the banks, critical in stabilizing channels and allowing meander development, to be modelled. This in turn requires a better understanding of flood-plain dynamics than we now have. At a much larger scale, though, alluvial-architecture models like those of Leeder (1978), Allen (1978, 1979), and Bridge and Leeder (1979) can be viewed as selective-deposition models for sand-mud systems. They predict that basins subsiding rapidly relative to the rate of channel avulsion selectively retain flood-plain muds in a manner analogous to that described here for gravels, although the mechanism is completely different.

There are other, more immediate uses for better developed versions of the models presented here. If a reliable quantitative or semi-quantitative relation between stratigraphic thickening and accelerated rate of grain-size change can be established, then horizontal grain-size profiles might be used to estimate thickness or subsidence-rate variations where section is missing, unexposed, or for which subsurface data are not available. Such estimates might also be useful in correlating unfossilferous gravel units. Alternatively, the form of grain-size variation, corrected for known effects of subsidence-rate variation, might provide useful information on source-area lithology or weathering regime.

More generally, one of the most important results of this study has been to show how, by coupling subsidence and sedimentation, an understanding of critical aspects of sediment mechanics in modern environments can be used to model large-scale features of sedimentary basins. It is striking that in spite of their extreme simplicity the models developed here are limited even more by lack of data against which they can be tested than by lack of understanding of the processes involved. It is yet more striking that the data that are available were all published at least twenty years ago. I hope that the work presented here will at least provide some incentive for more of the kind of thoroughgoing studies of alluvial basins that students of Francis Pettijohn carried out in the central Appalachians in the 1950s and 1960s. Without comprehensive, basic measurements like these it is difficult to see how quantitative basin-filling models of any kind can be tested and refined.

Acknowledgments. I thank J.S. Bridge, G.V. Middleton, J. Thorne, J.B. Southard, M. Covey, and K. Kleinspehn for thoughtful, constructive comments on this paper, and Susan Swanson for typing the manuscript. Financial support was provided by the ARCO junior-faculty program and NSF grant EAR-87-07041.

Appendix

To derive an approximate value for the partition function in a gravel/sand alluvial system, we begin with the simple sediment transport law of Meyer-Peter and Müller (1948):

$$Q_s = \frac{8}{(s-1)g}(\tau - \tau_c)^{3/2} \qquad (A1)$$

where Q_s is volume flux of sediment per unit width and time, s the specific gravity of the sediment, g gravitational acceleration, τ the kinematic boundary shear stress (force per unit area divided by fluid density), and τ_c the critical value of τ for the initiation of sediment motion.

If the gravel fraction in the bed p is small, it is given by:

$$p = \frac{\dfrac{d(fQ_s)}{dx}}{\dfrac{dQ_s}{dx}} \qquad (A2)$$

where Q_s is the volume flux of sediment and f and p refer to gravel fraction in the bed and flow, respectively. Note that long-term net aggradation is associated with spatial (as opposed to temporal) change in shear stress. Based on (A1) the relative change in the transport rate of the gravel is:

$$\frac{1}{fQ_s}\frac{d(fQ_s)}{dx} = \frac{1}{\tau - \tau_{cg}} \cdot \frac{3}{2}\frac{d\tau}{dx} \qquad (A3)$$

where τ_{cg} is the critical shear stress for the gravel. An analogous relation using τ_{cs}, the critical shear stress for the sand, holds for the sand fraction. Using (A2) and (A3), and $f \ll 1$, the partition coefficient is:

$$J = \frac{f}{p} = \frac{\tau - \tau_{cg}}{\tau - \tau_{cs}} \qquad (A4)$$

In gravel-bed rivers with noncohesive banks, Parker (1978) has shown that bank erosion acts to keep the bed shear stress τ during channel-forming floods adjusted to a value about 20% greater than the critical value for general bed-load motion. In the present case this critical value corresponds to τ_{cg}. Thus:

$$J = \frac{0.17}{(1 - 0.83 \, \tau_{cs}/\tau_{cg})}$$

If the grain-size of the gravel is large compared with that of the sand, then $\tau_{cs}/\tau_{cg} \ll 1$ in spite of equal-mobility effects (e.g., Andrews 1983) and $J \cong 0.17$.

References

ABBOTT, P.L. and PETERSON, G.L. (1978) Effects of abrasion durability on conglomerate clast populations: Examples from Cretaceous and Eocene conglomerates of the San Diego area, California. Journal Sedimentary Petrology 48:31–42.

AHNERT, F. (1970) Functional relationships between denudation, relief, and uplift in large mid-latitude drainage basins. American Journal Science 268:243–263.

ALLEN, J.R.L. (1978) Studies in fluviatile sedimentation: An exploratory quantitative model for the architecture of avulsion controlled alluvial suites. Sedimentary Geology 21:129–147.

ALLEN, J.R.L. (1979) Studies in fluviatile sedimentation: An elementary geometrical model for the connectedness of avulsion-related channel sand bodies. Sedimentary Geology 24:253–267.

ANDREWS, E.D. (1983) Entrainment of gravel from naturally sorted riverbed material. Geological Society America Bulletin 94:1225–1231.

ARMSTRONG, F.C. and ORIEL, S.S. (1965) Tectonic development of Idaho-Wyoming thrust belt. American Association Petroleum Geologists Bulletin 49:1847–1866.

BEAUMONT, C. (1981) Foreland basins. Geophysical Journal Royal Astronomical Society 65:291–329.

BLISSENBACH, E. (1964) Geology of alluvial fans in semiarid regions. Geological Society America Bulletin 5:175–190.

BRADLEY, W.C. (1970) Effect of weathering on abrasion of granitic gravel, Colorado River (Texas). Geological Society America Bulletin 81:61–80.

BRADLEY, W.C., FAHNESTOCK, R.K., and ROWEKAMP, E.T. (1972) Coarse sediment transport by flood flows on Knik River, Alaska. Geological Society America Bulletin 83:1261–1284.

BRIDGE, J.S. and DIEMER, J.A. (1983) Quantitative interpretation of an evolving ancient river system. Sedimentology 30:599–623.

BRIDGE, J.S. and LEEDER, M.R. (1979) A simulation model of alluvial stratigraphy. Sedimentology 26:617–644.

BRIERLY, G.J. and HICKIN, E.J. (1985) The downstream gradation of particle sizes in the Squamish River, British Columbia. Earth Surface Processes Landforms 10:597–606.

CROWELL, J.C. (1982) The Violin Breccia, Ridge basin, southern California. In: Crowell, J.C. and Link, M.H. (eds) Geologic History of Ridge Basin, Southern California. Pacific Section Society Economic Paleontologists Mineralogists, pp. 89–98.

ECKIS, R. (1928) Alluvial fans in the Cucamonga district, southern California. Journal Geology 36:224–247.

ENSLEY, R.A. and VEROSUB, K.L. (1982) Biostratigraphy and magnetostratigraphy of southern Ridge Basin, central Transverse Ranges, California. In: Crowell, J.C. and Link, M. (eds) Geologic History of Ridge Basin, Southern California. Pacific Section Society Economic Paleontologists Mineralogists, pp. 13–24.

FORCHE, F. (1935) Stratigraphie und paläogeographie des Buntsandsteins im Umkreis der Vogesen. Mitteilungen Geologischen Staatsinstitut Hamburg 15:15–55.

GORDON, E.A. and BRIDGE, J.S. (1987) Evolution of Catskill (Upper Devonian) river systems: Intra- and extrabasinal controls. Journal Sedimentary Petrology 57:234–249.

GROTZINGER, J.P. and McCORMICK, D.S. (1988) Flexure of the Early Proterozoic lithosphere and the evolution of Kilohigok Basin (1.9 Ga), Northwest Canadian Shield. In: Kleinspehn, K.L. and Paola, C. (eds) New Perspectives in Basin Analysis. New York: Springer-Verlag, pp. 405–430.

HELLER, P.L., WINSLOW, N.S., and PAOLA, C. (1986) Sedimentation and subsidence across a foreland basin: Observations and results from the western interior (abstract). Geological Society America 99th Annual Meeting Abstracts Programs, p. 634.

JANSEN, J.M.L. and PAINTER, R.B. (1974) Predicting sediment yield from climate and topography. Journal Hydrology 21:371–380.

JOHNSON, G.D., JOHNSON, N.M., OPDYKE, N.D., and TAHIRKHELI, R.A.K. (1979) Magnetic reversal stratigraphy and sedimentary tectonic history of the Upper Siwalik Group, eastern Salt Range and southwestern Kashmir. In: Farah, A. and De Jong, K.A. (eds)

Geodynamics of Pakistan. Quetta: Geological Survey Pakistan, pp. 149–165.

JORDAN, T.E. (1981) Thrust loads and foreland basin evolution, Cretaceous, western United States. American Association Petroleum Geologists Bulletin 65: 2506–2520.

JORDAN, T.E., FLEMINGS, P.B., and BEER, J.A. (1988) Dating thrust-fault activity by use of foreland-basin strata. In: Kleinspehn, K.L. and Paola, C. (eds) New Perspectives in Basin Analysis. New York: Springer-Verlag, pp. 307–330.

KOMAR, P.D. (1987) Selective grain entrainment by a current from a bed of mixed sizes: A reanalysis. Journal Sedimentary Petrology 57:203–211.

KRUMBEIN, W.C. (1941) The effects of abrasion on the size, shape and roundness of rock fragments. Journal Geology 49:482–520.

KUENEN, P.H. (1956) Experimental abrasion of pebbles 2. Rolling by current. Journal Geology 64:336–368.

LAWTON, T.F. (1986) Compositional trends within a clastic wedge adjacent to a fold-thrust belt: Indianola Group, central Utah, U.S.A. In: Allen, P.A. and Homewood, P. (eds) Foreland Basins. International Association Sedimentologists Special Publication 8:411–423.

LEEDER, M.R. (1978) A quantitative stratigraphic model of alluvium, with special reference to channel deposit density and interconnectedness. In: Miall, A.D. (ed) Fluvial Sedimentology. Canadian Society Petroleum Geologists Memoir 5, pp. 587–596.

McKENZIE, D. (1978) Some remarks on the development of sedimentary basins. Earth Planetary Science Letters 40:25–32.

MECKEL, L.D. (1967) Origin of Pottsville conglomerates (Pennsylvanian) in the central Appalachians. Geological Society America Bulletin 78:223–258.

MEYER-PETER, E. and MÜLLER, R. (1948) Formulas for bed-load transport. International Association Hydraulic Structures Research 2nd meeting Stockholm Appendix 2:39–64.

MINCH, J.A. (1979) The Late Mesozoic – Early Tertiary framework of continental sedimentation, northern Peninsular Ranges, Baja California, Mexico. In: Abbott, P.L. (ed) Eocene Depositional Systems, San Diego, California. Pacific Section Society Economic Paleontologists Mineralogists, pp. 43–67.

PARKER, G. (1978) Self-formed straight rivers with equilibrium banks and mobile bed. Part 2. The gravel river. Journal Fluid Mechanics 89:127–146.

PETTIJOHN, F.J. (1957) Sedimentary Rocks, 2nd edition. New York: Harper and Row, 718 pp.

PLUMLEY, W.J. (1948) Black Hills terrace gravels: A study in sediment transport. Journal Geology 56:526–577.

POTTER, P.E. and PETTIJOHN, F.J. (1977) Paleocurrents and Basin Analysis, 2nd edition. Berlin: Springer-Verlag, 425 pp.

READ, W.A. and DEAN, J.M. (1982) Quantitative relationships between numbers of fluvial cycles, bulk lithological composition and net subsidence in a Scottish Namurian basin. Sedimentology 29:181–200.

ROEHL, J.W. (1962) Sediment source areas, delivery ratios and influencing morphological factors. International Association Scientific Hydrology Publication 59:202–213.

RUST, B.R. and KOSTER, E.H. (1984) Coarse alluvial deposits. In: Walker, R.G. (ed) Facies Models, 2nd edition. Geoscience Canada Reprint Series 1, pp. 53–69.

SCHLEE, J. (1957) Upland gravels of southern Maryland. Geological Society America Bulletin 68:1371–1410.

SCHULTHEIS, N.H. and MOUNTJOY, E.W. (1978) Cadomin Conglomerate of western Alberta – a result of Early Cretaceous uplift of the Main Ranges. Bulletin Canadian Petroleum Geology 26:297–342.

SCHUMM, S.A. and STEVENS, M.A. (1973) Abrasion in place: A mechanism for rounding and size reduction in rivers. Geology 1:37–40.

SHAW, J. and KELLERHALS, R. (1982) The Composition of Recent Alluvial Gravels in Alberta River Beds. Alberta Research Council Bulletin 41, 151 p.

STANLEY, K.O. and WAYNE, W.J. (1972) Epeirogenic and climatic controls of Early Pleistocene fluvial sediment dispersal in Nebraska. Geological Society America Bulletin 83:3675–3690.

STEER, B.L. and ABBOTT, P.L. (1984) Paleohydrology of the Eocene Ballena Gravels, San Diego County, California. Sedimentary Geology 38:181–216.

TRÜMPY, R. (1980) Geology of Switzerland. Part A: An Outline of the Geology of Switzerland. Basel: Wepf and Co., 104 p.

WILCOCK, P.R. (1987) Bed-load transport of mixed-size sediment. Unpublished Ph.D. dissertation. Cambridge, Mass: Massachusetts Institute of Technology, 205 p.

WILSON, L. (1973) Variations in mean annual sediment yield as a function of mean annual precipitation. American Journal Science 273:335–359.

WILTSCHKO, D.V. and DORR, J.A. Jr. (1983) Timing of deformation in Overthrust Belt and foreland of Idaho, Wyoming, and Utah. American Association Petroleum Geologists Bulletin 67:1304–1322.

YEAKEL, L.S. (1962) Tuscarora, Juniata and Bald Eagle paleocurrents and paleogeography in the central Appalachians. Geological Society America Bulletin 73:1515–1540.

12

Relict Back-Arc Basins: Principles of Recognition and Possible New Examples from China

Kenneth J. Hsü

Abstract

The Junggar, Tarim, and Qaidam Basins are commonly considered to be cratonic blocks surrounded by orogenic belts and have thus been called intermontane basins. I propose that those basins be compared to the Black Sea and Caspian Basins, and suggest that Junggar was formed during the Carboniferous Period, and Tarim and Qaidam during the Permian Period, as back-arc basins behind volcanic arcs on the southern active margin of Paleozoic Asia. The relatively thin crust and the presence of very large positive magnetic anomalies under the Mesozoic and Cenozoic sediments of those basins indicate that their deepest depressions are floored, at least in part, by oceanic rocks. The oldest sediments in those basins are very likely marine shales. After arc-continent collisions during late Paleozoic and Triassic time, the basins became partially enclosed, and the euxinic sediments in those partially restricted basins could well be the source beds of the crude oils found recently in major oil fields of those basins. Junggar, Tarim, and Qaidam became inland basins with continental sedimentation after their communications to open sea were severed by the rising mountain chains. Isostatic basin subsidence permitted the accumulation of thick Mesozoic and Paleogene sediments before tectonic rejuvenation along Neogene faults.

Introduction

Many sedimentary basins are depressions underlain by thin crust. Except for regions under active compression, basin subsidence is related to isostastic adjustment (e.g., Hsü 1958). Phrased in modern terminology, the origin of sedimentary basins is to be sought either in downward flexure of the lithosphere on active plate-margins, or in intra-plate crustal thinning as a consequence of lithospheric stretching (Hsü 1982). In addition to the Airy type of isostastic adjustment, basin subsidence can also be related to mantle-density change (Hsü 1965), i.e., "thermal isostasy" because of the cooling of lithosphere (Oxburgh 1982).

Using this theoretical approach, the sedimentary basins of the world can be grouped into three categories, namely:

1. Basins formed by crustal compression.
2. Basins formed by crustal extension related to lithospheric stretching.
3. Basins formed by crustal extension related to lateral movement of crustal blocks, also known as pull-apart basins.

I presented this genetic classification orally in 1985 at Xining, China; it seemed that all basins could find their pigeonhole somewhere in the scheme. It was a rude awakening, however, when a young geologist in the audience of the Qinghai Provincial Geological Survey asked me about the origin of the Qaidam Basin.

The Qaidam Basin is situated near the northern edge of the Tibetan Plateau, bounded by the Altun and Qilian Mountains on the north and by the Kunlun Mountains on the south (Fig. 12.1). The basin occupies approximately 120,000 km² and the thickness of basin fill has been estimated to range from 8 to 15 km. Whereas the thick Neogene sediments have been deposited in actively subsiding basins bounded by active strike-slip faults, there did not seem to be an adequate explanation for the cause of subsidence during the Mesozoic Era. It is commonly believed that the basin is underlain by Precambrian metamorphic and granitic basement with its Sinian and Paleozoic sedimentary cover (Zhang et al. 1984). In the Chinese classification, the Qaidam,

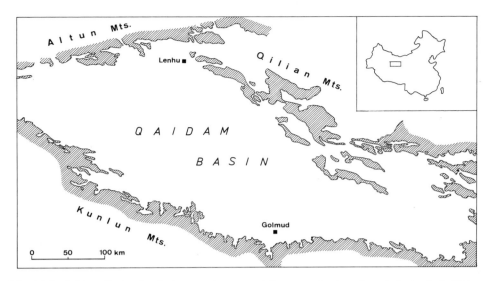

Fig. 12.1. Qaidam Basin. See index map in upper right corner for location of the basin in China.

Tarim, and Junggar Basins of northwest China have all be called "intermontane" basins, with their basin floors depressed during early Mesozoic compressive deformation (Zhang et al. 1984).

The mountains surrounding the large basins are not orographic features like those in the Basin and Range Province of western North America, where the relief has resulted from block faulting. The mountain ranges of northwest China are each an orogenic belt, apparently formed by collision of allochthonous continental blocks. The basins had been in existence long before the more recent tectonic reactivation along Cenozoic faults. They had subsided rapidly during Mesozoic time, when continental deposits, 5 to 15 km thick, were deposited; the Mesozoic subsidence could not be related to the Neogene faulting.

After a short lecture by the young geologist on the geology of Qaidam, I became aware of a similarity in the tectonic framework of the Qaidam Basin to that of the Black Sea. The one obvious difference is, of course, that the young euxinic basin hosts an inland sea some 2 km deep, whereas Qaidam is continental. However, the Black Sea would be filled up by another 3 km of sediments, which, judged on the basis of its Neogene sedimentation rate (Hsü 1978), could be deposited in another 30 million years. Considering that the Qaidam Basin is more than 100 million years older than the Black Sea, it is not difficult to imagine that the Qaidam had once been a feature like the Black Sea before it was completely overwhelmed by terrestrial sedimentation. Since the

Black Sea is a relict back-arc basin (Hsü 1978; Hsü *et al.* 1977), it is possible to postulate that the Qaidam Basin was also formed by lithospheric stretching behind an active island arc. In fact, several puzzles that could not be explained by previous hypotheses have found ready solution with the new postulate.

The origin of the Junggar, Tarim, and Qaidam Basins is a problem of not only academic interest, it is of great concern to exploration geologists because major hydrocarbon accumulations are expected to be found in each basin.

Relict Back-Arc Basins

Geologists of the fixistic school thought that marginal seas owed their origin to oceanization of continental crust by chemical processes at the crust-mantle boundary (Beloussov 1962), but the speculation has never been supported by geophysical evidence. The plate-tectonics theory postulates rifting and sea-floor spreading behind island arcs to explain the genesis of marginal seas (Karig 1970), and the postulate has been confirmed by geophysical and deep-sea-drilling investigations (Karig *et al.* 1975).

Not all back-arc basins are floored by oceanic crust. In the initial stage of development, such as the Aegean Sea north of the island arc of Pelopennesus-Crete-Rhodos (Cretan Arc), a back-arc basin is underlain by a stretched and thinned continental crust (Le Pichon 1982). In a more advanced stage of evolution, such as the Tyrrhenian Sea behind the Calabrian Arc, oceanic rocks are emplaced in the deepest parts of the basin (Hsü 1977). With further widening through sea-floor spreading, a back-arc basin can become a small ocean basin, such as the Philippine Sea (Karig et al. 1975).

A back-arc basin is commonly eliminated when the frontal arc collides with a continent or with another arc. Turbidites deposited in those basins are thrust over a carbonate or siliciclastic platform as flysch nappes or as the matrix of ophiolitic melanges. Back-arc basins may, however, survive arc-arc or arc-continent collisions. The South China Sea is a good example of such a relict basin (Fig. 12.2) displaying high heat flow typical of young crust.

The back-arc basin under the South China Sea was formed by sea-floor spreading from middle Oligocene to Early Miocene time (32–17 Ma). A strip of continental crust, called the Reed Bank/Calamian Arc or microcontinent (Taylor and Hayes 1980), was separated from mainland China to form a nonvolcanic arc south of the South China Basin. Its origin was thus comparable to that of the Aegean Basin today (cf. Reed Bank/Calamian Arc and Cretan Arc), except that the back-arc spreading of the South China Basin is sufficiently advanced that its abyssal plain is underlain by oceanic crust. The southeast-facing arc was separated from a northwest-facing Palawan volcanic arc by oceanic crust, which was consumed by subduction processes during Paleocene and Early Miocene time culminating in collision of the Reed Bank/Calamian and the Palawan Arc during Middle Miocene time. The South China Sea changed from an active to an inactive or relict back-arc basin as a consequence of that collision (Taylor and Hayes 1980). The subsequent history of the South China Sea was one of continued subsidence and sedimentation. Sediments from the Pearl and Red Rivers have built a continental terrace, and within probably not more than 100 million years, the South China Basin will fill with sediment to become a "continental basin" behind the Palawan Orogenic Belt. The Celebes and the Sulu Basins were also once back-arc basins formed by active sea-floor spreading, but are now trapped behind collisional orogenic belts (Fig. 12.2; Cardwell et al. 1980; Lee and McCabe 1986).

The Balearic Basin of the western Mediterranean is another relict back-arc basin, formed by sea-floor spreading during late Oligocene and Miocene time when Corsica and Sardinia were rifted from Europe

to form a nonvolcanic arc (Hsü 1977). The basin ceased to be active when this arc collided with Italy during the Middle Miocene Orogeny (before the Tyrrhenian Basin came into existence in late Neogene time).

Relict back-arc basins should have basement structures similar to those of active back-arc basins, bounded by a passive margin on one side and an arc margin on another. The arc margin of a relict basin should, however, have been involved in arc-arc or arc-continent collision to form the orogenic belt on the present border of the basin. If the arc was volcanic, its former existence could be deduced on the basis of the products of back-arc volcanism.

Relict back-arc basins are commonly underlain by ocean crust formed by sea-floor spreading. They may, however, be underlain by thinned continental crust, if back-arc rifting ceased before the sea-floor-spreading phase began. In either case, deep-marine sediments should be present at the bottom of the basinal sequence of a relict back-arc basin. The basin subsidence should be very rapid during the early stage because of the increase in the mantle density (Hsü 1965) after back-arc spreading ceased. The existence of an initially deep depression should further permit the existence of a very thick and isostatically adjusted sequence of infilling sediments.

Relict back-arc basins are easily distinguished from foreland basins by their basement structures and by their sedimentary histories. Foreland basins, such as the Molasse Basin of the Alpine System, are compressional structures, distinctively different from basins formed by back-arc extension. The tectonic unit in the mountains adjacent to the foreland basins is commonly the foreland folded belt, not a former island arc. The earliest "post-orogenic" sediments in a foreland basin (after collision) are molasse deposits, whereas deep-marine sediments are usually deposited contemporaneously with arc-volcanism.

Relict back-arc basins can be distinguished from pull-apart basins by their shape, their basement structures and by their tectonic history. Pull-apart basins are long, narrow and deep basins, bounded by strike-slip faults on both sides. Relict back-arc basins can be crescentic like the active Mariana Basin, or trapezoidal like the inactive South China Basin. Although some relict back-arc basins are now bounded by strike-slip faults, they should have been bounded, at the time of their formation, by extensional structures typical of a passive continental margin on one side and by those behind an island arc

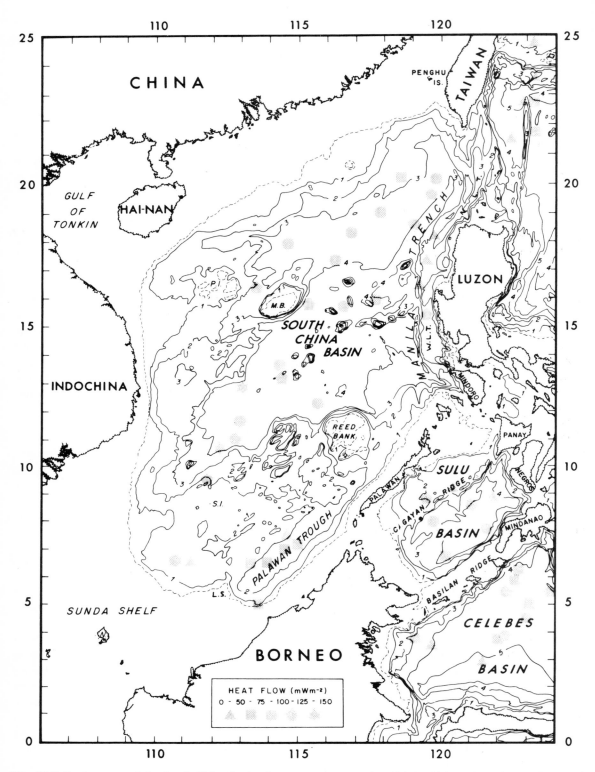

Fig. 12.2. Bathymetry of the South China Basin (from Taylor and Hayes 1980). Bathymetric contours in km. The heat flow values typical of young, inactive back-arc basins are commonly slightly higher than those of normal oceanic crust.

on the other. Furthermore, products of arc volcanism should be absent in basins where strike-slip faulting (rather than subduction) is the dominant mode of tectonics.

Using those criteria, the Black Sea and the Caspian Basins have been recognized as relict back-arc basins. Sediments deposited in the western part of the Cretaceous Black Sea Basin, now deformed and present in Bulgaria, provide evidence that the basin was created by rifting and by sea-floor spreading north of a Cretaceous island-arc (Hsü et al. 1977). This small ocean basin was trapped behind the mountains of Anatolia after the Tethys Ocean (south of the arc) was eliminated by an arc-continent collision during the Alpine orogenesis (Hsü et al. 1977; Hsü 1978). The Caspian Sea may be another Cretaceous back-arc basin north of another volcanic arc in Anatolia and is now trapped behind the same Tethyan orogenic system.

Using the Tethyan examples as analogs, I propose that the Junggar, Tarim, and Qaidam Basins of northwest China are relict back-arc basins of the Paleotethys system. They were formed during late Paleozoic and/or very early Mesozoic time by sea-floor spreading behind volcanic island arcs on the southern edge of the Asiatic continent. They were trapped behind a newly risen orogenic belt after arc-continent or arc-arc collision. The subsidence of those back-arc basins was caused by both mantle-density change and sedimentary loading.

Tectonic Evolution of North and Northwest China

The tectonics of north China are dominated by a history of interaction of the North China Blocks and the Siberian Craton. The collision of the Sino-Korean, the Alaxian, and the North Qaidam Blocks during a mid-Paleozoic (Caledonian) orogeny resulted in the Qilian and Helan deformed belts (Fig. 12.3). The collision of this composite North China Block and Siberia during the late Paleozoic ("Hercynian" or "Variscan") orogeny produced the broad and arcuate deformed belt that extends from Tianshan to Yinshan to the Great Xiangganling Mountains. At the end of Paleozoic time, North China was situated on the southern margin of the Asiatic continent and was bounded on the south by a north-dipping subduction zone. With consumption of the Paleotethys during Triassic time, the North and South China Blocks collided along the Qinling Axis (Fig. 12.3).

The geology of northwest China has been interpreted on the basis of the same model of mobile zones between stable blocks: the Junggar, Tarim, and Qaidam Basins are assumed to be three cratonic blocks (Yang et al. 1986). The Tianshan Mountains were formed after the collision of Junggar and Tarim Blocks, and the Kunlun Mountains were formed after the collision of the Tarim/Qaidam with the North Tibet Block (Fig. 12.3).

The postulate that the Junggar, Tarim, and Qaidam Basins are stable blocks does not provide adequate explanation for the cause and timing of the basin subsidence during the Mesozoic Era. Regions underlain by normal continental crust should be uplifted to form mountains and plateaus when they undergo compressive deformation, except in regions where the crust is downwarped to form foreland basins. The Permo-Triassic Junggar, Tarim, and Qaidam were not foreland basins. If they had been continental blocks overriding underthrusted continental crust, those regions should have been uplifted, similar to the modern Tibetan Plateau. Instead, the basin floors subsided rapidly, and the rates of synorogenic and post-orogenic subsidence were, on the average, approximately 0.1 m/10^3 yr during early Mesozoic time (Lanzhou 1981). The assumption that the Junggar, Tarim, and Qaidam Basins are underlain by continental crust of normal thickness cannot explain such subsidence. The early Mesozoic rapid subsidence is, however, a logical consequence of basin-filling if the Junggar, Tarim, and Qaidam Basins were late Paleozoic back-arc basins underlain by oceanic or thinned continental crust.

Paleogeographic reconstructions suggest that North China was bounded by the Paleotethys Ocean in latest Paleozoic and earliest Mesozoic time (Sengör and Hsü 1984). North China was, however, fringed by marginal seas as South China is now. Eventually, the Permo-Triassic marginal basins were trapped, as relict back-arc basins behind orogenic belts, like the Sulu, Celebes, and South China Basins of the southwest Pacific today.

Junggar Basin

The Junggar Basin of northern Xinjiang, triangular in shape, is located between Altay Mountains to the north, Tianshan Range to the south, and Western

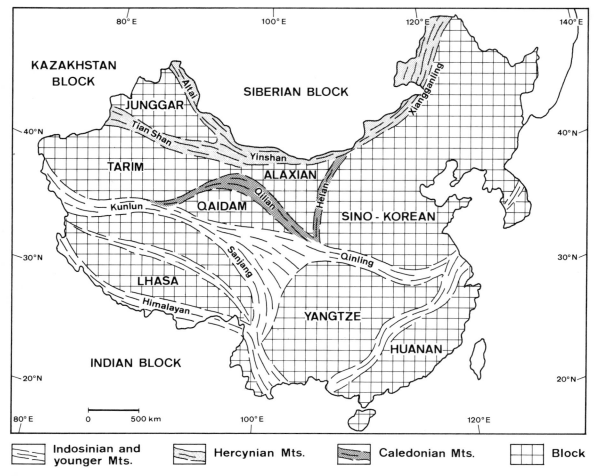

Fig. 12.3. Tectonic units of China. Junggar, Tarim, and Qaidam are shown as stable blocks, but they are probably underlain by oceanic crust. The unnamed block north of the Llasa Block is provisionally called the North Tibet Block.

Junggar Mountains to the northwest (Fig. 12.4). The basin has an area of 130,000 km² and is underlain by a sedimentary sequence more than 11 km thick (Fig. 12.5).

It is commonly assumed that the basin is underlain by Precambrian basement and that the collision of the Junggar and the Tarim Blocks took place during late Paleozoic time to form the Tianshan Range (Wang 1983). The lower Paleozoic strata of the Tianshan Range are mainly marine shales and sandstones, which apparently were deposited on the passive margin of a continental block. The upper Paleozoic strata include, however, thick volcanic flows, tuffs, breccias, and volcaniclastic rocks, typically found in an island-arc setting (Ulmishek 1984). The oldest strata encountered by deep drilling near

the northwestern margin of Junggar are of Carboniferous age (Fig. 12.6), consisting of dolomitic shale, tuff, and volcanic breccia. A facies change suggests that water depth increased eastward toward the basin center, where marine shales of Carboniferous age are likely to be present (Wang 1983). The Permian strata include turbidites and other coarse clastics in the Tianshan Range and on the periphery of the Junggar Basin, but they are mainly black shales in the central depression (Fig. 12.7).

The basement structures under the Junggar have been buried under a thick sedimentary sequence. The depth of the Moho under the Junggar basin is approximately 40 km (Huang *et al.* 1980). Since the sedimentary infill is more than 10 km thick, the thickness of the basement is less than 30 km. Such

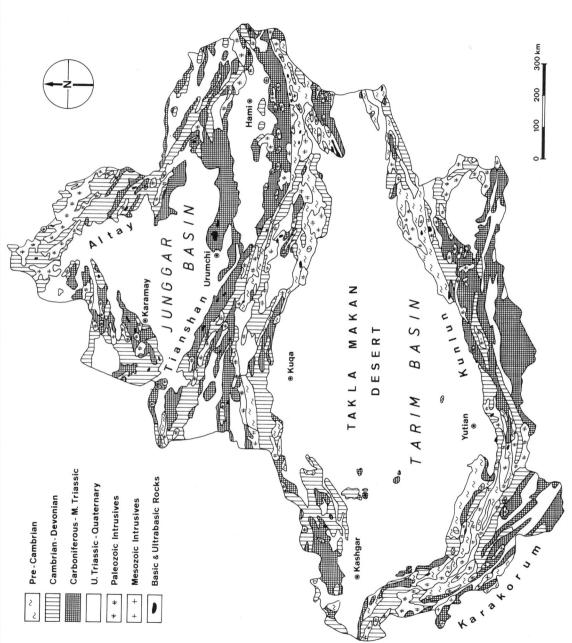

Fig. 12.4. Sketch geologic map of Xinjiang Province (modified after Anonymous 1973).

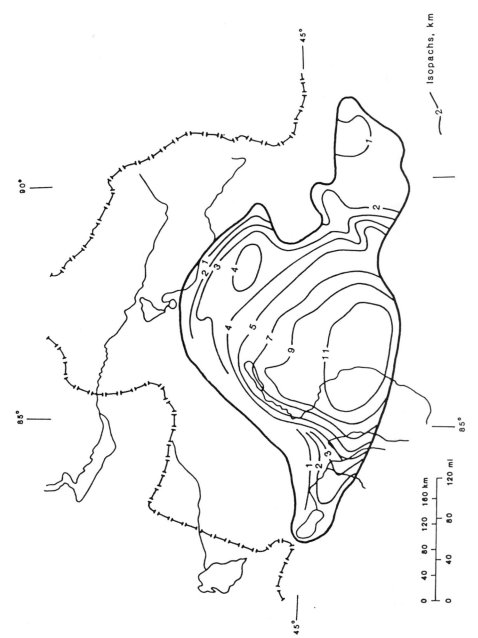

Fig. 12.5. Sediment-thickness in Junggar Basin. Isopach contours in km (modified after Ulmishek 1984). The dashed lines are the border of the Xinjiang province.

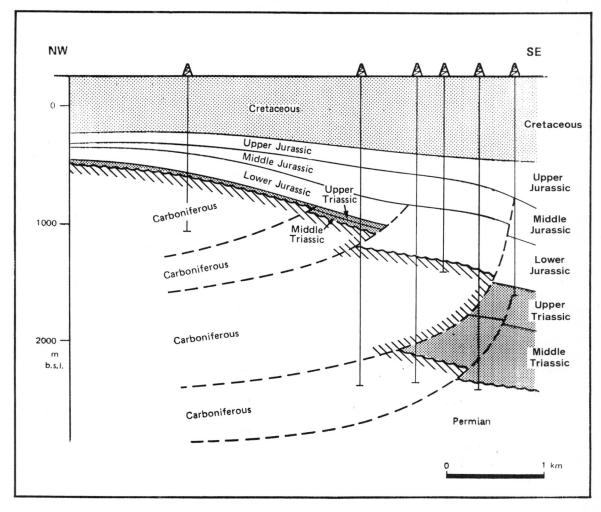

Fig. 12.6. Structural cross section, Karamay Field, Junggar Basin (after Ulmishek 1984).

an extended continental crust may underlie a passive continental margin or a back-arc basin. The presence of very thick flysch strata in the Tianshan and Bogda Mountains southeast of Junggar Basin suggests, however, that this Permian northern ocean margin was not a passive, but an active, continental margin. Furthermore, the existence of a late Paleozoic frontal arc south of the Junggar Basin is indicated by the presence of volcanic rocks of this age in the Tianshan Mountains. This volcanic arc separated the site of deep-marine sedimentation in the Junggar Basin from the open ocean. The basin was thus, by definition, a back-arc basin.

The presence of a large positive magnetic anomaly in the basin center is evidence for the occurrence of oceanic rocks (Zhang 1983), but the crustal thickness suggests that the crust under the

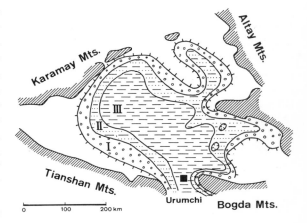

Fig. 12.7. Permian sediment-facies distribution, Junggar Basin (modified after Wang 1983). I = Coarse clastic facies. II = Siltstone facies. III = Marine shale facies.

Junggar is largely continental. The Aegean Basin may be a modern analog of such a youthful back-arc basin: The crust under the Aegean has been thinned by extension, but it is still largely continental, although oceanic basement is present under the Cretan Basin just north of the island of Crete (Hsü 1977).

The Permian back-arc basin of Junggar was trapped behind a collision zone. During Late Permian time, the northern Tianshan volcanic arc collided with continental basement, which now crops out in southern Tianshan (Fig. 12.3) to form this east-west trending mountain range. The connection of Junggar Basin to the open sea became more restricted (see Fig. 12.6), and was completely severed during Late Triassic time.

The Junggar Basin was transformed into a large inland lake during Jurassic time, comparable to the modern Caspian Sea (Wang 1983). The open ocean on the other side of the Tianshan Range, the Paleotethys, was not very distant and the climate of the Junggar region was temperate and humid. With the eventual shoaling of the lake, the basin became a huge coal swamp. The Jurassic coal reserve near Urumchi is reportedly one of the largest of the world.

After the Paleotethys between Xinjiang and northern Tibet was eliminated during the early Mesozoic Indosinian Orogeny (Sengör and Hsü 1984), the open ocean of Neotethys was more distant so that the prevailing climate became increasingly arid in the Junggar region. The Upper Jurassic and Lower Cretaceous formations are thus mainly red alluvial, lacustrine, and eolian deposits.

The rate of basin subsidence slowed considerably during Cretaceous time, when the limit of isostatic loading was apparently approached. The Upper Cretaceous and Paleogene sediments are relatively thin. Tectonic activity resumed, however, during late Cenozoic time. The Neogene "collapse" was triggered by strike-slip faulting, which owed its origin to the compression induced by the continued northward movement of the Indian subcontinent against mainland Asia (Molnar and Tapponnier 1975). In the Urumchi Basin, a narrow pull-apart basin just north of Tianshan, the Neogene sediments are 5 km thick (Ulmishek 1984).

Tarim Basin

The Tarim Basin between Tianshan and Kunlun Mountains is the largest in China, 560,000 km² in extent. Its size and shape are comparable to the Sea of Japan. The Tarim basement is deeply buried. Precambrian metamorphics and Paleozoic sediments are exposed in the mountain ranges surrounding the Tarim basin. They also crop out or are buried at shallow depth on an elevated ridge, the Central Tarim Ridge, within the basin (Figs. 12.3, 12.4, 12.8).

The Tarim Basin is assumed to be underlain by continental crust (Anonymous 1973). The average depth of the Moho under the Tarim Basin, as estimated from gravimetric data, is only approximately 40 km (Wang 1983). The average basement thickness should thus be less than 25 km, because the maximum sediment thickness is more than 15 km. Considering that the basement topography is probably rough, as under the young, back-arc Tyrrhenian Basin, the presence of thinner, or oceanic, crust in local troughs cannot be excluded.

A most curious feature in the Tarim region is the Central Tarim Ridge, an arc-shaped basement ridge in the Tarim Basin with the apex of the arc facing south (Figs. 12.8, 12.9, 12.10). The ridge, which stands some 15 km above the adjacent depressions, extends from Kashgar in the west to Qargan in the east for a distance of some 1,000 km, and seems to have been offset by a wrench fault into two segments. The basin is thus divided into two depressions, and both contain more than 15 km of sediment (Fig. 12.8). The Precambrian and Paleozoic sedimentary cover of the ridge is, as a rule, shallowly buried, and crops out locally (see Anonymous 1973). This basement structure is also defined by magnetic anomalies, which are strongly negative, not positive as are anomalies associated with basement elevations in many other basins (Fig. 12.9).

Some Chinese scientists have suggested that the Paleozoic stratigraphy of the Tarim Basin is similar to those of the Kunlun Mountains and of the Central Tarim Ridge, which were deposited on shallow marine carbonate platforms (Anonymous 1973; Wang 1983). There may indeed have been a broad carbonate platform between the Tianshan and Kunlun Mountains in early Paleozoic time, but the presence of the shallow marine sediments on the ridge does not necessarily imply that similar sediments are present in the basinal depressions as well, especially if the depressions were formed by back-arc sea-floor spreading.

The Permo-Carboniferous strata of the Kunlun Mountains and Tarim Basin are mainly volcaniclastic sediments and black shales (Fig. 12.11). Batholithic intrusions of late Paleozoic age form the backbone of the Kunlun Mountains (Fig. 12.4). Permian volcanics are present in the Kunlun Mountains

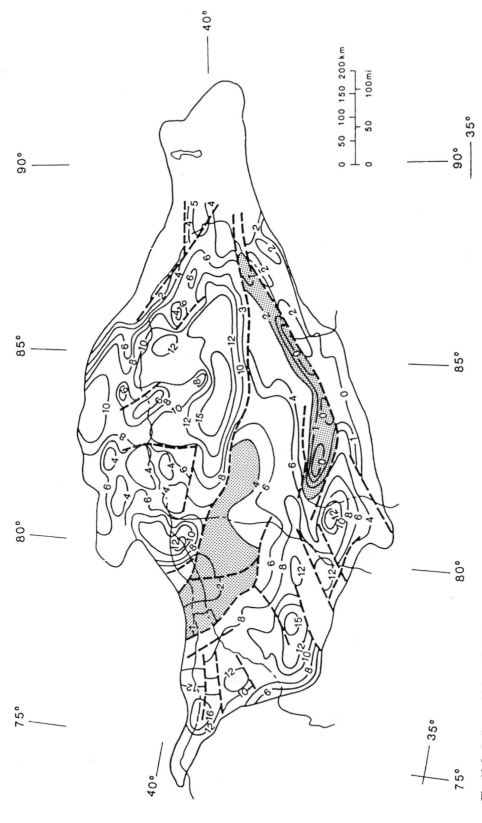

Fig. 12.8. Sediment thickness in Tarim Basin. Isopach contours in km (modified after Ulmishek 1984). The stippled area of the Central Tarim Ridge is buried under 4 km of sediment in the western segment and under 2 km in the eastern segment.

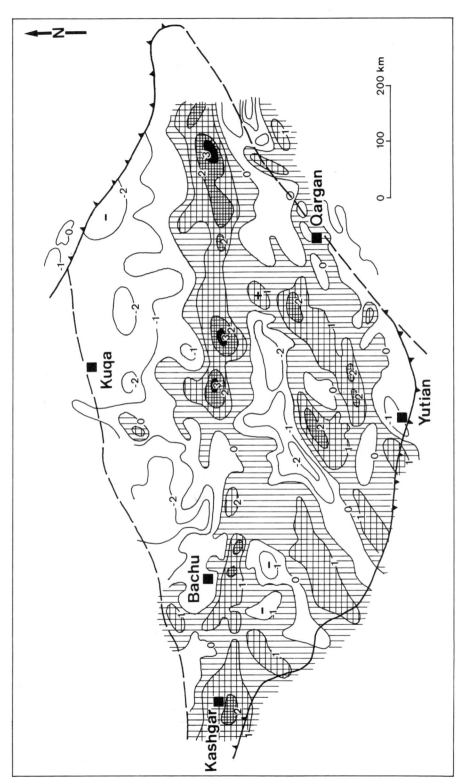

Fig. 12.9. Magnetic anomalies, Tarim Basement; contours in 100 gammas (modified after Zhang 1983).

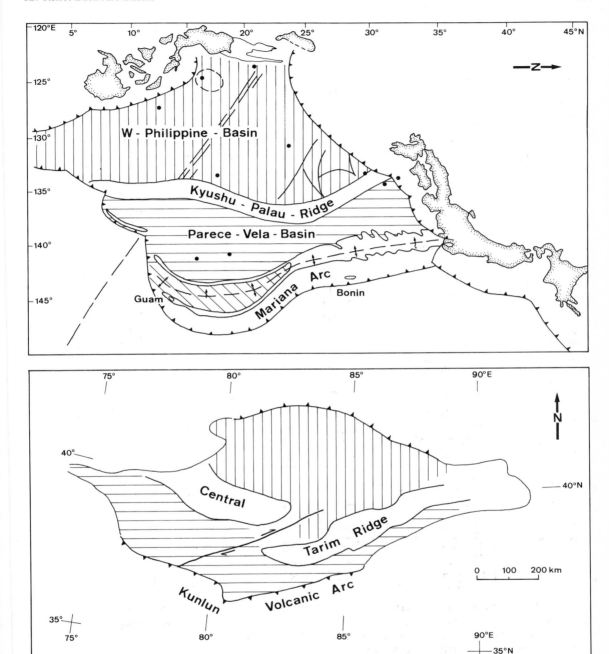

Fig. 12.10. Back-arc depressions of Philippine Sea (upper figure) and relict back-arc depressions of Tarim (lower figure) for comparison. Vertically striped depression is behind the remnant arc, and horizontally striped basin is in front of the remnant arc but behind the frontal arc. There is no equivalent of Mariana Trough (diagonal stripes) in the Tarim Basin.

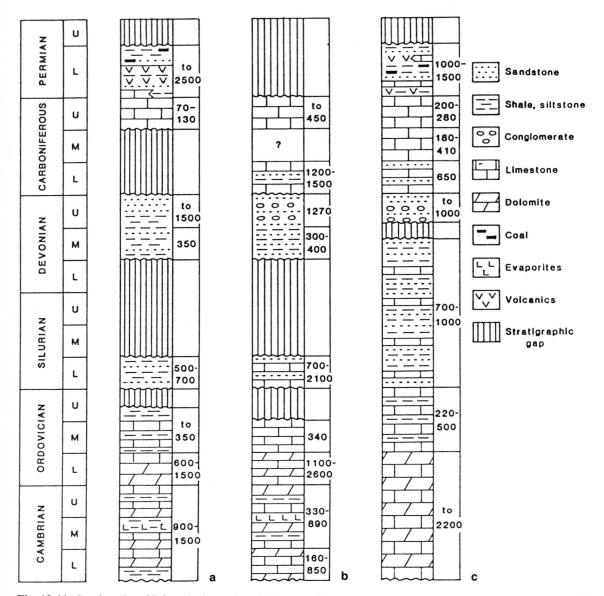

Fig. 12.11. Stratigraphy of Paleozoic formations in Tianshan and Kunlun Ranges (modified after Ulmishek 1984). Numbers to the right of the profiles indicate the range of thickness in metres of each stratigraphic interval. a = Jal-pin area, northwest of Tarim; b = Kuluketage Range, northeast of Tarim; c = Northern slope of Kunlun Mountains, south of Tarim.

and the Tarim Basin (Anonymous 1973; Ulmishek 1984). Deep-marine Permo-Triassic turbidites are found in the Karakorum Range, south of Kunlun (Fig. 12.4). All those rock types are typical of active plate margins. The Permian volcanics in Kunlun and in Tarim were the product of arc-volcanism and/or back-arc volcanism, whereas the Kunlun batholiths were the roots of volcanoes on an active plate margin. South of the Kunlun Arc, in deep-sea trenches

or in fore-arc basins, the Permo-Triassic turbidites formed accretionary wedges. The stratigraphic evidence thus permits a paleogeographic reconstruction of a Permo-Carboniferous island arc, with its apex facing south. The Tarim Basin north of the arc is thus, by definition, a back-arc basin, formed by rifting of continental crust and sea-floor spreading behind the arc. The black shales and volcanics in Tarim, ranging in age from Carboniferous to Triassic, were the deep-marine deposits of this back-arc depression.

The interpretation that the Tarim is a relict back-arc basin is supported by the fact that unusually large positive magnetic anomalies are present under the Tarim Basin (Fig. 12.9). The northern anomaly belt is in places more than 100 km wide and extends for more than 1,000 km in an east-west direction under the Taklimakan Desert; the maximum value exceeds 350 gammas. The southern belt is more than 500 km long, and is interrupted locally by negative anomalies. These large positive anomalies indicate the presence of mafic or ultramafic rocks in Tarim basement (Zhang 1983). Chinese scientists postulated that those rocks are exotic blocks of ophiolite in the melange of an orogenic belt, marking the position of suturing of the northern and southern halves of the Tarim basins during Precambrian deformation of the basement (Wang 1983; Zhang 1983). Ophiolitic melanges are commonly found, however, in mountains, where the crust is thickened by continental collision, not in basins underlain by thin crust. The postulate of a melange belt under the Tarim Basin also fails to explain the correlation of the magnitude of the anomaly with the depth of the Tarim basement. I propose, therefore, that the positive magnetic anomalies indicate the presence of oceanic crust under the Tarim sedimentary cover. The rifting of the back-arc depressions of the Tarim Basin had apparently proceeded far enough to enter the sea-floor-spreading stage when mafic and ultramafic rocks were emplaced. The depressions floored by ocean crust, according to isostatic considerations (Hsü 1958) should also have become the sites of maximum sedimentation, resulting in the positive correlation between the magnetic anomaly and the sediment thickness in Tarim Basin (cf. Figs. 12.8, 12.9).

The presence of continental basement on the Central Tarim Ridge is in accordance with the observation that the ridge is defined by a negative magnetic anomaly (Fig. 12.9). An arcuate belt of continental crust separating a back-arc basin into two parts has been called a remnant arc, such as the Kyushu-Palau Ridge between the West Philippine and Parela-Vela Basins (Karig *et al.* 1975). The Central Tarim Ridge has the geometry of just such a remnant arc, and the Permian back-arc basin complex of Tarim may have an actualistic analog in the Philippine Sea (Fig. 12.10). The frontal arc along the Kunlun trend is comparable to the modern Mariana Arc. The Central Tarim Ridge, which separates the northern and the southern depressions of Tarim, is equivalent to the modern Kyushu-Palau Ridge.

The Kunlun Arc collided with North Tibet during the Indosinian deformation of Late Triassic or Early Jurassic age. This orogenic movement eliminated the Paleotethys and caused the rise of the Kunlun and Karakorum Ranges. The Tarim Basin was trapped behind this Paleotethys orogenic belt, as the Black Sea was trapped behind the Tethys System.

The postulate that the Tarim depressions are underlain by thin crust of late Paleozoic/early Triassic age explains the sedimentary history of the basin. The Mesozoic subsidence, permitting the accumulation of a sedimentary sequence more than 10 km in basin center, was isostatic. The basin depth was already considerable when the basin became an inactive back-arc basin. Subsidence ensued because of an increase in mantle density caused by lithospheric cooling (Hsü 1965). The basin floor was depressed farther by the sedimentary load filling the deep basin (Hsü 1982).

The basin was completely isolated from the open sea shortly after the arc-continent collision such that the very thick Jurassic sediments were deposited mainly in alluvial, lacustrine, and swampy environments under warm and humid climatic conditions (Fig. 12.12). After the limit of isostatic subsidence was approached, the rate of sedimentation slowed so that the Cretaceous and Paleogene strata of Tarim Basin are relatively thin, mainly terrestrial red beds with shallow marine strata present in the western part of the basin.

The basin was reactivated during Neogene time, when thick variegated beds of alluvial and lacustrine origin were deposited in several marginal troughs. The pre-Neogene sedimentary history of Tarim Basin was, however, typical of that predicted for the sedimentary fill of a relict back-arc basin.

Qaidam Basin

The Qaidam Basin is situated near the northern edge of the Tibetan Plateau, bounded by the Altun and Qilian Mountains on the north and by the Kunlun Range on the south (Fig. 12.1). The basin area is about 120,000 km², and the thickness of basin fill has been estimated to range from 8 to 15 km.

One proposed view is that the basin is underlain by Precambrian metamorphic and granitic basement

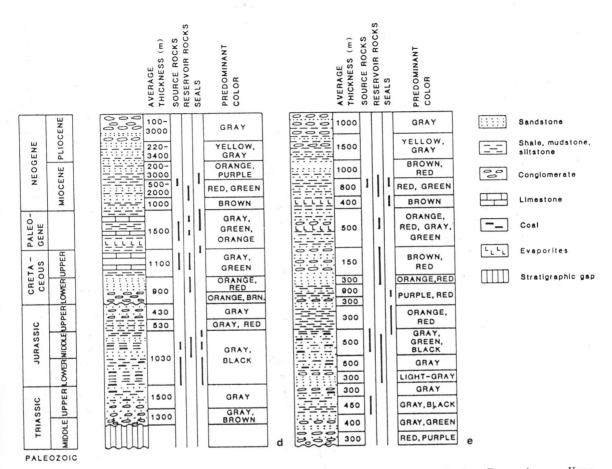

Fig. 12.12. Mesozoic stratigraphy of Tarim Basin. d = Southwestern Depression; e = Northern Depression near Kuqa. Information based upon wells drilled on the periphery of the basin (modified after Ulmishek 1984).

with a Sinian and Paleozoic sedimentary cover (Zhang *et al.* 1984). The Qilian Mountains were accreted to North China during the Caledonian Orogeny in late Silurian time (Figs. 12.3, 12.13). The nature of the basement under the Paleozoic sediments in the central depression of the Qaidam is, however, unknown.

The Devonian and Carboniferous strata in the mountains around Qaidam are continental and shallow marine in origin, and were apparently deposited on the passive margin of the Paleotethys Ocean. Late Paleozoic batholiths are, however, widely distributed in the Kunlun Mountains, as are Permian volcanics and Triassic turbidites. Their occurrences indicate that the passive margin was converted into an active margin before Permo-Triassic time.

The geology of the Qaidam basin is thus similar to that of the Tarim Basin. If the Tarim was a Permo-Triassic back-arc basin, the Qaidam may also have had a similar origin. I propose that both the Tarim and Qaidam Basins survived the Late Triassic arc-contintent collision that eliminated the Paleotethys, and that the pair of basins was trapped behind the Indosinian orogenic belt of Kunlun like the Black and Caspian Seas as a pair behind the Alpine orogen of Anatolia.

The idea that the genesis of the Qaidam Basin was related to subduction on an active continental margin has been vaguely suggested by Chinese scientists (e.g., Wang 1983). Recent geophysical data indicate that granitic crust is absent under the axial trough of basin (Ulmishek 1984). Thus the proposed back-arc

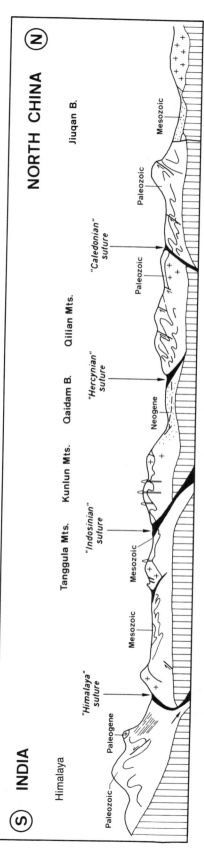

Fig. 12.13. Structural cross section across western China, showing the accretion of micro-continents and/or of island arcs to the Asiatic mainland since Paleozoic time. Qaidam is one of the two back-arc basins trapped behind the Kunlun Range after the Indosinian Orogeny; the other is the Tarim Basin (not shown here).

rifting had apparently proceeded far enough to permit local emplacement of oceanic crust under the Qaidam Basin.

The Mesozoic and early Paleogene subsidence history of Qaidam is very poorly known. A deep well on the northern margin (near Lenhu; Fig. 12.1) penetrated 138 m of Quaternary, 961 m of Neogene, 1,612 m of upper Paleogene, 2,544 m of lower Paleogene, 2,858 m of Jurassic, and 2,883 m of pre-Jurassic sediments, but a more than 7,500 m deep well (Han-2) in the basin center but did not penetrate beyond Miocene strata (Zhang 1983). The hypothesis that the Qaidam is a relict back-arc basin explains only the pre-Neogene subsidence history. Late Cenozoic tectonics created a Neo-Qaidam Basin, in which very thick Neogene and Quaternary sediments were deposited (Song and Liao 1982).

Hydrocarbon Potential in Relict Back-Arc Basins

The postulate that the Junggar, Tarim, and Qaidam Basins are relict back-arc basins is based upon the consideration of comparative tectonics and is supported by the available geological and geophysical data. Such an interpretation explains the timing and magnitude of subsidence in those basins.

Numerous oil fields have been found in the Junggar, Tarim, and Qaidam Basins, but none was economically significant until the Karamay field of the Junggar Basin was recently discovered. The Karamay field has an estimated reserve of 10 billion barrels (Ulmishek 1984). Huang and Li (1982) proposed lacustrine source beds for the Karamay oil, before geochemical analyses indicated a marine origin of the oil. The postulate that the Junggar was once a partially enclosed back-arc basin like the Black Sea is in accordance with the presence of thick Permo-Carboniferous euxinic black shales under the Mesozoic and Cenozoic sediments of the Junggar Basin and they are probably the main source beds for Junggar petroleum.

Petroleum exploration in the Tarim Basin has a history similar to that in the Junggar Basin. Production from shallow Cenozoic and Mesozoic pools began after 1949, but no major find was made until the recent discovery near Kuqa. The petroleum was found in Paleozoic carbonates and has a chemistry indicative of a marine origin (Hu and Zhang 1982).

The exposed Paleozoic rocks are, however, not good source beds (Huang and Li 1982). I propose that the oil came from Permo-Triassic black shales, deposited when the Tarim was a partially enclosed back-arc basin like the modern Black Sea. If so, the hydrocarbon potential of the Tarim is enormous. If, for example, 2.5 km of the more than 15 km of Tarim sediments were source beds, and their average hydrocarbon content was 4,000 ppm, the reserve of Tarim would be of the order of 10 billion tons, or 70 billion barrels.

The exploration of the Qaidam has focused on prospects in the Mesozoic and Cenozoic lacustrine sediments. None of the discoveries so far qualifies as a giant field. The hydrocarbon potential of the basin is, however, much greater if the Qaidam is a relict back-arc basin with early accumulation of thick euxinic black shales.

The question of whether the giant "intermontane" basins of northwest China were once partially enclosed back-arc basins can eventually be resolved by deep seismic reflection profiling and by deep drilling. Considering the extraordinary economic returns at stake, I expect such exploration ventures in the not too distant future.

Conclusions

Large sedimentary basins surrounded by mountains in China have been considered to be intracratonic basins. The cause of subsidence of these Chinese basins has always been a puzzle. I propose in this paper that the Junggar, Tarim, and Qaidam Basins were not in an intracratonic setting during late Paleozoic and early Mesozoic time, but were then marginal to the Asiatic continent. Geological and geophysical data support the postulate that these basins were relict back-arc basins.

In general, ancient back-arc basins are thought not to survive deformation during arc-arc or arc-continental collisions and not to be preserved as sites of post-orogenic sedimentation. The geology of the South China Sea clearly proves that a back-arc basin can survive collisional deformation. The basement structures and the inferred sedimentary histories of relict back-arc basins provide criteria for their recognition. Some relict back-arc basins contain large petroleum reserves. Rich hydrocarbon occurrences have been found, for example, in the Black

Sea and Caspian regions. The discovery of a giant field in the Junggar Basin provides further confirmation of the enormous hydrocarbon potential in relict back-arc basins.

References

ANONYMOUS (1973) Geologic Atlas of China, with Explanatory Notes (in Chinese). Beijing: Peoples' Publishing House, 149 p.

BELOUSSOV, V.V. (1962) Basic Problems in Geotectonics. New York: McGraw-Hill, 820 p.

CARDWELL, R.K., ISACKS, B.L., and KARIG, D.E. (1980) The spatial distribution of earthquakes, focal mechanism solutions, and subducted lithosphere in the Philippine and northeastern Indonesian islands. American Geophysical Union, Geophysics Monograph 23: 1–36.

HSÜ, K.J. (1958) Isostasy and a theory for the origin of geosynclines. American Journal Science 256:306–327.

HSÜ, K.J. (1965) Isostasy, crustal thinning, mantle changes and disappearances of ancient land masses. American Journal Science 263:97–109.

HSÜ, K.J. (1977) Tectonic evolution of the Mediterranean basins. In: Nairn, A.E.M., Kanes, W.H., and Stehli, F.G. (eds) The Ocean Basins and Margins, v. 4. The Eastern Mediterranean. New York: Plenum Press, pp. 29–75.

HSÜ, K.J. (1978) Stratigraphy of the lacustrine sedimentation in the Black Sea. In: Ross, D.A. et al. Initial Reports of the Deep Sea Drilling Project 42, Part 2. Washington, DC: United States Government Printing Office, pp. 509–524.

HSÜ, K.J. (1982) Geosynclines in plate-tectonic settings: Sediments in mountains. In: Hsü, K.J. (ed) Mountain Building Processes. London: Academic Press, pp. 3–12.

HSÜ, K.J., NACHEV, I.K., and VUCHEV, V.T. (1977) Geologic evolution of Bulgaria in light of plate tectonics. Tectonophysics 40:245–256.

HU, B. and ZHANG, J. (1982) Carbon-isotope composition of crude oils from Tarim Basin (in Chinese). Collection Petroleum Geology Articles 6:130–135.

HUANG, D. and LI, J. (1982) Genesis of Oil and Gas in Sediments of Continental Facies of China (in Chinese). Beijing: Petroleum Industry Publishing House, 355 p.

HUANG, J., REN, Y., JIANG, C., ZHANG, Z., and QIN, D. (1980) The Geotectonic Evolution of China. Beijing: Scientific Publishers, 124 p.

KARIG, D. (1970) Ridges and basins of the Tonga-Kermadec Island Arc System. Journal Geophysical Research 75:239–254.

KARIG, D., INGLE, J.C. et al. (1975) Initial Reports of the Deep Sea Drilling Project, Volume 31. Washington, DC: United States Government Printing Office, 927 p.

LANZHOU RESEARCH INSTITUTE OF GEOLOGY (1981) Genesis, Maturation and Migration of Hydrocarbons in Continental Sediments of China (in Chinese). Lanzhou: Peoples' Publishing Society Gansu Province, 269 p.

LEE, C.S. and McCABE, R. (1986) The Banda-Celebes-Sulu Basin: A trapped piece of Cretaceous-Eocene oceanic crust? Nature 322:51–54.

LE PICHON, X. (1982) Land-locked ocean basins and continental collision: The eastern Mediterranean as a case example. In: Hsü, K.J. (ed) Mountain Building Processes. London: Academic Press, pp. 201–211.

MOLNAR, P. and TAPPONNIER, P. (1975) Cenozoic tectonics of Asia: Effects of a continental collision. Science 189:419–426.

OXBURGH, E.R. (1982) Heterogeneous lithospheric stretching in earth history of orogenic belts. In: Hsü, K.J. (ed) Mountain Building Processes. London: Academic Press, pp. 85–94.

SENGÖR, A.M.C. and HSÜ, K.J. (1984) The Cimmerides of eastern Asia: History of the eastern end of Palaeo-Tethys. Memoires Societe Geologique France 147:139–167.

SONG, J. and LIAO, J. (1982) Structural characteristics and petroliferous regions in Chaidamu (Tsaidam) Basin. Acta Petrolei Sinica, Supplement for 1982, pp. 14–23.

TAYLOR, B. and HAYES, D.W. (1980) The tectonic evolution of the South China Basin. American Geophysical Union, Geophysics Monograph 23:77–88.

ULMISHEK, G. (1984) Geology and Petroleum Resources of Basins in Western China. Springfield, MA: United States National Technical Information Service, 131 p.

WANG, S. (1983) Petroleum Geology of China (in Chinese). Beijing: Petroleum Industry Publishing House, 348 p.

YANG, Z., CHANG, Y., and WANG, H. (1986) Geology of China. Oxford: Oxford University Press, 303 p.

ZHANG, Y. (1983) Geophysical and tectonic characteristics of large petroliferous basins of China (in Chinese). In: Zhu, X. (ed) Tectonic Evolution of Mesozoic and Cenozoic Basins of China. Beijing: Scientific Publishing House, pp. 39–47.

ZHANG, Z.M., LIOU, J.G., and COLEMAN, R.G. (1984) An outline of the plate-tectonics of China. Geological Society America Bulletin 95:364–370.

13

Synorogenic Sedimentation and Subsidence in a Plio-Pleistocene Collisional Basin, Eastern Taiwan

Neil Lundberg and Rebecca J. Dorsey

Abstract

The Coastal Range of eastern Taiwan exposes an accreted Miocene arc and forearc terrane that is overlain by a 5 to 6 km thick sequence of collision-derived Plio-Pleistocene sedimentary rocks. When the Luzon arc collided with the Chinese continental margin, the forearc basin became a collisional basin [here informally named the Coastal Range Collisional Basin (CRCB)], caught between the growing mountain belt and the dying arc. In the CRCB sequence, basal mudstones, olistostromes, and pebbly mudstones are overlain by turbidites (mainly thin-bedded fan-fringe and basin-plain facies) and tempestites (widespread shallow-marine, wave-influenced storm deposits). Local channel- and canyon-fill sequences contain abundant cobble-boulder conglomerate, and upper Pleistocene cobbly braided-stream deposits overlie the marine rocks with angular unconformity, dating uplift of the basin. The present Luzon Trough south of Taiwan, which should provide a modern analog for marine deposits of the CRCB, is notably lacking in broad shallow shelf areas corresponding to widespread shallow-water facies of the Plio-Pleistocene sequence. This suggests that early stages of the collision produced less dramatic subsidence than does the present propagation of the collision to the south.

The CRCB experienced increasing rates of subsidence and sedimentation prior to uplift and incorporation into the orogenic belt, similar to patterns seen in well known foreland basins. Backthrusts and backfolds within the rear flank of the Central Mountains apparently drove the orogenic load relatively eastward toward the CRCB, producing dramatic subsidence before the basin was uplifted. The early stages of the arc-continent collision possessed a regional depositional symmetry that may reflect a deeper symmetry of tectonic processes. Linear orogenic belts, in any case, should be expected to shed detritus laterally in both directions into basins deepened by tectonic loading.

Introduction

Sedimentary sequences deposited in basins that flank orogenic belts provide an important record of orogenic history and processes. Such stratigraphic records are particularly valuable in the case of presently active orogenic belts, for which the nature and history of the orogenic interior is well known, allowing correlation between tectonic processes and resulting patterns of sedimentation. In this paper we examine a thick sequence of sedimentary rocks exposed in the Coastal Range of eastern Taiwan, which was deposited during the ongoing arc-continent collision between the Luzon island arc and the Asian continental margin.

Arc-continent collisions provide a mechanism for continental growth through the accretion of exotic island arcs to continental margins. Considerable study has been devoted to arc-continent collisions, generally in ancient orogenic belts for which orogenic processes must be inferred. Notable examples include the Taconic orogeny of the Appalachians (Rowley and Kidd 1981; Stanley and Ratcliffe 1985) and the amalgamation of various displaced terranes onto the western margin of North America (Coney et al. 1980; Jones et al. 1982). Many of the displaced terranes of the western Cordillera have been identified as having constituted volcanic-arc terranes for at least part of their pre-accretion history (Coney et al. 1980; Silver and Smith 1983). An example involving an apparently simple arc terrane is the Early Cretaceous collision of the intraoceanic Togiak-Koyukuk arc with North America in southwest and west-central Alaska (Box 1985). In exam-

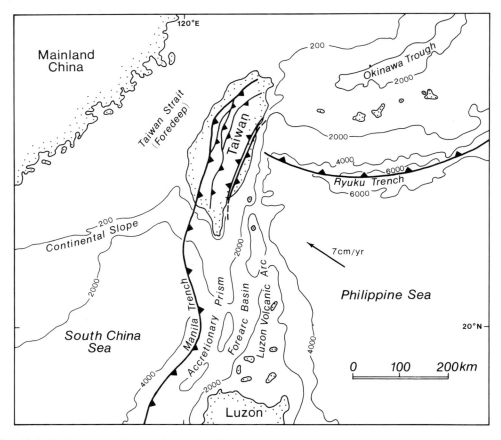

Fig. 13.1. Bathymetry and tectonic setting of Taiwan and surrounding regions. Contours are in metres.

ples older than Late Cretaceous, paleobathymetry is difficult to constrain, and often even biostratigraphic data are sparse. Sedimentary basins flanking the orogens are typically preserved, however, and the stratigraphic records may be compared to modern examples.

In order to understand complex collages constructed along continental margins with long histories of repeated convergence, it is helpful to study the complex festoons of island arcs and microcontinents of the southwest Pacific (Silver and Smith 1983; Hamilton 1985). Modern arc-continent collisions in this region include the Banda arc-Australia collision, the subject of recent marine work by a number of workers (Hussong *et al.* 1983; Karig *et al.* 1984; Breen *et al.* 1986), and the collision in Taiwan, which has progressed far enough that the collision zone is emergent and can be studied and sampled directly.

The orogenic belt in Taiwan is flanked by a sedimentary basin on each side, imparting a deposi-

tional symmetry to the collision zone. The western Taiwan foredeep is a continental foreland basin, whereas the eastern Coastal Range exposes a collisional basin, in which orogenic deposits overlie the accreted arc and forearc terrane. The collisional basin of eastern Taiwan, here informally named the Coastal Range Collisional Basin (CRCB), originated as the forearc basin of the intraoceanic Luzon volcanic arc. During the ensuing collision with the Chinese continental margin, the associated accretionary wedge grew into a large fold-and-thrust belt. As the collision progressed, the forearc basin evolved into a collisional basin caught between the growing mountain belt to the west and the dying arc to the east. In this contribution we report on lithofacies analysis and sedimentation and subsidence patterns in the CRCB. These data indicate that the development, filling, and ultimate fate of this basin have been controlled by tectonic processes in the orogenic belt in ways very similar to those operative in the paired foreland basin to the west.

Tectonic Setting and Regional Geology

In Taiwan the Luzon volcanic arc, perched on the leading edge of the Philippine Sea plate, is overriding the Asian continental margin toward the northwest (Fig. 13.1). The resulting oblique collision is closing the South China Sea Basin from the north, like a giant zipper, and building the island of Taiwan as an overgrown accretionary prism. Plate convergence is about 70 m/10³ yr directed to the northwest (Seno 1977), and the rate of southward propagation of the collision with respect to the arc is about 85 m/10³ yr (Suppe 1984).

The present orogeny in Taiwan began at about 4 Ma (Suppe 1980). Horizontal shortening of at least 100 to 200 km in northern and central Taiwan has been accommodated principally by slip on thrust faults and related folding within a wedge-shaped zone above a basal decollement (Suppe 1980, 1981). The mountain belt has achieved (or, in the south, is developing toward) a steady-state geometry (Suppe 1981), in which the regional surface slope is controlled by the rock strength, decollement friction, and fluid pressure (Davis et al. 1983; Dahlen et al. 1984). The mountain belt is dominated by west-vergent faults and folds, but the eastern flank of the Central Range contains significant east-vergent backthrusts and backfolds (Stanley et al. 1981; Suppe 1981).

This orogenic event has loaded the surrounding lithosphere, forming or deepening the flanking basins, and it has also provided the highlands from which detritus was shed to fill these basins. The principal sediment-source terranes are represented by rocks exposed in the Central Range today (Fig. 13.2). The Central Range is made up of Eocene to Miocene low-grade metasedimentary rocks of the Slate Formation and pre-Tertiary metamorphic rocks of the Tailuko and Yuli belts. The Slate Formation and pre-Pliocene sedimentary rocks of the western Taiwan foothills represent uplifted and deformed clastic deposits of a rifted-margin sequence. These were mudstones and sandstones that accumulated on the passive Chinese continental margin following Paleogene rifting that formed the South China Sea Basin. Rocks of the western Taiwan foothills and the Slate Formation have been metamorphosed to zeolite and lower greenschist facies during the young and ongoing collision. Pre-Tertiary metamorphic rocks of the Central Range represent the continental basement upon which the rifted-margin clastic wedge accumulated. This basement

complex consists of schist, marble, gneiss, migmatite, amphibolite, metabasalt, and metachert, and was metamorphosed to upper greenschist to amphibolite facies during the Nanao orogeny, at approximately 70 to 90 Ma (Ernst 1983). This event was accompanied by granitic intrusions, and apparently reflects an episode of pre-rifting Andean-type subduction beneath the Chinese continental margin. During the Neogene arc-continent collision these rocks were reburied, deformed, and affected by regional retrograde metamorphism at intermediate conditions, roughly 350 to 475°C and 4 kb (Liou 1981; Lo and Wang-Lee 1981; Ernst 1983).

The Collisional Basin of Eastern Taiwan

Sedimentary sequences preserved in the CRCB contain a valuable record of the early arc-continent collision, as documented by Chi et al. (1981). The Coastal Range exposes the accreted arc terrane, overlain in part by the olistostromal Lichi Melange, which includes blocks of the East Taiwan Ophiolite. Sediment deposited on arc rocks in the CRCB also includes fine to coarse clastic detritus clearly derived from a continental source terrane, the ancestral Central Range. The CRCB is bounded on the west by the Longitudinal Valley, site of a major active fault and the recognizable suture between volcanic-arc and continental crust.

The CRCB originated as a forearc basin, but apparently owes the great thickness of its sedimentary fill to tectonic loading in the nearby mountain belt. Stratigraphic sections in the CRCB are at least 6 km in composite thickness, whereas forearc basins of intraoceanic arc systems rarely exceed 1 to 2 km in thickness (Lundberg 1983). Tectonic loading apparently provided the subsidence necessary for their accumulation. The CRCB experienced an overall history of basin subsidence and filling very similar to that of classic foreland basins. The arc edifice constituted an efficient sediment barrier on the eastern side of the CRCB.

The CRCB has been strongly affected by tectonic compression during Plio-Pleistocene orogeny. The sequences of the CRCB have been themselves thrust and folded, and the Luzon arc terrane has undergone significant horizontal shortening. Quantitative estimates of shortening are hampered by lack of lateral continuity of stratigraphic units, but a crude estimate can be made by assuming a pre-collision width

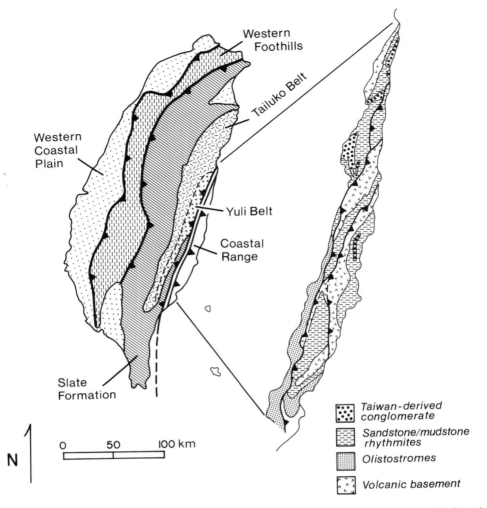

Fig. 13.2. Generalized bedrock geology of Taiwan and geologic sketch map of the Coastal Range showing distribution of principal lithologies. Western Coastal Plain and Western Foothills are part of the western Taiwan foredeep; Slate Formation, Tailuko Belt, and Yuli Belt make up the Central Range (metamorphic belt); Coastal Range represents accreted arc and forearc terrane of the Philippine Sea plate and overlying strata of the CRCB.

of the arc and forearc terrane similar to that south of Taiwan today, beyond the propagating tip of the collision. Between Taiwan and Luzon the combined width of the arc and forearc basin is roughly 100 km, as compared to the 30 km width of the Coastal Range (including possible extensions offshore and beneath the Longitudinal Valley), suggesting as much as 70 km of horizontal shortening. Ultimately these rocks were incorporated into the orogenic belt, with the present Longitudinal Valley remaining as a tectonic (and topographic) separation between the principal mountain range (Central Range) and the uplifted and deformed remnants of the CRCB and the underlying arc terrane (Coastal Range).

The arc basement of the Coastal Range is made up of volcanic and volcaniclastic rocks of the Miocene Tuluanshan Formation. Volcanic rocks are primarily andesite and basaltic andesite, agglomerate, and bedded tuff. Sedimentary rocks include volcanic-clast conglomerate, tuffaceous sandstone, volcaniclastic thin-bedded turbidites and shallow-marine tempestites, and arc-fringing reefal limestone. Diorite and dike rocks are also represented in the Chimei Igneous Complex. Whole-rock K-Ar dates of andesite range as old as 22 Ma (Ho 1969). Pyroclastic and arc-derived epiclastic deposition continued undiluted through Miocene time, and active arc volcanism continued, sporadically, well into Pleistocene time, based on metre-thick ash beds interbedded in Pleistocene orogenic sediments of the southern CRCB (Chi *et al.* 1981).

Stratigraphy

Marine sedimentary rocks of the CRCB are mainly Plio-Pleistocene in age, with basal strata possibly as old as latest Miocene (Chi *et al.* 1980; 1981). The normally bedded marine strata have been divided into two different stratigraphic schemes. Most workers follow the usage of Hsu (1956, 1976), who defined two lithostratigraphic units, the Takangkou Formation, comprising black shale and conglomerate, and the Chimei Formation, made up of alternating thin beds of sandstone and shale. Much of the sequence mapped as Takangkou Formation is made up of rhythmically interbedded sandstone and mudstone, however, which are very similar to the Chimei strata. This problem, coupled with the time-transgressive nature of the Takangkou/Chimei contact, has led to confusion in usage. The other stratigraphic scheme, suggested by Teng (1980), separates the early, arc-derived strata (Fanshuliao Formation) from the later Taiwan-derived strata (Paliwan Formation). These two groups of rocks are intimately interbedded near the base of the section, however, with abundant siltstones and very fine sandstones of equivocal provenance, making application of this scheme difficult at best. We consider it is most practical to group all normally bedded marine deposits above the Tuluanshan volcanic and volcaniclastic sequence into one formation, which for consistency should be called the Takangkou Formation.

The Lichi Melange is an olistostromal accumulation of lower Pliocene scaly mudstone, containing blocks of mainly Miocene age as well as interbeds of coherently bedded strata (Liou *et al.* 1977; Chi *et al.* 1981; Page and Suppe 1981). Detailed mapping by Min (1984) has shown that chaotic rocks of the Lichi Melange intertongue with normally bedded lower Pliocene deposits of the Takangkou Formation, as defined above.

The Pinanshan Conglomerate is a subaerial deposit of late Pleistocene age (Chi *et al.* 1981). It is exposed in the Longitudinal Valley, between the southern end of the Coastal Range and the Central Range, and is composed of clasts from both mountain ranges (Page and Suppe 1981).

Lithofacies

Sequences of the CRCB comprise four principal lithofacies: 1) mudstone; 2) olistostromes; 3) sandstone/mudstone rhythmites; and 4) Taiwan-derived conglomerate (Fig. 13.3). All four lithofacies are represented in the Takangkou Formation; the Lichi Melange contains principally the olistostromal lithofacies, and the Pinanchan Conglomerate is a late-stage Taiwan-derived conglomerate. These syncollisional deposits in most localities unconformably overlie volcanic-clast conglomerates or shallow-water reefal limestones.

Volcanic-Clast Conglomerate

A distinctive volcanic-clast conglomerate commonly forms the upper unit of the Tuluanshan (volcanic and volcaniclastic) basement complex. Clast lithologies are dominantly volcanic rocks of intermediate composition, with minor amounts of coralline limestone and mafic volcanic and plutonic rocks. These conglomerates are typically clast-supported and form massive beds up to 2 m thick, ranging from poorly sorted and ungraded to moderately well sorted and reverse to normally graded. Clasts are rounded to well rounded, and range in size up to 50 cm across (mean diameter 10–15 cm). The matrix of these conglomerates is calcite-cemented, poorly sorted volcaniclastic sandstone.

We interpret these thick-bedded volcanic-clast conglomerates to have accumulated in a near-shore, shallow-marine environment on the flank of the Luzon arc, prior to the onset of arc-continent collision. This is supported by regional lateral interfingering and local interbedding with shallow-water, arc-apron carbonates containing abundant corals and oysters in life position. The clasts are well rounded, but their coarseness and generally non-resistant lithologies suggest that transport was relatively limited.

Mudstone

A thick mudstone interval, 200 to 400 m thick, forms a widespread deposit at the base of the CRCB sequence (Fig. 13.4). Dark gray, massive to laminated silty mudstones generally contain widely spaced 3 to 5 cm thick interbeds of siltstone to fine-grained sandstone, and typically exhibit slump folds

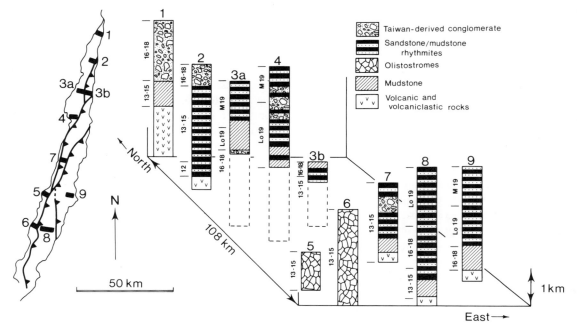

Fig. 13.3. Stratigraphic relations and lithologies in studied sections of the CRCB (shown as the inset in Fig. 13.2). Nannofossil zones noted at left of each column are from Chi *et al.* (1981), Chi (unpublished data), and recent field work. Stratigraphic sections have been only schematically restored to their original (pre-thrust) positions. Sec-

tions: 1 = Yenliaokeng; 2 = Shuilien-Fanshuliao; 3a = Feng Pin west; 3b = Feng Pin east; 4 = Hsiukuluanchi; 5 = Fuli; 6 = Kuanshan; 7 = Loholi; 8 = Tungho-Matagida; 9 = Chengkung. These sections were also studied by Chi *et al.* (1981), who describe lithologies, biostratigraphy, and structural geology.

Fig. 13.4. Outcrop photo of mudstone with several thin (~3 cm) sandstone beds, from lower portion of Tungho section (see Fig. 13.3 for location).

on a scale of metres to hundreds of metres. In some places the mudstone also includes pebbly mudstone, poorly sorted muddy-matrix conglomerate, and olistostromes. Illite and chlorite dominate the clay mineralogy of this mudstone, indicating a continental source in the orogenic belt (Buchovecky 1986).

We interpret this basal mudstone, which is laterally equivalent to the Lichi Melange, as a slope deposit that formed on the west margin of the CRCB. The sharp transition from volcanic-clast conglomerate and reefal limestone to mudstone and olistostrome marks an abrupt decrease in the energy of depositional processes, from high-energy currents of a shallow marine environment to the quiet settling of hemipelagic mud in a slope setting, although the slope mud was prone to slumping. This contact also represents a dramatic compositional shift from undiluted arc-derived sediment to large volumes of detritus shed off the collisional accretionary wedge, apparently marking the onset of collision-related orogenesis.

Olistostromes

Stratigraphically chaotic deposits of submarine slides in the CRCB include the well known Lichi Melange and numerous smaller bodies within more coherent stratigraphic sections. The Lichi Melange is restricted to the southern- and westernmost Coastal Range (Fig. 13.3), and was interpreted by Page and Suppe (1981) as an olistostromal unit deposited in moderately deep water on the western flank of the forearc basin. The smaller bodies are slide deposits composed of blocks of sandstone, marl, and tuff in a matrix of mudstone. Pebbly mudstones and pebbly sandstones are also common in the lower portions of sections, with well rounded pebbles and cobbles set in a very poorly sorted sandy to muddy matrix. Pebbles and cobbles are composed of sedimentary, metasedimentary, ophiolitic, and volcanic rocks. The pebbles and cobbles were clearly subaerially derived from the growing orogenic highlands of proto-Taiwan, and the matrix-supported pebbly mudstones and sandstones were deposited by mass-flow processes on or at the base of unstable slopes.

Sandstone/Mudstone Rhythmites

Rhythmically interbedded sandstones and mudstones make up the bulk of sedimentary rocks of the CRCB, forming stratigraphic intervals up to at least 3,000 m thick. These thin- to medium-bedded rhythmites have been interpreted uniformly as turbidites by previous workers on the basis of common flute casts, Bouma sequences, and graded beds (Teng 1979; Chi et al. 1981). Our detailed observations in a number of sections reveal, however, that a large portion of thin sandstone beds of the CRCB are tempestites, deposited by storm-generated currents above storm wave base. The distinction between the two can be quite subtle. At an outcrop scale, both are characterized by rhythmic alternations of sandstone and mudstone, with sandstone beds generally showing normal grading and erosive bases. Outcrops typically do not show organized vertical sequences in which sandstone beds thicken or thin upward; rather, bed thicknesses tend to vary irregularly within relatively narrow bounds. Descriptions of typical turbidites and tempestites of the CRCB follow, along with criteria useful for their distinction.

Turbidite sandstone beds show partial to complete Bouma sequences indicative of deposition from gradually waning unidirectional currents. Sharp erosional bases commonly show sole marks, including flute and groove casts, load casts, and flame structures. Turbidites typically comprise normally graded, vertical sequences of planar-laminated, medium-grained sand (T_b Bouma subdivision) overlain by unidirectional climbing-ripple and/or convolute-laminated, medium- to fine-grained sand (T_c), which grades into either planar-laminated muddy silt (T_d) or directly into laminated to massive silty mud (T_e). Massive, coarse to very coarse sand (T_a) commonly occurs at the bases of turbidite beds thicker than about 30 cm. Most turbidites of the CRCB are thin bedded, with sandstone bed thicknesses from 3 to 40 cm, averaging 10 to 20 cm, and with sandstone:mudstone ratios ranging from 1:4 to 1:1 (Fig. 13.5). Medium-bedded turbidites are relatively minor, and range from 40 to 100 cm thick (averaging 50–70 cm) with sandstone:mudstone ratios from 1:1 to 3:1.

Turbidity currents were active in transporting clastic detritus to the CRCB, but details of the dispersal system have not yet been resolved. It is likely that a submarine fan system provided detritus from the collision belt to the north, but a fan geometry has not yet been identified. The turbidite sequences studied thus far in the CRCB are composed principally of widespread, uninterrupted sections of distal, thin-bedded turbidites characteristic of fan-fringe and basin-plain environments.

Fig. 13.5. Outcrop photo of thin-bedded rhythmites of the Coastal Range Collisional Basin. This example is of lower Pleistocene turbidites from near the top of the Hsiukuluanchi section (see Fig. 13.3 for location), but at this scale they are indistinguishable from thin-bedded tempestites.

Rhythmites of the CRCB that we interpret to have been deposited (or redeposited) by storm-generated currents are typically laterally continuous and thinly bedded, with 2 to 20 cm thick sandstone beds alternating with laminated silty mudstone intervals 5 to 30 cm thick. Sedimentary structures that are most useful in recognizing tempestites of the CRCB are: 1) undulating (wavy) erosional bases; 2) basal planar laminae that grade laterally and vertically into low-angle cross-laminae (possibly hummocky cross-stratification: Harms *et al.* 1982); 3) internally complex, symmetric to asymmetric wave-generated ripple cross-laminae; and 4) typically distinct to sharp wave-rippled tops of sandstone beds (Fig. 13.6). This combination of erosional and bedding structures is commonly attributed to rapid input and deposition of sandy sediment above storm wave base in a muddy shelf environment, during peak and waning stages of major storm events (Johnson 1978; Walker 1984).

Intimate interbedding of turbidites and tempestites in several sections suggests that some of the turbidites of the CRCB were triggered by storm events that affected shallower (tempestite) parts of the basin (Walker 1984). We believe, however, that this mechanism for generating turbidity currents in offshore regions is probably restricted to sedimentary basins with shallow-shelf areas significantly steeper

than those of most modern continental shelves (Swift and Niedoroda 1985).

Taiwan-Derived Conglomerate

Taiwan-derived conglomerates of the CRCB are made up of a (typically) poorly sorted mixture of well rounded pebbles, cobbles, and boulders up to about 1 m across (Fig. 13.7). Clast compositions are dominated by sedimentary and metasedimentary rocks derived from the growing mountain belt of proto-Taiwan, but also include subordinate amounts of volcanic and ophiolitic detritus. Conglomerates range from clast supported to matrix supported, with single large boulders not infrequently "floating" in beds of poorly sorted pebble conglomerate. Beds of cobble-boulder conglomerate are commonly several metres thick, ranging up to 7 m, and range from strongly lenticular with downcutting erosional bases to tabular and planar based. Associated facies include medium- to thick-bedded pebble conglomerate and thin- to thick-bedded turbidites, all of which contain abundant evidence for rapid subaqueous deposition, dewatering, slumping, and sliding. We interpret these conglomerates and associated facies to represent rapid sedimentation from high-density turbidity currents in upper-fan channels at the base of a steep regional slope.

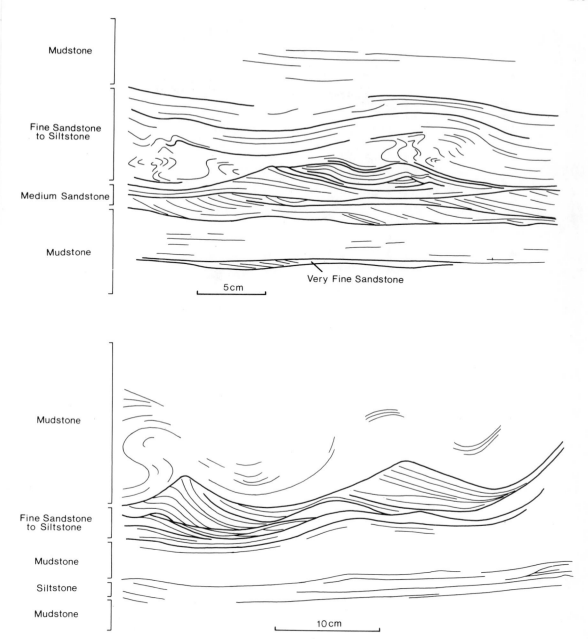

Fig. 13.6. Two examples of wave-generated ripple cross-laminae in tempestites of the Coastal Range Collisional Basin. Note wavy bases, distinct tops, and complex internal ripple laminae.

Fig. 13.7. Outcrop photo showing Taiwan-derived cobble-boulder conglomerate of the Shuilien section (see Fig. 13.3 for location). This outcrop is in a relatively well stratified interval of near-shore deposits; most conglomerates in the Coastal Range Collisional Basin are poorly stratified and medium to thick bedded.

Basin History and Paleogeography

Pre-collisional volcaniclastic deposits preserved in the Coastal Range accumulated in the forearc basin of an intraoceanic subduction zone during latest Miocene time, and were derived from the Luzon volcanic arc and shallow-water carbonates of the arc apron. Initial collision with the Chinese continental margin in earliest Pliocene time expanded the accretionary wedge into a major subaerial fold and thrust belt (Fig. 13.8). This terrane provided mixed ophiolitic and sedimentary detritus that was transported eastward into the defunct forearc basin, or CRCB, to form the Lichi Melange, pebbly mudstones, small olistostromes, and the widespread basal mudstone unit. As collision progressed, the CRCB subsided in a complex fashion and was filled by thick successions of turbidites, tempestites, and conglomerates. Sediment was derived mainly from the growing mountain belt of proto-Taiwan, which was progressively unroofed: sedimentary lithic fragments gave way to slate-grade fragments, followed in turn by higher-grade metasedimentary detritus (Dorsey 1985, in press). The CRCB has been raised above sea level since 0.5 Ma, followed by subaerial sedimentation of alluvial braided-stream deposits of the Pinanchan Conglomerate. This unit represents the final stage of basin filling, and dates uplift and incorporation of the CRCB into the orogenic belt.

Modern Analog

The best modern analog for Plio-Pleistocene sedimentation in the CRCB is found in the Luzon Trough (forearc basin) immediately south of Taiwan (Fig. 13.1). Because the collision is propagating southward, it should be possible to select an appropriate position (latitude) across the Luzon Trough as a modern analog for a given transect of the CRCB at any one time. This is corroborated by preliminary paleocurrent data from Plio-Pleistocene turbidites, which indicate that transport was mainly toward the south, as is suggested by the present physiography of the Luzon Trough. This situation allows a very detailed first pass at a paleogeographic model with which to compare results of onland study of lithofacies.

To a first approximation, the lithofacies summarized in the preceding section can be fitted easily into the present suite of environments suggested by bathymetry south of Taiwan, with one glaring exception. For significant periods of time during the Plio-Pleistocene, broad regions of relatively shallow water existed between the emergent, actively growing mountain belt to the west and the chain of dying arc volcanoes to the east. In contrast, water depth south of the Coastal Range at present increases rapidly to greater than 3,000 m in the Luzon Trough, with a narrow shelf extending over no more than several hundred km^2.

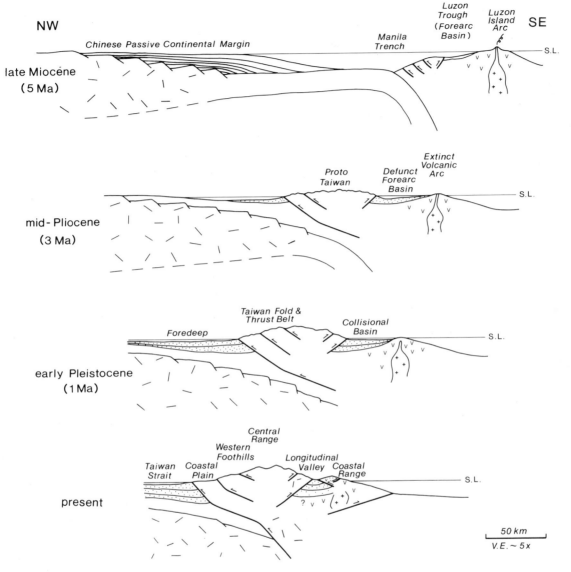

Fig. 13.8. Schematic diagram depicting evolution of the Taiwan collision.

All onland sections thus far studied in detail contain thick sections (1,500 m or more) of shallow-water, wave-influenced deposits. This indicates that a balance was maintained between rates of basin subsidence and sediment accumulation, such that water depth remained within a narrow range (~50–200 m) during deposition of a thickness of sediment that exceeds this depth by an order of magnitude. Furthermore, a shallow marine environment must have covered an area substantially larger than that documented by the sections preserved and exposed today. An area calculated for the present distribution of Plio-Pleistocene tempestites preserved in the CRCB is roughly 80 × 30 km, for an apparent minimum estimate of 2,400 km² of shelf during Plio-Pleistocene time. The discrepancy between modern and inferred Plio-Pleistocene bathymetry is enigmatic, and implies that initial phases of the collision produced slower subsidence than does the present propagation of the collision. This may be due in part to the smaller overall mass of the early orogenic load and/or buoyant, young ocean crust of the South China Sea.

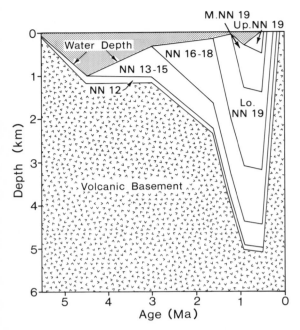

Fig. 13.9. Preliminary geohistory diagram, based on nannofossil ages by Chi (Chi *et al.* 1981, and unpublished data), paleobathymetry estimated from lithofacies analysis, and measured thicknesses. Backstripping calculated at each step using standard depth-porosity relationships computed for rapidly deposited sandstone and mudstone of the U.S. Gulf Coast (Xiao, H.B., personal communication 1986).

Subsidence History

A preliminary composite vertical section based on the best studied sections in the southern CRCB (Matagida and Chengkung; see Fig. 13.3) is 5.2 km thick. Basal mudstones, pebbly mudstones, and muddy turbidites of early Pliocene age pass upward into thin-bedded turbidites of early to late Pliocene age. Up section these are interbedded with and gradationally overlain by tempestites of late Pliocene to early Pleistocene age. Since early Pleistocene time (from about 0.5 Ma) the entire section has been uplifted; the basal contact and volcanic basement rocks are exposed well above sea level.

Methodology

We calculated a preliminary geohistory diagram (Fig. 13.9) based on this composite section, coupled with conservative estimates of paleobathymetry

based on lithofacies analysis, and assumed standard depth-porosity relationships computed for rapidly deposited sandstones and mudstones of the U.S. Gulf Coast (Xiao, H.B., personal communication 1986). We disregarded eustatic changes in sea level because the sediment thicknesses are sufficiently great that sea-level changes would superimpose only minor fluctuations on our computed curve. We also assumed that volcanic "basement" has not compacted, although these rocks contain a (generally small) percentage of volcaniclastic deposits that are subject to significant compaction.

The two major uncertainties in this analysis are: 1) the construction of the composite section; and 2) our paleobathymetric estimates. In defense of the first point, orogenic belts rarely expose complete stratigraphic sections without intervening faults; we combined two neighboring sections in the south using detailed biostratigraphy, but they are separated by a thrust fault. As a check we constructed a composite section using the two best-studied sections in the northern CRCB and achieved very similar results. Our estimates of water depth are generally well constrained for shallow-water marine deposits, but less so for slope mudstones, related olistostromes and pebbly mudstones, and turbidites. These may have accumulated in quite deep water; certainly there are abundant deep-water sedimentary rocks in some parts of the CRCB, as well as the compelling evidence of present bathymetry of the Luzon Trough. In the absence of unequivocal evidence of deep-water conditions we assigned these a maximum depth of 1,000 m. Using depths substantially greater than 1,000 m results in an unlikely subsidence history in which rapid subsidence alternates with basin uplift.

Discussion

An initial phase of subsidence is inferred from the transition from shallow-marine bioclastic limestones and associated volcanic-clast conglomerates (the forearc phase, Fig. 13.8) to mudstones and turbidites. Following this subsidence the basin filled to shallow marine conditions. Rapid subsidence (~ 0.8 m/10^3 yr) began at about 3 Ma, matched by equally rapid deposition, and these rates both increased dramatically, to about 5 m/10^3 yr in early Pleistocene time (about 1.6 to 0.9 Ma). Rates of deposition after this time period are poorly constrained due to uncertainty over how much of the upper part of the section may have been removed by erosion. Uplift since early Pleistocene time can be assigned in a minimum average rate, however, of approximately 5 m/10³ yr based on the subaerial deposition of the Pinanshan Conglomerate.

The dramatic increase in subsidence rates before this basin was uplifted is intriguing. A substantial portion of the subsidence was due to the weight of the sediment deposited (Fig. 13.10), but the residual tectonic subsidence still shows a significant increase in early Pleistocene time. This pattern is similar to that documented for foreland basins formed on continental crust in which the orogenic load approaches the site of deposition (Beaumont 1981; Kominz and Bond 1986), and it appears that the same happened here on the opposite (arcward) side of the orogenic belt. Backthrusts and backfolds at the rear (arcward) flank of the Central Range of Taiwan (Ernst 1977; Stanley et al. 1981; Suppe 1981) apparently caused the orogenic load to approach the CRCB to the east, producing dramatic subsidence before the basin itself was incorporated into the orogenic belt (Fig. 13.8).

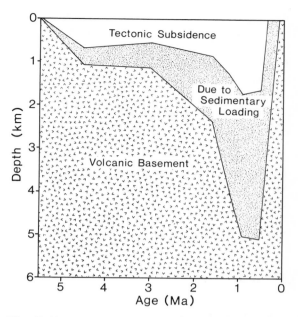

Fig. 13.10. Subsidence due to loading by sediment and water versus residual tectonic subsidence. Note increased rate of subsidence in early Pleistocene time (1.6 to 0.9 Ma).

Comparison with the Western Taiwan Foredeep

Orogenic sedimentation in the western Taiwan foredeep bears a number of key similarities to that in the CRCB of eastern Taiwan. The onset and subsequent evolution of the arc-continent collision controlled sedimentation in both basins, producing changes in rates of sedimentation, in lithologic composition, and in lithofacies patterns. Sedimentation rates increased dramatically as collision progressed. In the western foredeep, sedimentation rates (uncorrected for differential compaction) of 0.05 to 0.30 (averaging 0.15) m/10³ yr increased during the collision to a maximum of 3 m/10³ yr (Chang and Chi 1983). In the CRCB, pre-collision sedimentation rates of 0.10 to 0.40 m/10³ yr increased to about 5 m/10³ yr. The initial uplift of the proto-Taiwan mountain belt is recorded in the western foredeep by the first appearance, in lower Pliocene deposits, of abundant slate fragments (Lee 1963) and reworked nannofossils (Chang and Chi 1983) derived from the young orogen. The initial orogeny is recorded in the CRCB as the basal illite-rich mudstones (Buchovecky 1986) dated at about 4.5 Ma (Chi, W.R., personal communication 1986), and by the Lichi Melange and related deposits (Page and Suppe 1981). Progressive unroofing of the orogenic belt is

indicated by reverse trends in illite crystallinity in both basins (Manius et al. 1985; Buchovecky 1986) and by sandstone composition (Dorsey in press). The western foredeep shallowed with time as the orogen approached: early orogenic sedimentation was dominated by deep-water deposition of mud, followed by shallow marine, then deltaic, and finally by sandy and cobbly braided-river facies (Covey 1984a). The CRCB accumulated significant thicknesses of coarse-grained deep-water deposits, but also shallowed overall. Early deep-water mud deposition was followed by deposition of thin-bedded turbidites, followed by shallow-marine storm sedimentation with local incision of conglomerate-filled channels, and finally deposition of braided-stream conglomerates.

Finally, subsidence histories and subsequent deformation of the fill of the two basins have been similar. Geohistory diagrams from both the western foredeep (Covey 1984b) and the CRCB reveal dramatic increases in rates of both subsidence and sedimentation followed by rapid uplift as the basin margins were incorporated into the growing orogenic belt. The western Taiwan foredeep has been continually encroached upon by the westward-migrating orogenic belt, as the deformation front of the fold-and-thrust zone incorporates the easternmost foredeep deposits into the mountain belt, to be recycled back into younger foredeep strata. The CRCB was apparently likewise encroached upon by eastward-verging thrusts and folds from the west. It was later incorporated into the mountain belt by a still poorly understood sequence of mainly west-vergent, but also east-vergent, thrust faults. This comparison of the western foredeep with the eastern CRCB indicates a depositional symmetry that may reflect a deeper symmetry of tectonic processes in the Plio-Pleistocene orogeny in Taiwan. This is difficult to reconcile with the present dominance of westward vergence in the Taiwan orogen (Suppe 1980), and provides a focus for future research.

Conclusions

Lithofacies analysis of the Coastal Range Collisional Basin of eastern Taiwan reveals an overall shallowing-upward trend in depositional environments, although in detail the vertical sequences are much more complex. Basal mudstones with local olistostromes and/or pebbly mudstones deposited in slope and base-of-slope environments overlie a sharp, unconformable contact with underlying volcanic and volcaniclastic rocks of the Luzon arc. These mudstones and related rocks are overlain by sandy, thin-bedded turbidites, which comprise very thick intervals characteristic of fan-fringe and basin-plain environments. Mid-fan channel deposits and suprafan-lobe sequences are minor and present only locally. The turbidites in turn pass up gradationally into widespread deposits of shallow-marine, wave-influenced tempestites, and locally back into thin-bedded turbidites with canyon-fill deposits containing abundant cobble-boulder conglomerate. Following uplift above sea level in late Pleistocene time, cobbly braided-stream deposits accumulated on top of the older sedimentary rocks with angular unconformity. The offshore region immediately south of Taiwan, in which the arc-continent collision is propagating, provides an appropriate modern analog for marine deposits now exposed in the Coastal Range, with the exception that a satisfactory modern setting is not present for widespread shallow-water facies preserved in the onshore Plio-Pleistocene sections.

Analysis of the subsidence history of the Coastal Range Collisional Basin reveals a pattern of increasing rates of subsidence and sedimentation prior to dramatic uplift, very similar to that seen in well known foreland basins. This suggests that significant horizontal shortening along backthrusts at the nominal "rear" of the orogenic belt caused tectonic loading of the rearward collisional basin, accompanied by increased sedimentation, before this basin was itself incorporated into the orogenic belt. Our preliminary analysis implies that early stages of the Taiwan orogen possessed a component of regional symmetry, in both tectonic and depositional processes, that heretofore has not been appreciated. This aspect should be helpful in developing a better understanding of the mechanics of crustal interaction during arc-continent collision.

Acknowledgments. We thank NSF (EAR-8510656) and the Shell Foundation for funding. John Suppe, Chi Wen-Rong, Mike Covey, C.M. Wang-Lee, C.S. Ho, Chris Fong, and many other Chinese colleagues provided helpful discussions, and the Central Geological Survey kindly provided logistical support. We thank Chi Wen-Rong for access to unpublished data and for discussions on biostratigraphy, and Jay Namson, Eli Silver, and Jim Steidtmann for helpful comments. We also thank Karen Kleinspehn and Chris Paola for organizing a very successful symposium and for performing the yeoman's task of editing this volume.

References

BEAUMONT, C. (1981) Foreland basins. Geophysical Journal Royal Astronomical Society 65:291–329.

BOX, S.E. (1985) Early Cretaceous orogenic belt in northwestern Alaska: Internal organization, lateral extent, and tectonic interpretation. In: Howell, D.G. (ed) Tectonostratigraphic Terranes of the Circum-Pacific Region. Circum-Pacific Council for Energy and Mineral Resources Earth Science Series 1:137–145.

BREEN, N.A., SILVER, E.A., and HUSSONG, D.M. (1986) Structural styles of an accretionary wedge south of the island of Sumba, Indonesia, revealed by SeaMARC II side scan sonar. Geological Society America Bulletin 97:1250–1261.

BUCHOVECKY, E.J. (1986) Clues to the Development of the Taiwan Orogen from Mudstones of the Southern Coastal Ranges. Unpublished Senior Thesis, Princeton University, 31 p.

CHANG, S. and CHI, W.R. (1983) Neogene nannoplankton biostratigraphy in Taiwan and the tectonic implications. Petroleum Geology Taiwan 19:93–147.

CHI, W.R., NAMSON, J., and MEI, W.W. (1980) Calcareous nannoplankton biostratigraphy of Neogene sediments exposed along the Hsiukuluanchi in the Coastal Range, eastern Taiwan. Petroleum Geology Taiwan 17:74–87.

CHI, W.R., NAMSON, J., and SUPPE, J. (1981) Stratigraphic record of plate interactions in the Coastal Range of eastern Taiwan. Geological Society China Memoir 4:491–530.

CONEY, P.J., JONES, D.L., and MONGER, J.W.H. (1980) Cordilleran suspect terranes. Nature 288: 329–333.

COVEY, M. (1984a) Lithofacies analysis and basin reconstruction, Plio-Pleistocene western Taiwan foredeep. Petroleum Geology Taiwan 20:53–83.

COVEY, M. (1984b) Sedimentary and Tectonic Evolution of the Western Taiwan Foredeep. Unpublished Ph.D. Thesis, Princeton University, 152 p.

DAHLEN, F.A., SUPPE, J., and DAVIS, D. (1984) Mechanics of fold-and-thrust belts and accretionary wedges: Cohesive Coulomb theory. Journal Geophysical Research, 89:10,087–10,101.

DAVIS, D., SUPPE, J., and DAHLEN, F.A. (1983) Mechanics of fold-and-thrust belts and accretionary wedges. Journal Geophysical Research 88:1153–1172.

DORSEY, R.J. (1985) Petrography of Neogene sandstones from the Coastal Range of eastern Taiwan: Response to arc-continent collision. Petroleum Geology Taiwan 21:184–215.

DORSEY, R.J. (in press) Provenance evolution and unroofing history of a modern arc-continent collision: Evidence from petrography of Plio-Pleistocene sandstones, eastern Taiwan. Journal Sedimentary Petrology.

ERNST, W.G. (1977) Olistostromes and included ophiolitic debris from the Coastal Range of eastern Taiwan. Geological Society China Memoir 2:97–114.

ERNST, W.G. (1983) Mineral paragenesis in metamorphic rocks exposed along Tailuko Gorge, Central Mountain Range, Taiwan. Journal Metamorphic Geology 1:305–329.

HAMILTON, W. (1985) Subduction, magmatic arcs, and foreland deformation. In: Howell, D.G. (ed) Tectonostratigraphic Terranes of the Circum-Pacific Region. Circum-Pacific Council for Energy and Mineral Resources Earth Science Series 1:259–262.

HARMS, J.C., SOUTHARD, J.B., and WALKER, R.G. (1982) Structures and Sequences in Clastic Rocks. Society Economic Paleontologists Mineralogists Short Course Notes 9, 249 p.

HO, C.S. (1969) Geological significance of potassium-argon ages of the Chimei Igneous Complex in eastern Taiwan. Geological Survey Taiwan Bulletin 20:63–74.

HSU, T.L. (1956) Geology of the Coastal Range, eastern Taiwan. Geological Survey Taiwan Bulletin 8:39–63.

HSU, T.L. (1976) The Lichi Melange in the Coastal Range framework. Geological Survey Taiwan Bulletin 25: 89–95.

HUSSONG, D., SILVER, E., KARIG, D., and BLACKINGTON, J.G. (1983) SeaMARC II mapping of convergent plate margin features in eastern Indonesian waters. EOS Transactions, American Geophysical Union 64:239.

JOHNSON, H.D., and BALDWIN, C.T. (1986) Shallow siliciclastic seas. In: Reading, H.G. (ed) Sedimentary Environments and Facies. New York: Elsevier, pp. 229–282.

JONES, D.L., SILBERLING, N.J., GILBERT, W., and CONEY, P.J. (1982) Character, distribution, and tectonic significance of accretionary terranes in the central Alaska Range. Journal Geophysical Research 87:3709–3717.

KARIG, D.E., KLEMPERER, S., BARBER, A.J., CHARLTON, T., and HUSSONG, D.M. (1984) Collision tectonics across the Banda Arc at Timor. EOS Transactions American Geophysical Union 65:1088.

KOMINZ, M.A. and BOND, G.C. (1986) Geophysical modelling of the thermal history of foreland basins. Nature 320:252–256.

LEE, P.J. (1963) Lithofacies of the Toukoshan-Cholan Formation of western Taiwan. Geological Society China Proceedings 6:41–50.

LIOU, J.G. (1981) Recent high CO_2 activity and Cenozoic progressive metamorphism in Taiwan. Geological Society China Memoir 4:551–581.

LIOU, J.G., LAN, C.Y., SUPPE, J., and ERNST, W.G. (1977) The East Taiwan Ophiolite: Its Occurrence, Petrology, Metamorphism and Tectonic Setting. Minerals and Resources Service Organization Special Report 1, Taipei, Taiwan, 212 p.

LO, C.H. and WANG-LEE, C. (1981) Mineral chemistry in some gneissic bodies, the Hoping-Chipan area, Hualien, eastern Taiwan. Geological Society China Proceedings 24:40–55.

LUNDBERG, N. (1983) Development of forearcs of intraoceanic subduction zones. Tectonics 2:51–61.

MANIUS, W.G., COVEY, M., and STALLARD, R. (1985) The effects of provenance and diagenesis on clay content and crystallinity in Miocene through Pleistocene deposits, southwestern Taiwan. Petroleum Geology Taiwan 21:173–185.

MIN, L. (1984) Geology of Juiyan Area, Coastal Range, Eastern Taiwan (in Chinese). Unpublished Bachelor's Thesis, National Taiwan University, 54 p.

PAGE, B.M. and SUPPE, J. (1981) The Pliocene Lichi Melange in Taiwan: Its plate tectonic and olistostromal origin. American Journal Science 281:193–227.

ROWLEY, D.B. and KIDD, W.S.F. (1981) Stratigraphic relationships and detrital composition of the Medial Ordovician flysch of western New England: Implications for the tectonic evolution of the Taconic orogeny. Journal Geology 89:199–218.

SENO, T. (1977) The instantaneous rotation vector of the Philippine Sea plate relative to the Eurasian plate. Tectonophysics 42:209–226.

SILVER, E.A. and SMITH, R.B. (1983) Comparisons of terrane accretion in modern southeast Asia and the Mesozoic North American Cordillera. Geology 11: 198–202.

STANLEY, R.S., HILL, L.B., CHANG, H.C., and HU, H.N. (1981) A transect through the metamorphic core of the central mountains, southern Taiwan. Geological Society China Memoir 4:443–473.

STANLEY, R.S. and RATCLIFFE, N.M. (1985) Tectonic synthesis of the Taconian orogeny in western New England. Geological Society America Bulletin 96: 1227–1250.

SUPPE, J. (1980) A retrodeformable cross section of northern Taiwan. Geological Society China Proceedings 23:46–55.

SUPPE, J. (1981) Mechanics of mountain building in Taiwan. Geological Society China Memoir 4:67–89.

SUPPE, J. (1984) Kinematics of arc-continent collision, flipping of subduction, and back-arc spreading near Taiwan. Geological Society China Memoir 6:21–33.

SWIFT, D.J.P. and NIEDORODA, A.W. (1985) Fluid and sediment dynamics on continental shelves. In: Tillman, R.W., Swift, D.J.P., and Walker, R.G. (eds) Shelf Sands and Sandstone Reservoirs. Society Economic Paleontologists Mineralogists Short Course Notes 13:47–133.

TENG, L.S. (1979) Petrographical study of Neogene sandstones of the Coastal Range, eastern Taiwan (1. Northern Part). Acta Geologica Taiwanica 20:129–155.

TENG, L.S. (1980) Lithology and provenance of the Fanshuliao Formation, northern Coastal Range, eastern Taiwan. Geological Society China Proceedings 23: 118–129.

WALKER, R.G. (1984) Shelf and shallow marine sands. In: Walker, R.G. (ed) Facies Models. Geological Association Canada, Geoscience Canada Reprint Series 1:141–170.

14

Paleogeography and Tectonic Evolution Interpreted from Deformed Sequences: Principles, Limitations, and Examples from the Southwestern United States

J. Douglas Walker

Abstract

Paleogeographic analysis is an important way to reconstruct tectonic evolution. In many areas of interest, strata to be analyzed are deformed and metamorphosed. Although interpretations based on data collected from disrupted rocks are commonly poorly constrained relative to analysis of undeformed sequences, useful information can often be recovered.

Paleogeographic reconstructions within orogenic belts are more difficult due to metamorphism and deformation because: 1) material is uplifted and lost to erosion; and 2) the internal structure of sequences is altered. Interpretations are based on whatever features are still recognizable in the rocks and, when available, lithostratigraphic, chronostratigraphic, and/or biostratigraphic correlation with surrounding areas. Care must be taken to correlate rocks correctly; to this end, marker sequences, rather than marker beds, provide the most reliable tools.

Interpretation of the tectonic evolution of the Mojave Desert, southwestern United States, is based on the analysis of rocks that are presently exposed as metamorphosed and deformed roof pendants. Upper Precambrian to Paleozoic sequences are continuous with the Cordilleran miogeocline and show southwesterly paleogeographic trends. Correlation of these rocks is based on marker sequences. A fundamental event occurred in Permo-Triassic time that resulted in a change in the orientation of paleogeographic trends. This event was probably related to the change from a passive to a subducting margin along the western Cordillera.

Introduction

Understanding the tectonic evolution of an area is an important goal of basin analysis. This goal is often reached by relating paleogeographic reconstruc-

tions, derived using techniques of basin analysis, to tectonic setting. This process becomes increasingly difficult when rocks have undergone moderate to extreme deformation and metamorphism. The aim of this contribution is to describe some of the general methods, and their limitations, for using paleogeographic reconstructions to infer tectonic evolution in disrupted areas. Two examples, from the Mojave Desert, southwestern United States, form the basis for discussing how these methods can be applied to disrupted sequences.

Several aspects of regional tectonics can be studied using basin-analysis techniques and paleogeographic reconstructions. First, the tectonic elements of an area during the past can be identified. For example, an area can be identified as a passive margin based on careful paleogeographic reconstructions, or deformational events can be recognized from characteristic, synorogenic deposits. A second goal is to determine the timing of tectonic events. In the examples above, we would like to define the times of rifting and deformation. Third, an understanding of spatial and temporal variations in tectonic settings can be gained; the question of whether rifting was diachronous along a margin, and the time-space patterns of deformation are important in the examples cited. Lastly, provenance studies can be used to determine the possible source region of a basin fill. This is especially important in constraining the spatial and temporal relations between displaced or juxtaposed tectonic terranes and microplates, and also the possible points of origin of these crustal elements (Howell and Jones 1984; Schermer *et al.* 1984).

Taken together, the aspects described above lead to an understanding of the tectonic evolution of a region. The process of interpreting paleogeography and tectonic evolution becomes much more difficult and tenuous when the study area has undergone deformation and metamorphism. Kay (1945) stated in his classic paper on paleogeographic and palinspastic maps that paleogeographic reconstruction represents concepts or interpretations based on an incomplete geologic data set. Deformation and metamorphism further limit the data set available to build paleogeographic reconstructions, commonly making reconstructions and tectonic models derived from disrupted sequences much more speculative and interpretive. Useful and definitive interpretations of tectonic evolution can often be made, however, in spite of structural disruption and metamorphism.

Methods

Many different methods can be used successfully to decipher the paleogeographic significance of deformed and metamorphosed sequences. A full review of these methods is beyond the scope of this contribution. Instead, I provide a summary of methods used in the study of Mojave Desert rocks, and make reference to related methods and their possible application to other types of disruption.

Information that can be obtained from any sequence of deformed rocks falls into two general categories: 1) that which can be derived from the rocks alone; and 2) that which can be interpreted using data projected from undeformed sequences. The former deals with original rock features that are retained despite metamorphism and deformation. Sedimentary structures, thickness, primary mineralogy, and biostratigraphic data are commonly resolvable relict features. Determination of original features can be done in a variety of ways. Bed thickness and fossil morphology can be recovered by reversing finite strain in the rock (Ramsey and Huber 1983, and references therein). This depends less on the degree of finite strain than on the ability to define correctly the amount of strain. Even in rocks with shear strains of 2 or greater it is possible to get useful information provided the amount of finite strain is well known. This technique is difficult to apply to sequences containing mixed lithologies where strain is inhomogeneously distributed, and in areas of heterogeneous strain. Petrographic study of mineralogy and texture commonly allows the determination of original sediment compositions if pseu-

domorphs or relict textures are present. One example is that of graywackes where lithic grains are transformed into clay matrix in sandstones. Dickinson (1970) outlines clear criteria that can be used to distinguish between "pseudomatrix" derived from lithic fragments, detrital matrix, and cement. Often, sedimentary and biogenic structures can be characterized if the rock is not too severely strained. Structures such as cross-bedding, however, may become completely overprinted at shear strains above 2 or 3. (This quantity for shear was determined by computer simulations of deformation and subjective appraisal of when structures were unrecognizable.) Many microfossils, such as conodonts and radiolaria, can be recovered from deformed Phanerozoic rocks (e.g., Irwin et al. 1978; Hill 1985; Boucot and Rumble 1987; Walker 1987).

The second method is to project information into the deformed sections from undeformed areas by lithostratigraphic correlation. The pitfalls and shortcomings associated with lithostratigraphic correlation of undeformed sequences apply to a similar practice in deformed rocks, but to a much greater degree. For example, it is easier to correlate deformed sequences incorrectly because both internal structure and the succession of strata may be disturbed. In addition, similar lithologies of entirely different ages can be mistakenly linked because age and stratigraphic position are undefined. A common method in lithostratigraphy is to identify distinctive marker beds to use as tie-lines (Miall 1984). In deformed and metamorphosed areas, marker units are often inadequate because they can be easily lost, obscured, or duplicated. The use of a *marker sequence* is usually necessary: A distinctive succession of strata or association of rock types provides the most reliable way to correlate sequences. The use of marker sequences is shown below for examples from the Mojave Desert. Care should be taken not to ignore variations within the marker sequence when using this method. Changes in attribute ratio (e.g., sand/shale), bedding thickness, or other features may be interpretable. Also, the disappearance of key units may record unconformities.

Sequence correlation does not lend itself to deciphering diachronous events unless chronostratigraphic data are also available. This leads to another problem in correlation: the comparison of radiometric data with biostratigraphic ages. Often the ages of sequences in deformed areas are constrained by radiometric ages of cross-cutting or interbedded igneous rocks. If absolute ages are not available for correlative undeformed sequences, then large errors in correlation are probable. This problem arises from large uncertainties in calibrating the absolute and relative time scales. The magnitude of errors associated with these assignments is commonly about 5 to as much as 20 million years for Mesozoic and Paleozoic time. For examples of the magnitude of errors calculated for different time periods and a comparison of differences in calibrations see Harland *et al.* (1982), Palmer (1983), and P.F. Carr *et al.* (1984).

The techniques used in analyzing disrupted sedimentary sequences depend on the degree of metamorphism and deformation, as does the quality of information that can be gained. Probably the worst case is that of a regionally metamorphosed and deformed sedimentary sequence that has been heavily intruded and that has no undeformed reference sections (i.e., granite-gneiss terrain). Analysis here may consist largely of major and/or trace element geochemical studies to infer protoliths (see Eriksson *et al.* 1988: this volume). The ability to make useful interpretations about paleogeography and tectonics is not ensured even if such a sequence can be traced into less deformed protoliths. In cases of extreme disruption, protolith interpretation may be impossible. Often the only interpretations that can be made from these rocks concern the presence and gross nature of the sedimentary basin. This information itself may be sufficient, however, to constrain the position and extent of plate margins and original tectonic environment (e.g., passive-margin versus foreland) in highly disrupted areas. An example is the presence and significance of metasedimentary rocks in granite-greenstone belts (Windley 1984).

Less-deformed and less-metamorphosed sections are still difficult to decipher, but can yield substantially more information, especially if data from undeformed sequences are available. Rocks at upper greenschist and amphibolite grade metamorphism have undergone severe mineralogical changes; other features such as sedimentary structures and bed thickness may also be obscured by structural deformation. Many sedimentary features are still interpretable despite disruption. Protolith, stratigraphic succession, and even fossil ages can still be determined in many cases (Fig. 14.1; Hill 1985, and references therein; examples below). Care must be taken not to overinterpret these rocks. Cleavage or other deformational fabrics should not be confused with bedding, especially in fine-grained rocks.

Fig. 14.1. Crinoid stems in marble, eastern California. Deformation has sheared crinoid stems into parallelism.

Failure to recognize transposed bedding can result in misinterpretation of stratigraphic sequences and facing directions. Transposed bedding is recognized by locating fold hinges or finding adjacent beds with opposite facing. Facing indicators should also be carefully evaluated. One strategy of study in the highly deformed and metamorphosed rocks is to examine carefully fold noses and other areas of lower strain. Here features that may be obliterated in attenuated fold limbs can still be recognized.

Information that is more useful for correlations and reconstructions is easier to collect from rocks at lower grades of metamorphism (greenschist and lower) and levels of deformation. Special techniques, other than palinspastic restoration, are often of secondary importance. Original mineralogy is typically well preserved. Framework composition of sandstones often can be directly interpreted, although fine-grained rocks and carbonates may still be recrystallized despite the lower metamorphic grade. The main limitation on interpretations at lower metamorphic grade is often the amount and style of structural disruption.

An important area where extreme structural disruption occurs at relatively low metamorphic grade is in melanges. Sedimentological features are well preserved in competent layers and on a bed-by-bed scale. Larger-scale aspects of rock geometry, facies, and stratigraphic succession are often destroyed. Important interpretations of tectonics and paleogeography can, however, be made from melanges. Original mineralogy can be described for provenance determination. The assemblage of exotic material in a melange (material derived from a downgoing plate) can yield important paleogeographic and tectonic constraints. For example, exotic fragments with significantly different fossil or mineralogical assemblages from the host rock indicate large displacements for the subducted slab (Schermer *et al.* 1984). Fossils are often well preserved in melanges. The age of the melange matrix and intact blocks can hold keys to the timing of melange formation and hence timing of subduction and related deformation.

It is best to base correlations and reconstructions in disrupted areas on rocks that are most resistant to deformation and metamorphic/metasomatic changes. Siliciclastics, especially sandstones and conglomerates, often survive deformation fairly intact, and can form the main parts of marker sequences. Examples of this are given below for upper Precambrian and lower Cambrian rocks of the Mojave Desert. Resolution can be lost when siliciclastics are interbedded on a small scale with pelites or carbonates: Such sequences are easily transposed because of the ductility contrasts between these rocks. Although carbonates are susceptible to recrystallization that destroys original features, they are very important because they often contain the best fossil records.

Examples

Two examples in which tectonics can be deciphered by paleogeographic reconstruction of deformed and metamorphosed rocks are provided by upper Precambrian to lower Mesozoic rocks of the Mojave Desert, southern California (Fig. 14.2). The Mojave Desert lies in the southern continuation of the northwest-trending, Mesozoic Sierra Nevada Batholith. Because of voluminous intrusive rocks, pre-Jurassic strata generally occur as isolated, metamorphosed

(greenschist-amphibolite grade), and deformed roof pendants with areas of a few km^2 to several tens km^2. Pre-Jurassic rocks are sedimentary and minor volcanic and intrusive units (Burchfiel and Davis 1981).

Two questions arise about the tectonic evolution of the southern United States Cordillera: 1) do important paleogeographic or tectonic changes occur in the Precambrian/Paleozoic rifted margin and miogeocline as they are followed southwestward into the Mojave Desert; and 2) when did the trend of the margin change from northeast to northwest? Analysis of roof pendants in the Mojave Desert gives surprisingly detailed answers to both of these questions.

Late Precambrian and Paleozoic

Upper Precambrian and Paleozoic rocks of the western United States provide a classic example of a passive continental margin sequence (Stewart and Poole 1974). Miogeoclinal and cratonal strata can be traced through Utah and Nevada and into easternmost California (Fig. 14.2). An important tectonic question is: What happens to this sequence as it is traced farther west into the Mojave Desert area of southern California? The answer is in the deformed and intruded sequences of the Mojave Desert (Fig. 14.2).

The Upper Precambrian and Lower Cambrian sections contain many of the attributes outlined above for analysis of disrupted sequences. The facies trends, rock sequences, and facies geometries are well known from undeformed sequences (Stewart and Poole 1974). There are also very well defined marker sequences (and also marker beds) in the upper Precambrian (Stewart 1970) and Paleozoic (Stone *et al.* 1983) strata. The sequences contain distinctive sedimentary structures in particular units (such as *Skolithos* tubes in quartzite). Because of these features the sequence has been correlated and paleogeography reconstructed with some certainty for the northern and central part of the margin.

Upper Precambrian and Lower Cambrian rocks in the eastern Mojave Desert consist of siliciclastics with intercalated carbonates. A very distinctive sequence in these rocks consists of, in ascending order, the Wood Canyon Formation, Zabriskie Quartzite, and Carrara Formation (Fig. 14.3; Stewart 1970). The Wood Canyon Formation contains conglomeratic sandstone at its base, sandstone and shale in its middle, and is capped by interbedded

Fig. 14.2. Generalized trends of upper Precambrian and Paleozoic miogeoclinal and cratonic strata in the Mojave Desert region. Data taken mainly from Brown (1986) and Burchfiel and Davis (1972, 1981). Facies and isopach trends show clear northeast-southwest orientations. Lower part of figure shows the outcrop distribution of upper Precambrian and Paleozoic rocks (black) in the Mojave Desert, and 100 m isopach of Zabriskie Quartzite.

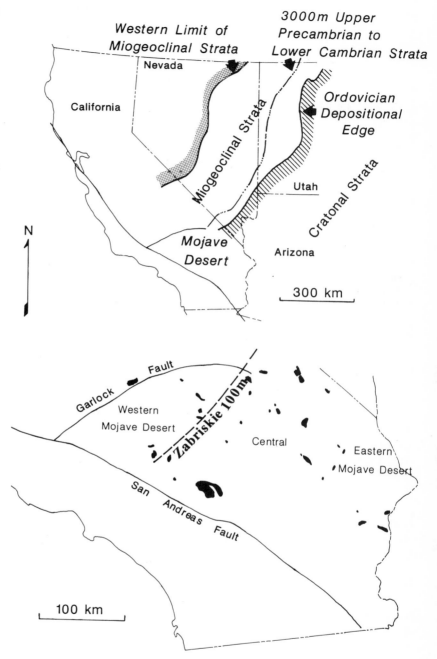

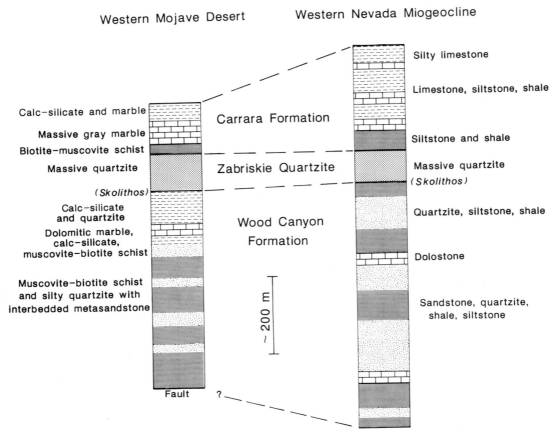

Fig. 14.3. Correlation diagram for Lower Cambrian rocks. Undeformed section (right) is from the Specter Range of western Nevada (Burchfiel 1964; Stewart 1970); deformed section (left) is composite for the western Mojave Desert (Stewart and Poole 1975; Miller 1981).

shale and dolostone. The Zabriskie Quartzite is a vitreous, light-colored sedimentary quartzite that usually contains well preserved *Skolithos* burrows. The overlying Carrara Formation consists of interbedded shale and limestone.

In the Mojave Desert, this part of the upper Precambrian/Cambrian section has been widely used for correlation. Stewart and Poole (1975) first recognized this distinctive sequence in deformed and metamorphosed sections in the western part of Mojave Desert (Fig. 14.3). These occurrences led them to suggest that the miogeocline was continuous across the Mojave Desert. Recent work supports this interpretation (Dunne 1977; Miller 1981). Because the Zabriskie Quartzite is such a mechanically competent unit, it has been used as the basis for isopach maps. Isopach trends show a dominantly northeast orientation and the unit thickens to the northwest (Fig. 14.2; Burchfiel and Davis 1981). Other units of

mixed lithologies have undergone inhomogeneous finite strain and hence are of questionable use in isopach studies.

Paleozoic strata are also locally exposed in the Mojave Desert (Dunne 1977; Miller 1981; Brown 1983, 1984; Stone *et al.* 1983). These rocks are much more deformed and recrystallized than the Upper Precambrian and Lower Cambrian strata. This difference arises because the younger rocks consist primarily of carbonates that are mechanically weak and easily recrystallized. For this reason, paleogeographic reconstructions based on these rocks are more tenuous. In the discussion below, descriptions of lithologies come from well preserved sections in southern Nevada and easternmost California. These rocks are useful in analyzing the Paleozoic paleogeography of the Mojave Desert, because there are preserved sections available for correlation, and they comprise a distinctive sequence, which is discussed below. For the most complete recent summary of the extent and correlation of these rocks in the Mojave Desert, see Brown (1986).

Several features have been used to correlate sequences and reconstruct paleogeography from these deformed rocks. The lower part of the section consists of banded dolostone and silty dolostone. In metamorphosed sections, this rock becomes banded dolomitic marble. Brown (1983) showed that this section could be correlated on both its physical characteristics (banded nature and succession of lithologies) and major element chemistry (it is distinct from other Paleozoic and Precambrian carbonates). This sequence is apparently continuous across the Mojave Desert. Mississippian to Pennsylvanian rocks consist of massive limestone and thin-bedded limestone, respectively. This couplet, often in excess of 1,000 m thick, forms a distinctive pair even in strongly deformed and metamorphosed areas and is mappable across the Mojave Desert. Correlations are further enhanced by preserved marker beds within these units. A distinctive cherty limestone occurs within the Mississippian section. When metamorphosed, this unit becomes an easily recognized calcite marble with calc-silicate pods. Correlation of this Mississippian unit, however, could not be made without first correlating the overall sequences, because cherty carbonates are present elsewhere in the section. In addition, some of the correlations outlined above are supported by sparse but diagnostic biostratigraphic data (Brown 1984).

Hence several important conclusions about the westward extent of the Cordilleran miogeocline can be made. The miogeocline was apparently continuous across the Mojave Desert, without major changes in depositional environments of miogeoclinal strata across this region. Miogeoclinal sedimentation was uninterrupted at least through Pennsylvanian time, because no units are added to or missing from the Mojave sequence when compared with well preserved sections to the east. Therefore, the tectonic evolution of the Mojave Desert region was essentially the same as that of the main part of the Cordilleran margin to the northeast through Pennsylvanian time.

Permian and Triassic

An important conclusion of the discussion above is that the Cordilleran passive margin continued south-westward across the Mojave Desert during Paleozoic time. A fundamental event(s) altered this picture so that by Jurassic time a northwest-trending convergent margin had developed in the Mojave area (Hamilton and Myers 1966). The question that immediately arises is how and when this change occurred: What was the Permian and Triassic tectonic evolution of the southwestern part of the Cordilleran margin? The answer lies in Permian and Triassic rocks of the Mojave Desert, which provide another example of how to obtain paleogeographic insight from deformed and metamorphosed rocks (Fig. 14.4).

Lower Permian rocks in the eastern Mojave Desert are the westward continuation of the miogeoclinal platform carbonates (Stone et al. 1983). Lower Permian strata apparently extend westward across the Mojave Desert, although they are only locally exposed. They are directly dated in the western Mojave Desert; other sections are correlated based on lithostratigraphy. Upper Permian sedimentary rocks are exposed only in the eastern part of the Mojave Desert (Stone et al. 1983). Upper Permian rocks are absent over most of the Mojave Desert except in its western part where Upper Permian volcanic and intrusive rocks are exposed (Miller 1981; M.D. Carr et al. 1984). Intrusion here was preceded by deformation and metamorphism. Relations of miogeoclinal rocks to this deformation and intrusion have been best described by Miller (1981) for an area in the western Mojave Desert, where a monzonite pluton dated at 242 ± 2 Ma (U-Pb zircon, J.E. Wright, personal communication 1983) intrudes deformed Paleozoic strata. Removal or nondeposition of Upper Permian rocks over the rest of the Mojave may be related to uplift during this event.

The age of the monzonite pluton emphasizes another problematic aspect of stratigraphy and reconstruction of paleogeography in deformed areas, namely inaccuracies and lack of resolution in calibration of the absolute and relative geologic time scales. Whether this pluton is Permian or Triassic is difficult to constrain; the age of the Permian-Triassic boundary is especially uncertain because of its poor calibration (placed between 237 to 250 Ma; Harland et al. 1982; Palmer 1983; P.F. Carr et al. 1984). In the case outlined here, biostratigraphic data are available to constrain better the geologic age of intrusion and deformation. The deformed rocks and pluton are overlain unconformably by lower Triassic rocks (Miller 1981; Walker 1987). Thus the intru-

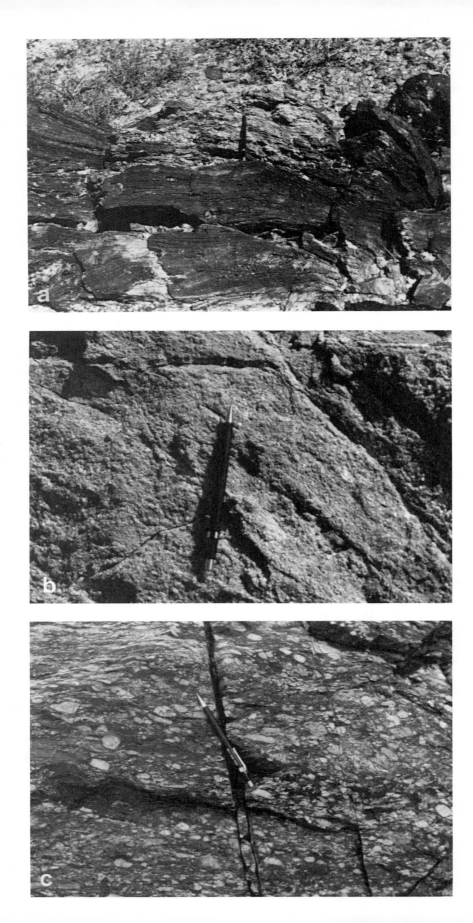

sion and disruption are probably of latest Permian or earliest Triassic age.

Lower Triassic rocks are locally exposed across the Mojave Desert and consist of a distinctive package of calc-silicate hornfels with intercalated marbles and metaconglomerate (Fig. 14.4; Walker 1987). These rocks are dated by conodonts in the western Mojave Desert (Walker 1987). In the eastern Mojave Desert Lower Triassic rocks are dated by macrofauna and correlated with confidence with well dated Lower Triassic rocks (Stone *et al.* 1983). Correlation of intervening deformed sections is based on the distinctive hornfels-marble-metaconglomerate sequence. Hence, it is best to interpret the monzonite intrusion and deformation in the western Mojave Desert as Late Permian in age because: 1) the deformed sections are overlain by Triassic rocks; and 2) a Late Permian age would correspond with and possibly be related to the hiatus in deposition to the east.

Paleogeographic maps constructed for Late Permian and Early Triassic time are presented in Figures 14.5 and 14.6. Data and interpretations for areas outside of the Mojave Desert are based on the work of Stewart *et al.* (1972), Poole and Wardlaw (1978), Peterson (1980), Lewis *et al.* (1983), Stone *et al.* (1983), M.D. Carr *et al.* (1984), Schweickert and Lahren (1984), and Speed (1984). These data permit the interpretation that the miogeocline continued across the Mojave Desert through Early Permian time and had been overprinted with a new trend in Early Triassic time. Late Permian deformation and intrusion record the initiation of events leading to this change in the western Mojave Desert. Reconstruction of Permian rocks shows a pattern that apparently differed from that of the earlier Paleozoic (Figs. 14.2, 14.5). Intrusion, uplift, and emergence signaled the change in paleogeographic trend and tectonic setting. This inference cannot be made with absolute certainty, however, because rocks in the northwestern part of the Mojave Desert (near Garlock Fault, Fig. 14.5) may not have been in

their present position at that time (Burchfiel and Davis 1981). Depositional overlap of these rocks did not occur until Early Triassic time. Paleogeographic reconstruction of the Lower Triassic rocks better constrains the tectonic evolution of this area (Fig. 14.6). Facies trends of the Lower Triassic rocks are oriented northwesterly (Walker 1987), based on preserved features of the sedimentary section. Conglomerate and sand become more common in the section westward, and sand/shale or clastic/chemical ratios show northwesterly trends. Conglomerate also becomes more common and coarser to the west. These data are used for interpretations because features such as sand/shale ratio or clast size should persist despite deformation and intrusion. All of the rocks discussed above are of shallow to marginal marine origin based on locally preserved crossbedding, mudcracks, burrows, and other environmental indicators (Fig. 14.7).

A full discussion of events leading to this change in paleogeography is beyond the scope of this chapter. Davis *et al.* (1978) and Burchfiel and Davis (1981) interpreted the change to have resulted from truncation of the southwestern extension of the Cordillera. They envisioned the mechanism to be sinistral strike-slip faulting that created a northwest trend to the margin in the Mojave Desert. This interpretation has been supported by additional data from the Mojave (Walker *et al.* 1984; Burchfiel *et al.* 1985; Walker 1987) and surrounding regions (Stone and Stevens 1984).

The examples above show that important details of the tectonic evolution of the southwestern United States can be understood by study of deformed and metamorphosed rocks. The conclusions made are based on a very limited and imperfect data set (see Figs. 14.2, 14.5, and 14.6 for distribution of rocks used in reconstructions), and thus have large uncertainties. The paleogeographic trends and timing of events obtained, however, are sufficiently detailed to be very useful in understanding the evolution of the Cordillera.

◄

Fig. 14.4. Deformed and metamorphosed Permian and Lower Triassic rocks: a = Deformed calc-silicate hornfels in Triassic sequence. Pencil for scale is 15 cm long. b = Coarsely recrystallized marble in Permian sequence. Pen-cil is 15 cm long. Protolith is horizontally laminated on a scale of 20–30 mm. c = Conglomerate in Triassic sequence. Note stretching of pebbles and inhomogeneous strain. Pencil is 15 cm long.

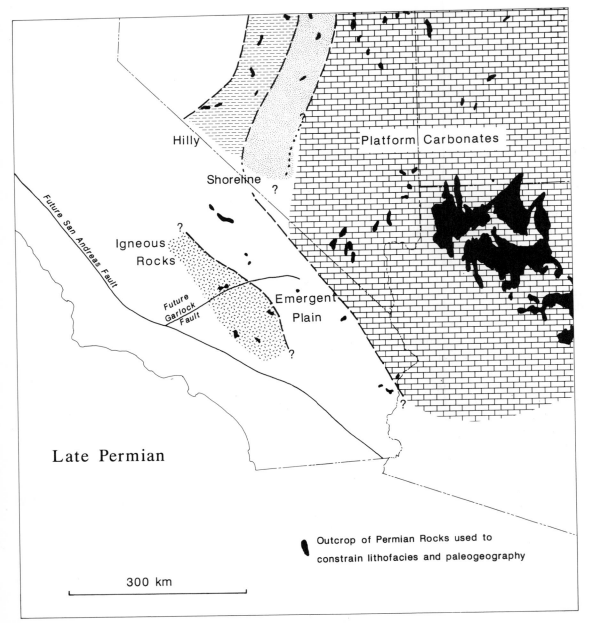

Fig. 14.5. Generalized Late Permian paleogeography and lithofacies of the southwestern United States. Hilly area was present in area of Antler Highland. Platform carbonates from the miogeocline and craton end in the eastern Mojave Desert. A continuous igneous belt is shown to trend northwesterly in the western Mojave Desert. Rocks in the northwestern Mojave Desert, around the Garlock Fault, may not have been in this position in Late Permian time; if these rocks are allochthonous the igneous belt may not be continuous. Regardless of the original position of these rocks, igneous rocks in the western Mojave Desert record a clear departure from earlier paleogeographic trends.

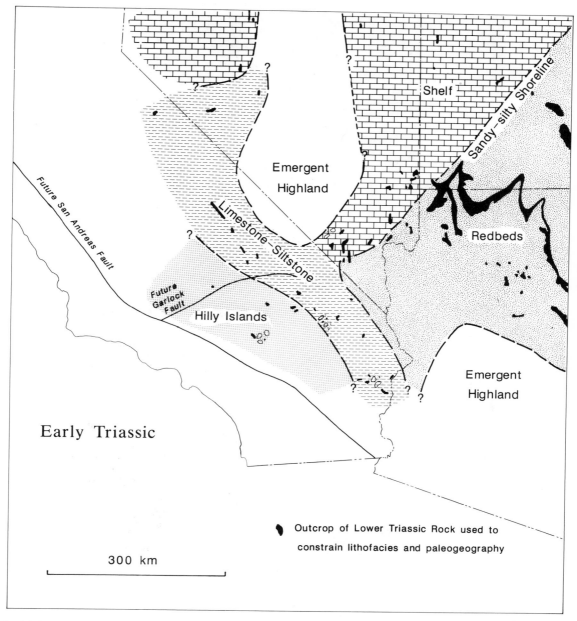

Fig. 14.6. Generalized Early Triassic paleogeography and lithofacies of the southwestern United States. The limestone-siltstone facies and marine rocks deposited around the hilly islands constitute the Lower Triassic overlap sequence on top of rocks in the northwestern Mojave Desert.

Fig. 14.7. Well preserved features in disrupted Lower Triassic rocks. a = Preserved climbing-ripple laminae in calc-silicate hornfels. Conodonts recovered from sur- rounding rocks indicate temperatures of 400 °C. b = Mud-cracks in calc-silicate hornfels.

Conclusions

Paleogeographic and basin analysis studies should be attempted in deformed areas. Even in highly disrupted areas, the study of deformed and metamorphosed rocks can lead to important conclusions regarding the tectonic evolution of a region. The reliability of such conclusions depends on how much information can be obtained from the rocks. Critical factors in analysis of disrupted rocks include: 1) the amount and style of deformation and metamorphism; 2) the areal distribution of remaining strata; and 3) presence of undeformed sequences to use for correlation.

One of the most basic tools for the study of disrupted sequences is lithostratigraphic correlation. For more certainty in correlation, marker sequences rather than marker beds must be used. When carefully applied, this method provides useful paleogeographic data.

Roof pendants of Phanerozoic rocks in the Mojave Desert area provide good examples of the study of disrupted sequences. A detailed history of the southern continuation of the Cordilleran miogeocline can be interpreted from these rocks. This history records miogeoclinal sedimentation through Pennsylvanian time, and a reorientation of the newly active margin in Late Permian or earliest Triassic time.

Acknowledgments. This study benefitted greatly from discussions with B.C. Burchfiel, M.D. Carr, C. Paola, and B.R. Wardlaw. K.A. Eriksson, K. Kleinspehn, C. Paola, and E.T. Wallin provided helpful reviews of the manuscript. Support for this work was provided by a National Science Foundation graduate fellowship, and NSF grant EAR-8314161, awarded to B.C. Burchfiel.

References

BOUCOT, A.J. and RUMBLE III, D. (1987) Comment on "Remarkable fossil locality: Crinoid stems from migmatite of the Coast Plutonic Complex, British Columbia." Geology 14:631.

BROWN, H.J. (1983) Possible Cambrian miogeoclinal strata in the Shadow Mountains, western Mojave Desert, California (Abstract). Geological Society America Abstracts Programs 15:413.

BROWN, H.J. (1984) Paleozoic carbonate stratigraphy of the San Bernardino Mountains, San Bernardino County, California. Geological Society America Abstracts with Programs 16:456.

BROWN, H.J. (1986) Stratigraphy and paleogeographic setting of Paleozoic rocks in the northern San Bernardino Mountains, California. In: Kooser, M.A. and Reynolds, R.E. (eds) Geology Around the Margins of the Eastern San Bernardino Mountains. Publication Inland Geological Society 1:105–115.

BURCHFIEL, B.C. (1964) Precambrian and Paleozoic stratigraphy of the Specter Range quadrangle, Nye County, Nevada. American Association Petroleum Geologists Bulletin 48:40–56.

BURCHFIEL, B.C. and DAVIS, G.A. (1972) Structural framework and evolution of the southern part of the Cordilleran Orogen, western United States. American Journal Science 272:97–118.

BURCHFIEL, B.C. and DAVIS, G.A. (1981) Mojave Desert and environs. In: Ernst, W.G. (ed) The Geotectonic Development of California. Englewood Cliffs, NJ: Prentice Hall, pp. 217–252.

BURCHFIEL, B.C., WALKER, J.D., and SPEED, R.C. (1985) Correlations of Paleozoic to early Mesozoic deformations and sequences, western Nevada to the Mojave Desert. Geological Society America Abstracts with Programs 17:345.

CARR, M.D., CHRISTIANSEN, R.L., and POOLE, F.G. (1984) Pre-Cenozoic geology of the El Paso Mountains, southwestern Great Basin, California — a summary. In: Lintz J. Jr. (ed) Western Geological Excursions, Vol. 4. Reno, Nevada: Department Geological Sciences, Mackay School Mines, pp. 84–93.

CARR, P.F., JONES, B.G., QUINN, B.G., and WRIGHT, A.J. (1984) Toward an objective Phanerozoic time scale. Geology 11:272–277.

DAVIS, G.A., MONGER, J.W.H., and BURCHFIEL, B.C. (1978) Mesozoic construction of the Cordilleran "collage," central British Columbia to central California. In: Howell, D.G. and McDougall, K.A. (eds) Mesozoic Paleogeography of the Western United States. Pacific Coast Paleogeography Symposium 2, Pacific Section, Society Economic Paleontologists Mineralogists, pp. 1–32.

DICKINSON, W.R. (1970) Interpreting detrital modes of graywacke and arkose. Journal Sedimentary Petrology 40:695–707.

DUNNE, G.C. (1977) Geology and structural evolution of Old Dad Mountain, Mojave Desert, California. Geological Society America Bulletin 88:737–748.

ERIKSSON, K.A., KIDD, W.S.F., and KRAPEZ, B. (1988) Basin analysis in regionally metamorphosed and deformed early archean terrains: Examples from southern Africa and western Australia. In: Kleinspehn, K.L. and Paola, C. (eds) New Perspectives in Basin Analysis. New York: Springer-Verlag, pp. 371–404.

HAMILTON, W. and MYERS, W.B. (1966) Cenozoic tectonics of the western United States. Reviews Geophysics 4:509–549.

HARLAND, W.B., COX, A.V., LLEWELLYN, P.G., PICKTON, C.A.G., SMITH, A.G., and WALTERS, R. (1982) A Geologic Time Scale. Cambridge, England: Cambridge University Press, 131 p.

HILL, M.L. (1985) Remarkable fossil locality: Crinoid stems from migmatite of the Coast Range Plutonic Complex, British Columbia. Geology 13:825–826.

HOWELL, D.G. and JONES, D.L. (1984) Tectonostratigraphic terrane analysis and some terrane vernacular. Stanford University Publications Geological Sciences 18:6–9.

IRWIN, W.P., JONES, D.L., and KAPLAN, T.H. (1978) Radiolarians from pre-Nevadan rocks of the Klamath Mountains, California and Oregon. In: Howell, D.G. and McDougall, K.A. (eds) Mesozoic Paleogeography of the Western United States. Pacific Coast Paleogeography Symposium 2, Pacific Section, Society Economic Paleontologists Mineralogists, pp. 303–310.

KAY, M. (1945) Paleogeographic and palinspastsic maps. American Association Petroleum Geologists 29:426–450.

LEWIS, M., WITTMAN, C., and STEVENS, C.H. (1983) Lower Triassic marine sedimentary rocks in east-central California. In: Marzolf, J.E. and Dunne, G.C. (eds) Evolution of Early Mesozoic Tectonostratigraphic Environments-Southwestern Colorado Plateau to Southern Inyo Mountains. Utah Geological Mineral Survey Special Studies 60, Guidebook part 2, pp. 50–54.

MIALL, A.D. (1984) Principles of Sedimentary Basin Analysis. New York: Springer-Verlag, 490 p.

MILLER, E.L. (1981) Geology of the Victorville region, California. Geological Society America Bulletin, part II, 92:554–608.

PALMER, A.R. (compiler) (1983) The decade of North American geology 1983 geologic time scale. Geology 11:503–504.

PETERSON, J.A. (1980) Permian paleogeography and sedimentary provinces, west central United States. In: Fouch, T.D. and Magathan, E.R. (eds) Paleozoic Paleogeography of the West-Central United States. Rocky Mountain Paleogeography Symposium 1, Rocky Mountain Section, Society Economic Paleontologists Mineralogists, pp. 271–292.

POOLE, F.G. and WARDLAW, B.R. (1978) Candelaria (Triassic) and Diablo (Permian) Formations in southern Toquima Range, central Nevada. In: Howell, D.G. and McDougall, K.A. (eds) Mesozoic Paleogeography of the Western United States. Pacific Coast Paleogeography Symposium 2, Pacific Section, Society Economic Paleontologists Mineralogists, pp. 271–276.

RAMSEY, J.G. and HUBER, M.I. (1983) The Techniques of Modern Structural Geology Volume 1: Strain Analysis. London: Academic Press, 307 p.

SCHERMER, E.R., HOWELL, D.G., and JONES, D.L. (1984) The origin of allochthonous terranes: Perspectives on the growth and shaping of continents. Annual Reviews Earth Planetary Science 12:107–131.

SCHWEICKERT, R.A. and LAHREN, M.M. (1984) Extent of the Antler and Sonoma belts, sutures, and transcurrent faults in eastern Sierra Nevada, California. Geological Society America Abstracts with Programs 16:648.

SPEED, R.C. (1984) Paleozoic and Mesozoic continental margin collision zone feature: Mina to Candelaria, Nevada, transverse. In: Lintz, J. Jr. (ed) Western Geological Excursions, Vol. 4. Reno, Nevada: Department Geological Sciences, Mackay School Mines, pp. 66–80.

STEWART, J.H. (1970) Upper Precambrian and Lower Cambrian Strata in the Southern Great Basin California and Nevada. United States Geological Survey Professional Paper 620, 206 p.

STEWART, J.H. and POOLE, F.G. (1974) Lower Paleozoic and uppermost Precambrian Cordilleran miogeocline, Great Basin, western United States. In: Dickinson, W.R. (ed) Tectonics and Sedimentation. Society Economic Paleontologists Mineralogists Special Publication 22, pp. 28–57.

STEWART, J.H. and POOLE, F.G. (1975) Extension of the Cordilleran miogeocline belt to the San Andreas Fault, southern California. Geological Society America Bulletin 86:205–212.

STEWART, J.H., POOLE, F.G., and WILSON, R.F. (1972) Stratigraphy and Origin of the Moenkopi Formation and Related Strata in the Colorado Plateau Region. United States Geological Survey Professional Paper 691, 195 p.

STONE, P., HOWARD, K.A., and HAMILTON, W. (1983) Correlation of metamorphosed Paleozoic strata of the southeastern Mojave Desert Region, California and Arizona. Geological Society America Bulletin 94:1135–1147.

STONE, P. and STEVENS, C.H. (1984) Stratigraphy and depositional history of Pennsylvanian and Permian rocks in the Owens Valley-Death Valley region, eastern California. In Lintz, J. Jr. (ed) Western Geological Excursions, Vol. 4. Reno, Nevada: Department Geological Sciences, Mackay School Mines, pp. 94–119.

WALKER, J.D. (1987) Permian to Middle Triassic rocks of the Mojave Desert. Arizona Geological Society Digest 18:1–14.

WALKER, J.D., BURCHFIEL, B.C., and WARDLAW, B.R. (1984) Early Triassic overlap sequence in the Mojave Desert: Its implications for Permo-Triassic tectonics and paleogeography (Abstract). Geological Society America Abstracts Programs 16:685.

WINDLEY, B.F. (1984) The Evolving Continents. Chichester, England: John Wiley and Sons Ltd., 399 p.

15
Sedimentary Basins in the Context of Allochthonous Terranes

Karen L. Kleinspehn

Abstract

Sedimentary basins have been increasingly regarded as mobile entities throughout the development of the geological sciences. The recognition of allochthonous terranes leads to the concept that the history of some sedimentary basins may be partly or entirely independent of the lithospheric plate in which they now occur. Allochthonous terranes are thought to be former oceanic plateaus, including sedimentary cover. These sediment accumulations are here named "inverse basins." Inverse basins are marine or nonmarine sedimentary accumulations that record sedimentation on surfaces of high topographic relief relative to the surrounding areas. Allochthonous terranes are recognized mainly on the basis of biostratigraphic, lithostratigraphic, paleomagnetic and other geophysical data. Inverse basins associated with allochthonous terranes are of two classes: 1) those with fill that predates or is coeval with amalgamation or accretion; and 2) those that record post-amalgamation or post-accretion displacement and tectonics. Understanding terrane history prior to accretion depends primarily on recognition of latitudinally dependent magnetic, sedimentologic, and paleontologic features in the sediments of class-1 basins, posing challenging multi-disciplinary problems to the basin analyst.

Introduction

Throughout the history of the geological sciences, sedimentary basins have been increasingly viewed as mobile entities. The earliest stratigraphers simply viewed sedimentary strata as having been deposited in precisely the position of present exposure. Vertical motion of sedimentary basins was incorporated into the geosynclinal concepts of Hall (1859), whereas post-depositional horizontal motion of sedimentary basins was recognized later as a product of compressional (Dana 1873) and extensional tectonics. Approximately eighty years later, the advent of plate tectonics brought the revolutionary view that not only is horizontal motion part of basin history, but that the magnitude of possible motion is global, as sedimentary basins are transported within moving lithospheric plates (Hess 1962; Dietz 1963; Mitchell and Reading 1969; Dewey and Bird 1970). More recently, recognition of allochthonous terranes now incorporated into orogenic belts (Monger and Ross 1971; Coney *et al.* 1980; Schermer *et al.* 1984; Howell 1985) leads to the concept that some sedimentary basins may shift tectonically from one lithospheric plate to another and that their earlier history can be independent of that of the lithospheric plate in which they now occur.

This paper focuses on allochthonous sedimentary basins that originate in oceanic settings or in continental margins and on basins that are superimposed over fault boundaries between older allochthonous terranes. Such basins commonly include those associated with volcanic arcs, with continental fragments within ocean basins, or with transcurrent fault zones. Although this paper does not concern autochthonous basins in continental interiors, their development may also be remotely related to allochthonous terranes. For example, epicratonic basins or rift basins may develop preferentially over the boundaries between Precambrian allochthonous terranes. Alternatively, subsidence in foreland basins may be a response to the accretion of allochthonous terranes (e.g., Covey 1986).

The purpose of this paper is to review, using selected examples, the evidence for allochthonous basins and how analysis of their sedimentary fill can be used to constrain their history of mobility and subsequent accretion to a new lithospheric plate.

Allochthonous Terranes and Sedimentary Basins

A sedimentary basin is a mappable accumulation of primarily sedimentary deposits. Many sedimentary basins include a significant component of volcanic strata. Earlier definitions impose geometric constraints on basins (wedge-shaped—Conybeare 1979; a circular depression or a linear crustal downwarp—Bates and Jackson 1980), but a more suitable definition is independent of geometry. Although basins superimposed on continental interiors, continental margins, oceanic crust, and volcanic arcs are included in published basin classifications (Dickinson 1974; Klemme 1975; Bally and Snelson 1980; Kingston et al. 1983), basins associated with nonvolcanic oceanic plateaus are neglected in such classifications, although many sedimentary basins preserved in orogenic belts were part of former oceanic plateaus. Modern oceanic plateaus, as defined by Nur and Ben-Avraham (1982a,b), include continental fragments, over–thickened oceanic crust, and volcanic arcs, and have variable thicknesses of sedimentary cover, either strictly marine or a combination of marine and nonmarine deposits. Although continental fragments of Indian-subcontinent size are not usually regarded as oceanic plateaus, a continuum exists between such subcontinents and small submarine plateaus with continental lithosphere, as observed in modern oceans. The relative role of transtension versus orthogonal rifting in producing these continental fragments is unclear. Oceanic plateaus are sites of subsidence and sediment accumulation, but probably because they are not depressions relative to the surrounding bathymetry, they have been overlooked in global classifications of sedimentary basins. I propose the term "inverse basins" for these sedimentary accumulations, marine or nonmarine, that record sedimentation on surfaces of high topographic relief relative to the surrounding areas. These oceanic plateaus and their sedimentary cover (inverse basins) become amalgamated and accrete to continents along active plate margins, and thus they are precursors to allochthonous terranes (Ben-Avraham et al. 1981; Schermer et al. 1984).

An allochthonous terrane is a fault-bounded crustal body, which may be defined on the basis of its distinctive, internally consistent stratigraphy and whose position and dimensions result from tectonic processes usually involving displacements of hundreds to thousands of kilometres (Berg et al. 1972; Howell 1985). Such terranes are also called tectonostratigraphic terranes (Jones et al. 1983; Howell 1985), displaced terranes (Jones et al. 1977; Irving 1979), exotic terranes (Howell 1980; Vedder et al. 1983; Zen 1983), accreted terranes (Silberling and Jones 1983), and suspect terranes (Coney et al. 1980; Williams and Hatcher 1982; Williams 1984), but hereafter will be referred to simply as allochthonous terranes (Schermer et al. 1984). Over 350 allochthonous terranes are proposed for the modern Circum-Pacific region alone (Howell 1985), although it is still unclear what fraction of these might represent former oceanic plateaus that were continental fragments versus plateaus of oceanic affinity.

Convergent or transpressional tectonics may result in the collision of two or more oceanic plateaus to form a composite terrane. This process, called terrane amalgamation, is distinct from terrane accretion in which an individual or a composite terrane collides with and welds to a continental margin (Jones et al. 1983). The recognition that terrane accretion along the peripheries of continents is a mechanism by which continents grow (Ben-Avraham et al. 1981; Schermer et al. 1984) is an embellishment on Dana's (1873) observation that continents are marked by the concentric addition of orogenic belts.

Terranes are classified as: 1) dominantly sedimentary with a coherent stratigraphy; 2) dominantly sedimentary with a disrupted, chaotic stratigraphy; or 3) dominantly metamorphic with metamorphic grade of upper greenschist facies or higher (Jones et al. 1983; Howell et al. 1985). Sedimentary terranes, cohesive or disrupted, are the most abundant type in mountain belts (Fig. 15.1) and may include a large component of volcanic strata. The distinction between coherent and disrupted sedimentary terranes is somewhat arbitrary as a continuum exists between mildly deformed strata and melange. Nevertheless, these two types of terranes dominate the allochthonous margin of western North America. Because oceanic plateaus are difficult to subduct, their upper portion may be severed from their basement during subduction (Ben-Avraham et al. 1981), thus preferentially preserving their sedimentary cover, the inverse basins. Alternatively, if a closing ocean basin is floored by young, warm oceanic lithosphere, the ophiolitic crust and its overlying sedimentary-volcanic cover may be obducted onto a continental margin where it would be preserved as a

Fig. 15.1. Map showing distribution of sedimentary, disrupted sedimentary, and metamorphic allochthonous terranes in the Western Cordillera of North America. Terrane boundaries are from Silberling and Jones (1984) and interpretations of the type of allochthonous terrane are from Silberling and Jones (1984) and Howell (1985). Q is the allochthonous terrane Quesnellia and C is the Cache Creek Terrane.

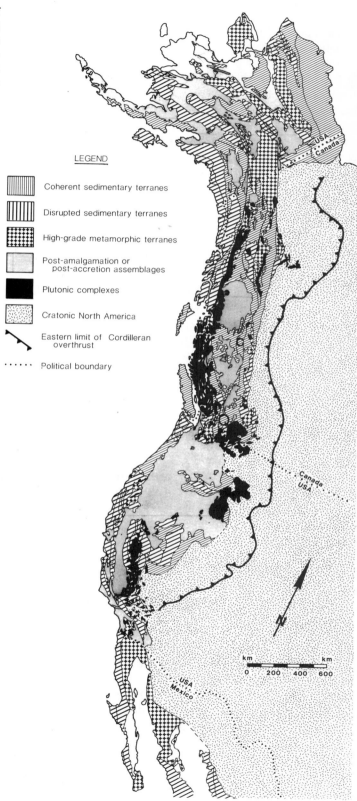

LEGEND

Coherent sedimentary terranes

Disrupted sedimentary terranes

High-grade metamorphic terranes

Post-amalgamation or
    post-accretion assemblages

Plutonic complexes

Cratonic North America

Eastern limit of Cordilleran
    overthrust

Political boundary

coherent or disrupted sedimentary terrane. The basement of sedimentary terranes may be fragments of continental margins, oceanic crust, or volcanic arcs (Nur and Ben-Avraham 1982a,b) but, whatever the basement, it is the sedimentary cover of the former oceanic plateaus that is exposed extensively in orogenic belts and is the primary focus of terrane studies. Consequently, continental growth via terrane accretion might be viewed as growth by the accretion of genetically unrelated sedimentary basins. Syn-accretionary or post-accretionary metamorphism may mask the original sedimentary nature of a terrane resulting in the metamorphic terranes as classified by Howell et al. (1985). Alternatively, metamorphic terranes may be the basement (commonly Precambrian) of continental oceanic plateaus from which the inverse-basin strata have been removed by erosion.

Sedimentary basins associated with allochthonous terranes can be grouped into two classes: 1) allochthonous basins in which present distribution of strata is limited to a single terrane; their basin fill was deposited on oceanic plateaus before or during amalgamation, or prior to or during accretion; or 2) allochthonous and autochthonous basins that post-date amalgamation or accretion such that portions of the strata overlap terrane boundaries. Inverse basins are preserved in orogenic belts as class-1 basins. Recognition of a terrane depends on data from a class-1 (allochthonous, single-terrane) basin; and delineation of a terrane's history of mobility depends on data from both classes of basins. By definition, class-1 basins are fault-bounded, whereas the second class may lack bounding faults. Vertical profiles through a given basin may record the transition of that basin from class 1 to class 2.

Recognition of Allochthonous Terranes

Allochthonous terranes, regardless of their origin, are identified and distinguished from adjacent terranes or autochthonous continental crust primarily on the basis of biostratigraphic, lithostratigraphic, and paleomagnetic or other geophysical data. The source of such data is commonly sedimentary basins of class 1 as defined above. Ideally, all four lines of evidence should be used jointly to identify a terrane, but in practice only one or two lines of evidence are often employed.

In order to satisfy rigorously the criteria discussed below, and thus qualify as allochthonous terranes, crustal blocks must have been far removed from the continents in which they now occur. Thus, it is reasonable to regard allochthonous terranes as former oceanic plateaus. If fault-bounded crustal blocks have merely undergone local displacement and reshuffling along a continental margin, it is not likely that they will meet these criteria for allochthonous terranes. As more detailed data become available, some of the Circum-Pacific terranes identified by Howell (1985) may be regarded as of more local origin rather than strictly allochthonous in terms of the criteria below.

Biostratigraphic Evidence

The concept of allochthonous terranes became a popular hypothesis as a result of biostratigraphic research in the Canadian segment of the North American Cordillera (Monger and Ross 1971). The concept was further amplified there and elsewhere in the North American Cordillera (Coney et al. 1980; Jones et al. 1983; Monger 1984) and in other orogenic belts (Howell 1980; Churkin and Trexler 1981; Nur and Ben-Avraham 1982a; Williams and Hatcher 1982; Schermer et al. 1984; Sengör 1984; Howell 1985).

To use paleontological data to recognize allochthonous terranes, fossils from within the terrane are compared to coeval fossils of the same taxonomic order from the surrounding area. In order to qualify as an allochthonous terrane, faunal and floral assemblages must be exotic relative to those across the boundary faults of the terrane. Differences between faunal and floral assemblages that can be explained in terms of structurally telescoped biofacies are inadequate to delineate an allochthonous terrane. In general, benthic organisms yield less ambiguous data than planktonic or nektonic organisms. For example, various planktonic organisms, although deposited in the same lithofacies, may have lived at two different depths within the same water column and thus constitute two different biofacies. Because skeletons of planktonic organisms are subject to long-distance post-mortem transport, such skeletons from two initially adjacent biofacies may be deposited in two widely separated areas and be misinterpreted as unrelated exotic organisms.

In the study of Monger and Ross (1971), benthic fusulinaceans of Permian age from areas in autochthonous North America were compared to those from what is now known as the Cache Creek Terrane (Fig. 15.1). Because their specimens were collected from algal-bearing carbonates, the possible depositional environments are limited to the photic zone, and clearly they were not comparing deep-marine fauna with shallow-marine specimens. By comparing specimens that not only had a benthic habit but occupied a similar ecological niche in a carbonate depositional setting, they minimized the chance that differences were attributable to different biofacies.

Lithostratigraphic Evidence

Boundaries of allochthonous terranes are marked by lateral discontinuities where coeval lithostratigraphic sequences at the formation level or greater cannot be correlated across a boundary fault, and differences between those sequences cannot be explained as simple lithofacies changes. These differences persist upward through a stratigraphic section until two terranes amalgamate or a terrane accretes to a continent, at which point their lithostratigraphic histories become similar. The fault boundary may then be obscured by development of a class-2 basin (discussed above) containing overlap sequences (Monger 1984) or successor basin sediments (*sensu* King 1966) that are deposited on both a terrane and its new neighbor(s).

Deposition of such overlapping strata (class-2 basins) is used to date terrane amalgamation or accretion. In addition, the terrane or its neighbor may shed identifiable detritus into the overlapping (class-2) basin (e.g., Eisbacher *et al.* 1974). Initial deposition of such distinctive detritus is used to document the proximity of the terrane or continent to its neighbor, and to date the time of uplift of the terrane or continental margin, which is inferred to be associated with deformation occurring during amalgamation or accretion.

The use of lithostratigraphic relationships to date amalgamation or accretion is complicated by the fact that terranes gradually close on one another at rates controlled by relative plate motions. Thus pre-amalgamation or pre-accretion deposits may record increasingly similar lithostratigraphic histories whose differences become progressively easier to explain in terms of different lithofacies within the same sedimentary basin.

The use of provenance data to date terrane amalgamation or accretion needs to be tempered by the understanding that: 1) dispersal systems may not be arranged in a pattern whereby paleocurrent analysis will suggest the immediate presence, temporally or spatially, of a terrane or its new neighbor, despite prior amalgamation or accretion; and 2) substantial uplift and unroofing of a terrane may occur prior to the appearance of detritus having a unique lithologic signature that is recognizable in sediments of class-2 basins. Geochemical, isotopic, or geochronometric studies of detritus may provide recognizable signatures in cases where petrologic data are ambiguous. In both cases, a geologically significant lag time may pass before lithostratigraphic evidence for terrane amalgamation or accretion is perceivable.

Paleomagnetic Evidence

Allochthonous terranes are recognized where paleomagnetic data for coeval rocks on both sides of a terrane boundary yield discordant paleomagnetic pole positions reflecting independent tectonic translation, rotation, or a combination of both (Irving 1979). Reliable paleomagnetic results depend upon oriented samples that contain the stable component of a remnant magnetization, that are dated with accuracy, and are collected from outcrops in which paleohorizontal planes are determined with certainty. Measuring paleohorizontal attitudes at the time magnetization was acquired is more ambiguous in plutons than in volcanic or sedimentary strata. Thus, allochthonous sedimentary basins of class 1 are a reliable source of paleomagnetic data used in terrane identification. Inclination data are used to constrain paleolatitudes, whereas differences in declination are less useful in recognizing allochthonous terranes.

Other Geophysical Data

Modern oceanic plateaus display a spectrum of basement types (Ben-Avraham *et al.* 1981). Basement thickness may be preserved upon incorporation into an orogenic belt, if the plateau acts as a rigid body. Alternatively the basement may be thickened during compressional deformation or may be thinned as the lower portion is subducted while the upper part of the plateau is accreted to a continent. The crustal density or thickness of a terrane may be inferred from gravity, aerial magnetic, or deep-seismic reflection surveys. Large, abrupt changes in crustal densities or thicknesses may mark terrane boundaries within an orogenic belt and should be particularly apparent in a collage of terranes where a terrane with crust of continental thickness lies outboard of a terrane(s) with crust of typical oceanic density and thickness. Steep crustal density or thickness gradients can be used to identify terranes and map their boundaries (e.g., Mortimer 1985). These techniques offer the potential to map subsurface terrane boundaries despite the presence of post-accretion overlap strata.

Mobility of Allochthonous Terranes

Allochthonous terranes are distinguished from simple fault-bounded crustal blocks by their magnitude of displacement. No lower or upper limit exists on the magnitude of offset required for a terrane to be considered allochthonous; the criteria of noncorrelatable lithostratigraphic and biostratigraphic relations described earlier must be met, and to do so commonly involves hundreds or thousands of kilometres of plate motion. The site of origin of most terranes remains obscure. Some terranes now identified in deformed continental margins, such as Quesnellia (Fig. 15.1), occur within the same lithospheric plate in which they are thought to have originated (Howell *et al.* 1985), whereas others, such as the Cache Creek Terrane (Fig. 15.1) are now part of another lithospheric plate (Monger 1984). Both of these terranes clearly were oceanic plateaus. Baja California serves as an example of an incipient oceanic plateau with sedimentary basins and underlying basement that initially were part of the North American Plate but, due to the Miocene development of a spreading ridge in the Gulf of California, are now incorporated into the Pacific Plate. With continued spreading, Baja California would undergo thermal subsidence to become a submarine oceanic plateau. With future development of a subduction complex at that plate boundary, Baja California could eventually be reincorporated into the North American Plate.

Offset along the bounding faults of allochthonous terranes potentially can be quantified using the same techniques as for minor faults (e.g., Kleinspehn 1985) in relying on displaced: 1) marker beds or time surfaces; 2) sequences of strata; 3) sedimentary, volcanic or metamorphic facies; 4) igneous intrusions; or 5) pre-existing structures such as faults or fold axes. These techniques are particularly useful for

defining post-accretionary faulting and disruption of terranes within a continental margin. However, in most cases, pre-accretion terrane mobility involves multiple changes in sense of motion over distances that are so large that terranes are now isolated from their initial neighbors and these techniques are not useful.

The fill of allochthonous basins should display predictable characteristics as a function of meridional or latitudinal displacement, and these characteristics may provide useful quantitative estimates of terrane displacement. Criteria for estimating the magnitude of displacement include:

1. Paleomagnetic evidence (both marine and nonmarine deposits)
2. Latitudinal or meridional faunal and floral geographic provinces (both marine and nonmarine deposits)
3. Meridional or latitudinal zones of sedimentation associated with upwelling (marine deposits only)
4. Latitudinally zoned lithofacies (both marine and nonmarine deposits)
5. Changes in subsidence rate and water depth associated with motion of the oceanic plateau away from a spreading ridge (marine deposits)

Paleomagnetic Measure of Displacement

Paleomagnetic studies permit quantification of large-scale latitudinal displacements as well as terrane rotation. Inclination of paleomagnetic poles, when corrected for tectonic tilting, provides a measure of displacement relative to the paleomagnetic field for coeval rocks of the autochthonous continent. That such data suggest displacements of hundreds to thousands of kilometres (Irving 1979) provided the final impetus in the development of the concept of allochthonous terranes. Ambiguous interpretations result for portions of the geologic record for which paleomagnetic inclinations cannot be uniquely correlated to a specific polarity reversal of the earth's magnetic field. In such cases, only the magnitude and not the hemisphere (north or south) of the resultant paleolatitude can be determined (Irving *et al.* 1985).

Determination of longitudinal displacement is derived from detailed analysis of differences in apparent polar-wander paths between a terrane and the continent to which it is accreted (Gordon *et al.* 1984; Livermore *et al.* 1986). The paucity of estimates of longitudinal displacement for terranes is explained by the absence of such polar wander data, which is in turn due to the structural complexity of many terranes.

Paleobiologic Measures of Displacement

Faunal and floral assemblages reflect syn-depositional climatic conditions and therefore indicate approximate paleolatitude. Thus paleobiologic data are useful not only in identifying allochthonous sedimentary basins, but also in determining their approximate paleolatitude. The paleolatitude of a fossil assemblage is compared to the paleolatitude predicted by plate-tectonic reconstructions to estimate the magnitude of displacement of the allochthonous basin relative to the autochthonous continental margin in which the basin now occurs (Monger 1984). Biofacies commonly display a north-south climatic zonation (Tipper 1981) or meridional biogeographic provinciality (Monger and Ross 1971; Miller and Wright 1987), which when disrupted can serve as a measure of displacement. Such measures are complicated because tropical shallow-marine faunal provinces are approximately twice as extensive areally as temperate or boreal provinces (Schopf 1980) and the dimensions of latitudinal zones vary from one organism to another and with time (e.g., Valentine 1973; McKenzie *et al.* 1980). Thus, the amount of displacement necessary for biogeographic differences to be perceivable varies. Climatically controlled provinciality is most marked in nonmarine organisms, shallow-marine organisms, or surface-dwelling pelagic organisms. Shallow-water fossils are a common component of strata on modern oceanic plateaus (e.g., McKenzie *et al.* 1980), and are potentially useful in measuring large displacements of ancient inverse basins (Schlanger 1981).

Upwelling Zones as Measures of Displacement

Upwelling in oceans is the product of prevailing winds or the divergence of surface waters (Kennett 1982). Wind-driven upwelling zones are concentrated on the eastern margins of modern ocean basins because of the Coriolis force. Given the persistent north-south orientation of the continental margin along the eastern margin of Panthalassa, it is reasonable to assume that zones of upwelling occurred there throughout much of Mesozoic and

Cenozoic time (Parrish and Curtis 1982). Oceanic plateaus whose displacement history carried them into such zones of upwelling may record sedimentation influenced by upwelling in the form of phosphoritic, organic-rich, or glauconitic strata if water depths were appropriate.

In the Pacific Ocean, latitudinal zones of upwelling associated with diverging surface waters at the equator (1–2° N or S) produce bands of high phytoplankton productivity that deepen the calcite compensation depth (CCD), resulting in zones of carbonate-rich pelagic sediments with high accumulation rates (Tracey *et al.* 1971). A few degrees north or south of the equator (>5°) the CCD shallows and the carbonate content of the surface sediments gradually decreases until siliceous sediments dominate. Farther away from the equator (>30°, Schopf 1980) red clay constitutes the surface sediments of the deep Pacific floor (Tracey *et al.* 1971). Because the upwelling is related to the Coriolis force near north-south oriented ocean margins, this latitudinal lithologic zonation may well have been present since Late Paleozoic time, although its width and axis have evidently shifted through time (Winterer *et al.* 1973; van Andel 1974). If similar zones existed in ancient ocean basins, as is particularly likely in Panthalassa, vertical stacking of these deep-marine lithofacies could be used to trace the meridional motion of allochthonous deep-sea basins across these latitudinal bands.

Latitudinally Zoned Lithofacies as Measures of Displacement

For both subaerial and submarine oceanic plateaus, siliciclastic sediments within their inverse basins may reflect displacement across latitudinal zones of climatically controlled lithofacies. The zones may express climatic control on weathering in the source area, the composition of meteoric water, or on depositional processes. For example, the Permo-Triassic northward migration of the Indian subcontinent is recorded by increasing maturation of alluvial sandy detritus as India passed from an arid subpolar region to a humid subtropical setting (Suttner and Dutta 1986). Analogous compositional trends may be anticipated in far-travelled oceanic plateaus.

Not only may framework composition in sands reflect climatic change but early silicate cements may also serve as indicators of paleoclimate. In the same Permo-Triassic sandstones from migrating

India, latitudinal zones of aridity were marked by ground water with high ionic concentration producing early cements of smectite, chlorite, and rare laumontite (Dutta and Suttner 1986). In contrast, kaolinite and quartz formed authigenically in zones of high precipitation. Again, displaced oceanic plateaus could display similar trends.

Climatically controlled nonmarine lithofacies such as evaporites, aeolian deposits, or glaciogenic deposits can be used to estimate the magnitude of meridional terrane displacement. To do so, facies observed in allochthonous basins would be compared with facies predicted for the paleolatitude of adjacent autochthonous basins. Ice-rafted debris indicative of glaciation (Poore 1981) or wind-blown dust suggesting arid conditions (Schopf 1980) are not restricted to nonmarine depositional environments, but are also deposited on the elevated but submarine surfaces of oceanic plateaus and thus may serve as indicators of paleolatitude. In aeolian dust, the concentration of illite increases with increasing latitude, whereas kaolinite is most abundant in equatorial sediments and chlorite has a constant concentration with a slight increase at the highest latitudes (Schopf 1980).

In determining paleolatitude using lithofacies, one must exercise caution in that orographic controls on climate may be misinterpreted as latitudinal controls, particularly for facies of humid versus arid settings.

Shelf sediments show a general trend from a biogenic origin at the equator, to authigenic minerals (if present) in tropical latitudes, to river-derived at mid-latitudes to glacial (if present) at the highest latitudes (Schopf 1980). Deep-ocean deposits are also zoned, with the sequence carbonate sediments-aeolian grains-siliceous oozes-glaciogenic debris observed with increasing latitude. Deep-marine deposition on modern oceanic plateaus that passed through the equatorial zone is marked by accelerated sediment accumulation rates that decrease rapidly away from the equator (Winterer *et al.* 1973; van Andel 1974).

Trends in Subsidence Rates and Paleo-Water Depth as a Measure of Displacement

As oceanic lithosphere and any associated oceanic plateaus move away from a spreading ridge they subside according to the exponential thermal subsidence curve for cooling oceanic lithosphere (Parsons and Sclater 1977). If sedimentation rates are less than subsidence rates, water depths over an oceanic plateau increase with time, but at a decreasing rate. This progressive change may be recorded by a transition from shallow-water or terrestrial sediments to deeper water facies in an inverse basin (Schlanger 1981). If the subsidence history of an allochthonous inverse basin can be determined, it may be possible to estimate the distance that the basin was displaced relative to a spreading center.

All of these concepts for determining paleolatitude and displacement history are complicated by the interaction between atmospheric and oceanic circulation, tectonism, and distribution of the continents. These general concepts are presented here in an attempt to stimulate their application to the study of allochthonous terranes, although the information derived from them may be only semi-quantitative.

Conclusions

The concept that exotic "microcontinents" may serve as a source of sediment is not new, nor is the hypothesis that such microcontinents change location and dimensions through time. Since the presence of "Appalachia" and "Cascadia" was first proposed (Chamberlin 1882; Schuchert 1910), such concepts have played a principal role in our attempts to understand the evolution of continental margins. Some "lost" microcontinents of the last century are in fact not lost, but appear as, for example, "Pacifica," on recent maps of reconstructed allochthonous terranes (Nur and Ben-Avraham 1983). Although allochthonous terranes are not identified per se in all continental margins, their presence is inferred in many by the usage of the term "microcontinent." For example, models for the evolution of Tethys include mobile microcontinents that now occur as fault-bounded, coherent stratigraphic packages along the Tethyan belt (Dewey *et al.* 1973; Dercourt *et al.* 1986). Because the term "microcontinent" restricts the basement of an oceanic plateau to that of continental lithosphere, it omits other types of basement. Thus the term "allochthonous terrane" is a preferable, more widely applicable usage.

Early evidence for "lost" microcontinents is from paleocurrent data gathered as part of basin analyses (Schuchert 1910; Potter and Pettijohn 1977). Modern analyses of basins associated with allochthonous terranes continue to yield data critical to deciphering the complex magnitude and timing of terrane motion, the timing and location of terrane amalgamation, the timing of terrane accretion, and any post-accretionary fragmentation of allochthonous terranes. Provenance, paleobiologic, paleoclimatic, or paleomagnetic data from sediments deposited on an amalgamated, composite terrane may provide, in the absence of volcanic deposits, the only record of amalgamation far removed from a continental margin as well as the only record of post-amalgamation motion of the composite terrane.

Although the concept of allochthonous terranes was conceived through classical stratigraphic studies, such studies alone are inadequate to define rigorously allochthonous terranes or their history of mobility. Analysis of allochthonous terranes clearly poses challenging, multi-disciplinary problems to the basin analyst.

Acknowledgments. David Howell, Harold Williams, Chris Paola, and Eric Mohring are thanked for their insightful comments on earlier versions of this manuscript.

References

BALLY, A.W. and SNELSON, S. (1980) Realms of subsidence. In: Miall, A.D. (ed) Facts and Principles of World Petroleum Occurrence. Canadian Society Petroleum Geologists Memoir 6, pp. 9–75.

BATES, R.L. and JACKSON, J.A. (eds) (1980) Glossary of Geology. Falls Church, Virginia: American Geological Institute, 749 p.

BEN-AVRAHAM, Z., NUR, A., JONES, D., and COX, A. (1981) Continental accretion: From oceanic plateaus to allochthonous terranes. Science 213:47–54.

BERG, H.C., JONES, D.L., and RICHTER, D.H. (1972) Gravina-Nutzotin belt—Tectonic significance of an upper Mesozoic sedimentary and volcanic sequence in southeastern Alaska. United States Geological Survey Professional Paper 800D, pp. 1–24.

CHAMBERLIN, T.C. (1882) Geology of Wisconsin, v. 4. Madison, Wisconsin: Wisconsin Geological Survey, 779 p.

CHURKIN, M. and TREXLER, J.H. (1981) Continental plates and accreted oceanic terranes in the Arctic. In: Nairn, A.E.M., Churkin, M., and Stehli, F.G. (eds) The Ocean Basins and Margins, The Arctic Ocean. New York: Plenum Press, pp. 1–20.

CONEY, P.J., JONES, D.L., and MONGER, J.W.H. (1980) Cordilleran suspect terranes. Nature 288:329–333.

CONYBEARE, C.E.B. (1979) Lithostratigraphic Analysis of Sedimentary Basins. New York: Academic Press, 555 p.

COVEY, M. (1986) The evolution of foreland basins to steady state: Evidence from the western Taiwan foreland basin. In: Allen, P.A. and Homewood, P. (eds) Foreland Basins. International Association Sedimentologists Special Publication 8, pp. 77–90.

DANA, J.D. (1873) On some results of the earth's contraction from cooling, including a discussion of the origin of mountains and the nature of the earth's interior. American Journal Science, series 3, 6:161–171.

DERCOURT, J., ZONENSHAIN, L.P., RICOU, L.E., KAZMIN, V.G., LE PICHON, X., KNIPPER, A.L., GRANDJACQUET, C., SBORTSHIKOV, I.M., GEYSSANT, J., LEPVRIER, C., PECHERSKY, D.H., BOULIN, J., SIBUET, J.C., SAVOSTIN, L.A., SOROKHTIN, O., WESTPHAL, M., BAZHENOV, M.L., LAUER, J.P., and BIJU–DUVAL, B. (1986) Geologic evolution of the Tethys belt from the Atlantic to the Pamirs since the Lias. Tectonophysics 123:241–315.

DEWEY, J.F. and BIRD, J.M. (1970) Mountain belts and the new global tectonics. Journal Geophysical Research 75:2625–2647.

DEWEY, J.F., PITMAN, W.C., RYAN, W.B.F., and BONNIN, J. (1973) Plate tectonics and the evolution of the Alpine System. Geological Society America Bulletin 84:3137–3180.

DICKINSON, W.R. (1974) Plate tectonics and sedimentation. In: Dickinson, W.R. (ed) Tectonics and Sedimentation. Society Economic Paleontologists Mineralogists Special Publication 22, pp. 1–27.

DIETZ, R.S. (1963) Collapsing continental rises: An actualistic concept of geosynclines and mountain building. Journal Geology 71:314–333.

DUTTA, P.K. and SUTTNER, L.J. (1986) Alluvial sandstone composition and paleoclimate, II. Authigenic mineralogy. Journal Sedimentary Petrology 56:346–358.

EISBACHER, G.H., CARRIGY, M.A., and CAMPBELL, R.B. (1974) Paleodrainage pattern and late-orogenic basins of the Canadian Cordillera. In: Dickinson, W.R. (ed) Tectonics and Sedimentation. Society Economic Paleontologists Mineralogists Special Publication 22, pp. 143–166.

GORDON, R.G., COX, A., and O'HARE, S. (1984) Paleomagnetic Euler poles and the apparent polar wander and absolute motion of North America since the Carboniferous. Tectonics 3:499–537.

HALL, J. (1859) Description and figures of the organic remains of the lower Helderberg Group and the Oriskany Sandstone. Natural History of New York, Part 6, Palaeontology, volume 3. Albany, NY: Geological Survey New York, 532 p.

HESS, H.H. (1962) History of ocean basins. In: Engel, A.E.J., James, H.L., and Leonard, B.F. (eds) Petrologic

Studies: A Volume to Honor A.F. Buddington. Geolooical Society America, pp. 599–620.

HOWELL, D.G. (1980) Mesozoic accretion of exotic terranes along the New Zealand segment of Gondwanaland. Geology 8:487–491.

HOWELL, D.G. (ed) (1985) Tectonostratigraphic Terranes of the Circum-Pacific Region. Houston: Circum-Pacific Council for Energy and Mineral Resources, Earth Sciences Series Number 1, 585 p.

HOWELL, D.G., JONES, D.L., and SCHERMER, E.R. (1985) Tectonostratigraphic terranes of the Circum-Pacific region. In: Howell, D.G. (ed) Tectonostratigraphic Terranes of the Circum-Pacific Region. Houston: Circum-Pacific Council for Energy and Mineral Resources, Earth Sciences Series Number 1, pp. 3–30.

IRVING, E. (1979) Paleopoles and paleolatitudes of North America and speculations about displaced terranes. Canadian Journal Earth Sciences 16:669–694.

IRVING, E., WOODSWORTH, G.J., WYNNE, P.J., and MORRISON, A. (1985) Paleomagnetic evidence for displacement from the south of the Coast Plutonic Complex, British Columbia. Canadian Journal Earth Sciences 22:584–598.

JONES, D.L., HOWELL, D.G., CONEY, P.T., and MONGER, J.W.H. (1983) Recognition, character and analysis of tectonostratigraphic terranes in western North America. In: Hashimoto, M. and Uyeda, S. (eds) Accretion Tectonics in the Circum-Pacific Regions. Boston: D. Reidel Publishing Company, pp. 21–35.

JONES, D.L., SILBERLING, N.J., and HILLHOUSE, J. (1977) Wrangellia—a displaced terrane in northwestern North America. Canadian Journal Earth Sciences 14:2565–2577.

KENNETT, J. (1982) Marine Geology. Englewood Cliffs, NJ: Prentice-Hall, Inc., 813 p.

KING, P.B. (1966) North American Cordillera. In: Tectonic History and Mineral Deposits of the Western Cordillera, Canadian Institute Mining Metallurgy, Special Volume 8, pp. 1–25.

KINGSTON, D.R., DISHROON, C.P., and WILLIAMS, P.A. (1983) Global basin classification system. American Association Petroleum Geologists Bulletin 67: 2175–2193.

KLEINSPEHN, K.L. (1985) Cretaceous sedimentation and tectonics, Tyaughton-Methow Basin, southwestern British Columbia. Canadian Journal Earth Sciences 22:154–174.

KLEMME, H.D. (1975) Giant oil fields related to their geologic setting—a possible guide to exploration. Bulletin Canadian Petroleum Geology 23:30–66.

LIVERMORE, R.A., SMITH, A.G., and VINE, F.J. (1986) Late Palaeozoic to early Mesozoic evolution of Pangaea. Nature 322:162–165.

McKENZIE, J., BERENOULLI, D., and SCHLANGER, S.O. (1980) Shallow-water carbonate sediments from the Emperor Seamounts: Their diagenesis and paleo-

geographic significance. In: Jackson, E.D., Koisumi, I. et al. Initial Reports of the Deep Sea Drilling Project 55. Washington, DC: United States Government Printing Office, pp. 415–455.

MILLER, M.M. and WRIGHT, J.E. (1987) Paleogeographic implications of Permian Tethyan corals from the Klamath Mountains, California. Geology 15:266–269.

MITCHELL, A.H.G. and READING, H.G. (1969) Continental margins, geosynclines and ocean floor spreading. Journal Geology 77:629–646.

MONGER, J.W.H. (1984) Cordilleran tectonics: A Canadian perspective. Bulletin Société Géologique France 26:255–278.

MONGER, J.W.H. and ROSS, C.A. (1971) Distribution of fusulinaceans in the western Canadian Cordillera. Canadian Journal Earth Sciences 8:259–278.

MORTIMER, N. (1985) Structural and metamorphic aspects of Middle Jurassic terrane juxtaposition, northeastern Klamath Mountains, California. In: Howell, D.G. (ed) Tectonostratigraphic Terranes of the Circum-Pacific Region. Houston: Circum-Pacific Council for Energy and Mineral Resources, Earth Science Series Number 1, pp. 201–214.

NUR, A. and BEN-AVRAHAM, Z. (1982a) Displaced terranes and mountain building. In: Hsü, K.J. (ed) Mountain Building Processes. New York: Academic Press, pp. 73–84.

NUR, A. and BEN-AVRAHAM, Z. (1982b) Oceanic plateaus, the fragmentation of continents, and mountain building. Journal Geophysical Research 87:3644–3661.

NUR, A. and BEN-AVRAHAM, Z. (1983) Break-up and accretion tectonics. In: Hashimoto, M. and Uyeda, S. (eds) Accretion Tectonics in the Circum-Pacific Regions. Boston: D. Reidel Publishing, pp. 3–18.

PARRISH, J.T. and CURTIS, F.L. (1982) Atmospheric circulation, upwelling, and organic-rich rocks in the Mesozoic and Cenozoic Eras. Palaeogeography, Palaeoclimatology, Palaeoecology 40:31–66.

PARSONS, B. and SCLATER, J.C. (1977) An analysis of the variation of ocean floor bathymetry and heat flow with age. Journal Geophysical Research 82:803–827.

POORE, R.Z. (1981) Temporal and spatial distribution of ice-rafted mineral grains in Pliocene sediments of the North Atlantic: Implications for Late Cenozoic climatic history. In: Warme, J.E., Douglas, R.G., and Winterer, E.L. (eds) The Deep Sea Drilling Project: A Decade of Progress. Society Economic Paleontologists Mineralogists Special Publication 32, pp. 505–516.

POTTER, P.E. and PETTIJOHN, F.J. (1977) Paleocurrents and Basin Analysis. New York: Springer-Verlag, 425 p.

SCHERMER, E.R., HOWELL, D.G., and JONES, D.L. (1984) The origin of allochthonous terranes: Perspectives on the growth and shaping of continents. Annual Review Earth Planetary Sciences 12:107–131.

SCHLANGER, S.O. (1981) Shallow-water limestones in oceanic basins as tectonic and paleoceanographic indicators. In: Warme, J.E., Douglas, R.G., and Winterer, E.L. (eds) The Deep Sea Drilling Project: A Decade of Progress. Society Economic Paleontologists Mineralogists Special Publication 32, pp. 209–226.

SCHOPF, T.J.M. (1980) Paleoceanography. Cambridge, MA: Harvard University Press, 341 p.

SCHUCHERT, C. (1910) Paleogeography of North America. Geological Society America Bulletin 20:427–606.

SENGÖR, A.M.C. (1984) The Cimmeride Orogenic System and the Tectonics of Eurasia. Geological Society America Special Paper 195, 82 p.

SILBERLING, N.J. and JONES, D.L. (1983) Paleontological evidence for northward displacement of Mesozoic rocks in accreted terranes of the western Cordillera (abstract). Victoria, BC: Program with Abstracts, Annual Meeting, Geological Association Canada, p. A62.

SILBERLING, N.J. and JONES, D.L. (eds) (1984) Lithotectonic terrane map of the North American Cordillera. United States Geological Survey Open-File Report 84–523.

SUTTNER, L.J. and DUTTA, P.K. (1986) Alluvial sandstone composition and paleoclimate, I. Framework mineralogy. Journal Sedimentary Petrology 56:329–345.

TIPPER, H.W. (1981) Offset of an upper Pliensbachian geographic zonation in the North American Cordillera by transcurrent movement. Canadian Journal Earth Sciences 18:1788–1792.

TRACEY, J.I. et al. (1971) Initial Reports of the Deep Sea Drilling Project 8. Washington, DC: United States Government Printing Office, 1037 p.

VALENTINE, J.W. (1973) Evolutionary Paleoecology of the Marine Biosphere. Englewood Cliffs, NJ: Prentice-Hall, Inc., 511 p.

VAN ANDEL, T.H. (1974) Cenozoic migration of the Pacific Plate, northward shift of the axis of deposition, and paleobathymetry of the central Equatorial Pacific. Geology 2:507–510.

VEDDER, J.G., HOWELL, D.G., and McLEAN, H. (1983) Stratigraphy, sedimentation, and tectonic accretion of exotic terranes, southern Coast Ranges California. In: Watkins, J.S. and Drake, C.L. (eds) Continental Margin Geology. American Association Petroleum Geologists Memoir 34, pp. 471–496.

WILLIAMS, H. (1984) Miogeoclines and suspect terranes of the Caledonian-Appalachian Orogen: Tectonic patterns in the North Atlantic region. Canadian Journal Earth Sciences 21:363–380.

WILLIAMS, H. and HATCHER, R.D. (1982) Suspect terranes and accretionary history of the Appalachian orogen. Geology 10:530–536.

WINTERER, E.L., EWING, J.I. et al. (1973) Initial Reports of the Deep Sea Drilling Project 17. Washington, DC: United States Government Printing Office, 930 p.

ZEN, E. (1983) Exotic terranes in the New England Appalachians — limits, candidates, and ages: A speculative essay. In: Hatcher, R.D., Williams, H., and Zietz, I. (eds) Contributions to the Tectonics and Geophysics of Mountain Chains. Geological Society America Memoir 158, pp. 55–81.

16
Dating Thrust-Fault Activity by Use of Foreland-Basin Strata

Teresa E. Jordan, Peter B. Flemings, and James A. Beer

Abstract

Foreland basins form at the sides of thrusted mountain belts because of flexure of the lithosphere under a load. In retroarc and in some peripheral foreland basins, the entire flexure results from the load of thrust sheets. Thus the subsidence history of the basin is an indirect measure of the history of thrusting. Although the texture and migrations of facies of the basin fill certainly respond to thrust motions, they also depend on the other independent controls (climate, the lithology of the thrust belt, and base level). Therefore it is unreliable to use textural history or facies migrations as indicators of thrust history.

Strata exposed at Huaco, Argentina, accumulated during Miocene and Pliocene time in the Bermejo foreland basin in response to deformation of the Precordillera thrust belt. Because of the availability of detailed chronological data and facies descriptions, we utilize the Huaco sequence as a test case of various approaches to estimating thrust history.

Tectonic subsidence histories, derived from well constrained stratal accumulation histories, are sensitive indicators of the thickening in the thrust belt, and are independent of local sedimentary processes. However, it is difficult to quantify the subsidence caused by the thrust loads. The chronology of an entire thrust belt can be better determined if the subsidence history is known at many locations in the basin.

The first appearance of diagnostic clasts of sand and gravel can demonstrate that thrusting had begun before the diagnostic grains appear, if the source units have known structural positions. Sands provide a more complete uplift history than do gravels because they are more widely distributed.

In the absence of absolute chronological control, variations in subsidence rate are sometimes estimated by comparing varying rock textures and geometries to simulations of deposition under conditions of variable subsidence rate. The appearance of conglomerates in a foreland basin sequence is probably the most common facies that is utilized as a tectonic indicator. Whereas at Huaco the textural and facies patterns help to refine the interpretations of thrust history that are based on subsidence rates and sedimentary petrology, the textural and facies information alone do not provide a reliable record of deformation.

Introduction

Foreland basins and thrust-faulted orogenic belts are genetically linked and together constitute the geologic record of an orogeny (Fig. 16.1). Because the basins are preferentially preserved, the foreland-basin strata often form the basis for analyzing the deformation history of the mountain belt. In particular, the times of motion on thrust faults are commonly deciphered from the sedimentary record, in combination with more direct dating techniques such as cross-cutting structural relations. Classic discussions of the methods and difficulties of dating thrust-fault activity were presented by Armstrong and Oriel (1965) and Oriel and Armstrong (1966). The purpose of this contribution is to evaluate the relative accuracy of several methods of dating thrust-fault activity.

The stratigraphic record of a foreland basin reflects a two-part hierarchy of independent controls. The first-order control is the regional subsidence pattern imposed by flexure of the lithospheric plate on which the basin is located. The load that causes the flexure can be of two types: 1) the thickening pile of thrust sheets in the thrust belt; and 2) subcrustal buoyancy contrasts or mantle dynamic loads. The rheology of the lithosphere and the relative contributions of the two types of load determine

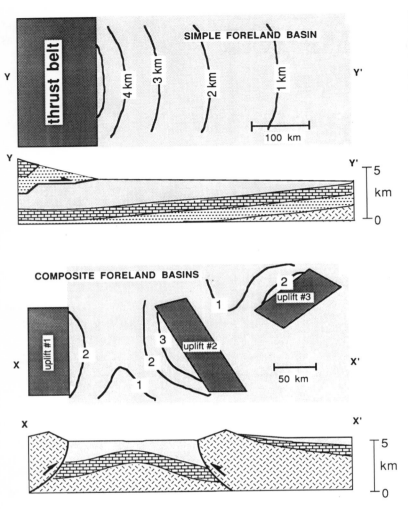

Fig. 16.1. Schematic maps and cross sections of two types of foreland basins and the mountain belts with which they are spatially and genetically associated. Contours show thickness of strata that are contemporaneous with shortening in the mountain belt. Note different scales of the two maps.

the waveform of the flexed basin, whereas the history of the loads controls the timing and magnitude of the subsidence. Second-order controls influence the character of the strata but cannot independently generate the basin. The three major second-order controls are the lithology of the thrust belt, climate, and eustatic sea-level controls on base level. For the purpose of establishing the timing of thrust faulting, the geologist's goal is to evaluate the characteristics of the strata that most closely relate to the first-order control.

We will consider *only* foreland basins in which the load is generated primarily by thrust faulting and where subcrustal loads are of little importance. This is the sub-category of foreland basins for which studies of the modern gravitational field or of the ancient foreland basin system have shown that the

thrust-related load is in flexurally controlled isostatic balance with its foreland. Documented examples show that these foreland basins are commonly retroarc basins (Dickinson 1974), in which the thrust system is antithetic to the subducting plate margin, such as the North American Cordillera and the Andes (e.g., Beaumont 1981; Jordan 1981; Lyon-Caen *et al.* 1985). At least one example of a peripheral foreland basin (i.e., synthetic with the subducting plate margin; Dickinson 1974) that is in balance with the thrust load is known: Karner and Watts (1983) showed that the foreland basin of the Himalaya is in gravitational balance with the topography of the mountains. Much more commonly, however, the load generated by peripheral thrust belts is not adequate to account for all of the flexure in their foreland basins. In that case, modelling shows that a combination of the thrust sheet load and an "extra" load account for the geometry of the basin. Examples include the Alpine Molasse basin, the Adriatic Sea/Apennine system, the Zagros Mountains and Persian Gulf Basin, and the Appalachian system (Karner and Watts 1983; Royden and Karner 1984; Snyder and Barazangi 1986; Stockmal *et al.* 1986). The "extra" load in these systems seems to be directly related to the subduction of a lower plate consisting of thin continental margin or oceanic lithosphere, or to dynamic effects of mantle convection (Royden and Karner 1984; Stockmal *et al.* 1986). The same principles of dating thrust faults apply to such peripheral foreland basins, but the interpretations of the basin are more complicated because of the possibility that the additional, independent load might also change through time. Thus we restrict our examples to the Andes, the North America Cordillera, and the Himalaya.

We evaluate the state of the art of dating thrust-fault motions based on the sedimentary record of the foreland basin, proceeding from methods that best estimate thrust history to less useful methods. To examine the foreland basin stratigraphy, the competing importance of lithology, climate, and sea-level changes (the second-order controls) must first be considered. Although the discussion of the accuracy of general methods of dating thrust faults applies to all basins, for brevity we use a single example, a Neogene foreland basin in the Andes Mountain system, for detailed illustration.

Clarification of the impact of thrust-fault motion on the stratigraphic record is essential to both stratigraphic and structural studies. The first goal is simply to dispense with circular reasoning. All too often, a thrust-belt geologist examines the published history of foreland-basin strata and suggests that a series of thrust faults moved at a given time because conglomerates interfinger in the basin at that time. A foreland-basin geologist, in search of an understanding of the controls on progradational patterns, then consults a paper describing the thrust belt and finds that a thrust fault moved at the time of a progradation, and suggests that the facies changes reflect deformation pulses. No new insight into basin controls can be derived from such circles. However, if the independent controls on a given section can be quantified, then the rheological properties of the lithosphere *and* the nature of the response of surficial processes to the tectonics can be evaluated. Advances in regional geologic studies of thrust belts also depend upon correctly dating thrust faults. For

instance, in the majority of cases, rates of motion on individual thrust faults are not known, nor is it known whether the net rate of shortening in a thrust belt is relatively continuous or discontinuous. Both questions could be addressed with better correlations of thrust activity to the strata in the basins.

Chronological Resolution

Whether the accuracy of the traditional means of dating thrust motions is adequate, or the weaknesses of those traditional methods (discussed here) are important, depends upon the goals of a given project. In pioneering studies of new regions, the traditional methods are adequate to suggest the beginning of deformation on time scales of 10^7 years. If the intent of the project is to establish a detailed thrust chronology, determine lithospheric rheology, or document depositional response to tectonic activity, that resolution is not adequate.

There are two basic approaches to improving the chronologic determination of thrust motion: direct and indirect dating. The only direct means of dating motion on a fault are those that involve material in the fault zone or rocks that have physical cross-cutting relations to the fault. Indirect methods use strata deposited at a distance from the fault.

The least ambiguous means for determining the time of motion on a thrust fault is by bracketing it between the youngest rock unit which is cut by the fault and the oldest unit that overlies or cuts the trace of the fault. Examples of the use of this technique in the Idaho-Wyoming thrust belt show that it is highly successful in the cratonic edge of the thrust system, where many of the near-surface fault relations are preserved (e.g., Royse *et al.* 1975; Dorr and Gingerich 1980, Wiltschko and Dorr 1983). In some cases the time of motion can be bracketed to within about 1 to 2 million years (e.g., La Barge Fault; Dorr and Gingerich 1980). In other cases, it cannot resolve an age to better than a 50 million year time span (e.g., Crawford-Meade thrust system; Wiltschko and Dorr 1983). The method is limited by the tendency for erosion to strip away critical overlapping relations, particularly in the interior of thrust belts where continued deformation tends to raise the older faults to elevations where they are increasingly accessible to erosion (Burbank and Reynolds 1988; Steidtmann and Schmitt 1988: this volume). Furthermore, the critical overlapping relations most commonly involve nonmarine strata and those commonly are difficult to date due to lack of preserved fauna.

An ambiguity also exists in the interpretation of cross-cutting relations because faults may have prolonged histories of continuous or discontinuous motion, as illustrated by the Absaroka Thrust of the Idaho-Wyoming thrust belt (Royse *et al.* 1975; Wiltschko and Dorr 1983). If a given major decollement system moves recurrently or over a long time span, but varying minor splays accommodate that motion for shorter times, then cross-cutting relations for a single splay do not fully describe motion on the major thrust system.

So while cross-cutting relations offer a straightforward means of dating a fault, there are two inherent difficulties. First, they are spatially limited to regions with low preservation potential, and second, the resolution of the timing is often poor even where the necessary rock relations are preserved.

The indirect dating approaches are discussed at length below. Beyond the question of the relative validity of the indirect approaches, their utility depends upon the resolution of the methods used to date the strata. Experience with dating thrust faults using cross-cutting relations implies that shortening across major faults occurs during spans of less than 10 million years (e.g., Wiltschko and Dorr 1983) and in some cases during less than 2 to 4 million years (Royse, F., personal communication 1986; G. Johnson *et al.* 1986; N.M.. Johnson *et al.* 1986; Burbank and Raynolds 1988: this volume). Therefore, the chronological resolution of the dating method must be *significantly* better than 10 million years if changes in the rate of subsidence caused by fault motion are to be discernible. In Phanerozoic marine foreland basins, this should not be difficult to attain. In nonmarine basins, abundant faunas with excellent biostratigraphic correlation to the absolute time scale will be adequate in some cases, but there are many examples of foreland-basin strata in which that information is not available (e.g., the Catskill delta of New York State, many of the fluvial strata of the Idaho-Wyoming thrust belt, and Neogene strata of the eastern Andes of Argentina). Elsewhere, volcanic ash beds prove adequate, if they are intercalated at short intervals in the section and can be dated with high-resolution absolute chronological techniques. A technique with excellent utility in Neogene nonmarine sections is magnetic-polarity stratigraphy (see Johnson *et al.* 1988; Burbank and Raynolds 1988: this volume); the isochrons that are established have resolutions on the order of 10^4 years (see discussion in Behrensmeyer and Tauxe 1982; Johnson *et al.* 1988: this volume). This technique has the advantage that the samples are of sedimentary rock that does not require exotic preservation of biological or volcanic materials. Magnetic-polarity stratigraphy is used in the Andean example that we present in this chapter.

Controls on Basin Stratigraphy

First-Order Controls: Tectonic Subsidence

Rheology

The mechanical link between the thrusted mountain system and the subsiding basin at its flank has been established at a general level. Modelling studies indicate that the lithosphere behaves as a rigid plate overlying a fluid mantle in response to loading over time scales greater than 10^5 years (e.g., Karner and Watts 1983). Crustal loads, such as those caused by thickening along thrust faults, are compensated regionally by flexure of the lithosphere (Fig. 16.2; Beaumont 1981; Jordan 1981).

The exact rheological model that most appropriately describes the response of the lithosphere to loading is controversial. Modern observation of the loaded lithosphere suggests that the displacement can be modelled as elastic deformation (e.g., Watts 1978; Caldwell and Turcotte 1979). The stratigraphic record of foreland basins provides the opportunity to examine the mechanical behavior of the lithosphere over longer durations. Examination of the rheology of the lithosphere based on foreland-basin strata has centered on possible viscous behavior of the lithosphere (Quinlan and Beaumont 1984), on lateral variations in lithospheric properties (Karner and Watts 1983; Stockmal *et al.* 1986), and on thermal history (Kominz and Bond 1986).

In spite of the differences between elastic, viscoelastic, and temperature-dependent rheological models, generalizations can be made about the response of the lithosphere to loading. In the short term, crustal thickening due to shortening in the mountain belt results in a deviation from isostatic equilibrium that is compensated by subsidence. However, because the lithosphere has lateral strength, the subsidence is not restricted to the thickened area (the case of Airy isostatic balance), but is instead distributed flexurally over the region adjacent to the mountain belt. The deflected

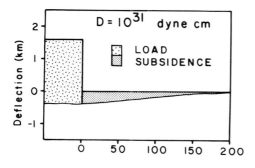

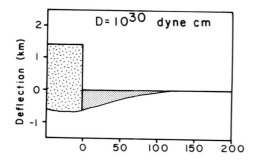

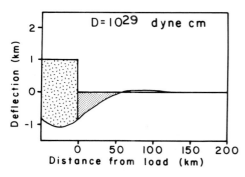

Fig. 16.2. Cross sections of the response of an infinite elastic plate to a load, illustrating the difference in amplitude (height) of the tectonic uplift in the mountains and the differences in wavelength (width) and amplitude of the tectonic subsidence in the basin that result from varying the elastic properties of the plate. The load, which is reasonable for a major thrust system, is 50 km wide and 2 km thick in each case. The elastic properties of the plate are expressed as its flexural rigidity (D). The displacements are appropriate to a crustal density of 2.7 gm/cm³ and mantle density of 3.3 gm/cm³.

lithosphere has a characteristic form (Fig. 16.2); the subsidence is largest proximal to the load, diminishes gradually as distance increases, and becomes a slight uplift (an outer bulge) at a distance from the load. The wavelength of this deformation and the maximum deflection for a given load depend only upon the rigidity of the lithosphere for a given structural model of the thrust plates. For elastic models the characteristic wavelength depends on the effective elastic thickness (stiffness, expressed as flexural rigidity) and varies widely in different basins. For example, estimated wavelengths determined for the United States Cretaceous Western Interior Basin and for the modern Himalaya and foreland basin span from 50 to 186 km, respectively (Jordan 1981; Karner and Watts 1983).

More problematic is the long-term behavior of the lithosphere due to loading. In the elastic model the mechanical properties of the plate do not change through time. In a viscoelastic model, the initial response of the lithosphere is similar to the elastic model but, through time, the rigidity of the lithosphere decreases. Because we examine here only the basin history that is contemporaneous with thrusting, the long-term behavior is not in question and an elastic model is adequate.

On the other hand, the short-term behavior of the lithosphere is critical to our topic. Theoretically, the isostatic compensation in response to loading, and the resultant flexure, is not instantaneous because of the high viscosity of the underlying asthenosphere. Studies of crustal rebound during deglaciation show that isostatic compensation occurs on time scales of 10^3 to 10^4 years (Turcotte and Schubert 1982). Thus, in comparison to the time scales of thrust activity (10^5 to 10^7 years) the lithosphere may be considered to deform instantaneously.

In summary, the response of the lithosphere to thrust-sheet loading is the creation of an asymmetric depression adjacent to the load. The wavelength and form of the resulting basin are dependent upon the stiffness of the lithosphere, which shows considerable variation in different sedimentary basins.

Crustal Thickening

Thrust loading is fundamental to triggering flexural subsidence in foreland basins. Thus it is important to understand the styles and rates at which the thrust loads form.

The thrust-load history, which we detect indirectly by examining the foreland basin strata, is an expres-

Table 16.1. Rates of shortening and thickening in thrusted mountain systems.*

Location	Duration (10^6yr)	Shortening (m/10^3yr)	Approximate thickening (m/10^3yr)	Reference
Central Precordillera, Argentina	8.5	6–7	0.2–0.3	Ortiz and Zambrano 1981 Allmendinger, R.A., personal communication 1985 N. Johnson et al. 1986
Subandean thrust belt, Argentina	3–5	15–30	0.5–0.8	Mingramm et al. 1979 Allmendinger et al. 1983 Sheffels et al. 1986
Idaho-Wyoming salient of North American Cordilleran thrust belt	80	1.7	0.05	Royse et al. 1975 Jordan 1981 Allmendinger, R.A., personal communication 1986
Southern Canadian Cordilleran thrust belt	100	2	0.05	Price and Mountjoy 1970 Price 1981 Price and Fermor 1985 Kominz and Bond 1986
(Individual thrust) Prospect Thrust	2	6.0	0.6–1.0	Royse et al. 1975 Royse, F., personal communication 1986

*Restricted to retroarc thrust systems.

sion of the thrust-thickening history. The thickening is directly dependent upon the shortening history because crustal volume is conserved during shortening. The *rate* of loading (thickening) is dependent on both the magnitude (volume) and duration of thrusting. The volume relates to the structural geometry of thrust belts, which is well known in many locations, and generalizations are possible (e.g., Dahlstrom 1970; Boyer and Elliott 1982). Essentially, two styles of thrust belts are common in parts of orogenic belts where the lithosphere is cold enough to generate flexural basins (Fig. 16.1). The most common are those with "thin-skinned" geometries (e.g., Appalachian Valley and Ridge or Canadian Cordilleran overthrust belt) where loading is distributed over a broad zone. Less common are the "basement uplifts" (e.g., Rocky Mountain Foreland style), where much thicker parts of the continental crust are involved and thus much thicker loads can be created. However, the areas subjected to loading by basement uplifts are smaller and more discontinuous than in thin-skinned thrust belts, resulting in basins that are geometrically distinct (Fig. 16.1; Chapin and Cather 1981; Hagen et al. 1985; Flemings et al. 1986).

The rates of shortening and thickening in some retroarc thin-skinned thrust belts are summarized in Table 16.1, with two examples from the Andes and

two from the North American Cordillera. To emphasize possible differences between rates appropriate to entire systems versus individual faults, Table 16.1 contains one example taken from the movement history of a single thrust in the North American Cordillera, for which cross-cutting relations provide tight constraints on the time of motion. For the North American thrust-system examples, the duration of thrusting is taken as the time span during which the foreland basin was actively subsiding, and all Mesozoic-Cenozoic shortening in the mountain system has been summed. In the Andean examples, the amount of shortening can be estimated only in the foothills thrust belts, so Table 16.1 lists timing information that is pertinent to those belts. Although there is a striking difference in the duration of shortening in the Neogene Andean thrust belts compared to the older North American examples, the upper four examples of Table 16.1 each involve the net deformation across a minimum of four thrust faults. The rates of shortening in some peripheral thrust belts are known to be much greater (e.g., Covey 1986).

Given the geometry of the thrust sheets, the rate of thickening can be estimated. The rates of thickening are .05 to .8 m/10^3 yr for the listed retroarc thrust systems and perhaps as great as 1 m/10^3 yr for the Prospect Fault alone (Table 16.1). These set the upper limit of rates of tectonic subsidence in proximal parts of their foreland basins. The actual rate of subsidence will be a fraction of the thickening rate where the fraction is determined by the flexural rigidity. A low flexural rigidity will result in a higher fraction of subsidence (lower fraction of uplift) than will a high flexural rigidity (Fig. 16.2), such that the rates of proximal basin subsidence should be about 20 to 40% of the thrust-thickening rates.

The crustal thickening by thrust faulting serves as the loading event that initiates and maintains subsidence in the foreland basin, and the rheology of the lithosphere determines the geometric form of the subsidence. Second-order factors control the mass redistribution from the mountain belt to the basin.

Second-Order Controls

The bedrock lithology of the thrust belt is an independent control on the texture of detritus entering the basin. The type of sediment load in a depositional system, which is a mixture of bed load, suspended load, and dissolved load, varies as a function of the source material. For example, a quartzite source area initially yields coarse-grained material because of its resistance to chemical weathering. In such lithologies, the spacing of joints and thickness of beds are prime controls on the size of the clasts that weather from it (Blatt 1982). Shales, however, yield fine-grained material because of their fissile nature and susceptibility to chemical weathering. In that the type of sediment load affects the channel geometry and nature of the transport process in an alluvial environment (Schumm 1968b), thrust-belt lithologies can affect facies relationships in the basin (e.g., Graham et al. 1986).

Climate, through its influence on vegetation, weathering processes, and hydrologic factors, is another independent variable in the system that contributes to a basin's stratigraphy. As an example, a thrust belt with limestone bedrock exposed in a tropical climate would result in largely dissolved load, whereas the same thrust belt exposed in an arid climate would result in conglomerates and sandstones. Although the effect of climatic changes through time on fluvial sediments has been documented in several Quaternary examples (e.g., Schumm 1968a) there is still disagreement about the role of climate in alluvial history (Karlstrom and Karlstrom 1987), and it is even more difficult to establish how climatic fluctuations are expressed in the ancient stratigraphic

record. Independent confirmation of such climatic fluctuations is rare.

Eustatic sea-level change is the third important independent control to consider. Because foreland basins occur, by definition, on continental crust they tend to be characterized by nonmarine or shallow-marine deposition. This is especially true of the foreland basins discussed here, which are in flexural balance with the thrust belt. It is much less true of peripheral foreland basins that are partly generated by a sub-thrust load. Of the foreland basins discussed here, whether a given basin is partly marine or completely nonmarine depends upon the thickness of the continental crust and eustatic sea level. If the basin is partly marine, the role of short-term eustatic cycles is obvious: shifting shoreline facies reflect the balance between rates of eustatic sea-level change, sediment supply, transport, and subsidence (e.g., Vail et al. 1977; Pitman 1978). It is not clear to what extent facies in nonmarine basins are affected by sea-level changes. Posamentier and Vail (in press), citing studies of alluviation in response to river dams, predict that the effect of a eustatic base-level change propagates the entire length of the drainage basin. The longitudinal profile of the stream is thought to shift in the same direction as the shoreline on a time scale similar to that of eustatic cycles. Snow and Slingerland (1987), in a numerical treatment of this problem, arrive at a similar conclusion. Leopold and Bull (1979), however, present evidence that suggests base-level changes have only local effects. In any case, whether or not a given foreland basin will respond to base-level changes depends upon the nature of intermediate base levels in its drainage system. Structurally controlled basins that have internal drainage will be insensitive to eustatic base-level variations.

Huaco Section, Bermejo Foreland Basin, Argentina

Thrust motion can be recognized in the stratigraphic record either through measurements of its first-order impact on the basin (i.e., subsidence) or through the impacts that thrust motion might have on the second-order controls. For instance, thrust motion might result in exposure of new lithologies, changing patterns of drainage basins and slopes, and changing discharge of streams and rates of supply of sediment to the basin (see also Burbank and

Raynolds 1988: this volume). Below we examine various approaches to trying to decipher the influence of thrust faulting on synorogenic strata and we illustrate these approaches using a sequence of Neogene strata from a retroarc foreland basin on the eastern flank of the Andes Mountains.

The nonmarine sequence exposed at Sierra de Huaco in San Juan Province, Argentina, is nearly 6,000 m thick. The strata crop out at the interface between the Precordillera thin-skinned thrust belt and the still-subsiding Bermejo River Valley (Figs. 16.3, 16.4); the Neogene strata and Holocene valley are referred to together as the Bermejo Basin. The age of the lower 5,400 m of exposed section spans the interval from about 14 to 2.4 Ma (Fig. 16.5).

The region of the Precordillera mountains was previously recognized as a thin-skinned thrust belt from its structural geometry (Fig. 16.4; e.g., Baldis and Chebli 1969; Ortiz and Zambrano 1981), but the time of thrusting has been investigated only recently, using magnetic-polarity stratigraphy of the synorogenic strata. Those studies began with the Huaco strata (Johnsson et al. 1984; N.M. Johnson et al. 1986) and are continuing with strata exposed

at other locations (e.g., Bercowski et al. 1986; Johnson et al. 1987; Reynolds et al. 1987).

The Huaco section is well suited to serve as a test of the dating methods that may be applied in other foreland-basin sequences because: 1) it is representative of ancient sequences in that major portions of the Bermejo Basin will probably be preserved, judging by its structural position and lateral continuity (Figs. 16.3, 16.4); 2) high-resolution chronological data exist, based on magnetic-polarity stratigraphy with dated volcanic ash interbeds; 3) several of the techniques that are discussed below can be applied independently at Huaco; and 4) preliminary data from other sites corroborate our interpretation of the Huaco strata.

At Huaco, the contact of the Neogene section with underlying Mesozoic units is buried, although probably no more than a few hundred metres of Cenozoic strata are covered. All of the formation boundaries are conformable in the region of the field study, although the contact between the upper two units (Rio Jachal and Mogna Formations) is discordant elsewhere (Ortiz and Zambrano 1981; N.M. Johnson et al. 1986). The section was studied in two

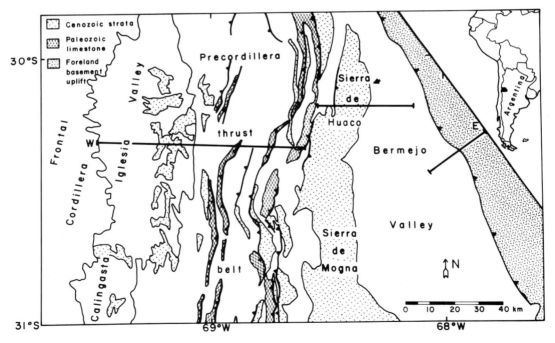

Fig. 16.3. Map of the Precordillera thrust belt and Bermejo foreland-basin system in west-central Argentina. Heavy lines are thrust faults with teeth directed toward the upper plate (see also Baldis and Chebli 1969; N.M. Johnson et al. 1986; Jordan and Gardeweg in press). Inset map of South America shows location of larger map. The synorogenic Cenozoic strata in the Sierra de Huaco (bold arrow) are derived from source areas to their west. The approximate location of the cross section (Fig. 16.4) is shown by line segments.

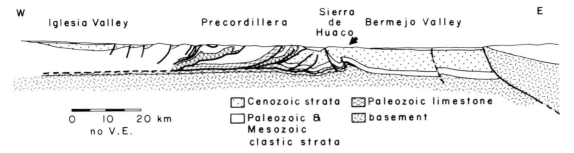

Fig. 16.4. Composite, generalized east-west oriented structural cross section of the Precordillera thrust belt and Bermejo foreland basin. Simplified from Ortiz and Zambrano (1981), Allmendinger, R.W. (personal communication 1986), and Jordan and Allmendinger (1986). Heavy arrow indicates location of stratigraphic profile in Figure 16.5.

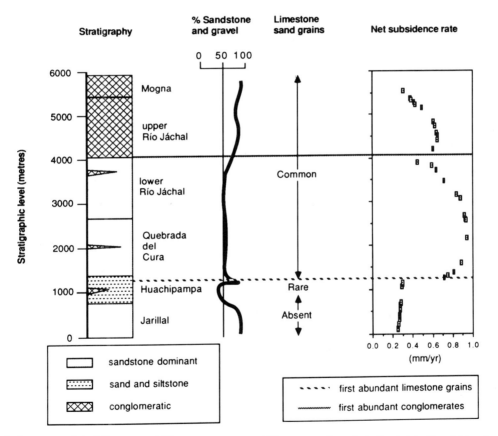

Fig. 16.5. Summary of the Miocene and Pliocene nonmarine clastic strata of the Sierra de Huaco. The left column shows the basic lithologic subdivisions and formations identified in the field. The content of sand plus gravel is shown in the middle column, and the presence or absence of limestone clasts is indicated in the right column. The box on the right shows the "net subsidence rate" as a function of the stratigraphic level. The rate was calculated by determining the slope of the decompacted accumulation curve (Fig. 16.6).

separate regions within the Sierra de Huaco, resulting in a composite section whose upper and lower parts are separated geographically by about 15 km. The lower and upper parts of the section overlap by more than 700 m of thickness and more than 500,000 years (N.M. Johnson *et al.* 1986).

Indirect Dating Methods

We first discuss two indirect methods of dating thrust-fault activity that are not subject to circular reasoning or ambiguity. Subsequently we discuss methods that are less reliable because they depend on facets of the strata produced by the complex integration of the second-order controls.

Subsidence History

Where cross-cutting relations with thrust faults do not exist, tectonic subsidence history is the best indicator of thrust history, as subsidence is controlled by flexure of basement beneath the load of thrust sheets in the mountain belt. The tectonic subsidence history is measured by examining the thickness of preserved strata. Thickness trends through time at a given point in the basin and lateral variations in thickness both provide information about activity in the thrust belt. The tectonic subsidence history at a single location in the basin allows identification of the times of motion on thrust faults although the location of the thrust(s) cannot be specified. If multiple sites are compared or if complementary data are available, however, then even the location of the thrust can be determined.

The thickness of strata in a basin is the result of four components that create space for deposition: 1) true tectonic subsidence, which is the motion that the basement would have undergone even if no sediments were deposited (Fig. 16.2); 2) subsidence caused by the load of the accumulating pile of sediment; 3) an apparent subsidence due to compaction of the underlying strata; and 4) an apparent gain or loss of space if the elevation at which deposition occurs changes through time (this is the result of using the depositional surface as the fixed frame of reference). If the goal is to examine structural activity in the thrust belt, only the tectonic subsidence is of interest.

"Backstripping" is the name given to the set of techniques by which the stratal thickness is subdivided into the individual components caused by sediment loading, compaction, and elevation change, so that the tectonic subsidence component can be isolated (Watts and Ryan 1976; Steckler and Watts 1978; Van Hinte 1978; Sclater and Christie 1980; Bond and Kominz 1984). In practice, backstripping a column of strata in a foreland basin to determine the tectonic subsidence history correctly is extremely complicated.

Decompacting the measured stratal thicknesses corrects for progressive compaction with increasing burial and for the dependence of compaction on lithology, e.g., muds and clast-supported sediments compact differently. Compaction curves for some basins have been compiled (e.g., Sclater and Christie 1980; Dickinson *et al.* 1987), but the extent to which they are appropriate to any other sedimentary sequence depends on the mixture of mechanical compaction, cementation, and dissolution that occurred in those strata. A number of workers have corrected for the compaction of syn-thrust strata in a variety of foreland basins (Allen *et al.* 1986; Kominz and Bond 1986; Cross 1986; Heller *et al.* 1986; Homewood *et al.* 1986). Figure 16.6 illustrates the importance of this procedure in interpreting the details of thrust timing, by comparing a curve from Huaco that has been corrected for stratal compaction to the same data from which compaction was not filtered. Because the sand to mud ratio was not constant through the column (Fig. 16.5) and because the upper part has never been deeply buried, the decompacted curve (open boxes, Fig. 16.6) reveals details of the history that are different than those suggested by the raw data (filled boxes, Fig. 16.6). The decompacted curve reveals that an increase in accumulation rate that occurred approximately 8.5 Ma was quite abrupt, whereas the uncorrected data imply a longer-term transition. Also, the decompacted curve better shows that the rate of stratal accumulation diminished after about 4 Ma. We will interpret these features in subsequent discussions.

In this example, the details of the compaction behavior were constrained by the sandstone textures and mudstone densities in the Huaco and neighboring sections. In the sandstones, grain packing varied as a function of depth and of the cementation history of each formation (Damanti, J.F., personal communication 1986). We derived analytical expressions for the empirical relations between compaction and depth, analogous to the approach of Sclater and Christie (1980), resulting in three expressions for sandstones, depending on their cementation history,

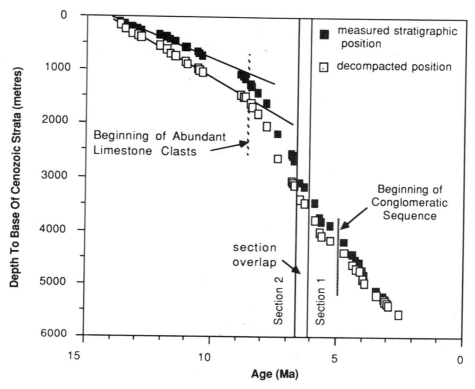

Fig. 16.6. Accumulation history of the Huaco strata. The black boxes indicate the stratigraphic position of magnetic-polarity reversals relative to their age; the curve formed from this information shows the rate of accumulation of fully compacted strata. The open boxes on the same diagram plot the decompacted position of the magnetic-polarity reversals; that curve indicates the rate of accumulation of strata, corrected for compaction that occurred after the deposition of each dated horizon. Thus, the open-box curve is a more correct description of the subsidence of the base of the section at each point in time. The field work and sampling at Huaco were accomplished in two overlapping sections; the base of section 1 and top of section 2 are indicated by the two vertical lines in the middle of the diagram. Chronological data are from Johnsson *et al.* (1984) and N.M. Johnson *et al.* (1986). The two line segments in the upper part of the diagram serve to emphasize the difference in the detailed form of the two curves in the interval between 8 and 9 Ma.

and one for mudstones. Each open box (Fig. 16.6) represents net stratal thickness at that point in time: Strata deeply buried beneath the surface at that time had already been significantly compacted; those at the surface had not yet compacted.

Another element of the compaction history that is much more problematical concerns the possible compaction of underlying strata that pre-date the foreland-basin phase. Those strata may continue to compact under the increasing burial load of the foreland-basin fill, thus creating space for more deposition. Whether or not that compaction should be treated during backstripping in the same manner as the compaction of foreland-basin strata depends on the individual basin. If there is a significant unconformity ($\approx 10^8$ years) between the pre-

foreland basin-strata and foreland-basin phase, the porosity and cementation history of the older strata may have been strongly influenced by time, temperature, and fluid history, resulting in rocks that are no longer sensitive to the mechanical effects of additional burial. In the Huaco example, the underlying units are Cambrian and Ordovician carbonates and Carboniferous to Triassic clastic units, whose compaction we have not included in Figure 16.6. The carbonates were probably fully cemented during the long interval of exposure near the surface during Paleozoic time (e.g., Schmoker 1984). Approximately one third of the 1,400 m of Upper Paleozoic and Mesozoic strata includes interbeds of shale and siltstone (di Paola and Marchese 1973) that might have compacted substantially, but it is pure speculation to estimate their degree of lithification at the beginning of the Miocene foreland-basin cycle. Kominz and Bond (1986) took into account the compaction of some pre-foreland-basin strata during the foreland-basin stage because there was not a significant unconformity beneath the foreland-basin fill.

Accounting for elevation changes during the deposition of a sequence is an important but difficult step in backstripping a column. The basin might have been starved and its elevation diminishing through time, or it might have been filling with strata at a rate exceeding the net subsidence of tectonic subsidence, compaction, and sediment loading so that its elevation increased through time, or the elevation might have been controlled by a drainage threshold into another linked basin. However, elevation at the time of deposition is difficult to constrain in a nonmarine basin. Even in marine strata, the precision of water-depth estimates can be relatively low, depending on the age of the sediment (Ingle 1975; Van Hinte 1978). Thus backstripped tectonic subsidence histories remain somewhat imprecise. In some basins the relative elevation changes are a small component of the overall subsidence history, but whether or not this is true must be determined from the available paleobathymetric/paleotopographic data.

The example from Huaco (Figs. 16.5, 16.6) has not been corrected for elevation, although the correction can be qualitatively evaluated. The upper 5,000 m of the section represent progradation of an alluvial fan system over a marginal playa (N.M. Johnson et al. 1986). Because the Holocene Bermejo Basin can be considered as a continuation of the Mio-Pliocene depositional system, the Holocene fans, which have about 250 m of relief, can be used to constrain the probable elevation changes during deposition of the strata. Thus an increase in elevation of 250 m is reasonable for the prograding section and, because of the relationship between topographic slope and grain size (Lane 1955), most of the elevation increase would have occurred during accumulation of the conglomerate facies (Figs. 16.5, 16.6; beginning about 5.0 Ma). If flexure caused by a load on the lithosphere creates an increase in elevation of about 60 to 80% of the thickness of the load, the 250 m elevation increase would be responsible for a maximum of 300 to 400 m of added strata. Thus, the net stratal accumulation (Fig. 16.6, open boxes) is partly related to elevation increase rather than to tectonic subsidence, and the rate of subsidence was less than illustrated (Fig. 16.5).

To date, the most difficult step of backstripping a column in a foreland basin is determining the component of the subsidence caused by the load of the basin-filling strata. It is common practice to make an Airy-isostasy estimate of this factor (e.g., Homewood et al. 1986; Biddle et al. 1986; Kominz and Bond 1986; Allen et al. 1986; Cross 1986; Heller et al. 1986), resulting in a tectonic subsidence curve that always has the same form as the "net subsidence" curve (corrected for only compaction and elevation). However, this is explicitly inaccurate in a foreland basin, which exists because subsidence is not accommodated by Airy behavior. The problem is that correcting for flexural subsidence due to the basin fill requires a full three-dimensional reconstruction of stratal thicknesses and ages and is computationally difficult if the thickness patterns prove to be complex in three dimensions. In the case of Huaco, the thickness patterns to the east and west are poorly known (in the direction perpendicular to the thrust belt). Therefore we have not attempted a flexural estimate of subsidence under the load of the strata. We do not show an Airy estimate of that component of subsidence (Figs. 16.5, 16.6) because it would not provide any additional information about the thrusting history but might be falsely interpreted to add validity to a so-called "tectonic subsidence" curve. Thus, even in parts of the section where there was no need for an elevation correction, we have calculated only the net subsidence of the top of the underlying Mesozoic units (Figs. 16.5, 16.6).

Based only on the shape of the "net subsidence rate" curve (Fig. 16.5), the history at Huaco of flexure under a load is intriguing. The abrupt increase in rate at the Huachipampa-Quebrada del Cura contact can be interpreted as a likely time of thrust-sheet loading. The remainder of the curve is difficult to interpret without supplementary information. The long interval of diminishing rate of subsidence (2,000 to 4,000 m in Fig. 16.5, or 7.5 to 5 Ma in Fig. 16.6) has several possible explanations. It could represent the difference in subsidence between the two geographically separated parts of the measured section (N.M. Johnson et al. 1986), although we do not favor that interpretation. It could represent continued thrust motion with either a lower rate, a different geometry, or in a different location. Finally, this high but decreasing accumulation rate could represent only the effects of mass redistribution from the already-thrusted mountain belt to the basin. In all likelihood thrust motion and mass redistribution by surficial processes could both have been active. Without three-dimensional knowledge of stratal thicknesses and flexural backstripping, we cannot determine what was the balance between these two processes. The next inflection in the subsidence-rate curve (Fig. 16.5) occurs at about 4,000 m. Although it is simple to interpret that some important event in the load history occurred at that point in time, it is difficult to describe the nature of the event. The discussion of the balance between thrust loading and depositional redistribution of the load could also apply in this case.

Unroofing History

Clast petrology is another useful and reliable method of determining thrust-movement history in some situations. The petrology of clasts that accumulate in the foreland basin records the erosion of the thrust-fault system, plus any other source areas that are more internal in the orogenic system and those on the distal margins of the basin. Lag times are important in this method of determining the time of motion on thrust faults: Deposition of any given clast is direct evidence that a rock of that lithology was exposed to erosion at some point in the geologic past. The lags are those related to erosion, to storage and transport in the fluvial system, and to possible recycling in the basin system.

However, if the bedrock lithology of the thrust system is known, then the evidence is quite direct: The presence of clasts in the basin that were exposed in the source area by motion on a specific thrust fault requires that the thrust fault moved before the stratum bearing those clasts was deposited (e.g., Armstrong and Oriel 1965; Johnson *et al.* 1985; Lawton, 1986). The likelihood that clasts will be of unambiguous parentage depends upon the bedrock lithology and upon the degree to which weathering has modified the bedrock. Again, a limestone unit in a thrust system may be a diagnostic characteristic within the regional geologic setting, but if the limestone is entirely dissolved rather than forming bed load in the basin, it will not be a useful clue to uplift of that thrust system.

The utility of clast lithology in structural interpretations is well illustrated by the Huaco section. In the Precordillera, the thrusted Paleozoic section consists entirely of siliciclastic strata with two important exceptions: in the Western Precordillera,

basalts are involved in the faulting, and in the Central Precordillera, the basal unit above the decollement is a thick carbonate (Figs. 16.3, 16.4). Documenting the presence of basaltic grains in the sandstones is difficult because other mafic volcanic detritus was supplied from more interior parts of the mountain system (Frontal Cordillera, Fig. 16.3). However, the limestone grains are readily identified in thin section among the sand population and, where present, indicate that the basal structural unit was locally exposed and thus must have been thrusted. Limestone grains are common in sandstones younger than about 8.5 Ma, coincident with the sudden increase in the rate of net subsidence (Fig. 16.5). Limestone grains are rare in the next lower 180 m of strata (Fig. 16.5; one or two limestone grains per thin section in a series of 5 samples). Limestone grains are absent below that horizon. In isolation, the interpretation of this interval (8.5 to ≈ 9.1 Ma) is ambiguous: The rare limestone grains could represent calcrete or multiply reworked grains. Within the context of other information about the section and region, however, this interval more likely represents the time that deformation of the limestone-thrust belt was just beginning, with only local exposures of the deeper structural levels.

Only the first appearance of distinctive lithologies is unambiguous. Subsequent occurrences of the same lithology can be due to either: 1) continued erosion of the same fault block or long-term storage of material derived from it in the hillslope systems; 2) renewed uplift of the source area on underlying and structurally unrelated thrust faults, as the thrust system locus of activity shifts; or 3) cannibalization of previously deposited foreland-basin strata that contained the clast lithology in question, which is a normal phenomenon on the side of the foreland basin adjacent to the thrust belt (e.g., Steidtmann and Schmitt 1988: this volume). For these reasons, the continued presence of limestone grains in the Huaco section does not provide much information about the structural history after 8.5 Ma.

Both sand and gravel reflect the bedrock composition of the thrust belt and both have been used in studies of thrust-fault timing (e.g., Armstrong and Oriel 1965; Royse *et al.* 1975; Johnson *et al.* 1985; N.M. Johnson *et al.* 1986; Lawton 1986; DeCelles *et al.* 1987). Any introduction of new clast lithologies of gravel size into the basin is probably accompanied by sand-sized fragments of similar material, so the first appearance of the diagnostic lithologies should be independent of the grain size. Sand-sized detritus has the great advantage that it is more widely distributed in the foreland basin than is coarser detritus. Gravel is usually restricted to steep basin margins and high-competence river systems. The basin margins are the least likely parts of the basin to be preserved, whereas ribbons of conglomerate in the distal channel systems may constitute a very low percentage of the basin fill. As discussed below, progradation of marginal conglomerate facies into the basin interior, where they are more likely to be preserved, depends on complex surficial processes. In total, the conglomerate horizons in a section tend to be very discontinuous samples of passing geologic time, and the first preserved occurrence of some diagnostic clast type might lag millions of years behind its first appearance in the basin system. Although sandstones are subject to the same progradational facies controls as are coarser clastic deposits, they represent a more faithful record of geologic events.

In the Huaco example (Fig. 16.5) conglomerates are rare in the lower part of the section and become abundant in the upper 1,500 m. The first conglomerates recognized were deposited at approximately 8.5 Ma. They are laterally discontinuous horizons and do not contain limestone clasts. The next unit that has scattered pebbles at the bases of some of the major sand bodies and a few thin conglomerate lenses, the Quebrada del Cura Formation, was deposited between about 8.2 and 6.8 Ma. Pebbles include limestone clasts. The facies in which conglomerate is common spans the time from 5.0 Ma to the top of the preserved section (the youngest dated horizon is about 2.4 Ma). In it the pebble-sized to locally boulder-sized clasts are rich in limestone. A reconnaissance examination of the section would probably have led to the interpretation that the limestones were exposed by thrusting beginning 5.0 Ma, and a more thorough study of the sparse gravels below might have indicated that thrusting of the limestone unit began during Quebrada del Cura time, but neither interpretation would have been as accurate as that based upon the sandstone compositions.

Conglomerates have advantages over sandstones in the ease of establishing the source rock of a given grain lithology. Where source rock sections are available for inspection, it is relatively straightforward to recognize the parent lithologies of the conglomerate (e.g., Graham et al. 1986; Lawton 1986).

The specific source lithology of sand-sized grains is considerably more difficult to interpret than of conglomerates (Dickinson 1970). Quartz and feldspar grains are not highly specific to their sources, whereas lithic clasts are of much greater utility. Which types of grains are preserved depends upon the bedrock lithology in the source area, the degree of weathering, abrasion during transport, and diagenesis. Heavy minerals can be used successfully to identify source units (e.g., Johnson et al. 1985; Morton 1985). More sophisticated methods of describing grain petrology (Füchtbauer 1974; Dorsey 1985; papers contained in Zuffa 1985) and chemistry, such as isotopic studies (see Heller et al. 1985; Heller and Frost 1988: this volume), can overcome some of the problems of identifying source areas.

Methods Requiring Facies Interpretations

Strata that accumulate in the foreland basin represent mass transferred from the shortened orogenic belt to a topographically lower region. Facies, their lateral changes, and their migrations through time are primary characteristics of the strata and, in many basins, a great deal of information already exists about facies distribution. However, attempts to date thrust-fault activity through interpretations of depositional facies are plagued by serious limitations. The basic problem is that facies are as much a function of complex surficial processes (with inherent lag times) as of thrust-related relief changes in the source area (see also Paola 1988: this volume).

Fluvial Simulation Models

Where chronological data are lacking, interpretations of the texture of strata have been used to estimate relative rates of subsidence. This approach is based on the results of fluvial simulation models that predict the ratio of preserved overbank sediments to channel sediments in a meandering river system as a function of subsidence rate (Allen 1978, 1979; Leeder 1978; Bridge and Leeder 1979). The important premise involved is that coarse-grained sediment is preferentially preserved in a fluvial system. Simulations suggest that the interaction of channel avulsion with subsidence results in different degrees of sand-body interconnectedness and different values of sand/shale ratios with changing subsidence rate. These predicted results have been used to interpret varying tectonic subsidence rates in ancient fluvial sequences (e.g., Kraus 1980; Blakey and

Gubitosa 1984). Severe limitations to the approach include the need to prove that the rate and grain size of sediment supplied to the system were constant and to establish that the stream geometries and avulsion frequency were constant throughout the time spanned by the section. In other words, the user must prove that subsidence was the only variable. We are unaware of any examples in which these conditions have been met.

Although the lower part of the section at Huaco is texturally similar to the system simulated by Allen (1978, 1979), Leeder (1978), and Bridge and Leeder (1979), it does not provide a perfect example because much of the section was deposited in arid, flash-flood dominated streams and fans (Beer *et al.* 1986) rather than in the meandering systems for which simulations were designed. Thus sand-body interconnectedness characteristics of the Huaco section are not pertinent to the models. However, the general philosophy behind preservation of varying ratios of channel to overbank deposits, or coarse-to-fine ratios, should be applicable. According to the model of Allen (1978), slower accumulation rates are envisioned to allow more time for reworking within the depositional environment and thus more opportunity for fine-grained materials to be flushed out of the system while coarse deposits are retained as channel deposits. Thus the coarse-to-fine ratio should increase as the rate of accumulation decreases. At Huaco, a comparison of strata deposited in the interval from about 8.2 to 10.4 Ma, the Huachipampa Formation, to the units overlying and underlying it implies that this premise is inaccurate (Fig. 16.5). The percentage of sand is lower in the Huachipampa than in the underlying unit and yet the rate of net subsidence is constant. One should note, however, that because the style of fluvial deposition changed notably between these two units (Beer *et al.* 1986, 1987), the condition of the simulation model that only subsidence rate may vary has been violated. Where there is a change in net subsidence rate (at 8.5 Ma, the Huachipampa-Quebrada del Cura contact) there is no change in fluvial style, yet the percentage of sand increases and, in clear contradiction to the model, the net subsidence rate also increases.

A different aspect of Bridge and Leeder's model (1979) might be applicable to Huaco strata. They suggest that at any given instant in a basin's development, if there is differential subsidence, channels will tend to shift to and flow parallel to the axis of maximum subsidence, causing that area to be rela-

tively coarser than areas of less subsidence. This phenomenon was substantiated in the field by Suttner (1969) and Read and Dean (1982), although their measure of subsidence rate (stratal thickness during a specific time interval) was not corrected for the effects of compaction. At Huaco the simultaneous increase in sandstone percentage and accumulation rate at the Huachipampa-Quebrada del Cura boundary might have been considered to reflect this type of behavior. However, facies and paleocurrent data (Beer *et al.* 1986, 1987) show that the drainage was directed perpendicular to structural strike, and thus probably perpendicular to differential subsidence, rendering this interpretation inappropriate.

Paola (1988: this volume) develops a mass-balance model of filling a subsiding alluvial basin with gravel and sand, a coarser alluvial system than was considered by Allen, Leeder, and Bridge. Paola (1988: this volume) develops several sedimentological criteria that should differentiate basins with rapid asymmetric subsidence (characteristic of foreland basins) from basins with slow, uniform subsidence. In principle, variations through time in the subsidence rate and geometry might result in a stack of strata with varying diagnostic sedimentological characteristics, and thus would be of use in detecting times of thrust motion.

In the middle of the Huaco section, there is no correlation between textures and the changing rate of net subsidence (Fig. 16.5), but in the upper part there are correlations that warrant consideration of Paola's (1988: this volume) models. The thick Rio Jachal Formation conglomerate is a local facies, not observed to the north or south of the area in which we studied the two sections. Because its distribution is restricted *parallel* to structural strike, it differs from the thick gravels that are restricted *perpendicular* to strike that Paola (1988: this volume) predicts might accompany rapid, asymmetric subsidence. It is plausible that the base of the Rio Jachal conglomerate represents a stream capture or some other abrupt change in drainage pattern, rather than being a specific consequence of subtle changes in subsidence history at Huaco. In contrast, the Mogna Formation conglomerate is a regional facies (Ortiz and Zambrano 1981), and might be like the thin, regional sheets of gravel that Paola predicts are characteristic of slow, regional subsidence. The Mogna conglomerate is not well constrained by chronological data (Johnsson *et al.* 1984); it began to accumulate during a time of decreasing subsidence rate (Fig. 16.5).

Facies Migration

There are many examples of the use of lateral facies changes and their migrations through time to date thrust movements. For example, appearances of pulses of conglomerate in a section are interpreted as indicators of new or renewed thrusting (e.g., Wiltschko and Dorr 1983). Several recent studies have concluded that peak thrust activity correlates to intervals of fine-grained strata (Beck and Vondra 1985; Bilodeau and Blair 1986; Heller et al. 1986). Prograding (Swift et al. 1985) or retrograding facies (Ettensohn 1985) are alternatively interpreted as correlating to times of thrusting. In partly marine basins, shifting shoreline positions are an aspect of the prograding/retrograding facies interpretation.

As has been discussed elsewhere with reference to eustatic sea-level controls on shoreline positions and facies tracts (e.g., Vail et al. 1977), the accumulation of the sediments in laterally migrating facies tracts results from the rate at which sediment is supplied to a given location compared to the rate at which space is created at that location. Sea-level changes, subsidence, and changes in the elevation of the depositional surface interact to create space for potential deposition. Thrust motions, which we wish to detect, play a role by: 1) controlling the rate of tectonic subsidence; 2) exposing varying bedrock lithology, which results in changes in weathering rate and detrital products, which in turn control transport mechanism and depositional profile; and 3) controlling local-relief distribution in the mountain system, which can influence both climate and drainage organization.

Thrust-fault activity includes a basinward shift in the locus of uplift in addition to changing degrees of uplift in one location. Clearly, on the time scale over which more than one of the major thrust systems develop, the basin margin and its sedimentary facies must migrate in space. For thrust belts that obey the rule that younger thrusts cut progressively toward the craton (Dahlstrom 1970), this results in a facies progradation. The difficulty is to determine on a shorter time scale, that is, the scale of increase and decline of activity on a single fault, what the facies migration pattern will be.

The rate of supply of sediment to the margin of the depositional basin is the integrated result of the rates of uplift, denudation, and transportation. Table 16.2

Table 16.2. Comparison of uplift and denudation rates.

Uplift rates (m/10^3yr)*		
Short term (Subandean thrust belt)	0.30–0.64	
Long term (Southern Canadian Cordillera)	0.03–0.04	
Denudation rates (m/10^3yr)		
Normal relief	0.01–0.1	
Steep relief †	0.1 –1.0	(N. Saunders and S. Young 1983)

*Calculated from thickening rate data in Table 16.1.
†Saunders and Young (1983) classified relief such that "steep refers to mountainous or other steeply dissected regions, or individual slopes above 25°."

is an attempt to compare the ranges of values that are possible for uplift and denudation. Uplift rates are derived from the rates of crustal thickening given in Table 16.1, making certain reasonable assumptions about the crustal response to loading. The range of "normal" flexural rigidities (10^{29} to 10^{31} dyne-cm) are such that, in general, about 60 to 80% of any increment of added crustal thickness remains above the original datum, whereas 20 to 40% subsides below the datum (Fig. 16.2). The denudation rates in Table 16.2 are from a recent comprehensive survey by Saunders and Young (1983); rates are for high relief and low relief areas under various climatic conditions. In comparison with the uplift rates, one sees that the denudation rates are of the same order of magnitude (10^{-2} to 10^{-1} m/10^3 yr), suggesting that three scenarios are possible: denudation rate may be less than, or equal to, or, for short durations, greater than uplift rate, depending on the topographic, lithologic, and climatic setting. The importance of transport rates as a link between denuding the source area and filling space in the basin is not fully understood. For example, sediment that is moved downslope by mass wasting may reside on the mountain hillslopes until headward stream erosion is sufficient to transport the excess sediment. Trimble (1975, 1977) indicates that agricultural practices have caused deposition of significant volumes of sediment within mountain stream systems. A change in the environmental conditions can result in flushing that sediment out of the system later, thus producing unsteady sediment supply to a basin although denudation itself may be steady. Parker (1985) suggests that such buffers in sedimentary systems may operate on times scales of geologic interest. Until such processes are better understood, the assumption that the rate at which material erodes on the basin margin is the same as the rate at which material is delivered to the basin is untested.

Facies migration patterns relate directly to the question of whether or not mountain belts attain steady-state conditions (no net increase or loss in relief over a long time period). If the mountain system reaches a steady state (e.g., Hack 1960; Suppe 1981; Ahnert 1984; Covey 1986), then the rates of uplift and denudation are equal. Whether or not motion on a thrust fault will temporarily destroy that equilibrium condition depends upon its rate of shortening and thickening. If the thrusting rate is slower than the capacity of the system to maintain an equilibrium, then denudation and transportation will maintain the topography at the original elevation, implying that there has been no net crustal thickening (erosion equalling structural thickening; Anastasio and De Paor 1985). Thus, in the basin, no added tectonic subsidence will occur, yet more material is supplied by erosion.

At the beginning of a cycle of mountain building, steady state cannot exist, or there would be no mountain belts (e.g., Schumm 1963; Ahnert 1970). Similarly, during geologically short intervals the rate of motion on individual thrust faults may exceed the capacity of erosion to maintain the steady state. In those cases, the crust will thicken causing uplift and the rate of denudation will increase in response to increased relief. If so, more material is supplied to the basin at the same time as more space is created by subsidence. Eventually, the rate of tectonic subsidence will decline as the period of activity on that fault (or set of faults) is passed or as a steady state of thickening and denudation is achieved in the mountain system. Thus, at some point in time the rate of supply of sediments exceeds the declining rate of subsidence.

Although the long-term competition between rates of uplift, denudation, and subsidence can be described, the net facies result cannot be. At both a basin-wide scale and a local scale, it is not possible to make general pronouncements about whether the rate of sediment accumulation exceeds the rate at which space is created. First, the magnitudes of uplift and subsidence are coupled through the rheology (Fig. 16.2). For instance, a high flexural rigidity (stiff lithosphere) creates more uplift and less subsidence than a low flexural rigidity (soft lithosphere). In that denudation rate is exponentially proportional to uplift rate (Ahnert 1970, 1984), soft lithosphere causes relatively greater subsidence in the proximal part of the basin *and* less sediment supply than will

stiffer lithosphere. As noted previously, the flexural rigidity of the lithosphere does not appear to be constant among the basins studied to date and therefore the balance of uplift and subsidence differs among basins. Second, the sediment supplied to the basin is not distributed on a uniform plane over the entire region of subsidence. Rather, depositional relief develops on the sides along which detritus enters the basin. The appropriate depositional slope is a function of grain size of the detritus in the system and discharge (e.g., Lane 1955; Snow and Slingerland 1987) and is not dependent upon the form of the subsiding basement beneath it. The depositional slope that is established will determine how the volume of sediment is distributed throughout the basin, and therefore it controls the rate of sediment accumulation at a local scale.

Even if the flexural properties of a mountain belt/basin pair are known, the balance between vertical motions and denudation is dependent upon bedrock lithology and climate, and the depositional relief is also dependent upon the climate (hydrology), lithology (texture), and base level. Thus it is impossible to generalize as to whether or not progradation will occur for any given increment of thrust thickening. A single episode of thrust motion might cause first progradation and then retrogradation of facies, or the reverse might occur. Therefore inverting the logic of this discussion and using a preserved facies migration to indicate tectonic activity, which is our goal, is probably nonunique at best and currently impossible. Forward models that can evaluate the facies responses to each of the independent variables (thrust motion, climate, lithology, and base level) are needed that will guide the stratigraphic interpretations (e.g., Karner 1986; Flemings and Jordan 1987; Snow and Slingerland, 1987; Paola 1988: this volume).

The Huaco sequence demonstrates that it is tenuous to interpret thrust timing through facies analysis. First, if the conglomerates are treated as evidence of increased depositional slope, and thus pulses of conglomerate are interpreted as indications of thrust fault activity (e.g., Wiltschko and Dorr 1983), the interpretation of first thrusting would be at odds with other information (discussion above and Figs. 16.5, 16.6). Second, if fine-grained facies are interpreted to represent times of active thrusting (initial subsidence rate presumed to be greater than sediment supply rate) (e.g., Beck and Vondra 1985; Bilodeau and Blair 1986), the contact between Jarillal and Huachipampa Formations (10 Ma) would be assigned a tectonic significance that is not indicated by the measured "net subsidence rate." Conversely, the time at which facies rapidly began to coarsen (but still lacking conglomerates) is about 8.5 Ma; this is the time at which the Huaco location began to subside rapidly. Third, perhaps the long-term facies progradation at Huaco parallels a long-term eastward migration of the Precordillera thrusts. However, the available information is insufficient to prove that the thrusts migrated monotonically to the east; other independent variables could have combined to produce the same progradational trend.

An unrelated complication to the use of foreland-basin facies migration patterns is the potential impact of distant or eustatic base-level controls. As discussed by Posamentier and Vail (in press), the entire continental depositional system could respond to eustatic sea-level changes. In some foreland basins this may not be a problem over long time scales because the basin may be structurally bounded and maintained at a regional base level (this might be particularly common in composite foreland basins, Fig. 16.1b). However, structural or erosional activity that either raises or breaches the point in the basin that controls the regional base level would have the same impact as a eustatic sea-level variation, acting as an external control on depositional systems in the basin. While it should be possible to detect eustatically controlled facies migration if the timing of the foreland basin strata can be compared to eustatic sea-level history, there will be no equally universal way to recognize the impact of regional base-level changes.

Table 16.3. Use of foreland-basin strata to date deformation.

Approach	Accuracy (speed of response)	Sensitivity to varying locations in the basin
Subsidence history	Little or no lag time	Moderate but systematic
Unroofing: sedimentary petrology of Sandstone Conglomerate	Short lag time; subsequent reworking	Moderate High
Facies migration	Lag time variable, depending on climate, lithology, and base level	High

Conclusions and Recommendations

To learn more about the dynamics of the interaction of thrust belts and foreland basins we must proceed with great caution in correlating events in the two parts of the system, or the interpretations will be based on circular reasoning. Cross-cutting relations provide dates of thrust motions that can be correlated by biostratigraphy or chronostratigraphy to the basin, if the relations define narrow time windows. The most reliable approaches to indirect correlations are those that are based on the mechanics that link the thrust shortening in the mountain belt to the basin. The sedimentary record of the foreland basin strata also strongly records the impact of the three second-order controls (climate,

lithology, eustatic base level) on sediment supply and distribution. Because their relative importances, feedback relations, and lag times are not well known, it can be difficult to distinguish in the stratigraphic record among thrust motion, climate change, sequential unroofing of new lithologies, and eustatic sea-level change. We suggest that there is a hierarchy of accuracy of the popular methods of dating motion on thrust faults (Table 16.3).

In a foreland basin that is *not* dynamically controlled by a subducting oceanic or thinned continental plate and where cross-cutting relations are unavailable, the most accurate indirect means of dating thrust activity should be by determining rates of tectonic subsidence from backstripped sedimentary sections and noting times of change of the rate. Because subsidence varies spatially within the basin, results from two or more sections complement one another but should be evaluated individually. The greater the chronological resolution of the dating method employed in the strata, the more detailed and accurate will be the dating of thrust motion. The Huaco data illustrate the success of this approach as well as its limitations. The limitations exist because of the difficulty of calculating a true tectonic subsidence rate, and because we are examining only one location. The section records the beginning of a phase of thrusting in the Precordillera (within the flexural wavelength of Huaco) at approximately 8.5 Ma. It is speculative to interpret the history of thrusting versus the effect of erosion and transportation of sediment load during the interval after 8.5 Ma: The subsidence curve probably includes an interesting record of the geometry of several faults, their rates of motion, and the shifting sediment load. Furthermore, the structural history before 8.5 Ma is not easily interpreted from the Huaco subsidence history: Accumulation of strata is obvious, but the cause of the subsidence is not. This information must be derived from other locations in more interior parts of the mountain belt (e.g., Reynolds *et al.* 1986, 1987).

In the absence of good chronological control with which to establish rates of subsidence, some general conclusions can be drawn about the times of deformation. The safest, most reliable correlation that can be drawn is that the beginning of a long-term cycle of subsidence in the foreland basin indicates the time at which the mountain belt began to form. Long-term migration of facies and the depocenter away from the mountain belt are expected until the mountain belt ceases shortening. At Huaco, this reasoning implies that the Andes began a phase of uplift before 14 Ma, and thrusting continued until more recently than 2.4 Ma.

Theoretically, clast petrology of the foreland basin sandstones is a reliable way to date the beginning of motions on individual thrust sheets. This will be useful only if enough is known about the structural geology of the thrust belt to know which bedrock lithologies occur in which thrust plates, and the interpretation will be vastly improved if the drainage patterns of the basin are established by facies and paleocurrent studies. First appearances of diagnostic clast types in the section are straightforward to interpret, although some unknown lag time must be assumed to have passed between the point in time when thrusting began on that plate and when the clasts were deposited in the preserved stratum. Subsequent occurrence of the same clast types cannot be interpreted as direct evidence that thrusting continued on that fault system. Sandstones are more difficult to work with petrographically than are conglomerates, but provide a much better sampling of detrital clast types supplied to the basin. At Huaco, the first appearance of limestone clasts in the sand population indicates that thrusting might have begun in the Central Precordillera about 9.1 Ma and that it clearly was active by 8.5 Ma, which predated conglomerates with limestone clasts.

Analysis of facies patterns and facies migrations through time is important in understanding basin evolution. Drainage patterns are particularly useful complementary information in attempts to interpret the strata in the foreland basin in terms of the thrust history of the basin margins. However, interpretation of the histories of specific thrusts based only on those observations should be avoided.

Interweaving the information at Huaco gained from "net subsidence rate" history, limestone-clast history, and only a very general interpretation of facies migration, produces a more complete thrust history than is possible from any one method alone. Based only on the information from Huaco, the cause of basin subsidence prior to 8.5 Ma could have been deformation anywhere outside of the Central Precordillera and Sierra de Huaco. The Huaco site might have been at a distance from the tectonic load that approximates the flexural wavelength of the basin, or it could have been even farther away, and owe its subsidence to the load of the basin fill that lay closer to the tectonic load. The limestone clasts clarify that the increase in accumulation rate corresponded to development of thrusts in the Central Precordillera. The continued high rate of "net subsidence" indicates that the thrusts in the Central Precordillera remained active until at least 7.5 Ma. Given the general trend to more proximal facies, it is reasonable to conclude that deformation after 8.5 Ma migrated eastward across the Central Precordillera toward the Huaco section. The beginning of a second cycle of increasing and then decreasing subsidence rate coincides with the base of the conglomerate facies in the Rio Jachal Formation. This coincidence suggests the possibility that at about 5.0 Ma the locus of active thrusting shifted to the set of thrusts immediately west of the Sierra de Huaco, although other explanations for both features are possible and their coincidence could be strictly fortuitous. Cross-cutting relations demonstrate that deformation continued to migrate eastward, deforming the Sierra de Huaco after 2.4 Ma (N.M. Johnson *et al.* 1986).

The underlying cause for subsidence in these foreland basins is the motion of thrust sheets, and thus it is reasonable to use the history of strata in the basin as a means of determining thrust history. However, the system that controls the facies distribution at any point in time includes rheology, climate, lithology, and eustatic base level, in addition to thrust history. Study of the interactions of the independent variables and their facies products should be a priority of research in foreland basins. However, with the present state of knowledge, it seems quite unlikely that we are able to accurately discern the role of one variable, thrust motion, among the facies products.

Acknowledgments. The authors thank Noye Johnson of Dartmouth College for his cooperation and inspiration in this study, John Damanti, Carol Lee Roark, Mike Gefell, and R.W. Allmendinger of Cornell for their assistance, and the editors, Chris Paola and Karen Kleinspehn, for their help, patience, and comments. The comments of Douglas W. Burbank, Tim Lawton, and Neil Lundberg improved the original manuscript. Financial support has been provided by NSF EAR-8206787 and 8418131 and the Shell Companies Foundation. Cornell's Institute for Study of the Continents (INSTOC) Contribution #60.

References

AHNERT, F. (1970) Functional relationships between denudation and uplift in large mid-latitude drainage basins. American Journal Science 268:243–263.

AHNERT, F. (1984) Local relief and the height limits of mountain ranges. American Journal Science 284:1035–1055.

ALLEN, J.R.L. (1978) An exploratory quantitative model for the architecture of avulsion-controlled alluvial suites. Sedimentary Geology 21:129–147.

ALLEN, J.R.L. (1979) Studies in fluviatile sedimentation: An elementary model for the connectedness of avulsion-related channel sand bodies. Sedimentary Geology 24:253–267.

ALLEN, P.A., HOMEWOOD, P., and WILLIAMS, G.D. (1986) Foreland basins: An introduction. In: Allen, P. and Homewood, P. (eds) Foreland Basins. International Association Sedimentologists Special Publication 8, pp. 3–12.

ALLMENDINGER, R.W., RAMOS, V.A., JORDAN, T.E., PALMA, M., and ISACKS, B.L. (1983) Paleogeography and Andean structural geometry, northwest Argentina. Tectonics 2:1–16.

ANASTASIO, D.J. and DE PAOR, D.G. (1985) Thrusting and sedimentation along an emergent thrust front: An example from the external Sierras of the southern Pyrenees, Spain. Geological Society America Abstracts with Programs 17:513.

ARMSTRONG, F.C. and ORIEL, S.S. (1965) Tectonic development of Idaho-Wyoming thrust belt. American Association Petroleum Geologists Bulletin 49:1847–1866.

BALDIS, B.A. and CHEBLI, G.A. (1969) Estructura profunda del area central de la Precordillera Sanjuanina. IV Jornadas Geologicas Argentinas 1:47–65.

BEAUMONT, C. (1981) Foreland basins. Geophysical Journal Royal Astronomical Society 65:291–329.

BECK, R.A. and VONDRA, C.F. (1985) Syntectonic sedimentation and Laramide basement thrusting, Rocky Mountain foreland, Wyoming, U.S.A. (abstract). Inter-national Association of Sedimentologists Symposium on Foreland Basins, Fribourg, Switzerland, p. 36.

BEER, J.A., JORDAN, T.E., and JOHNSON, N.M. (1986) Sedimentary characteristics as a function of sedimentary accumulation rate: A study of Miocene fluvial sediments, Bermejo Basin, Argentina (abstract). Society Economic Paleontologists Mineralogists, Annual Midyear Meeting, Raleigh, North Carolina, p. 8.

BEER, J.A., JORDAN, T.E., and JOHNSON, N.M. (1987) Velocidad de sedimentacion y ambientes sedimentarios asociados con el desarrollo mioceno de la Precordillera, Huaco, Provincia de San Juan, Republica Argentina. X Congreso Geologico Argentino, Actas 2, pp. 83–90.

BEHRENSMEYER, A.K. and TAUXE, L. (1982) Isochronous fluvial systems in Miocene deposits of Northern Pakistan. Sedimentology 29:331–352.

BERCOWSKI, F., BERENSTEIN, L., JOHNSON, N.M., and NAESER, C. (1986) Sedimentología, magnetoestratigrafía y edad isotópica del Terciario en Lomas de las Tapias, Ullum, Provincia de San Juan (Abstract). La Plata, Argentina, 1° Reunion Argentina de Sedimentología, Octubre, 1986, Actas, pp. 169–172.

BIDDLE, K.T., ULIANA, M.A., MITCHUM, R.M., JR., FITZGERALD, M.G., and WRIGHT, R.C. (1986) The stratigraphic and structural evolution of the central and eastern Magallanes Basin, southern South America. In: Allen, P. and Homewood, P. (eds) Foreland Basins. International Association Sedimentologists Special Publication 8, pp. 41–61.

BILODEAU, W.L. and BLAIR, T.C. (1986) Tectonics and sedimentation: Timing of tectonic events using sedimentary rocks and facies. Geological Society America Abstracts with Programs 18:542.

BLAKEY, R.C. and GUBITOSA, R. (1984) Controls of sandstone body geometry and architecture in the Chinle Formation (Upper Triassic), Colorado Plateau. Sedimentary Geology 38:51–86.

BLATT, H. (1982) Sedimentary Petrology. San Francisco: W.H. Freeman and Company, 564 p.

BOND, G.C. and KOMINZ, M.A. (1984) Construction of tectonic subsidence curves for the early Paleozoic miogeocline, southern Canadian Rocky Mountains: Implications for subsidence mechanisms, age of breakup, and crustal thinning. Geological Society America Bulletin 95:155–173.

BOYER, S.E. and ELLIOTT, D. (1982) Thrust systems. American Association Petroleum Geologists Bulletin 66:1196–1230.

BRIDGE, J.S. and LEEDER, M.R. (1979) A simulation model of alluvial stratigraphy. Sedimentology 26:617–644.

BURBANK, D.W. and RAYNOLDS, R.G.H. (1988) Stratigraphic keys to the timing of thrusting in terrestrial foreland basins: Applications to the northwestern Himalaya. In: Kleinspehn, K.L. and Paola, C. (eds)

New Perspectives in Basin Analysis. New York: Springer-Verlag, pp. 331–351.

CALDWELL, J.G. and TURCOTTE, D.L. (1979) Dependence of the thickness of the elastic oceanic lithosphere on age. Journal Geophysical Research 84:7572–7576.

CHAPIN, C.E. and CATHER, S.E. (1981) Eocene tectonics and sedimentation in the Colorado Plateau–Rocky Mountain area. In: Dickinson, W.R. and Payne, W.D. (eds) Relations of Tectonics to Ore Deposits in the Southern Cordillera. Arizona Geological Society Digest 14:173–198.

COVEY, M. (1986) The evolution of foreland basins to steady state: Evidence from the western Taiwan foreland basin. In: Allen, P. and Homewood, P. (eds) Foreland Basins. International Association Sedimentologists Special Publication 8, pp. 77–90.

CROSS, T.A. (1986) Tectonic controls of foreland basin subsidence and Laramide style deformation, western United States. In: Allen, P. and Homewood, P. (eds) Foreland Basins. International Association Sedimentologists Special Publication 8, pp. 15–39.

DAHLSTROM (1970) Structural geology in the eastern margin of the Canadian Rocky Mountains. Bulletin Canadian Petroleum Geology 18:332–406.

DECELLES, P., TOLSON, R.B., GRAHAM, S.A., SMITH, G.A., INGERSOLL, R.V., WHITE, J., RICE, R., MOXON, I., LEMKE, L., HANDSCHY, J.W., FOLLO, M.F., EDWARDS, D.P., CAVAZZA, W., CALDWELL, M., and BARGAR, E. (1987) Laramide thrust-generated alluvial-fan sedimentation, Sphinx Conglomerate, southwestern Montana. American Association Petroleum Geologists Bulletin 71:135–155.

DICKINSON, W.R. (1970) Interpreting detrital modes of graywacke and arkose. Journal Sedimentary Petrology 40:695–707.

DICKINSON, W.R. (1974) Plate tectonics and sedimentation. In: Dickinson, W.R. (ed) Tectonics and Sedimentation. Society Economic Paleontologists Mineralogists Special Publication 22, pp. 1–27.

DICKINSON, W.R., ARMIN, R.A., BECKVAR, N., GOODLIN, T.C., JANECKE, S.U., MARK, R.A., NORRIS, R.D., RADEL, R., and WORTMAN, A.A. (1987) Geohistory analysis of rates of sediment accumulation and subsidence for selected California basins. In: Ingersoll, R.V. and Ernst, W.G. (eds) Cenozoic Basin Development of Coastal California (Rubey volume 6). Englewood Cliffs, N.J.: Prentice-Hall, pp. 1–23.

DI PAOLA, E.C. and MARCHESE, H.G. (1973) Petrologia y litoestratigrafia de las sedimentitas Paleozoicas de Huaco, San Juan, Republica Argentina. Asociacion Geologica Argentina Revista 28:369–381.

DORR, J.A., Jr. and GINGERICH, P.D. (1980) Early Cenozoic mammalian paleontology, geologic structure, and tectonic history in the overthrust belt near La Barge, western Wyoming. Wyoming University Contributions to Geology 18, no. 2:101–115.

DORSEY, R.J. (1985) Petrography of Neogene sandstones, eastern Taiwan: Response to arc-continent collision. Geological Society America Abstracts with Programs 17:565.

ETTENSOHN, F.R. (1985) The Catskill Delta complex and the Acadian Orogeny, a model. In: Woodrow, D.L. and Sevon, W.D. (eds) The Catskill Delta. Geological Society America Special Paper 201, pp. 39–50.

FLEMINGS, P.B., JORDAN, T.E., and REYNOLDS, S.A. (1986) Flexural analysis of two broken foreland basins: The Late Cenozoic Bermejo Basin and the Early Cenozoic Green River Basin (Abstract). American Association Petroleum Geologists Bulletin 70:591.

FLEMINGS, P.B. and JORDAN, T.E. (1987) Synthetic stratigraphy of foreland basins (Abstract). EOS 68:419.

FÜCHTBAUER, H. (1974) Sediments and Sedimentary Rocks 1. Stuttgart: Schweizerbart'sche Verlagsbuchhandlung, 464 p.

GRAHAM, S.A., TOLSON, R.B., DECELLES, P.G., INGERSOLL, R.V., BARGAR, E., CALDWELL, M., CAVAZZA, W., EDWARDS, D.P., FOLLO, M.F., HANDSCHY, J.W., LEMKE, L., MOXON, I., RICE, R., SMITH, G.A., and WHITE, J. (1986) Provenance modelling as a technique for analyzing source terrane evolution and controls on foreland sedimentation. In: Allen, P. and Homewood, P. (eds) Foreland Basins. International Association Sedimentologists Special Publication 8, pp. 425–436.

HACK, J.T. (1960) Interpretation of erosional topography in humid temperate regions. American Journal Science 258-A:80–97.

HAGEN, E.S., SHUSTER, M.W., and FURLONG, K.P. (1985) Tectonic loading and subsidence of intermontane basins: Wyoming foreland province. Geology 13:585–588.

HELLER, P.L. and FROST, C.D. (1988) Isotopic provenance of clastic deposits: Application of geochemistry to sedimentary provenance studies. In: Kleinspehn, K.L. and Paola, C. (eds) New Perspectives in Basin Analysis. New York: Springer-Verlag, pp. 27–42.

HELLER, P.L., PETERMAN, Z.E., O'NEIL, J.R., and SHAFIQULLAH, M. (1985) Isotopic provenance of sandstones from the Eocene Tyee Formation, Oregon Coast Range. Geological Society America Bulletin 96:770–780.

HELLER, P.L., WINSLOW, N.S., and PAOLA, C. (1986) Sedimentation and subsidence across a foreland basin: Observations and results from the Western Interior. Geological Society America Abstracts with Programs, 18:634.

HOMEWOOD, P., ALLEN, P.A., and WILLIAMS, G.D. (1986) Dynamics of the Molasse Basin of western Switzerland. In: Allen, P. and Homewood, P. (eds) Foreland Basins. International Association Sedimentologists Special Publication 8, pp. 199–217.

INGLE, J.C. (1975) Paleobathymetric analyses of sedimentary basins. In: Current Concepts of Depositional Systems with Applications for Petroleum Geology, San Joaquin Geological Society Short Course, Chapter 11, pp. 1–12.

JOHNSON, A.T., JORDAN, T.E., JOHNSON, N.M., and NAESER, C. (1987) Cronologia y velocidad de sedimentacion en una secuencia volcaniclastica, Rodeo, Provincia de San Juan, Argentina. X Congreso Geologico Argentino, Actas 2, pp. 83–86.

JOHNSON, G.D., RAYNOLDS, R.G.H., and BURBANK, D.W. (1986) Late Cenozoic tectonics and sedimentation in the north-western Himalayan foredeep:I. Thrust ramping and associated deformation in the Potwar region. In: Allen, P. and Homewood, P. (eds) Foreland Basins. International Association Sedimentologists Special Publication 8, pp. 273–291.

JOHNSON, N.M., JORDAN, T.E., JOHNSSON, P.A., and NAESER, C.W. (1986) Magnetic polarity stratigraphy, age and tectonic setting of fluvial sediments in an eastern Andean foreland basin, San Juan Province, Argentina. In: Allen, P. and Homewood, P. (eds) Foreland Basins. International Association Sedimentologists Special Publication 8, pp. 63–75.

JOHNSON, N.M., SHEIKH, K.A., DAWSON-SAUNDERS, E., and McRAE, L.E. (1988) The use of magnetic-reversal time lines in stratigraphic analysis: A case study in measuring variability in sedimentation rates. In: Kleinspehn, K.L. and Paola, C. (eds) New Perspectives in Basin Analysis. New York: Springer-Verlag, pp. 189–200.

JOHNSON, N.M., STIX, J., TAUXE, L., CERVENY, P.J., and TAHIRKHELI, R.A.K. (1985) Paleomagnetic chronology, fluvial processes and tectonic implications of the Siwalik deposits near Chinji Village, Pakistan. Journal Geology 93:27–40.

JOHNSSON, P.A., JOHNSON, N.M., JORDAN, T.E., and NAESER, C.W. (1984) Magnetic polarity stratigraphy and age of the Quebrada del Cura, Río Jachal, and Mogna Formations near Huaco, San Juan Province, Argentina. Noveno Congreso Geologico Argentino, Actas III, pp. 81–96.

JORDAN, T.E. (1981) Thrust loads and foreland basin evolution, Cretaceous, western United States. American Association Petroleum Geologists Bulletin 65:2506–2520.

JORDAN, T.E. and ALLMENDINGER, R.W. (1986) The Sierras Pampeanas of Argentina: A modern analogue of Rocky Mountain foreland deformation. American Journal Science 286:737–764.

JORDAN, T.E. and GARDEWEG, M. (in press) Tectonic evolution of the late Cenozoic Central Andes (20°–33°S). In: Z. Ben-Avraham (ed) The Evolution of the Pacific Ocean Margins. New York: Oxford University Press.

KARLSTROM, E.T. and KARLSTROM, T.N.V. (1987) Summary of a GSA symposium: Late Quaternary alluvial history of the American West: Toward a process paradigm. Geology 15:88–89.

KARNER, G.D. (1986) On the relationship between foreland basin stratigraphy and thrust sheet migration and denudation (Abstract). EOS 67:1193.

KARNER, G.D. and WATTS, A.B. (1983) Gravity anomalies and flexure of the lithosphere at mountain ranges. Journal Geophysical Research 88:10449–10472.

KOMINZ, M.A. and BOND, G.C. (1986) Geophysical modelling of the thermal history of foreland basins. Nature 320:252–256.

KRAUS, M.J. (1980) Genesis of a fluvial sheet sandstone, Willwood Formation, northwest Wyoming. University Michigan Papers Paleontology, no. 24, pp. 87–94.

LANE, E.W. (1955) The importance of fluvial morphology in hydraulic engineering. Proceedings American Society Civil Engineers 81, paper 795:1–17.

LAWTON, T.F. (1986) Compositional trends within a clastic wedge adjacent to a fold-thrust belt: Indianola Group, central Utah, U.S.A. In: Allen, P. and Homewood, P. (eds) Foreland Basins. International Association Sedimentologists Special Publication 8, pp. 411–423.

LEEDER, M.R. (1978) A quantitative stratigraphic model for alluvium, with special reference to channel deposit density and interconnectedness. In: Miall, A.D. (ed) Fluvial Sedimentology. Canadian Society Petroleum Geologists Memoir 5, pp. 587–596.

LEOPOLD, L.B. and BULL, W.B. (1979) Base level, aggradation, and grade. Proceedings American Philosophical Society 123:168–201.

LYON-CAEN, H., MOLNAR, P., and SUAREZ, G. (1985) Gravity anomalies and flexure of the Brazilian Shield beneath the Bolivian Andes. Earth Planetary Science Letters 75:81–92.

MINGRAMM, A., RUSSO, A., POZZO, A., and CAZAU, L. (1979) Sierras Subandinas. In: Segundo Simposio de Geologia Regional Argentina: Cordoba, Academia Nacional de Ciencias, Volume 1, pp. 95–138.

MORTON, A.D. (1985) Heavy minerals in provenance studies. In: Zuffa, G.G. (ed) Provenance of Arenites. NATO Advanced Science Institutes Series. Dordrecht, Holland: D. Reidel Publishing Company, pp. 249–277.

ORIEL, S.S. and ARMSTRONG, F.C. (1966) Times of thrusting in Idaho-Wyoming thrust belt: Reply. American Association Petroleum Geologists Bulletin 50:2614–2621.

ORTIZ, A. and ZAMBRANO, J.J. (1981) La provincia geologica Precordillera Oriental. Octavo Congreso Geologico Argentino 3:59–74.

PAOLA, C. (1988) Subsidence and gravel transport in alluvial basins. In: Kleinspehn, K.L. and Paola, C. (eds) New Perspectives in Basin Analysis. New York: Springer-Verlag, pp. 231–243.

PARKER, R.B. (1985) Buffers, energy storage, and the mode and tempo of geologic events. Geology 13:440–442.

PITMAN, W.C. III (1978) Relationship between eustacy and stratigraphic sequences of passive margins. Geological Society America Bulletin 89:1389–1403.

POSAMENTIER, H.W. and VAIL, P.R. (in press) Eustatic controls on clastic deposition. In: Wilgus, C.K., Van Wagner, J., Mitchum, R., and Posamentier, H. (eds) Sea Level Research—An Integrated Approach. Society Economic Paleontologists Mineralogists Special Publication 41.

PRICE, R.A. (1981) The Cordilleran foreland thrust and fold belt in the southern Canadian Rocky Mountains. In: McClay, K.R. and Price, N.J. (eds) Thrust and Nappe Tectonics. Geological Society London Special Publication 9, pp. 427–448.

PRICE, R.A. and FERMOR, P.R. (1985) Structure section of the Cordilleran foreland thrust and fold belt west of Calgary, Alberta. Geological Survey of Canada Paper 84-14.

PRICE, R.A. and MOUNTJOY, E.W. (1970) Geologic structure of the Canadian Rocky Mountains between Bow and Athabasca rivers—a progress report. In: Wheeler, J.O. (ed) Structure of the Southern Canadian Cordillera. Geological Association Canada Special Paper 6, pp. 7–25.

QUINLAN, G.M. and BEAUMONT, C. (1984) Appalachian thrusting, lithospheric flexure and the Paleozoic stratigraphy of the eastern interior of North America. Canadian Journal Earth Sciences 21:973–996.

READ, W.A. and DEAN, J.M. (1982) Quantitative relationships between numbers of fluvial cycles, bulk lithological composition and net subsidence in a Scottish Namurian basin. Sedimentology 29:181–200.

REYNOLDS, J.H., JORDAN, T.E., and JOHNSON, N.M. (1987) Cronologia neogenica y velocidad de sedimentacion en la Cuenca de La Troya, La Rioja. X Congreso Geologico Argentino, Actas 2, pp. 109–112.

REYNOLDS, J., NICKELSEN, J.J., TABBUTT, K.D., JORDAN, T.E., and JOHNSON, N.M. (1986) Variation in sediment accumulation rate within the Central Andean foreland basin (16-10 Ma interval), San Juan and La Rioja Provinces, Argentina. Geological Society America Abstracts with Programs 18: 729.

ROYDEN, L. and KARNER, G.D. (1984) Flexure of the continental lithosphere beneath Apennine and Carpathian foredeep basins. Nature 309:142–144.

ROYSE, F., Jr., WARNER, M.A., and REESE, D.L. (1975) Thrust belt structural geometry and related stratigraphic problems, Wyoming-Idaho-northern Utah. Rocky Mountain Association Geologists Symposium, pp. 41–54.

SAUNDERS, I. and YOUNG, A. (1983) Rates of surface processes on slopes, slope retreat and denudation. Earth Surfaces Processes Landforms 8:473–501.

SCHMOKER, J.W. (1984) Empirical relation between carbonate porosity and thermal maturity: An approach to regional porosity prediction. American Association Petroleum Geologists Bulletin 68:1697–1703.

SCHUMM, S.A. (1963) Disparity between present rates of denudation and orogeny. United States Geological Survey Professional Paper 454-H, pp. H1–H13.

SCHUMM, S.A. (1968a) River adjustment to altered hydrologic regimen—Murrumbidgee River and paleochannels, Australia. United States Geological Survey Professional Paper 598, 65 p.

SCHUMM, S.A. (1968b) Speculations concerning paleohydrologic controls of terrestrial sedimentation. Geological Society America Bulletin 79:1573–1588.

SCLATER, J.G. and CHRISTIE, P.A.F. (1980) Continental stretching: An explanation of the post-mid-Cretaceous subsidence of the central North Sea Basin. Journal Geophysical Research 85:3730–3735.

SHEFFELS, B., BURCHFIEL, B.C., and MOLNAR, P. (1986) Deformational style and crustal shortening in the Bolivian Andes (Abstract). EOS 67:1241.

SNOW, R.S. and SLINGERLAND, R.L. (1987) Mathematical modeling of graded river profiles. Journal Geology 95:15–33.

SNYDER, D.B. and BARAZANGI, M. (1986) Deep crustal structure and flexure of the Arabian plate beneath the Zagros collisional mountain belt as inferred from gravity observations. Tectonics 5:361–373.

STECKLER, M.S. and WATTS, A.B. (1978) Subsidence of the Atlantic-type continental margin off New York. Earth Planetary Science Letters 41:1–13.

STEIDTMANN, J.R. and SCHMITT, J.G. (1988) Provenance and dispersal of tectogenic sediments in thin-skinned, thrusted terrains. Kleinspehn, K.L. and Paola, C. (eds) New Perspectives in Basin Analysis. New York: Springer-Verlag, pp. 353–366.

STOCKMAL, G.S., BEAUMONT, C., and BOUTILIER, R. (1986) Geodynamic models of convergent tectonics: The transition from rifted margin to overthrust belt and consequences for foreland-basin development. American Association Petroleum Geologists Bulletin 70:181–190.

SUPPE, J. (1981) Mechanics of mountain building and metamorphism in Taiwan. Geological Society China Memoir 4, pp. 67–89.

SUTTNER, L.J. (1969) Stratigraphic and petrographic analysis of Upper Jurassic-Lower Cretaceous Morrison and Kootenai Formations, southwest Montana. American Association Petroleum Geologists 53:1391–1410.

SWIFT, D.P.L., THORNE, J., and NUMMEDAL, D. (1985) Sequence stratigraphy in foreland basins: Inferences from the Cretaceous Western Interior. Offshore Technology Conference Paper 4846, pp. 47–54.

TRIMBLE, S.W. (1975) Denudation studies: Can we assume stream steady state? Science 188:1207–1208.

TRIMBLE, S.W. (1977) The fallacy of stream equilibrium in contemporary denudation studies. American Journal Science 277:876–887.

TURCOTTE, D.L. and SCHUBERT, G. (1982) Geodynamics: Applications of Continuum Physics to Geological Problems. New York: John Wiley and Sons, 450 p.

VAIL, P.R., MITCHUM, R.M., Jr., and THOMPSON, S., III (1977) Relative changes of sea level from coastal onlap. In: Payton, C.E. (ed) Seismic Stratigraphy— Applications to Hydrocarbon Exploration. American Association Petroleum Geologists Memoir 26, pp. 63–82.

VAN HINTE, F.E. (1978) Geohistory analysis—application of micropaleontology in exploration geology. American Association Petroleum Geologists Bulletin 62:201–222.

WATTS, A.B. (1978) An analysis of isostasy in the world's oceans, 1: Hawaiian-Emperor seamount chain. Journal Geophysical Research 83:5989–6004.

WATTS, A.B. and RYAN, W.B.F. (1976) Flexure of the lithosphere and continental margin basins. Tectonophysics 36:25–44.

WILTSCHKO, D.V. and DORR, J.A., Jr. (1983) Timing of deformation in overthrust belt and foreland of Idaho, Wyoming, and Utah. American Association Petroleum Geologists Bulletin 67:1304–1322.

ZUFFA, G.G. (ed) (1985) Provenance of Arenites. NATO Advanced Science Institutes Series. Dordrecht, Holland: D. Reidel Publishing Company, 408 p.

17
Stratigraphic Keys to the Timing of Thrusting in Terrestrial Foreland Basins: Applications to the Northwestern Himalaya

DOUGLAS W. BURBANK and ROBERT G.H. RAYNOLDS

Abstract

Precise determination of the timing and nature of deformational events in foreland basins is critically dependent on the interpretation of stratigraphic indicators of tectonic activity within a detailed chronologic framework. Frequently data from numerous stratigraphic sections must be synthesized in order to generate a clear definition of a structural event and its response within a sedimentary sequence. In terrestrial deposits, magnetostratigraphic studies can provide the temporal control necessary to establish a reliable and coherent synthesis of tectonism that is based on an amalgamation of the diverse responses to, and indicators of, tectonic activity found in geographically separated localities. Useful indicators in this type of analysis include: 1) classical indicators of tectonism, such as the timing of changes in paleocurrents, provenance, and facies, and the ages of unconformities; and 2) indicators based on magnetic data or precise chronologic control, such as the history of syn- and post-depositional tectonic rotation and changes in the rate of sediment accumulation. Three examples based on the interpretation of such stratigraphic data from the northwestern Himalayan foreland basin are used to illustrate diverse thrusting events during the past 5 million years, including initial thrust propagation, later episodes of reactivation, and large-scale out-of-sequence thrusting. These examples illustrate the high degree to which reliable specification of the sequence of thrust events that are closely spaced in time but are distributed across a broad region is critically dependent on the availability of detailed, time-controlled stratigraphic records.

Introduction

Situated along the margins of collisional mountain ranges, most foreland basins are characterized by a succession of transient depocenters that migrate generally outward from the orogenic axis (Raynolds and Johnson 1985; Puigdefabregas *et al.* 1986). This migration is modulated primarily by deformation along the proximal basin margin (Miall 1978; Homewood *et al.* 1986), where there is progressive encroachment of thrusting and associated folding during continuing convergence (e.g., Armstrong and Oriel 1965; Beaumont 1981; Jordan 1981). The systematic displacement of depocenters is a response to crustal loading by successive thrust sheets, but this large-scale trend does not necessarily reflect the complexity of local faulting and associated sedimentation in the more proximal localities.

The objectives of this paper are to describe several ways in which thrusting episodes affecting terrestrial sediments can be discerned in the stratigraphic record, to show how detailed time constraints can be placed on these tectonic events, and to illustrate how the synthesis of well dated, local records of tectonism and sedimentation facilitates the development of a more comprehensive view of regional tectonic patterns. Recent stratigraphic studies of the northwestern Himalayan foreland that have used magnetic-polarity stratigraphy (Opdyke *et al.* 1979; Frost 1979; Moragne 1979; G. Johnson *et al.* 1979; Raynolds 1980; Burbank 1982; N. Johnson *et al.* 1982, 1985) and fission-track dating (G. Johnson *et al.* 1982; Burbank 1982) to provide chronologic control are used here to depict the local response to nearby tectonism. When considered at a regional scale, these data indicate that thrust-front migration is not smooth and systematic, but rather is characterized by complex sequences of deformation, unevenly distributed in both time and space. These conclusions concerning the pace and sequence of tectonic events are only possible due to the availability of tight temporal constraints.

A Model for the Sedimentary Response to Thrusting

Along the margins of terrestrial foreland basins in compressional tectonic settings, thrust and wrench faulting and associated folding represent the dominant structural style. Although lateral ramps, tear faults, and attendant sedimentation can have a profound effect on local depositional patterns (Elliott 1985), this analysis concerns primarily the deformation and sedimentary response that appears to be directly related to thrust faulting.

Prior to the initiation of a new thrust in a terrestrial foreland basin (e.g., the peripheral basin of the Himalaya or the retroarc basin of the Andes), certain sedimentary conditions commonly prevail along the proximal basin margin (Fig. 17.1A). In this simplified model, an older thrust delimits the border of the most active foreland sedimentation. Behind a thrust that has generated a topographic high, both antecedent and consequent rivers drain the uplifted terrain. Along lengthy mountain fronts, many of the major river courses may be focused by structural reentrants from which they debouch on to the foreland basin (Eisbacher et al. 1974; Oberlander 1985). Virtually all of the rivers draining the older fault-bounded terrain build alluvial cones that coalesce along the basin margin and prograde toward the basin axis. Parallel to the mountain front, an elongate trough accommodates a largely longitudinal master drainage system (Eisbacher et al. 1974). In terrestrial foreland basins adjacent to rapidly uplifting ranges, bed-load-dominated rivers tend to characterize the proximal fluvial system (Johnson et al. 1979), and coarse-grained facies near the apices of the alluvial cones fine basinwards into sandy braid plains (Friend et al. 1985; Hirst and Nichols 1986).

Our model predicts that, when the thrust front steps out farther into the basin, a number of structural and depositional changes occur (Fig. 17.1B). Initially, the rate of sediment accumulation decreases as growing, thrust-cored anticlines cause the local rate of basin subsidence to diminish. Eventually, the anticline becomes a feature of positive relief, and the thrust itself breaks the surface and may expose pre-molasse bedrock (Fig. 17.1B). Ahead of the thrust, older molasse sediments are folded, eroded, and sometimes tectonically rotated. Sediments derived from the newly uplifted terrain and deposited unconformably on these folded strata prograde into the proximal portions of the basin.

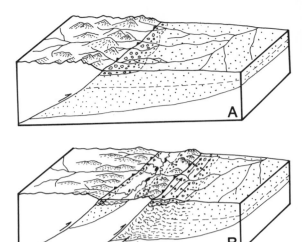

Fig. 17.1. Idealized model of the sedimentary and structural changes that could accompany a thrusting event along the proximal margin of a terrestrial foreland basin. Blackened areas denote lakes. Prior to the second thrusting event, the proximal basin margin is delineated by the older thrust that raised earlier molasse sediments and perhaps crystalline basement (A). Where streams debouch into the basin, alluvial cones prograde toward the basin axis and coalesce into a coarse alluvial apron. The trunk stream usually has a longitudinal orientation, subparallel to the mountain front. When a new thrust steps out into the proximal foreland (B), marked changes in the location and nature of sedimentation occur. An intermontane basin may develop behind the thrust, while ahead of the thrust, older molasse sediments are deformed, and coarse clastic sediments prograde from the newly thrusted terrain.

Behind the thrust, a new intermontane or piggyback basin is generated (Ori and Friend 1984). With thrusts of limited lateral extent, the river systems are diverted parallel to the thrust until they can exit along a lateral ramp (Elliott 1985). Behind thrusts of more regional extent, pre-existing fluvial systems frequently are partially ponded, and sluggish meandering or lacustrine depositional facies with a longitudinal orientation replace the previous braided system of a primarily transverse character (Burbank and Johnson 1983; Burbank and Tahirkheli 1985). Consequent drainage networks develop on the back slope of the thrusted terrain (Fig. 17.1B) and introduce detritus from new lithologies into the intermontane basin.

Stratigraphic evidence of the thrusting event can be preserved in both the proximal molasse and the new intermontane basin. This evidence includes unconformities, tectonic rotations, and changes in paleocurrents, provenance, facies, and sediment-accumulation rates.

Unconformities

Unconformities developed on folded molasse strata give clear indications of tectonic disruption (Fig. 17.2A). During prolonged intervals with multiple episodes of thrusting, progressive unconformities may develop adjacent to the thrust (Riba 1976; Anadon et al. 1986). Frequently, the hiatus represented by these unconformities diminishes rapidly toward more distal locations, so that a nearly continuous record of sedimentation, punctuated merely by scattered diastemic surfaces, characterizes the distal stratigraphic record. By dating the youngest strata below and the oldest strata above, these tectonic unconformities can provide ages constraining the timing of deformation. Whereas, close to the fault, the direct relationship between the unconformity and the faulting can be demonstrated clearly, it is also here that the greatest erosion of underlying strata occurs. For such cases, the age of tectonism cannot be closely constrained. Consequently, more distal localities are usually preferable for precisely dating an episode of faulting (Burbank and Raynolds 1984). However, to be useful, the tectonic significance of the distal unconformity must be unambiguous.

Within the intermontane basin, widespread unconformities are generated on the back side of the thrusted terrain, as consequent streams develop on the dip slope (Fig. 17.2A). This zone can be locally extended during thrusting (Platt 1986; Yin 1986), and renewed uplift can cause decoupling and gentle folding of newly deposited intermontane sediments (Burbank and Johnson 1983). Both the intensity of folding and the extent of the unconformities diminish basinward, such that uninterrupted deposition frequently prevails in the intermontane basin center, despite the major uplift along the distal basin margin. As in the molasse sediments on the distal side of the new thrust, constraining the age of the unconformity also helps to specify the time of deformation.

Rotations

Tectonic rotations of blocks of variable scales occur during compression, and these can be defined through statistical analysis of the magnetic data from single or multiple sites. At the largest scale

(100 to >1,000 km), differential convergence between two colliding continental masses or fragments may cause a steady, regional rotation about a vertical axis amounting to 1–2° per million years (Minster and Jordan 1978). At the small scale, localized shearing along a thrust front and associated tear faults may create extensive and complex rotations of small blocks (<1 km). However, neither of these two types of rotations is very useful for dating thrust events. The large-scale rotations cannot be tied to specific, local tectonic movements, and the small block rotations, although directly related to specific events, are usually difficult to date due to limited exposure and disrupted stratigraphy.

Between the largest and smallest scales, rotations of intermediate scale blocks (1–100 km) are particularly useful in defining thrust-belt deformation. As thrust sheets glide forward, they may undergo a differential rotation about a vertical axis due either to the specific geometry of the sliding mass and/or the terrain being overridden, or to the irregular character of the slippage surface, which varies in its ability to facilitate gliding (Fig. 17.2B; Seeber et al. 1981; Butler et al. 1987). Similarly, shear stresses exerted by an advancing thrust sheet can generate rotations of the strata that are being folded and eroded in front of the thrust (Fig. 17.2B). In each of these two cases, dating of the rotational events can help to delimit the history of thrust movement. Indeed, in particularly well dated successions, rotational studies can reveal a record of motion from initiation to completion that is more detailed than that discernable in the physical stratigraphic and structural record (e.g., Hornafius et al. 1986).

Facies, Provenance, Paleocurrents, and Sediment-Accumulation Rates

In front of a newly formed thrust sheet, higher-gradient streams frequently supplant the previous fluvial network in terrestrial foreland basins. The detrital load of these rivers usually increases such that they introduce coarser-grained sediments, including gravels in many proximal localities, into the terrain adjacent to the thrust (Fig. 17.2C; Eisbacher et al. 1974; DeCelles et al. 1987; Hirst and Nichols 1986). Consequently, an upward-coarsening succession, sometimes with abrupt basal contacts, may record the nearby thrusting events. The propagation of the coarse facies from an uplifted block is time transgressive, so that the first appearance of a

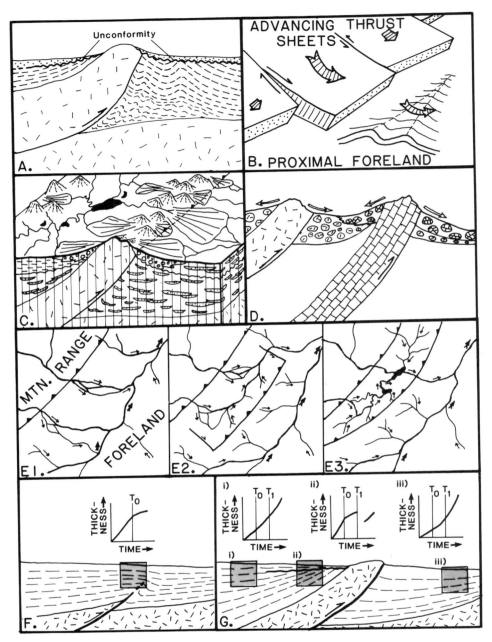

Fig. 17.2. Responses to thrusting recorded in the terrestrial sedimentary record. A: Unconformities frequently develop above the deformed molasse sediments in front of the thrust, as well as atop the uplifted sediments on the back side of the thrust. B: Differential tectonic rotations about a vertical axis can occur both in the advancing thrust sheet itself and in the deformed foreland in front of the thrust sheet. Syndeformational sedimentation may record this rotation, such that the ages of initiation and termination of the rotational event may also serve to constrain the duration of the thrusting event. C: Facies changes. Prior to thrusting, low-sinuosity alluvial deposits characterize the proximal foredeep. Following thrusting and uplift, the new source area can shed coarse clastic debris into both the newly formed intermontane basin and the foreland ahead of the thrust. In the center of the intermontane basins, sluggish fluvial and even lacustrine systems supplant the previous lower sinuosity river facies, whereas in the foreland, an abrupt and distinct upward coarsening trend can occur. D: Provenance. Newly thrusted and uplifted bedrock terrains can provide distinct clast assemblages that record the initial surficial exposure of the new source terrain. E: Paleocurrents. In response to thrusting, the previous longitudinal trunk stream and transverse tributaries (1) may be displaced basinward. If the new thrust is of limited lateral extent (2), rivers entering the

thrust-related facies at any given locality is unlikely to provide an accurate date for the tectonic event by itself. However, dating the coarse-grained influx at several localities along a line perpendicular to the thrust front localities can permit an assessment of the rate of transgression. This rate may be extrapolated back in time and space toward the thrust front to estimate the time of the inception of progradation from the thrusted terrain. Whereas in the Himalayan foreland most thrusting events are associated with coarse-grained sediments, in other foreland basins, such as in the northern Rockies (Beck and Vondra 1985; Winslow et al. 1985), fine-grained facies are sometimes localized adjacent to uplifts by thrusting events. Given both the potentially diverse depositional response to thrusting and the observation that coarse-grained facies may have nontectonic origins (e.g., glacial outwash), their genetic relationship to a given thrust must be evaluated before they can be used to date thrust activity.

Within the newly formed intermontane depression behind the thrust, two distinctive facies changes occur in our model (Fig. 17.2C). Vigorous uplift can readily pond transverse streams, such that lacustrine and high-sinuousity fluvial facies often replace less sinuous, higher-gradient stream deposits along the axial portions of the valley (Burbank and Johnson 1983). Secondly, rapid uplift promotes the development of coarse-grained deposits shed off the flank of the thrusted terrain and prograding toward the intermontane basin center (Fig. 17.2C; Steidtmann and Schmitt 1988: this volume).

Changes in provenance frequently accompany or precede these facies variations (Blatt 1967; Fuchtbauer 1967). Because thrusting inevitably disrupts previous drainage patterns, the ensuing fluvial reorganization can cause some compositional variability at many localities. Moreover, the newly thrusted terrain, which formerly lay below the prevailing depositional plain, can become a new source area for both the foreland and the intermontane basin (Fig. 17.2D). If this new source varies distinctively from the lithologies previously contributing to foreland deposition, then observable changes in provenance may be interpreted as a direct response to thrusting (Puigdefabregas et al. 1986; DeCelles et al. 1987). Because debris from newly exposed lithologies pervades all facies, provenance changes are often less time-transgressive than the consanguinous facies changes (see Jordan et al. 1988: this volume, for further discussion), and, therefore, the ages of provenance variations may more closely date specific tectonic events than do site-specific facies variations.

In terrestrial environments, it is inevitable that thrusting causes some drainage reorganization in the disrupted terrain. Consequently, important paleocurrent changes can provide distinctive records of thrusting events (Fig. 17.2E; e.g., Eisbacher et al. 1974; Burbank and Reynolds 1984). Whereas the course of major streams may be little altered by thrusting (Oberlander 1985), the pattern of many tributaries is frequently dramatically changed. The uplifted thrust terrain becomes a drainage divide for local streams. Longitudinal drainage develops along the axis of the intermontane basis, whereas in the deformed region in front of the thrust, synclinal axes and strike valleys may control the local fluvial network. The record of these paleocurrent changes provides useful insights into the topographic changes that accompanied thrusting.

The rate of net sediment accumulation in most foreland basins is modulated in part by the rate of subsidence of the basement in response to sediment and tectonic loading. Sea-level changes and the rate of sediment production exert additional, but often less important, control on sediment accumulation. Two contrasting responses to thrusting can occur in terrestrial foreland basins. In the incipient stages of development of a new thrust in a more distal posi-

---

◄

basin may be diverted subparallel to the thrust until they can exit along the lateral thrust ramp, whereas with more extensive thrusts (3), a largely centripetal drainage may develop throughout most of the intermontane basin. In either case, the thrusting causes a reversal of paleocurrents along the back side of the thrust. F: Changes in sediment-accumulation rates during the incipient growth of thrust-cored anticlines. Beginning at time T_0, antiformal growth causes a decrease in sediment-accumulation rates in the region where uplift is progressing. At time T_1, the thrust uplift exceeds the accumulation rate and the sequence is truncated. G: Sediment-accumulation rate variations during and following thrusting. Toward the center of the intermontane depression (i), rates may be nearly unperturbed by thrusting. Immediately behind the thrust (ii), sediment accumulation will be slowed and then terminated by erosion during thrusting. Subsequently, intermontane sediments can be accumulated (after T_1). Ahead of the thrust (iii), loading causes accelerated sediment accumulation during thrusting (T_0-T_1).

tion, the rate of subsidence locally diminishes as initial folding and the growth of thrust-cored anticlines commence (Fig. 17.2F). Hence, in the area where the new thrust will cut toward the surface, decelerating sediment-accumulation rates are expected (see Burbank and Raynolds 1984 and Raynolds and Johnson 1985 for discussion). Subsequently, after movement along a thrust has emplaced a new load on the crust (Fig. 17.2G), increased rates of subsidence and sediment accumulation are anticipated in areas ahead of the new thrust due to downward crustal flexure (see Beaumont 1981 and Jordan *et al.* 1988: this volume for discussion). The effects of thrusting on an intermontane or piggyback basin can be diverse (Steidtmann and Schmitt 1988: this volume). Renewed movement along a pre-existing thrust defining the distal edge of basin would be expected to diminish sediment-accumulation rates within much of the basin, whereas thrusting that leads to the initial structural definition of a basin could cause either an increase or decrease in accumulation rates as a function of position in the basin and the previous depositional history.

Chronologic Control of the Stratigraphic Record

Traditionally, biostratigraphic and radiometric ages have been used to define the timing of events seen in the stratigraphic record (see Miall 1984 for discussion). In terrestrial settings, the precision of biostratigraphic ages often is not high, because faunal stage boundaries are not reliably tied to a detailed chronometric scale (Harland *et al.* 1982). Moreover, the nonsynchrony of faunal changes and the uneven nature of the terrestrial fossil record dictate that many stage boundaries are time transgressive between widely separated localities. Therefore, biostratigraphic dates in terrestrial successions are likely to be both precise and accurate only where local stage boundaries are tied directly to an independent chronologic scale (e.g., Barry *et al.* 1982). Although radiometric dates can overcome this problem, material suitable for isotopic dating is unavailable in many stratigraphic successions.

Because magnetic reversals are globally synchronous phenomena and because the magnetic-polarity time scale is quite well established for Cenozoic and Mesozoic time (Mankinen and Dalrymple 1979; Harland *et al.* 1982; Berggren *et al.*

1985; Kent and Gradstein 1985), magnetic-polarity stratigraphy provides a dating tool of potentially high precision and accuracy. The critical task is to establish an unambiguous correlation between a locally observed succession of reversals and the global magnetic-polarity time scale. Radiometric dates on volcanic ashes or other igneous rocks intercalated in the stratigraphic sequence are helpful in generating a reliable correlation.

When a correlation is made successfully, each identified magnetic-reversal boundary introduces a time line into the local stratigraphy (e.g., Johnson *et al.* 1988: this volume). Interpolation between known reversals permits estimation of the age of events recorded by intervening strata. Although a precision of 50–90 Ka can be achieved in some situations (Badgley *et al.* 1986), more typically an uncertainty of 100–200 Ka is associated with most interpolated magnetostratigraphic dates for Neogene time. The precision decreases with greater antiquity and during intervals of less frequent reversals. In general, however, magnetostratigraphic studies provide a chronologic control that permits more precise dating of the stratigraphic record than has been common in the past, and they allow reliable correlations to be made between scattered localities exhibiting little or no faunal or lithological similarities (see Johnson *et al.* 1988: this volume). Such chronologic studies, when combined with stratigraphic observations, facilitate the development of a detailed regional synthesis.

Examples from the Northwestern Himalayan Foreland Basin

With the exception of the Indus and Bengal fans that flank the Indian subcontinent at a large distance from the Himalaya, deposition in the Himalayan foreland basin is exclusively terrestrial. This fact, in combination with both the extensive Cenozoic deformation and the excellent stratigraphic exposures in the northwestern Himalaya, makes this an appropriate area in which to apply previously described stratigraphic criteria to examine the character and chronology of thrusting events in a terrestrial foreland basin.

Because few individual localities provide all of the indicators that are necessary to date a deformational event reliably, data must be combined from numerous dated sequences. Since the mid 1970s, a large

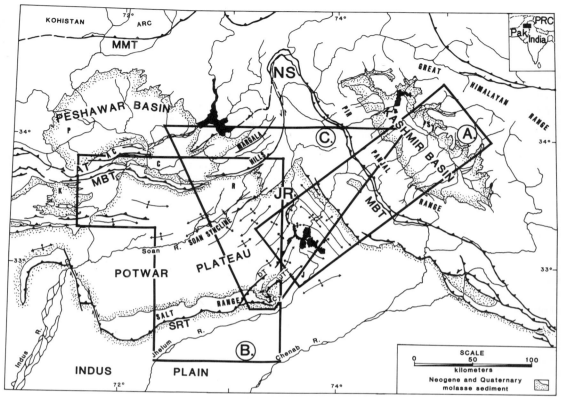

Fig. 17.3. Map of the northwestern Himalayan foreland basin, the southern margin of Kohistan, the southwestern margin of the Himalaya, and the major intermontane basins in the vicinity of the Northwest Syntaxis (NS). The major anticlinal axes in the deformed molasse sediments, as well as major thrust faults (barbed lines) and strike-slip faults in the region surrounding the Jhelum Re-entrant (JR) are delineated. Box A: Figure 17.4; Box B: Figure 17.7; Box C: Figure 17.10. Major thrust faults: Attock Thrust (AT); Main Boundary Thrust (MBT); Main Mantle Thrust (MMT); Salt Range Thrust (SRT). Other localities: Campbellpore (C); Jhelum (J); Kohat (K); Peshawar (P); Rawalpindi (R); Srinagar (S); Attock-Cherat Range (AC); Margala Hills (M).

number of magnetostratigraphic studies have been completed in the northwestern Himalayan foreland basin and adjacent intermontane basins (Fig. 17.3). The tectonic implications of some of these chronologic studies have been described by G. Johnson *et al.* (1979), N. Johnson *et al.* (1982, 1985), Burbank and Raynolds (1984), and Burbank *et al.* (1986). Data from this region are used here to demonstrate how stratigraphic and structural information can be integrated within a reliable chronologic framework to provide a synthesis of both local and regional tectonism. The examples serve: 1) to illustrate the initiation and subsequent recurrence of thrusting along a major fault system; 2) to delimit an interval of uplift that precedes by several millon years the previously accepted age of fault development in this region; and 3) to define out-of-sequence thrusting events on the

local scale. When considered in concert with relevant structural data, these examples from widely separated localities suggest both regional coherence in a chronologic and structural framework and potential causal linkages between disparate deformational events.

The Kashmir Basin and the Main Boundary Thrust (India)

The Kashmir Basin is one of the largest late Cenozoic intermontane basins in the Himalayan chain. It lies along the eastern limb of the Northwest Syntaxis (Fig. 17.3), where it is bounded to the northeast by the mountains of the High Himalaya and to the southwest by the Pir Panjal Range. The Pir

Panjal Range itself is bordered by the Main Boundary Thrust (MBT) system (Figs. 17.3 and 17.4), and it is the movement of these thrusts that has uplifted the Pir Panjal and formed the Kashmir Basin. Numerous stratigraphic sections in the Kashmir Basin and in the deformed molasse of the nearby Jhelum Re-entrant (Fig. 17.4) have been dated using magnetostratigraphy (Fig. 17.5; Johnson *et al.* 1979; Raynolds 1980; Burbank 1982; Raynolds and Johnson 1985).

Four lines of evidence can be used to delimit the initial structural development of the Kashmir Basin (Fig. 17.6). Within the basin itself, a distinctive lacustrine succession [part of the Karewa Group (Bhatt 1976)] contains a thick conglomeratic unit near its base. Magnetostratigraphic dating of the overlying strata at Hirpur (Burbank 1982) indicates that the top of the conglomerates dates from ~3.5 Ma. Although neither the conglomerates nor the abbreviated lacustrine strata underlying them have been dated using magnetics, average sediment-accumulation rates derived from the overlying sequence can be extrapolated through these updated units to yield estimated ages for them. Such an extrapolation indicates that deposition of the lowest lacustrine units had commenced by at least 4–4.5 Ma (Fig. 17.6). This sedimentation was a response to ponding of the fluvial system by initial uplift of the Pir Panjal (Burbank and Johnson 1983) and is typical of intermontane deposition in basins where antecedent rivers traverse newly formed, rapidly uplifting thrusts (Figs. 17.2C, 17.2E).

Additional evidence for deformation comes from the southwest of the Pir Panjal, where a major change in paleocurrents and provenance occurs between 4–5 Ma (Fig. 17.6). Most fluvial strata in the Jhelum Re-entrant prior to 5 Ma exhibit paleocurrents directed to the E and ENE and contain detritus dominated by "white" sandstones with abundant hornblende, granitic pebbles, and white quartzitic clasts (Raynolds 1982). These lithologies resemble those transported by the modern Indus River, and, by analogy, the ancient fluvial system is interpreted also to have drained the strongly uplifted and deeply dissected regions of the High Himalaya. Unlike the modern Indus, however, it appears that this ancient fluvial network joined the Ganges River and drained along the axis of the foredeep to the east (Raynolds 1982). By 4–5 Ma, this system is replaced by a fluvial network almost everywhere oriented to the south and characterized by "brown" sandstones containing volcanic and brown quartzite clasts and

lacking hornblende and granitic pebbles (Fig. 17.6). These lithologies are analogous to those of the modern Jhelum River, and, like the present-day Jhelum, the ancient system is interpreted to have drained the newly formed Pir Panjal Range.

The third line of evidence derives from the time-transgressive influx of conglomerates into the proximal foreland of the Jhelum Re-entrant (Fig. 17.6). Conglomerates carrying clasts typical of lithologies exposed in the Pir Panjal first appear at Sakrana and Rata-Dadial (Fig. 17.4) 2.5–3.0 Ma. The steady advance of this distinctive facies can be traced to the southwest at an average progradational rate of ~3 cm/yr (Burbank and Raynolds 1984; Raynolds and Johnson 1985). If this rate is extrapolated back through time and toward the MBT, it also suggests that the initiation of conglomeratic sedimentation on proximal alluvial cones commenced at about 4–5 Ma. Finally, although few of the sections in this region comprised rocks dated as older than ~4 Ma, the two sequences that are older (Ganda Paik and Basawa/Sanghoi Kas; Fig. 17.4) show distinctive increases in sediment-accumulation rates at ~5 Ma (Fig. 17.6). At Ganda Paik, this change also coincides with the disappearance of "white" sandstones and the inception of "brown" sandstones. These rate changes can be interpreted as a response to subsidence due to thrust loading along the MBT. Thus, facies and provenance changes, paleocurrent variations, and rates of sediment accumulation and facies migration gathered from numerous sections on either side of the MBT permit rather tight constraints to be placed on the initial stage of movement of this fault system.

Whereas the foregoing delineates beginning of thrusting along the MBT, how are renewed or accelerated movements along faults recorded in the sedimentary record? Once intermontane sedimentation is initiated in the piggyback basin, what changes occur that can be interpreted as responses to further thrusting? In Kashmir, two subsequent intervals of uplift can be delineated in the Pir Panjal Range based on data from the intermontane basin. Although a centripetal drainage undoubtedly prevailed throughout most of the basin's history, major conglomeratic influxes transgressed to the southwest across the dated localities on several occasions prior to 1.9 Ma (Fig. 17.6; Burbank and Johnson 1983). These depositional events reveal both: 1) the low relief that prevailed in the central and southern portions of the Kashmir Basin at that time; and 2) the continued importance of uplift along the northeastern margin

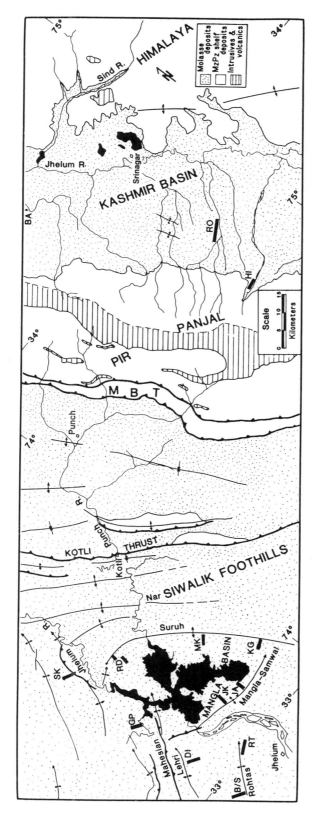

Fig. 17.4. Simplified geologic map extending from the northeastern margin of the Kashmir basin across the Main Boundary Thrust (MBT) to the Jhelum River in the eastern Potwar Plateau. The stippled pattern (molasse deposits) represents both Siwalik and Murree fluvial sediments, as well as the predominantly lacustrine Karewa sediments in Kashmir. Section locations are shown by black rectangles: Baramula (BA); Basawa/Sanghoi Kas (B/S); Dina (DI); Ganda Paik (GP); Hirpur (HI); Jari (JA); Jhel Kas (JK); Kas Guma (KG); Mawa Kaneli (MK); Rata-Dadial (RD); Rohtas (RT); Romushi (RO); Sakrana (SK). For location of area, see Box A in Figure 17.3.

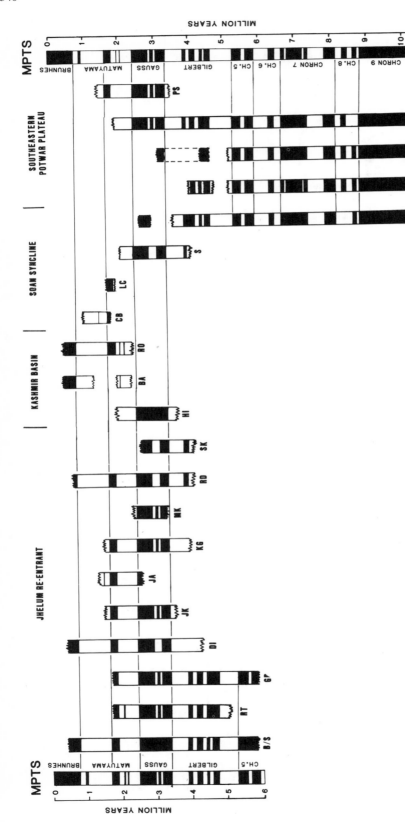

Fig. 17.5. Correlation diagram for the local magnetostratigraphies of the north-western Himalayan foreland and intermontane basins. The correlations between the local magnetostratigraphy and the magnetic-polarity time scale (MPTS) are assisted by fission-track dates on volcanic ashes, faunal data, and distinctive lithostratigraphic horizons that are correlative between nearby sections. Both fossil occurrences and the frequent presence of prominent ashes associated with the Gauss-Matuyama boundary aid in the recognition of identifiable chrons and sub-chrons at the local scale. The sections are arranged in the spatial sequence in which they would be encountered on a traverse 1) starting from the Jhelum River and ending in Kashmir [Basawa/Sanghoi Kas (B/S); Rohtas (RT); Ganda Paik (GP); Dina (DI); Jhel Kas (JK); Jari (JA); Kas Guma (KG); Mawa Kaneli (MK); Rata-Dadial (RD); Sakrana (SK); Hirpur (HI); Baramula (BA); Romushi (RO)]; and 2) starting at the Campbellpore Basin (CB) and ending in the Salt Range [Lei Conglomerate (LC); Soan (S); Kas Dovac (KD); Bhaun (BN); Tatrot/Andar (TA); Kotal Kund (KK); and Pind Savikka (PS)].

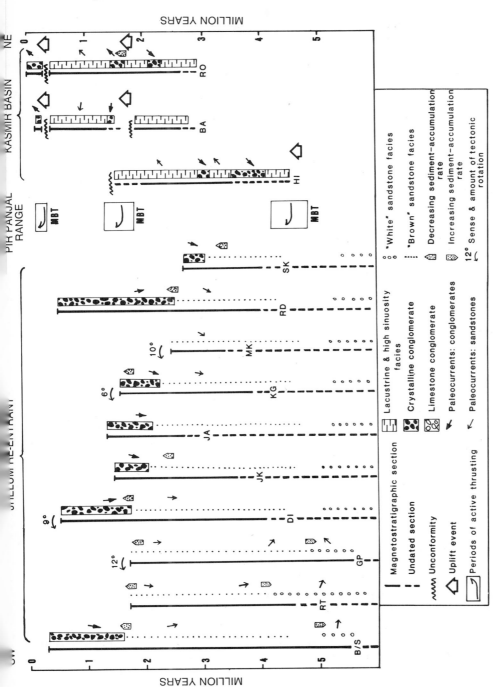

Fig. 17.6. Schematic representation of tectonic and sedimentologic events for the local sequences aligned along the Kashmir-to-Jhelum traverse. (For location, see Fig. 17.4). The solid vertical line represents the temporal duration of each section as interpreted from the magnetostratigraphic data. The amount of post-depositional rotation is determined from the magnetostratigraphic data from each section. Those sections showing no rotation are those whose mean magnetic orientation is not significantly different from geographic north and south. Each of the events (influx of conglomerates, initiation of tectonic uplift, erosion, changes in sediment-accumulation rates, and paleocurrent direction) is shown in its proper temporal position in each profile. Section abbreviations are the same as those for Figure 17.5. Concurrent changes in the tectonic indicators listed above are used to delineate the initiation of thrusting on the Main Boundary Thrust (MBT) between 4 and 5 Ma. Recurrent uplift events (at ~ 1.9 and 0.4 Ma) are similarly inferred from changes in sediment-accumulation rates, unconformities, and paleocurrent changes. Data are from Johnson *et al.* (1979), Raynolds (1980), and Burbank (1982).

of the basin. After 1.8 Ma, sediment-accumulation rates decline (Burbank *et al.* 1986), and subsequent paleocurrents, including those of conglomerates reflecting Pir Panjal lithologies, are directed towards the basin center. This rate decrease and paleocurrent reversal are interpreted to reflect renewed or accelerated uplift in the Pir Panjal. Along the flanks of the Pir Panjal, a syndepositional unconformity truncates strata dated at slightly less than 2 Ma at Baramula (Fig. 17.6). As seen in some other areas where deposition and deformation are occurring concurrently (e.g., Riba 1976), the angular discordance at Baramula diminishes rapidly away from the uplifted mountain front. This uplift appears to be synchronous with that inferred from sediment-accumulation rates and paleocurrent data gathered within the basin 60 km to the southeast at Romushi (Fig. 17.6), and it suggests an uplift event of regional scale. Whereas many of the sections southwest of the Pir Panjal display *decreasing* sediment-accumulation rates at about this same time (~1.8 Ma; Fig. 17.6), flexural models (e.g., Beaumont 1981) would suggest that additional thrusting along the MBT should have caused *accelerated* rates here. The observed rates suggest that thrust deformation was propagating farther into the foreland at this time (e.g., Fig. 17.2F), perhaps along the undated Kotli Thrust or more southerly thrust-cored anticlines (Fig. 17.4).

Although the data described here delimit three distinct episodes of uplift over the past 5 million years within the Pir Panjal Range, none of the sections used in this analysis is closer than 30 km to the MBT system (Figs. 17.3, 17.4). Immediately south of the former thrusts, sediments coeval with the thrusting have been eroded or overthrust, whereas to the north, the pre-Miocene bedrock is exposed. Thus, areas in close proximity to the thrusts reveal little concerning the history of faulting. Nonetheless, a rather detailed record of thrusting and uplift can be interpreted from the well dated stratigraphic successions in the more distant localities described here.

The Eastern Salt Range

How can differential rotations be used to date thrusting events, and can short-lived deformational events be discerned solely through rotational data? Stratigraphic data from the eastern Salt Range in Pakistan provide some answers to these questions.

The Salt Range marks the southern limit of prominent deformation in the region of the Northwest Syntaxis (Fig. 17.3). The range itself, comprising Phanerozoic bedrock and overlying late Cenozoic molasse sediments (Gee 1980), represents the southern termination of a large salt-lubricated detachment that underlies nearly all of the Potwar Plateau (Seeber and Armbruster 1979; Lillie *et al.* in press). To the south, modern molasse deposition continues unabated on the present Jhelum and Indus plains (Fig. 17.7). As the southernmost deformed zone, the Salt Range has traditionally been thought also to represent the most recent (Pleistocene) deformation in the region (Burbank and Raynolds 1984; Yeats *et al.* 1984; Lillie *et al.* in press; Butler *et al.* 1987). However, unconformities and both rotational and provenance data indicate that early Pliocene uplift, as well as Pleistocene deformation, occurred in several areas.

Two areas are described here: one from Bhaun, north of the central Salt Range, and one from Tatrot/Andar/Kotal Kund near the eastern terminus of the range (Fig. 17.7). The paleomagnetic record at Bhaun (Opdyke *et al.* 1979, 1982; N. Johnson *et al.* 1982) is interpreted to span an interval dated at ~9.5–4 Ma (Fig. 17.5). Observational data related to facies, provenance, rotation, and unconformities within the stratigraphic record provide clear evidence for tectonism toward the end of this time. First, distinctive conglomerates, comprising clasts of Eocene carbonates and of red granites derived from the upper Paleozoic Talchir glacial beds (G. Johnson *et al.* 1986), appear in the Bhaun succession for the first time at about 5 Ma (Fig. 17.8). Because the known source areas for the Talchir clasts are in the Salt Range (Gee 1945), this indicates that uplift had already exposed this nearby source area to the south of the Bhaun section (i.e., in a more distal location with respect to the orogenic axis). Second, the base of these conglomerates at 5 Ma marks an angular unconformity attributable to uplift. Third, the strata below the unconformity have been rotated counterclockwise more than 30° with respect to the expected magnetic axial-dipole direction, whereas the overlying, apparently Gilbert-aged sediments are essentially unrotated (Fig. 17.8; Opdyke *et al.* 1982). This differential rotation across an unconformity associated with an important provenance and facies change can be interpreted to result from thrusting and associated shearing at this time. The unrotated strata above the unconformity constrain this event to have occurred at ~5 Ma, and the

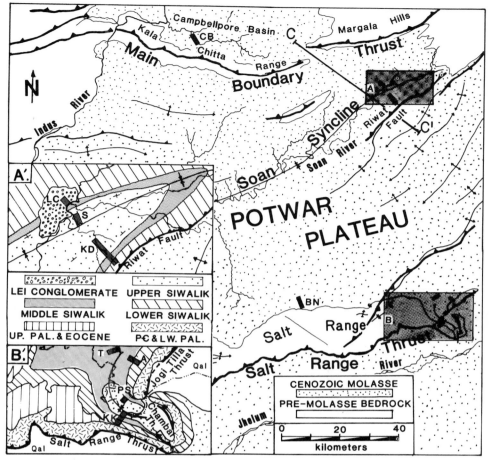

Fig. 17.7. Tectonic map indicating the major thrusts and folds in the vicinity of the Soan Syncline and the eastern Salt Range on the western side of the Jhelum Re-entrant (for location, see Box B, Fig. 17.3). The allochthonous Potwar Plateau (including the Soan Syncline) is riding above the Salt Range Thrust. Late Cenozoic molasse sediments are shown by the stippled pattern. The studied area of the Soan Syncline (A) between the MBT and the Riwat Fault is shown in inset Map A′. Magnetic sections: Kas Dovac (KD); Soan (S); Lei Conglomerate (LC). The position of the Campbellpore Basin (CB) section is shown in the upper left part of the main map. The studied area of the eastern Salt Range (B) is shown in inset map B′. Magnetic sections: Andar (A); Kotal Kund (KK); Pind Savikka (PS); Tatrot (T). Note that the Upper Paleozoic and Eocene strata (vertical stripes) include the Talchir boulder source area. The position of the Bhaun (BN) section in the central-eastern Salt Range is shown on the main map. Line C-C′ indicates the position of the cross section in Figure 17.9B. Salt Range Thrust modified from Yeats *et al.* (1984).

southerly source of the conglomerate indicates that uplift to the south defined a new piggyback basin at this time. This uplift probably resulted from early Pliocene movement along at least a segment of the Salt Range Thrust (Seeber *et al.* 1981). Rotational data (Opdyke *et al.* 1982) collected along the back slope of the present Salt Range Thrust indicate differences of rotation of up to 30° between localities less than 30 km apart. Such data imply inhomogeneous behavior of the advancing thrust

sheet, perhaps due either to segmentation (e.g., Fig. 17.2B) or to multiple deformational events affecting different portions of the thrusted terrain.

Whereas at Bhaun, rotational data across an unconformity help to define a thrusting event, farther east in the Salt Range, lack of differential rotation across an unconformity of both the same age and probable genesis helps to define a still younger episode of thrusting. The Jogi Tilla structure, a southeast-verging succession of thrusts that bring

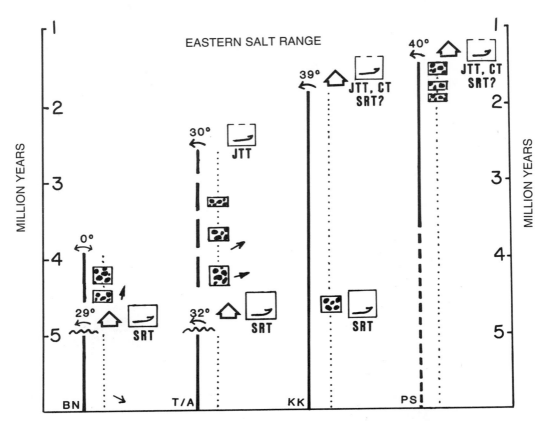

Fig. 17.8. History of deformation in the eastern Salt Range. Earliest movements occur at Bhaun (BN), where a differential rotation, unconformity, and facies change reflect early motion on the Salt Range Thrust (SRT). Data from Tatrot/Andar (T/A) reflect this same deformation. The second rotation of Tatrot/Andar occurs several million years later and is likely to be due to thrusting along the Jogi Tilla Thrust (JTT) and Chambal Thrust (CT) that is recorded by Pleistocene data at Kotal Kund (KK) and Pind Savikka (PS). Subsidiary motion along the SRT may also occur at this time. Data from Frost (1979), Opdyke et al. (1982), Johnson et al. (1986) and Burbank and Beck (in press). See Figure 17.6 for explanation of symbols.

the entire Phanerozoic succession to the surface (Gee 1980), trends northeast from the eastern termination of the Salt Range (Fig. 17.7). On the northern flank of this structure, molasse sediments older than 5.0 Ma at Andar Kas and Tatrot (Figs. 17.5, 17.7) are rotated over 30° and are unconformably overlain by strata dated at ~4.5–3.0 Ma (Fig. 17.8; Opdyke et al. 1979, 1982; N. Johnson et al. 1982; Barry et al. 1982). Similar to Bhaun, Salt Range-derived, Paleozoic Talchir clasts first appear just above this unconformity, and, at nearby Kotal Kund, Talchir clasts also first appear at ~5 Ma, although no observable unconformity is preserved. These data are interpreted to be responses to the same episode of Salt Range thrusting and uplift that affected the Bhaun sequence in the early Pliocene.

There is, however, no differential rotation across the unconformity at Tatrot/Andar, because the overlying strata are also rotated 30° (Fig. 17.8; Burbank and Beck in press). Consequently, this rotation at Tatrot/Andar cannot be attributed to early Pliocene thrusting, but must be due to a later event occurring sometime in the past 3 million years. Chronologic and rotational data from Kotal Kund (N.M. Johnson et al. 1982) and nearby Pind Savikka (Frost 1979) suggest when this later deformation occurred (Fig. 17.8). These sequences lie adjacent to the Jogi Tilla and the Chambal structures (an east southeast-directed succession of thrusts). Both sections extend to about 1.5–2.0 Ma, and both exhibit >30° rotations of the earliest Pleistocene and older strata. The location of all three sections with respect to the local thrusts suggests that the most likely explanation for their rotation and folding is that thrusting along both the Jogi Tilla and Chambal structures (Fig. 17.7) during Pleistocene time primarily generated the deformation observable today. Lesser movement along the Salt Range Thrust may also have been involved (Butler *et al.* 1987). In reconstructing the thrusting events in this region, the rotational data from several nearby sections, along with provenance, unconformity, and paleocurrent data, are critical to the interpretation of the deformational history. Without them, the sequence of Tatrot/Andar would probably be incorrectly interpreted as recording only a single thrusting event.

The Soan Syncline and Main Boundary Thrust (Pakistan)

Whereas structural cross-cutting relationships and stratal attitudes can often permit the sequence of thrusting to be discerned in imbricate fans of thrusts, the sequence of motion along widely separated thrusts is frequently difficult to resolve due to the absence of original cross-cutting structures or to the erosion of intervening strata following deformation. Chronologic data from magnetostratigraphy can provide the temporal control of the stratigraphic record needed to resolve these timing difficulties, even when no cross-cutting information is available. In this example from northern Pakistan, two major faults are spaced about 30 km apart. Despite this large separation, chronostratigraphic studies here can confine deformational episodes to narrow time windows and can clearly demonstrate that out-of-sequence thrusting has occurred.

The Soan Syncline lies about 20 km south of the MBT in Pakistan and is truncated along its southern margin by the Riwat Fault (Fig. 17.7). Chronologic control of the synclinal strata is derived from magnetostratigraphic studies by Moragne (1979), Raynolds (1980), and Burbank and Beck (unpublished data). The southernmost section at Kas Dovac spans an interval dated at ~ 10–2.4 Ma (Fig. 17.5). Strata older than ~ 3.5 Ma contain abundant Eocene limestone clasts carried by rivers flowing from source areas to the northwest. A syndepositional unconformity truncates the Kas Dovac section at ~ 3.5 Ma, and the overlying, finer-grained, Gauss-aged strata (Fig. 17.5) contain no limestone pebbles, but instead contain quartzites and volcanic clasts apparently

derived from the northeast (Fig. 17.9A). This unconformity dies out to the north, where sedimentation continued unabated throughout this interval. According to our model (Fig. 17.2A), the diminution of the depositional hiatus to the northwest should indicate uplift in the southeast. Whereas the strata below the unconformity are counterclockwise rotated an average of > 30° (Burbank and Raynolds 1984), the overlying sequence is essentially unrotated (Fig. 17.9A). Consequently, these data suggest that thrusting along the Riwat Fault along the southern margin of the present Soan Syncline at ~ 3.5 Ma (Fig. 17.9B) caused a rotation of the underlying strata and created an axial drainage (now occupied by the Soan River). This drainage pattern (Fig. 17.9C) could explain both the lower energy facies and the exclusion of limestone clasts that is observed in the supraconformable sequence on the northern flank of the Riwat structure.

The vertically dipping northern limb of the Soan Syncline comprises at least 6 km of molasse strata (Gill 1952) ranging from ~ 2.1 to > 9.5 Ma (Figs. 17.5, 17.9; Moragne 1979; Raynolds 1980). The entire sequence is overlain by the subhorizontal Lei conglomerate, the base of which is dated at ~ 1.9 Ma (Figs. 17.5, 17.9; G. Johnson *et al.* 1982). Thus, a strong compressional event between 2.1 and 1.9 Ma is seen to have terminated sedimentation at Soan and caused > 6 km of uplift and erosion (minimum rate of > 30 m/10^3 yr) prior to deposition of the Lei conglomerate 200,000 years later. This deformation is viewed as a response to large-scale motions on the MBT to the north (Figs. 17.7, 17.9). This interpretation is reinforced by the initiation at this time (Fig. 17.5) of low-energy intermontane sedimentation in the Campbellpore basin (Burbank 1982; G. Johnson *et al.* 1982), a piggyback basin defined by movement along the MBT (Fig. 17.9).

The well dated depositional sequences in the Soan region provide extraordinarily tight temporal constraints on the timing, sequence, and rate of deformation of thrusted strata. Clear definition of syndepositional unconformities along with tightly constrained ages on angular unconformities define discrete chronologic windows within which deformation occurred. Although it is perhaps rare that deformational events can be specified so clearly through stratigraphic analysis, the studies in the Soan area aptly illustrate the manner in which the time resolution of the magnetic time scale can bring structural events into sharp focus.

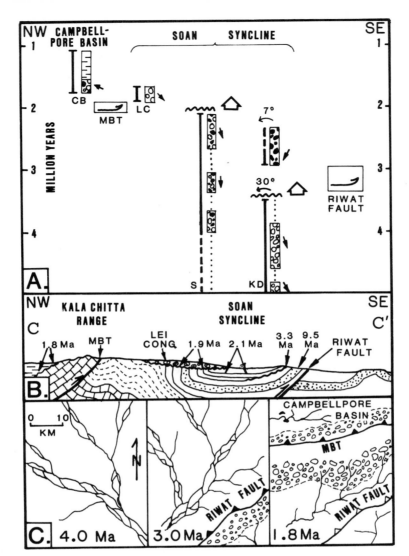

Fig. 17.9. History of deformation of Soan Syncline area. (See Fig. 17.7 for localities.) A: Temporal position of the relevant magnetostratigraphic sections and the stratigraphic and rotational data recorded at each section. Important points to note in the Kas Dovac (KD) section on the southeastern margin of the Soan Syncline are the presence of an unconformity constrained to an interval between 3 and 3.5 Ma, the exclusion of limestone-bearing clasts above this unconformity, and the rotation of the section approximately 30° counterclockwise. This unconformity is not seen in the Soan (S) section on the north limb of the syncline where sedimentation continued uninterruptedly until about 2.1 Ma. The overlying limestone-bearing conglomerates of the Lei Conglomerate (LC) entered the area at approximately 1.9 Ma. To the north of the MBT, sedimentation in the Campbellpore Basin (CB) commenced at ~1.8 Ma and was subsequently characterized by sluggish fluvial and lacustrine depositional environments. Symbols are the same as in Figure 17.6. Data are from Moragne (1979), Raynolds (1980), G.D. Johnson *et al.* (1982), and Burbank and Beck (unpublished data). B: Simplified geologic cross section that illustrates the relevant time constraints on the structural events. (Location of cross section is shown as line C-C′ in Fig. 17.7.) The initial thrusting on the Riwat Fault caused the rotation and synsedimentary unconformity observed at Kas Dovac and dated at about 3.3 Ma. This unconformity dies out to the north where sedimentation continued until approximately 2.1 Ma. Subsequently about 6 km of uplift and erosion took place in the 200,000 year interval prior to the commencement of deposition of the Lei Conglomerate at approximately 1.9 Ma. This deformational event is seen as a response to major movement along the MBT and is inferred to have initiated intermontane sedimentation in the Campbellpore Basin beginning at ~1.8 Ma. The data here indicate that out-of-sequence thrusting occurred in this area with the Riwat Fault becoming active about 1 million years prior to major motion on the MBT. C: Inferred changes in the drainage and sedimentation patterns that occurred at ~4.0, 3.0, and 1.8 Ma. Filled barbs on thrust indicate active faulting.

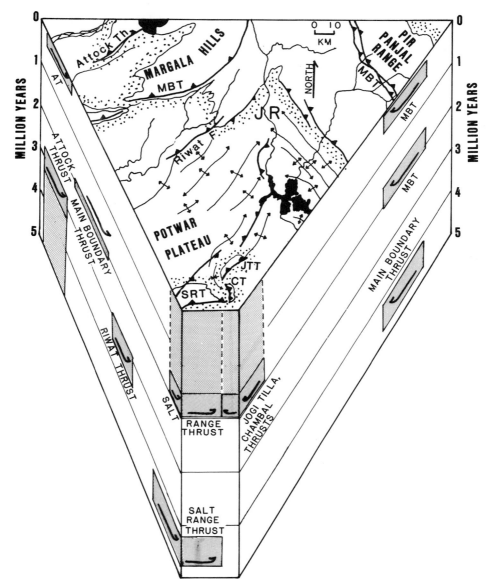

Fig. 17.10. Summary of fault motions in the study area (area C, Fig. 17.3). The vertical axis represents time (0–5 Ma). The locations of faults are determined from orthogonal projections onto the faces of the block. The movement history of the Attock Thrust (AT) is discussed in Burbank (1983) and Burbank and Tahirkheli (1985). The shaded boxes delineate the time and space domain over which each thrust is interpreted to have been active. The data suggest that thrusting does not systematically become younger southward, particularly in the eastern Potwar region, where the Salt Range and Riwat Thrusts each moved prior to major motion on the MBT along the Margala Hills and Kala Chitta Range.

Summary and Conclusions

These examples from the northwestern Himalaya foreland basin illustrate several ways in which stratigraphic data related to deformation can be interpreted within a reliable chronologic framework. The detailed history of thrusting that emerges from such studies provides new insights into the possible sequence and pattern of deformational events in a foreland basin. Previously it was thought that the thrusts described here developed from the Himalaya sequentially as deformation rippled systematically from north to south across the basin (Gansser 1964; Burbank and Raynolds 1984). These examples show that there have been multiple episodes of substantial movement along some thrust systems [e.g., the MBT in Kashmir (Figs. 17.5, 17.10)]. Moreover, the explicit thrust chronologies illustrated here demonstrate that in several cases substantial amounts of movement occurred on more southerly (distal) thrusts prior to movement on more northerly (proximal) thrusts (Fig. 17.10). The thrusts examined are spaced 30–100 km apart, and although they each apparently root into the same sole thrust (Lillie *et al.* in press), their large spatial separation means they should not be regarded as a single hanging-wall imbricate thrust fan. Instead, these data indicate that major out-of-sequence thrusting has occurred within the Himalayan foreland during the past 5 million years, a finding at considerable variance with the classical models of deformation in this region. Because cross-cutting relationships are not preserved between most of these widely spaced faults, such a finding is strongly dependent on examination of the stratigraphic indicators of thrusting within the chronologic framework derived from the magnetic studies.

The tight temporal constraints also permit an examination of the regional coherence and pattern of deformation. For example, in the northwestern Himalaya, one might now ask whether there is a causal linkage between the initial uplift of the Pir Panjal Range and the apparently synchronous deformation of the Salt Range, how the timing and amount of shortening in this region can be used to test various models for thrust-wedge development (e.g., Platt 1986; Yin 1986), and whether the very rapid deformation of the Soan Syncline could be a response to an encounter between the sole thrust and various down-to-the-north basement faults that underlie the northern margin of the Salt Range (Johnson *et al.* 1986; Lillie *et al.* in press). Chronologic studies of thrusting cannot answer these questions directly, but they can suggest temporal linkages that would otherwise be obscure, and they can provide a data base for assessing tectonic models.

In conclusion, paleomagnetic studies have provided an important key to the interpretation of regional tectonic and depositional histories. Many of the deformational events from the northwestern Himalaya described here have been identified previously on the basis of stratigraphic investigations. Magnetostratigraphic studies, however, permit recognition of additional tectonic events which were latent in the physical stratigraphy, but which are clearly defined by rotational magnetic data.

Moreover, the chronologic control provided by the magnetic studies creates a sound basis for synthesizing data from diverse localities within a temporal framework and generates a new level of precision in the description of specific events. As a consequence, the tectonic evolution of a region in both time and space can now be described more accurately than has previously been possible, and various models for the depositional and tectonic evolution of the region can be tested. The successes that have been achieved in recent years with magnetostratigraphic studies in terrestrial sediments are serving to dispel the skepticism that greeted early practitioners who adopted techniques previously reserved for studies of marine and igneous rocks. The potential of these techniques for examining both the timing of events and the rates of various processes in diverse tectonic and depositional environments has just begun to be tapped.

Acknowledgments. This research was supported by NSF grants EAR 7803639, EAR 8018779, and INT 8019373 to Dartmouth College and grants from the Richard E. Stoiber Field Fund, the Geological Society of America, Amoco International, Sigma Xi, Marathon Oil, and the Shell Foundation to D.W.B. We thank our colleagues at Dartmouth College, Peshawar University, and the Geological Survey of Pakistan, and we particularly appreciate the assistance of Gary Johnson, Noye Johnson, R.A.K. Tahirkheli, M. Khan, and G.M. Bhatt. Critical reviews of an earlier version of this manuscript by J. Steidtmann, C. Paola, R. Yeats, and K. Kleinspehn have enhanced this paper and are gratefully acknowledged.

References

ANADON, P., CABRERA, L., COLUMBO, F., MARZO, M., and RIBA, O. (1986) Syntectonic intraformational unconformities in alluvial fan deposits, eastern Ebro Basin margins (NE Spain). In: Allen, P.A. and Homewood, P. (eds) Foreland Basins. International Association Sedimentologists Special Publication 8, pp. 259–271.

ARMSTRONG, F.C. and ORIEL, S.S. (1965) Tectonic development of the Idaho-Wyoming thrust belt. American Association Petroleum Geologists Bulletin 49:1847–1866.

BADGLEY, C., TAUXE, L., and BOOKSTEIN, F.L. (1986) Estimating the error of age interpretation in sedimentary rocks. Nature 319:139–141.

BARRY, J.C., LINDSAY, E.H., and JACOBS, L.L. (1982) A biostratigraphic zonation of the middle and upper Siwaliks of the Potwar Plateau of northern Pakistan. Palaeogeography, Palaeoclimatology, Palaeoecology 37:95–130.

BEAUMONT, C. (1981) Foreland Basins. Geophysical Journal Royal Astronomical Society 65:291–329.

BECK, R.A. and VONDRA, C.F. (1985) Syntectonic sedimentation and Laramide basement faulting, Rocky Mountain Foreland, Wyoming, U.S.A. (Abstract). International Association Sedimentologists Symposium on Foreland Basins, Fribourg, Switzerland, p. 36.

BERGGREN, W.A., KENT, D.V., FLYNN, J.J., and VAN CONVERING, J.A. (1985) Cenozoic geochronology. Geological Society America Bulletin 96:1407–1418.

BHATT, D.K. (1976) Stratigraphical status of the Karewa Group of Kashmir, India. Himalayan Geology 6:197–208.

BLATT, H. (1967) Provenance determinations and recycling of sediments. Journal Sedimentary Petrology 37:1031–1044.

BURBANK, D.W. (1982) The Chronologic and Stratigraphic Evolution of the Kashmir and Peshawar Intermontane Basins, Northwestern Himalaya. Unpublished Ph.D. dissertation, Dartmouth College, Hanover, N.H., 291 p.

BURBANK, D.W. (1983) The chronology of intermontane-basin development in the northwestern Himalaya and the evolution of the Northwest Syntaxis. Earth Planetary Science Letters 64:77–92.

BURBANK, D.W. and BECK, R.A. (in press) Early Pliocene uplift of the Salt Range: Temporal constraints on thrust wedge development, northwest Himalaya, Pakistan. In: Malinconico, L.L. and Lillie, R.J. (eds) Tectonics and Geophysics of the Western Himalaya. Geological Society America Special Paper.

BURBANK, D.W. and JOHNSON, G.D. (1983) The Late Cenozoic chronologic and stratigraphic development of Kashmir intermontane basin, northwestern Himalaya. Palaeogeography, Palaeoclimatology, Palaeoecology 43:205–235.

BURBANK, D.W. and RAYNOLDS, R.G.H. (1984) Sequential late Cenozoic structural disruption of the northern Himalayan foredeep. Nature 311:114–118.

BURBANK, D.W., RAYNOLDS, R.G.H., and JOHNSON, G.D. (1986) Late Cenozoic tectonic and sedimentation in the northwestern Himalayan foredeep: II. Eastern limb of the Northwest Syntaxis and regional synthesis. In: Allen, P.A. and Homewood, P. (eds) Foreland Basins. International Association Sedimentologists Special Publication 8, pp. 293–306.

BURBANK, D.W. and TAHIRKHELI, R.A.K. (1985) The magnetostratigraphy, fission-track dating, and stratigraphic evolution of the Peshawar intermontane basin, northern Pakistan. Geological Society American Bulletin 96:539–552.

BUTLER, R.W.H., COWARD, N.P., HARWOOD, G.M., and KNIPE, R.J. (1987) Salt control on thrust geometry, structural style and gravitational collapse along the Himalayan mountain front in the Salt Range of northern Pakistan. In: O'Brien, J.J. and Lesche, I. (eds) Dynamical Geology of Salt and Related Structures. Austin, Texas: Academic Press, pp. 399–418.

DeCELLES, P.G., TOLSON, R.B., GRAHAM, S.A., SMITH, G.A., INGERSOLL, R.V., WHITE, J., SCHMIDT, C.J., RICE, R., MOXON, I., LEMKE, L., HANDSCHY, J.W., FOLLO, M.F., EDWARDS, D.P., CAVAZZA, W., CALDWELL, M., and BARGAR, E. (1987) Laramide thrust-generated alluvial-fan sedimentation, Sphinx Conglomerate, southwestern Montana. American Association Petroleum Geologists Bulletin 71:135–155.

EISBACHER, G.H., CARRIGY, M.A., and CAMPBELL, R.B. (1974) Paleodrainage pattern and late-orogenic basins of the Canadian Cordillera. In: Dickinson, W.R. (ed) Tectonics and Sedimentation. Society Economic Paleontologists Mineralogists Special Publication 22, pp. 143–166.

ELLIOTT, T. (1985) Foreland basins: The nature of the links between sedimentation and thrust-nappe tectonics in basin-filled successions (Abstract). International Association Sedimentologists Symposium on Foreland Basins, Fribourg, Switzerland, p. 51.

FRIEND, P.F., HIRST, J.P.B., NICHOLS, G.J., and VAN GELDER, A. (1985) Patterns of early Miocene fluvial deposition in the northern Ebro Basin, Spain (Abstract). Sixth European Regional Meeting International Association Sedimentology, Lleida, Spain, p. 171.

FROST, C.D. (1979) Geochronology and Depositional Environment of a Late Pliocene Age Siwalik Sequence Enclosing Several Volcanic Tuff Horizons, Pind Savikka Area, Eastern Salt Range, Pakistan. Unpublished A.B. Thesis, Dartmouth College, Hanover, N.H., 41 p.

FUCHTBAUER, H. (1967) Die Sandsteine in der Molasse Nordlich der Alpen. Geoligische Rundschau 56:266–300.

GANSSER, A. (1964) Geology of the Himalayas. London: Interscience, 289 p.

GEE, E.R. (1945) The age of the Saline Series of the Punjab and Kohat. Proceedings National Academy Science India 14:269–312.

GEE, E.R. (1980) Pakistan Geological Map: Salt Range Series (1:50,000), 6 sheets. Quetta: Geological Survey Pakistan.

GILL, W.D. (1952) The stratigraphy of the Siwalik Series in the northern Potwar, Punjab, Pakistan. Quarterly Journal Geological Society London 107:395–421.

HARLAND, W.B., COX, A.V., LLEWELLYN, P.G., PICKTON, S.A.G., SMITH, A.G., and WALTERS, R. (1982) A Geologic Time Scale. Cambridge, UK: Cambridge University Press, 128 p.

HIRST, J.P.P. and NICHOLS, G.J. (1986) Thrust tectonic controls on Miocene alluvial distribution patterns in southern Pyrenees. In: Allen, P.A. and Homewood, P. (eds) Foreland Basins. International Association Sedimentologists Special Publication 8, pp. 247–258.

HOMEWOOD, P., ALLEN, P.A., and WILLIAMS, G.D. (1986) Dynamics Molasse Basin of western Switzerland. In: Allen, P.A. and Homewood, P. (eds) Foreland Basins. International Association Sedimentologists Special Publication 8, pp. 199–217.

HORNAFIUS, S., LUYENDYK, B.P., TERRES, R.R., and KAMERLING, M.J. (1986) Timing and extent of Neogene tectonic rotation in the western Transverse Ranges, California. Geological Society America Bulletin 97:1476–1487.

JOHNSON, G.D., JOHNSON, N.M., OPDYKE, N.D., and TAHIRKHELI, R.A.K. (1979) Magnetic reversal stratigraphy and sedimentary tectonic history of the Upper Siwalik Group, Eastern Salt Range and Southwestern Kashmir. In: Farah, A. and DeJong, K.A. (eds) Geodynamics of Pakistan. Quetta, Pakistan: Geological Survey Pakistan, pp. 149–165.

JOHNSON, G.D., RAYNOLDS, R.G.H., and BURBANK, D.W. (1986) Late Cenozoic tectonic and sedimentation in the northwestern Himalayan foredeep: I. Thrust ramping and associated deformation in the Potwar region. In: Allen, P.A. and Homewood, P.(eds) Foreland Basins. International Association Sedimentologists Special Publication 8, pp. 273–291.

JOHNSON, G.D., ZEITLER, P., NAESER, C.W., JOHNSON, N.M., SUMMERS, D.M., FROST, C.D., OPDYKE, N.D., and TAHIRKHELI, R.A.K. (1982) The occurrence and fission-track ages of late Neogene and Quaternary volcanic sediments, Siwalik Group, northern Pakistan. Palaeogeography, Palaeoclimatology, Palaeoecology 37:63–93.

JOHNSON, N.M., OPDYKE, N.D., JOHNSON, G.D., LINDSAY, E.H., and TAHIRKHELI, R.A.K. (1982) Magnetic polarity stratigraphy and ages of Siwalik Group rocks of the Potwar Plateau, Pakistan. Palaeogeography, Palaeoclimatology, Palaeoecology 37:17–42.

JOHNSON, N.M., SHEIKH, K.A., DAWSON-SAUNDERS, E., and McRAE, L.E. (1988) The use of magnetic-reversal time lines in stratigraphic analysis: A case study in measuring variability in sedimentation rates. In: Kleinspehn, K.L. and Paola, C. (eds) New Perspectives in Basin Analysis. New York: Springer-Verlag, pp. 189–200.

JOHNSON, N.M., STIX, J., TAUXE, L., CERVENY, P.F., and TAHIRKHELI, R.A.K. (1985) Paleomagnetic chronology, fluvial processes, and tectonic implications of the Siwalik deposits near Chinji Village, Pakistan. Journal Geology 93:27–40.

JORDAN, T.E. (1981) Thrust loads and foreland basin evolution, Cretaceous, western United States. American Association Petroleum Geologists Bulletin 65:2506–2520.

JORDAN, T.E., FLEMINGS, P.B., and BEER, J.A. (1988) Dating thrust-fault activity by use of foreland-basin strata. In: Kleinspehn, K.L. and Paola, C. (eds) New Perspectives in Basin Analysis. New York: Springer-Verlag, pp. 307–330.

KENT, D.V. and GRADSTEIN, F.M. (1985) A Cretaceous and Jurassic geochronology. Geological Society America Bulletin 96:1419–1427.

LILLIE, R.J., JOHNSON, G.D., YOUSUF, M., ZAMIN, A.S.H., and YEATS, R.S. (in press) Structural development within the Himalayan foreland fold-and-thrust belt of Pakistan. In: Beaumont, C. and Tankard, A.J. (eds) Basins of Eastern Canada and Worldwide Analogs. Canadian Society Petroleum Geologists Special Volume.

MANKINEN, E.A. and DALRYMPLE, G.B. (1979) Revised geomagnetic polarity time scale for the interval 0–5 m.y. B. P. Journal Geophysical Research 84:615–626.

MIALL, A.D. (1978) Tectonic setting and syndepositional deformation of molasse and other nonmarine-paralic sedimentary basins. Canadian Journal Earth Sciences 15:1613–1632.

MIALL, A.D. (1984) Principles of Basin Analysis. New York: Springer-Verlag, 490 p.

MINSTER, J.B. and JORDAN, T.H. (1978) Present-day plate motions. Journal Geophysical Research 83:5331–5354.

MORAGNE, J.H. (1979) Magnetic Polarity Stratigraphy and Timing of Deformation for an Upper Siwalik Sedimentary Sequence, Soan Syncline, Pakistan. Unpublished B.A. Thesis, Dartmouth College, Hanover, NH, 31 p.

OBERLANDER, M. (1985) Origin of drainage transverse to structures in orogens. In: Morisawa, M. and Hack, J.T. (eds) Tectonic Geomorphology. Proceedings 15th Annual Binghamton Geomorphology Symposium, 1984. Boston, MA: Allen and Unwin, pp. 155–182.

OPDYKE, N.D., JOHNSON, N.M., JOHNSON, G.D., LINDSAY, E.H., and TAHIRKHELI, R.A.K. (1982) Palaeomagnetism of the Middle Siwalik Formations of Northern Pakistan and rotation of the Salt Range Decollement. Palaeogeography, Palaeoclimatology, Palaeoecology 37:1–15.

OPDYKE, N.D., LINDSAY, E.H., JOHNSON, G.D., JOHNSON, N.M., TAHIRKHELI, R.A.K., and MIZRA, M.A. (1979) Magnetic polarity stratigraphy and vertebrate palaeontology of the Upper Siwalik Subgroup of northern Pakistan. Palaeogeography, Palaeoclimatology, Palaeoecology 27:1–34.

ORI, G. and FRIEND, P.F. (1984) Sedimentary basins formed and carried piggyback on active thrust sheets. Geology 12:475–478.

PLATT, J.P. (1986) Dynamics of orogenic wedges and the uplift of high-pressure metamorphic rocks. Geological Society America Bulletin 97:1037–1053.

PUIGDEFABREGAS, C., MUNOZ, J.A., and MARZO, M. (1986) Thrust belt development in the eastern Pyrenees and related depositional sequences in the southern foreland basin. In: Allen, P.A. and Home-wood, P. (eds) Foreland Basins. International Association Sedimentologists Special Publication 8, pp. 229–246.

RAYNOLDS, R.G.H. (1980) The Plio-Pleistocene Structural and Stratigraphic Evolution of the Eastern Potwar Plateau, Pakistan. Unpublished Ph.D. dissertation, Dartmouth College, Hanover, N.H., 256 p.

RAYNOLDS, R.G.H. (1982) Did the ancestral Indus River flow into the Ganges drainage? University Peshawar Geological Bulletin 14:141–150.

RAYNOLDS, R.G.H. and JOHNSON, G.D. (1985) Rates of Neogene depositional and deformational processes, north-west Himalayan foredeep margin, Pakistan. In: Snelling, N.J. (ed) The Chronology of the Geological Record. Geological Society London Memoir 10, pp. 297–311.

RIBA, O. (1976) Syntectonic unconformities of the Alta Cardener, Pyrenees: A genetic interpretation. Sedimentary Geology 15:213–233.

SEEBER, L. and ARMBRUSTER, J. (1979) Seismicity of the Hazara arc in northern Pakistan: Decollement vs. basement faulting. In: Farah, A. and de Jong, K.A. (eds) Geodynamics of Pakistan. Quetta, Pakistan: Geological Survey Pakistan, pp. 131–142.

SEEBER, L., ARMBRUSTER, J., and QUITTMEYER, R.C. (1981) Seismicity and continental subduction in the Himalayan arc. In: Gupta, H.K. and Delaney, F.M. (eds) Zagros-Hindu Kush-Himalaya Geodynamic Evolution. American Geophysical Union Geodynamic Series 3, pp. 215–242.

STEIDTMANN, J.R. and SCHMITT, J.G. (1988) Provenance and dispersal of tectogenic sediments in thin-skinned, thrusted terrains. In: Kleinspehn, K.L. and Paola, C. (eds) New Perspectives in Basin Analysis. New York: Springer-Verlag, pp. 353–366.

WINSLOW, N.S., HELLER, P.L., and PAOLA, C. (1985) Enigmatic origin of lower Cretaceous gravel beds in Wyoming (Abstract). Society Economic Paleontologists Mineralogists Mid-year Meeting, Golden, Colorado, pp. 98–99.

YEATS, R.S., KHAN, S.H., and AKHTAR, M. (1984) Late Quaternary deformation of the Salt Range of Pakistan. Geological Society America Bulletin 95:958–966.

YIN, A. (1986) A mechanical model for a wedge-shaped thrust sheet (Abstract). EOS Transactions American Geophysical Union 67:1242.

18
Provenance and Dispersal of Tectogenic Sediments in Thin-Skinned, Thrusted Terrains

JAMES R. STEIDTMANN and JAMES G. SCHMITT

Abstract

The style of deformation in thin-skinned, thrusted terrains provides a setting for complex provenance and dispersal relations between source and sediment. Whereas it is commonly assumed that clastic wedges are shed by active thrusts and therefore can be used to date thrust motion, passive sources, formed when inactive terrain is transported over a ramp, may also generate tectogenic sediments. Furthermore, these sediments may be dispersed either synthetically or antithetically to tectonic transport. In some cases the clast composition of these deposits displays an inverted stratigraphy, reflecting unroofing in the source, while in other instances, blended clast compositions may form because of the simultaneous erosion of a thick stratigraphic section, complex geology in the source, or depositional amalgamation during sediment accumulation. Finally, during the evolution of a thrusted terrain, tectogenic sediments are tectonically transported, continually reworked, and cannibalized.

Case histories from the thrust belt of the western United States demonstrate these concepts. Some are rather clearcut examples while others are more circumstantial and are given as alternatives to prevailing interpretations. Although presented individually, each concept is an integral part of an uplift-erosion-transport-deposition system inherent to thin-skinned, thrusted terrains. An understanding of these concepts is a potentially powerful tool and should provide the basis for more fruitful studies of tectogenic sediments.

Introduction

Datable coarse clastic sediments shed by active faults have long been used to indicate time of major thrust events. For the thrust belt of the western United States, much of this work is reported in Armstrong and Oriel (1965), Oriel and Armstrong (1966), Armstrong (1968), Royse et al. (1975), Dorr et al. (1977), Wiltschko and Dorr (1983), and Bohannon (1984). In general the model assumed is that: 1) tectogenic sediments were shed as alluvial fans in front of rising thrust sheets; 2) the age of these sediments, usually determined from pollen and/or vertebrate fossils, indicates the time of motion on some fault now located in the direction from which the thrusts emanated; 3) except for minor reactivation, each fault had one period of major activity; and 4) this time of motion was the same along the entire length of the fault.

Where cross-cutting and overlapping relations between tectogenic deposits and associated faults can be well documented by field mapping or subsurface techniques, the resolution of fault timing is limited only by the precision of the sediment ages. Where cross-cutting and overlapping relations are ambiguous or absent, circumstantial evidence is used. This less diagnostic approach involves lines of reasoning based on paleocurrent and/or compositional evidence. That is, a fault is considered to have been active at a certain time if coarse clastics of that age came from the appropriate direction and have a clast composition that present geology suggests could have been supplied during motion on that thrust. This approach, even in the simplest of geologic settings, is likely to give poorly constrained, ambiguous, or erroneous results. In the internally complex setting of thin-skinned thrusted terrains where the horizontal component of tectonic transport is relatively great, intramontane basins are detached and transported, tectogenic sediments are tectonically transported and continually cannibalized, and where ramping and associated folding, imbrication, and duplexing are the rule, attributing coarse clastics to specific thrust events based solely on composition and paleotransport direction is tenuous at best. Nevertheless, dating thrust faults by this technique, together with the more clear-cut approach using overlapping and cross-cutting relations, has resulted in an informal dogma for the western United States thrust belt that major thrusts young in the direction of tectonic transport.

Concepts

Introduction

Recent analyses (Steidtmann 1985; Schmitt 1986) of our earlier work (e.g., Dorr *et al.* 1977; Schmitt 1982, 1985; Hurst and Steidtmann 1986) and that of others (e.g., Armstrong and Oriel 1965; Royse *et al.* 1975; Crawford 1979; Wiltschko and Dorr 1983) suggest that some of the source-sediment relations assumed in the above-mentioned model do not take into account several structural controls common to thin-skinned thrusting. Specifically, ramping can: 1) generate passive sediment sources that shed sediments but are only indirectly related to fault motion; 2) form paleoslopes toward the interior of the thrust belt causing sediment to be transported in a direction opposite to that of tectonic transport; and 3) amalga-

mate source-rock compositions where layered rocks are steeply tilted during tectonic transport over a ramp, thus inhibiting the formation of unroofing sequences. Furthermore, any two or all three of these source-sediment relations may occur together, producing a deposit in which the true tectonic significance is difficult or impossible to decipher without more definitive evidence.

Active and Passive Sediment Sources

In thin-skinned thrusting, much of the vertical component of tectonic transport, and thus much of the structural and topographic relief, is the direct result of ramping, where the thrusts cut up section through resistant units. Where a thrust ramps to the surface and forms a topographic high (A, Fig. 18.1), an active source is formed that may or may not shed tectogenic sediment suitable for dating the fault motion. At the same time, however, terrain composed of older, inactive thrusts (B, Fig. 18.1) in the hanging wall of the active thrust is also transported over ramps and may form topographic highs. We call these terrains passive sources if they supply sediment to intramontane basins, but dispersal of these sediments dates motion on the active fault (A, Fig. 18.1) and not faulting in the interior of the thrust belt. Misinterpretation of sediments derived from

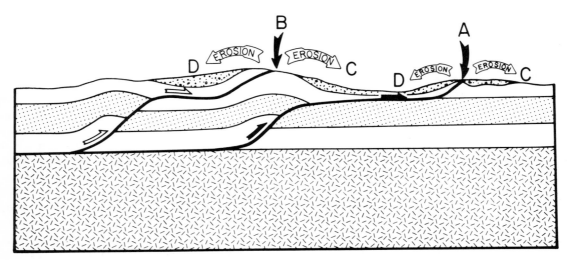

Fig. 18.1. Schematic diagram demonstrating the structural settings for active (A) and passive (B) sources, and synthetic (C) and antithetic (D) dispersal in a thin-skinned, thrusted terrain. Closed arrows on faults indicate motion related to generation of the tectogenic sediments. Open arrows indicate sense of motion on inactive older faults.

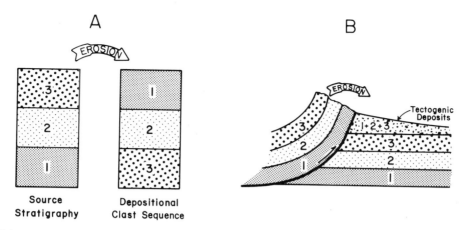

Fig. 18.2. Schematic diagram demonstrating structural settings for the formation of a classic unroofing sequence with inverted stratigraphy (A) and that for blended clast compositions (B).

passive sources can result in attributing younger motion to faulting than was actually the case.

Synthetic and Antithetic Sediment Dispersal

Transport of tectogenic sediments in the same direction as tectonic transport (C, Fig. 18.1) is the commonly assumed setting for the clastic-wedge/thrust association. Although this may be the most common setting for thrust-derived clastics, it is by no means the only one. Where hanging-wall beds dip toward the interior of the thrust belt, paleoslope, and therefore sediment dispersal, is opposite to tectonic transport (D, Fig. 18.1). We refer to this source-transport relation as antithetic dispersal, following the structural terminology common to thrusted terrains. Similarly, sediment transport in the same direction as tectonic transport is referred to as synthetic dispersal.

Inverted and Blended Clast Composition

In settings where distinct source-rock compositions are eroded sequentially, as in the case of predominantly vertical uplift of a stratigraphic section, unroofing sequences are commonly formed in the resultant clastic wedge (Fig. 18.2A). This erosional inverted clast stratigraphy can provide valuable information about the evolving source and aid in the identification of specific source areas (e.g., Dutcher *et al.* 1986). However, in thin-skinned, thrusted terrains, where horizontal transport dominates, layered

rocks consisting of different lithologies are commonly exposed to erosion together as they pass up and over a ramp (Fig. 18.2B) providing a "conveyor belt" of several rock types that are blended during erosion and dispersal. Tectogenic deposits thus formed may show no unroofing sequences but contain the same blended clast composition throughout, sometimes for relatively great thickness. While such a setting does not produce particularly detailed information about the source, it may indicate that the source was formed by tectonic transport over a ramp. Furthermore, the complex structure and uplift histories of thin-skinned, thrusted terrains can result in combinations of inverted sequences and blended clast compositions that vary laterally and reflect extremely local provenance controls (Hurst 1984). Finally, in distal deposits where floodplains are constructed of sediments derived from large areas, clast composition is blended by fluvial processes.

Cannibalism and Tectonic Transport of Proximal Deposits

Even though clastic wedges with synthetic transport may be typical of deposition at a thrust front, their preservation potential, and thus their potential for dating faulting events, is not necessarily good. This stems from the fact that, for major faults, time of thrusting youngs in the direction of tectonic transport. Thus, except for the deposits associated with the youngest (frontal) fault, clastic wedges are subsequently uplifted on the hanging wall of younger thrusts, transported tectonically, and cannibalized by erosion. This phenomenon probably explains the paucity of tectogenic coarse clastics in thin-skinned, thrusted terrains and has led some workers (e.g., Shuster and Steidtmann 1987; Burbank and Raynolds 1988: this volume) to examine more distal sediments, still preserved in front of the thrust belt, for evidence of the time of thrust-sheet emplacement.

Applications

General Statement

The following discussion identifies specific cases that illustrate the concepts of active and passive sources, synthetic and antithetic dispersal, inverted, blended, and complex clast compositions, cannibalization and tectonic transport of proximal deposits, and combinations thereof. They are presented to document that, in fact, these phenomena do occur in thin-skinned thrusted terrains. Some are rather clear-cut examples. Others are circumstantial and are presented as an alternative to prevailing interpretations. In neither case, however, do we intend to imply that any specific provenance determination or dating of any particular faulting event is incorrect, but only that a correct interpretation of these tectogenic deposits may involve previously unrecognized complications.

Active and Passive Sources

Deposits shed by active thrust sources represent the commonly assumed source-sediment relation used for dating thrust events. The lower Eocene Conglomerate Member of the Wasatch Formation in southwestern Wyoming (Oriel 1962) and the laterally equivalent Lookout Mountain Conglomerate (Dorr and Steidtmann 1977; Dorr et al. 1977) demonstrate these relations quite well. The lack of textural maturity and clast composition of these deposits show they were derived from the hanging wall of the frontal thrust immediately to the west. Unconformities within the Wasatch and cross-cutting relations with younger faults in the Lookout Mountain Conglomerate indicate that the faults were active during deposition (Dorr et al. 1977).

In contrast, the Fossil Basin in the thrust belt of southwestern Wyoming (Fig. 18.3) was being filled with sediments from at least one passive source while the basin itself was being transported on the hanging wall of the Absaroka Thrust system (Hurst and Steidtmann 1986). Such basins are variously referred to as detached, piggyback (Ori and Friend 1984), or thrust-sheet-top (Atkinson and Elliott 1985) basins. During early Eocene time, the Fossil Basin was filled from both the east and west when coarse, clastic wedges formed along the basin margins. Work by Hurst and Steidtmann (1986) shows that the coarseness and distribution of these tectogenic sediments, referred to as the Tunp Conglomerate, indicate that they were shed from sources generated by fault motion on both sides of the basin.

Along the eastern margin of the Fossil Basin, clast compositions and the westward intertonguing of the Tunp Conglomerate with finer-grained basin facies of the Wasatch Formation indicate that the Tunp had its source on the hanging wall of the Absaroka Thrust just east of the Fossil Basin. Overlap of folds associated with Absaroka thrusting by Paleocene Evanston Formation strata makes it quite clear, however, that there was no motion on the Absaroka fault after middle Paleocene time (Oriel and Tracey 1970), and therefore that active uplift of its leading edge could not have triggered deposition of the lower Eocene Tunp. However, the Darby Thrust, farther east, was active during early Eocene time (Lamerson 1982) and it has been suggested by Hurst and Steidtmann (1986) that the leading edge of the Absaroka Thrust sheet was uplifted, by passive rotation over a ramp in the Darby Thrust (Fig. 18.3), to form a sediment source for the Tunp Conglomerate.

The Upper Cretaceous Hams Fork Conglomerate Member of the Evanston Formation in northeastern Utah and southwestern Wyoming (Fig. 18.4) was probably also derived from a passive source. The Hams Fork contains abundant, distinctive, upper Precambrian quartzite clasts (Oriel and Tracey 1970; Crawford 1979) that must have been derived from source rocks carried only in the hanging wall of the Paris-Willard Thrust (Royse et al. 1975; Oriel and Platt 1980). However, motion on the Absaroka Thrust, and not the Paris-Willard, has been shown to be the appropriate age to form a Upper Cretaceous topographic high. According to Oriel and Tracey (1970), major motion along the Absaroka Thrust shed and overrode the Upper Cretaceous Hams Fork Conglomerate Member of the Evanston Formation. Undeformed Paleocene strata of the upper portion of the Evanston Formation overlap the Absaroka Thrust and limit its latest motion as pre-Paleocene.

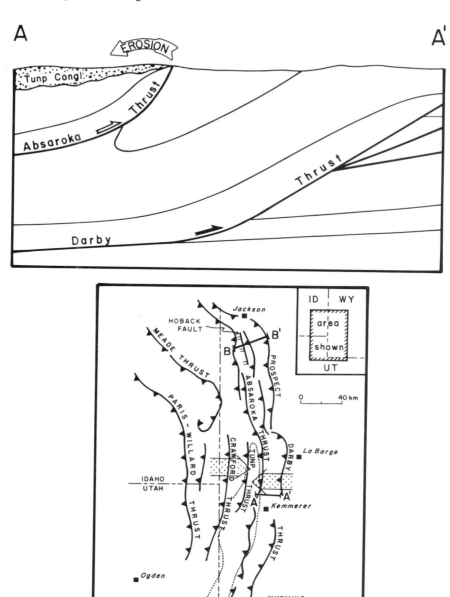

Fig. 18.3. Location map and schematic cross section of eastern side of the Fossil Basin. Extent of the basin indicated by dotted line. Large dotted arrows indicate sediment dispersal. Cross section shows that the leading edge of the Absaroka Thrust was rotated over a ramp in the younger Darby Thrust, generating a passive source. Tecto- genic sediments were transported antithetically to form the Tunp Conglomerate on the east side of the basin. Closed arrows on faults indicate motion related to genera- tion of the tectogenic sediments. Open arrows indicate sense of motion on inactive older faults. Modified from Lamerson (1982).

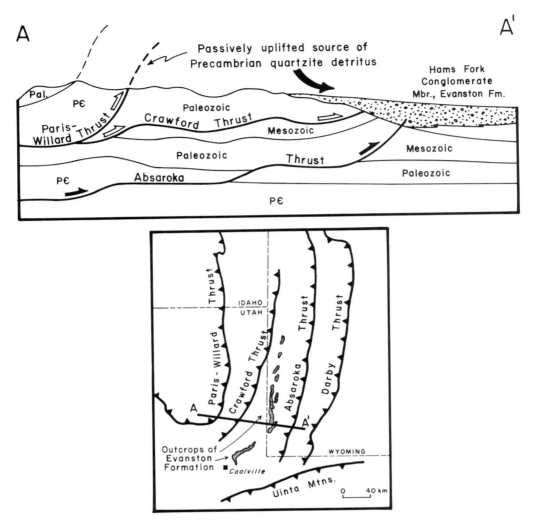

Fig. 18.4. Map and schematic cross section showing the structural relations for passive uplift of the Paris-Willard Thrust plate during Absaroka Thrust motion and the resulting generation of Hams Fork Conglomerate. Closed arrows on faults indicate motion related to generation of the tectogenic sediments. Open arrows indicate sense of motion on inactive older faults. Structural interpretation from Royse et al. (1975).

Crawford (1979) noted these same source area-provenance relations and suggested that initial motion on the Paris-Willard Thrust was synchronous with Late Cretaceous Hams Fork deposition. This interpretation places initial Paris-Willard Thrust movement at a significantly younger date than other interpretations (Armstrong and Oriel 1965; Wiltschko and Dorr 1983) and requires concurrent major motion along both the Paris-Willard and Absaroka Thrusts. In addition, it is in conflict with observations that the Paris-Willard Thrust is folded by transport over an Absaroka Thrust footwall ramp (Royse et al. 1975; Bruhn et al. 1983).

Rather than calling upon interpretations involving recurrent thrust motion or major changes in the sequence of thrust development, we believe that passive uplift is a more logical explanation for deriving the Hams Fork Conglomerate. Regional structural interpretations show that the Paris-Willard Thrust was folded into a broad antiformal feature (Royse et al. 1975) as it was passively transported over a ramp in the Absaroka Thrust (Fig. 18.4; Bruhn et al. 1983). Thus, the highland terrain on the Paris-Willard Thrust plate was once again uplifted and could have served as the source for quartzite clasts in the Hams Fork. Dispersal of Hams Fork sediments may therefore record motion on the younger Absaroka Thrust rather than the fortuitous reactivation of the older Paris-Willard Thrust system during Absaroka Thrust motion.

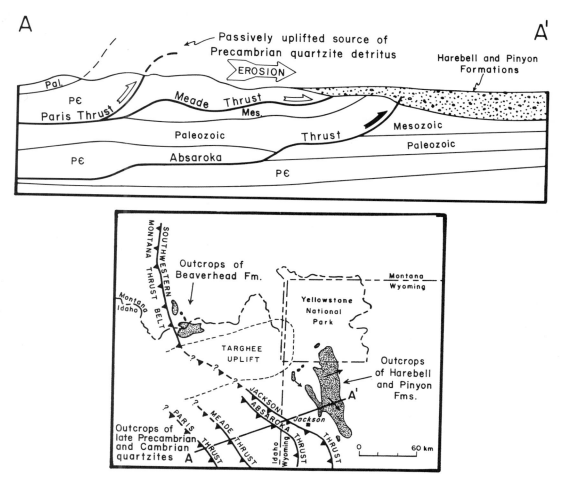

Fig. 18.5. Map and schematic cross section showing the structural setting for the Harebell and Pinyon Formations. A source in Precambrian metasedimentary rocks on the Paris Thrust sheet was formed when it was passively uplifted over a ramp in the Absaroka Thrust. Closed arrows on faults indicate motion related to generation of the tectogenic sediments. Open arrows indicate sense of motion on inactive older faults. Modified from Lindsey (1972) and Royse *et al.* (1975). Arrows in the Harebell-Pinyon outcrop area on the map indicate the dispersal directions of the conglomerates.

Quartzite cobbles and boulders in the Upper Cretaceous Harebell Formation and uppermost Cretaceous and Paleocene Pinyon Conglomerate, in the Jackson Hole area of northwestern Wyoming (Fig. 18.5) may also have been derived from passive uplift of the Paris Thrust sheet. Lindsey (1972) concluded that paleocurrent evidence, clast composition, and facies distribution indicate long-distance transport from western and northwestern sources in Idaho and Montana, including source rocks in the Belt Supergroup in Idaho and Montana, and Precambrian and Cambrian quartzites in southeastern Idaho. Love (1972) claimed a much nearer source in what he called the Targhee Uplift (Fig. 18.5), a source that was north and northwest of the

Teton Range, and is now buried beneath the eastern Snake River Plain.

Neither of these interpretations particularly appeal to us. The long-distance transport of Lindsey (1972) is difficult to reconcile with the quantity and caliber of conglomerate, and there is no other corroborating evidence for the huge, now-buried quartzite source suggested by Love (1972). Instead, we favor a passive source in the hanging wall of the Paris-Willard Thrust similar to that postulated above for the origin of the Hams Fork Conglomerate to the south. We suggest that erosion and transport of quartzite clasts were triggered when Precambrian metasedimentary rocks in the inactive Paris sheet were transported up and over a ramp in the Absaroka

Thrust (Fig. 18.5). Admittedly, the evidence for a passive source here is not as convincing as that for the Hams Fork but, given the data available and the similarities to the Hams Fork case, it is worth considering.

Synthetic and Antithetic Dispersal

As with active sources, synthetic transport is commonly assumed for dating thrust events, and the Wasatch and equivalent units of the western Wyoming Thrust Belt are excellent examples. Because these units are associated with the youngest, and therefore easternmost thrusts, they are still well preserved and their geometry and paleocurrent indicators show transport to the east, synthetic to thrusting.

Perhaps the most obvious example of antithetic dispersal is the Tunp Conglomerate on the east side of the Fossil Basin (Fig. 18.3). Sparse paleocurrent evidence and westward intertonguing with fine-grained basin center deposits (Hurst and Steidtmann 1986) leave little doubt that these sediments were shed westward down the hanging-wall dip slope of the eastwardly transported Absaroka Thrust plate.

Another example of antithetic dispersal is given by Mann (1974) in which thrust-derived Eocene Wasatch Formation conglomerates on the west side of the Wasatch Mountains were transported from southeast to northwest, opposite to the direction of thrusting.

Inverted and Blended Clast Compositions

The Upper Cretaceous conglomerate on Little Muddy Creek in southwestern Wyoming is a proximal, alluvial fan-delta deposit (Schmitt et al. 1986) that exhibits well developed, inverted clast stratigraphy (Royse et al. 1975). Sandstone clasts of the Lower Cretaceous Aspen Formation are present at the base and are overlain by a sequence of conglomerate in which the first appearance of progressively older clast lithologies occurs at successively higher stratigraphic intervals. Upsection the following sequence of first appearances of distinctive clasts occurs: Lower Cretaceous Bear River Formation sandstone, Lower Cretaceous Ephraim Formation conglomerate, Jurassic Twin Creek limestone, Jurassic Nugget Formation sandstone, and Triassic Woodside Formation red siltstone and Thaynes Formation limestone (Royse et al. 1975).

These sediments are preserved over an area of only about 2.5 km². Here, deposition accompanied uplift during the earlier of two movements along the Absaroka Thrust (Fig. 18.6) when sedimentary rocks on the basinward limb of the leading-edge anticline were sequentially eroded.

The Upper Cretaceous Sphinx Conglomerate of southwest Montana demonstrates both blending of clast compositions internally and an overall inverted clast stratigraphy. Clast counts by DeCelles et al. (1987) show that as motion along the Scarface Thrust continued, large portions of the stratigraphic column were exposed to erosion on the hanging wall (Fig. 18.7). The almost simultaneous exposure of several stratigraphic units, ranging in age from Mississippian through Cretaceous, is evidenced by the intimate association of numerous clast types in the Sphinx alluvial fan deposits. Dispersal of detritus from this portion of the upper-plate stratigraphic section occurred as strata were brought to the surface on a thrust and simultaneously exposed to erosion (Fig. 18.7). Strata were also exposed to erosion along minor thrust splays, resulting in even more blending of clast compositions. However, over the entire thickness of the Sphinx, there is a general trend in composition from Mesozoic clasts near the base to Paleozoic clasts at the top. The Sphinx crops out over an area similar in size to that of the above-mentioned conglomerate on Little Muddy Creek, and both are interpreted as the deposit of a single alluvial fan. Thus, it appears that unroofing sequences may be discernable on the scale of one fan where a hanging-wall anticline is breached by erosion.

At the other extreme, unroofing sequences form on a much larger scale and over several to tens of millions of years. Lawton (1985, 1986) notes this for the Cretaceous Indianola Group of central Utah where an upward trend from Paleozoic carbonate clasts to Precambrian quartzite clasts reflects initial erosion of an active thrust plate (Canyon Range Thrust) followed by erosion of a much younger, active thrust (Pavant Thrust) carrying an older, inactive, thrusted terrain.

In contrast, the work of Jefferson (1982) indicates that, in the Cedar Hills area of central Utah, deposits that are laterally equivalent to those described by Lawton (1985) contain no inverted clast stratigraphy. Clast-count data reveal sub-equal proportions of limestone and quartzite throughout the stratigraphic section. This led Jefferson to conclude that Indianola streams tapped a structurally complex

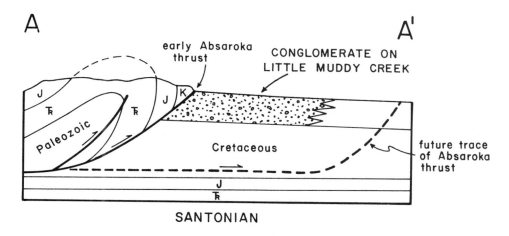

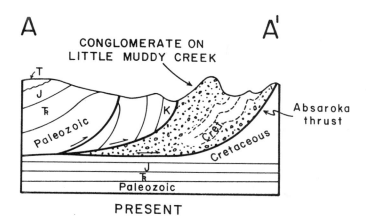

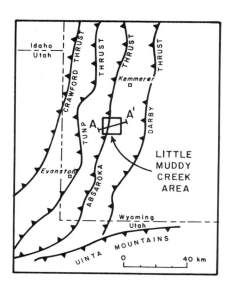

Fig. 18.6. Map and schematic cross sections showing the structural setting for deposition of the conglomerate on Little Muddy Creek during Santonian time, and the present structural setting related to subsequent motion on the Absaroka Thrust (after Royse and Warner 1987).

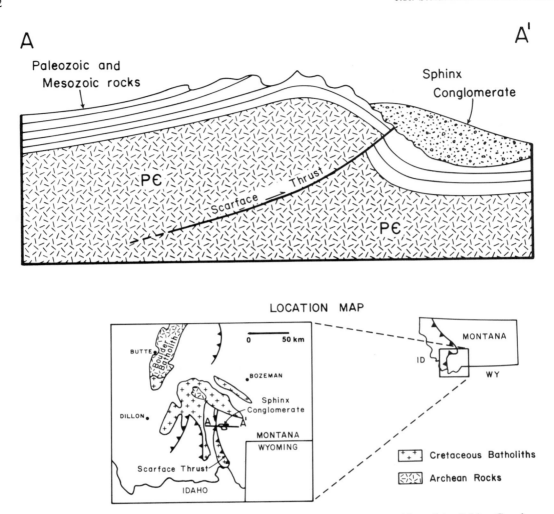

Fig. 18.7. Map and schematic cross section showing the structural setting for deposition of the Sphinx Conglomerate (after DeCelles *et al.* 1987).

stratigraphic section producing mixed cobble lithologies throughout the section, and therefore that simple unroofing of a source area did not occur (Jefferson 1982). Thus, both large-scale unroofing sequences and blended clast compositions may be recognizable within one stratigraphic unit in different areas and indicate distinct, local source-sediment relations.

One of the best documented examples of blended clast compositions is in the Upper Cretaceous Echo Canyon Conglomerate of northeastern Utah. Clast-count data indicate source rocks ranging from Cambrian to Jurassic age, with Mississippian to Jurassic clasts most common (Crawford 1979). However, there is no discernable variation in clast composition with stratigraphic position. Hence, the Echo

Canyon conglomerates were dispersed from a terrain characterized by simultaneous exposure of many stratigraphic units. Royse *et al.* (1975) attributed the origin of the Echo Canyon to motion along the Crawford Thrust as ramping of the upper plate fed Mississippian through Jurassic strata into the surface of erosion together.

Blended clast compositions also may arise where tectogenic sediments from a variety of specific source areas are mixed in a fluvial system, particularly in the more distal braid-plain settings down depositional slope from alluvial fans. Jefferson (1982) suggested that braid-plain processes during transport and deposition of the above-mentioned Indianola Group sediments may have mixed detritus from numerous sources, thus making it difficult to

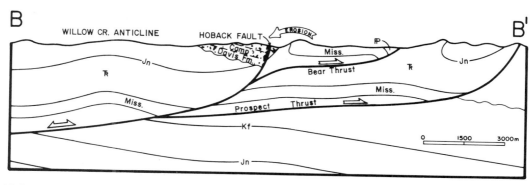

Fig. 18.8. Cross section showing the structural setting of the Miocene-Pliocene Camp Davis Formation along the Hoback listric normal fault. Location of cross section (B-B′) is shown on map in Figure 18.3 (modified from Royse *et al.* 1975).

correlate these tectogenic deposits with motion on a particular thrust.

The complex structural relations characteristic of thin-skinned, thrusted terrains may also influence compositional trends in the sediment fill of younger, post-thrusting basins superposed on older thrusted terrain. This influence is demonstrated in late Tertiary, listric-normal, fault-bounded basins that formed in the western thrust belt, probably in response to Basin and Range extension. This extension resulted in the reactivation of thrusts as listric normal faults with accompanying rapid subsidence and consequent rapid deposition.

One might expect that classic unroofing sequences should readily develop in such a setting. In fact, in other tectonic settings where the structural relief between basins and adjacent ranges was great, such as those along the flanks of the Laramide Beartooth uplift of southwestern Montana, there are obvious inverted clast stratigraphies (Fleuckinger 1970; Jobling 1974). However, listric-normal, fault-bounded basins in thin-skinned, thrusted terrains often do not possess such evidence. Rather, because of the inherently complex structural geometry of earlier, thrusted terrain adjacent to the younger basins, a diverse array of stratigraphic units may be eroded simultaneously.

One example of such a basin fill is the Miocene-Pliocene Camp Davis Formation of northwestern Wyoming (Fig. 18.8). The Camp Davis was deposited as a thick (1,700 m) sequence of conglomerate, sandstone, and freshwater limestone in response to motion on the Hoback Listric-Normal Fault (Fig. 18.8; Dorr *et al.* 1977). Those authors reported inverted clast stratigraphy but more recent,

detailed examination of clast distributions shows no clear-cut unroofing sequence in the Camp Davis (Olson and Schmitt 1986). Rather, a chaotic distribution of clast compositions provides evidence of concurrent erosion of numerous stratigraphic units in the hanging walls of both the Bear and Prospect Thrusts. Thus, during motion on the Hoback Normal Fault, progressive headward erosion of canyons in the uplifted Hoback Fault-block tapped older, thrusted terrain giving rise to the complex distribution of clast lithologies.

Cannibalism and Tectonic Transport of Proximal Deposits

The most compelling evidence for cannibalism of proximal tectogenic deposits is their general paucity throughout the thrust belt (Fig. 18.9). Admittedly, later erosion of topographically higher terrains has removed much of the evidence, but even where fine-grained tectogenic clastics are still present, their coarser, proximal counterparts commonly are not. This suggests that the coarse clastics were continually reworked as they were uplifted and eroded on the hanging walls of subsequent thrusts. It should also be pointed out that this process results in the transport of coarse debris by tectonic, rather than hydraulic, processes. In addition to the overall lack of proximal, tectogenic deposits, there is excellent evidence, from the few instances where they are preserved, that their destruction in thrusted terrains is common. The thrust-generated alluvial-fan deposits of the above-mentioned Sphinx Conglomerate contain reworked early-cemented clasts of

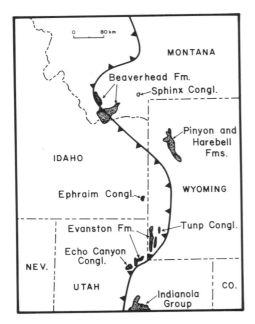

Fig. 18.9. Generalized map of a portion of the fold-thrust belt in the western United States showing the distribution of outcrops of coarse (i.e., cobble and boulder), proximal tectogenic deposits (modified from Wilson 1970).

conglomerate indicating that proximal portions of alluvial fans were cannibalized (DeCelles *et al.* 1987). This erosion and redeposition was due to continued basinward thrust propagation and renewed uplift of previously deposited gravelly deposits.

Cannibalization is also evident in the conglomerate on Little Muddy Creek in southwestern Wyoming (Fig. 18.6). After the sediment was shed from early uplift on the Absaroka Thrust, motion on a younger Absaroka Thrust splay to the east folded the conglomerate in the hanging-wall anticline (Vietti 1974). Subsequent erosion removed many of the younger stratigraphic units from this anticline, leaving the conglomerate preserved only in the west-dipping limb of the anticline (Royse and Warner 1987). Thus, a significant portion of the conglomerate on Little Muddy Creek was cannibalized by thrust-induced deformation, uplift, and erosion.

Summary and Conclusions

Although these concepts have been presented individually for the sake of clarity, each is an integral part of an uplift-erosion-transport-deposition system

genetically and intimately related to the style of deformation in thin-skinned, thrusted terrains. It should be anticipated that these provenance and dispersal relationships will occur in complex but predictable combinations. For example, because passive sources, antithetic transport, and blended clast compositions may all be related to ramping at depth, it is likely that a single deposit will demonstrate all three. Furthermore, cannibalization and tectonic transport of sediment are simply different manifestations of clastic-wedge development during thrust propagation that youngs in the direction of tectonic transport. In this setting, alluvial fans at the thrust front are destined to become intramontane deposits that are susceptible to erosion as they are uplifted and tectonically transported during the formation of younger thrusts. Finally, detached basins, which form and fill while being transported on thrust sheets, are appropriate depositional sites for sediments that display antithetic dispersal, passive sources, and blended clast compositions. Recognition of these relations is critical to meaningful interpretations of tectogenic sediments. At the same time, these concepts should serve as a starting point for a new and better understanding of source-sediment relations in thin-skinned, thrusted terrains.

Acknowledgments. This work on tectogenic sediments in the thrust belt of the western United States was supported by Chevron, U.S.A. and Amoco Production Company, both of Denver, Colorado. The authors benefitted greatly from discussions with Art Berman, Frank Royse, and Katey Sippel.

References

ARMSTRONG, F.C. and ORIEL, S.S. (1965) Tectonic development of Idaho-Wyoming thrust belt. American Association Petroleum Geologists Bulletin 49:1847–1866.

ARMSTRONG, R.L. (1968) Sevier orogenic belt in Nevada and Utah. Geological Society America Bulletin 79:273–288.

ATKINSON, C.D. and ELLIOTT, T. (1985) Patterns of sedimentation in the south Pyrenean basin: A tectono-sedimentary model for molasse accumulation in a thrust-sheet-top basin. International Association Sedimentologists Symposium on Foreland Basins, Fribourg, Program and Abstracts, p. 32.

BOHANNON, R.G. (1984) Mesozoic and Cenozoic tectonic development of the Muddy, north Muddy and northern Black mountains, Clark County, Nevada. Geological Society America Memoir 157, pp. 125–148.

BRUHN, R.L., PICARD, M.D., and BECK, S.L. (1983) Mesozoic and early Tertiary structure and sedimentology of the central Wasatch Mountains, Uinta Mountains and Uinta basin. Utah Geological Mineralogical Survey Special Studies 59, pp. 63–105.

BURBANK, D.W. and RAYNOLDS, R.G.H. (1988) Stratigraphic keys to the timing of thrusting in terrestrial foreland basins: Applications to the northwestern Himalaya. In: Kleinspehn, K.L. and Paola, C. (eds) New Perspectives in Basin Analysis. New York: Springer-Verlag, pp. 331–351.

CRAWFORD, K.A. (1979) Sedimentology and Tectonic Significance of the Late Cretaceous-Paleocene Echo Canyon and Evanston Synorogenic Conglomerates of the North-Central Utah Thrust Belt. Unpublished MS thesis, University of Wisconsin, Madison, 143 p.

DECELLES, P.G., TOLSON, R.B., GRAHAM, S.A., SMITH, G.A., INGERSOLL, R.V., WHITE, J., RICE, R., MOXON, I., LEMKE, L., HANDSCHY, J.W., FOLO, M.F., EDWARDS, D.P., CAVAZZA, W., CALDWELL, M., and BARGAR, E. (1987) Laramide thrust-generated alluvial-fan sedimentation, Sphinx Conglomerate, southwestern Montana. American Association Petroleum Geologists Bulletin 71:135–155.

DORR, J.A. JR., SPEARING, D.R., and STEIDTMANN, J.R. (1977) Deformation and Deposition between a Foreland Uplift and an Impinging Thrust Belt: Hoback Basin, Wyoming. Geological Society America Special Paper 177, 82 p.

DORR, J.A. JR. and STEIDTMANN, J.R. (1977) Stratigraphic-tectonic implications of a new, earliest Eocene mammalian faunule from central western Wyoming. Wyoming Geological Association Guidebook, 29th Annual Field Conference, pp. 327–337.

DUTCHER, L.A.F., DUTCHER, R.R., and JOBLING, J.L. (1986) Stratigraphy, sedimentology and structural geology of Laramide synorogenic sediments marginal to the Beartooth Mountains, Montana and Wyoming. In: Garrison, P.B. (ed) Geology of the Beartooth Uplift and Adjacent Basins. Billings, Montana: Montana Geological Society Guidebook, pp. 33–52.

FLEUCKINGER, L.A. (1970) Stratigraphy, Petrography and Origin of Tertiary Sediments off the Front of the Beartooth Mountains, Montana-Wyoming. Unpublished Ph.D. Dissertation, Pennsylvania State University, 249 p.

HURST, D.J. (1984) Depositional Environment and Tectonic Significance of the Tunp Member of the Wasatch Formation, Southwest Wyoming. Unpublished MS Thesis, University of Wyoming, 115 p.

HURST, D.J. and STEIDTMANN, J.R. (1986) Stratigraphy and tectonic significance of the Tunp conglomerate in the Fossil basin, southwest Wyoming. Mountain Geologist 23:6–13.

JEFFERSON, W.S. (1982) Structural and stratigraphic relations of Upper Cretaceous to Lower Tertiary orogenic sediments of the Cedar Hills, Utah. Utah Geological Association Publication 10, pp. 65–80.

JOBLING, J.L. (1974) Stratigraphy, Petrography and Structure of the Laramide (Paleocene) Sediments Marginal to the Beartooth Mountains, Montana. Unpublished Ph.D. Dissertation, Pennsylvania State University, 102 p.

LAMERSON, P.R. (1982) The Fossil Basin and its relationship to the Absaroka Thrust system, Wyoming and Utah. In: Powers, R.B. (ed) Geologic Studies of the Cordilleran Thrust Belt. Denver: Rocky Mountain Association Geologists Cordilleran Thrust Belt Studies, pp. 279–340.

LAWTON, T.F. (1985) Style and timing of frontal structures, thrust belt, central Utah. American Association Petroleum Geologists Bulletin 69:1145–1159.

LAWTON, T.F. (1986) Compositional trends within a clastic wedge adjacent to a fold-thrust belt: Indianola Group, Central Utah, U.S.A. In: Allen, P.A. and Homewood, P. (eds) Foreland Basins. International Association Sedimentologists Special Publication 8, pp. 411–423.

LINDSEY, D.A. (1972) Sedimentary Petrology and Paleocurrents of the Harebell Formation, Pinyon Conglomerate and Associated Coarse Clastic Deposits, Northwestern Wyoming. United States Geological Survey Professional Paper 734-B, 68 p.

LOVE, J.D. (1972) Harebell Formation (Upper Cretaceous) and Pinyon Conglomerate (Uppermost Cretaceous and Paleocene), Northwestern Wyoming. United States Geological Survey Professional Paper 734-A, 54 p.

MANN, D.C. (1974) Clastic Laramide Sediments of the Wasatch Hinterland, Northeastern Utah. Unpublished Master's thesis, Salt Lake City, University of Utah, 113 p.

OLSON, T.J. and SCHMITT, J.G. (1986) Sedimentary evolution of the Miocene-Pliocene Camp Davis Formation, northwestern Wyoming. Geological Society America Abstracts with Programs 18:400–401.

ORI, G.G. and FRIEND, P.F. (1984) Sedimentary basins formed and carried piggyback on active thrust sheets. Geology 12:475–478.

ORIEL, S.S. (1962) Main body of Wasatch Formation. American Association Petroleum Geologists Bulletin 46:2161–2173.

ORIEL, S.S. and ARMSTRONG, F.C. (1966) Times of thrusting in Idaho-Wyoming thrust belt—Reply. American Association Petroleum Geologists Bulletin 50:2614–2621.

ORIEL, S.S. and PLATT, L.B. (1980) Geologic map of the Preston 1 × 2 degree quadrangle, southeastern Idaho and western Wyoming. United States Geological Survey Map I-1127.

ORIEL, S.S. and TRACEY, J.I. (1970) Uppermost Cretaceous and Tertiary Stratigraphy of Fossil Basin, Southwestern Wyoming. United States Geological Survey Professional Paper 635, 53 p.

ROYSE, F., WARNER, M.A., and REESE, D.L. (1975) Thrust belt structural geometry and related stratigraphic problems, Wyoming-Idaho-northern Utah. In: Bolyard, D.W. (ed) Symposium on Deep Drilling Frontiers in the Central Rocky Mountains. Denver: Rocky Mountain Association Geologists, pp. 41–54.

ROYSE, F. and WARNER, M.A. (1987) Little Muddy Creek area, Lincoln County, Wyoming. Geological Society America Decade North American Geology Guidebook, pp. 213–216.

SCHMITT, J.G. (1982) Origin and Sedimentary Tectonics of the Upper Cretaceous Frontier Formation Conglomerates in the Wyoming-Idaho-Utah Thrust Belt. Unpublished Ph.D. Thesis, University of Wyoming, 252 p.

SCHMITT, J.G. (1985) Synorogenic sedimentation of Upper Cretaceous Frontier Formation conglomerates and associated strata, Wyoming-Idaho-Utah thrust belt. Mountain Geologist 22:5–6.

SCHMITT, J.G. (1986) Sequence and timing of structural development in Wyoming-Idaho-Utah thrust belt: What do provenance studies of foredeep basin conglomerates indicate? (Abstract) American Association Petroleum Geologists Bulletin 70:1054.

SCHMITT, J.G., ANDERSON, D., DIEM, D., HAZEN, D., LLOYD, J., MILLER, E., OLSON, T.J., SALT, K., SINGDAHLSEN, D., STINE, A., and TROMBETTA, M. (1986) Fan-delta deposition of the Upper Cretaceous Little Muddy Creek Conglomerate, southwestern Wyoming. Geological Society America Abstracts with Programs 18:410.

SHUSTER, M.W. and STEIDTMANN, J.R. (1987) Fluvial-sandstone architecture and thrust-induced subsidence, northern Green River basin, Wyoming. In: Ethridge, F.G., Flores, R.M., and Harvey, M.D. (eds) Recent Developments in Fluvial Sedimentology. Society Economic Paleontologists Mineralogists Special Publication 39, pp. 279–286.

STEIDTMANN, J.R. (1985) Structural controls on composition and distribution of the tectogenic sediments in the Wyoming thrust belt and foreland. Society Economic Paleontologists Mineralogists, Program with Abstracts, p. 86.

VIETTI, J.S. (1974) Structural Geology of the Ryckmann Creek Anticline Area, Lincoln and Uinta Counties, Wyoming. Unpublished MS Thesis, University of Wyoming, 102 p.

WILSON, M.D. (1970) Upper Cretaceous-Paleocene synorogenic conglomerate of southwestern Montana. American Association Petroleum Geologists Bulletin 54:1843–1867.

WILTSCHKO, D.V. and DORR, J.A. JR. (1983) Timing of deformation in overthrust belt and foreland of Idaho, Wyoming and Utah. American Association Petroleum Geologists Bulletin 67:1304–1322.

Part IV
Precambrian Basins: Introduction

RICHARD W. OJAKANGAS

Precambrian rocks constitute about 20% of the Earth's land surface. Many of these rock units are metasedimentary, ranging upward in metamorphic grade to granulite facies, and many have been highly deformed as well. "Seeing through" the metamorphism and deformation, in order to accomplish basin analysis, is the challenge for workers in Precambrian terranes, especially in those of Archean and early Proterozoic age.

Early on, Precambrian rocks were commonly avoided because of metamorphism, deformation, and their lack of fossils. A few early workers (e.g., A.C. Lawson in Ontario and J.W. Gruner in northeastern Minnesota) demonstrated that coherent Archean stratigraphic sequences were indeed mappable in these areas.

Francis J. Pettijohn conducted sedimentological studies and mapped in Archean basins in Ontario, and published a series of papers on this work between 1934 and 1943. His classic culminating paper, published in 1943 with the deceptively simple title of "Archean Sedimentation," was an early example of basin analysis. He recognized the graywacke-greenstone association and compared the graywackes to Phanerozoic flysch sediments in eugeosynclines. He realized that lithologic similarity between different volcanic-sedimentary belts does not imply the same age. He emphasized the lack of quartzite and limestone in the Archean basins. He contrasted orogenic "flysch-type" sedimentary facies with epeirogenic "platform-type" sedimentary facies, and noted that they were mutually exclusive. In his summary, he stated (1943a, p. 968) "The similarity of the Archean deposits to those of the Pleistocene glaciation, pointed out at numerous places in the paper, does not mean that the deposits

of the Archean are glacial. The similarity, striking as it is, proves the essential aquatic nature of the Archean deposits. . . . The similarity follows from the most fundamental common factor in the origin of the deposits of the two ages—i.e., the rapid rate of accumulation. . . . It thus appears that the Archean sediments are fundamentally no different from orogenic sediments of other ages."

Pettijohn's interest in Precambrian rocks never waned, and he returned decades later to one of the basins to apply modern sedimentological concepts and techniques of basin analysis (Walker and Pettijohn 1971). For example, the "varved slates" and "megavarves" of his 1943a paper were reinterpreted as turbidites. Pettijohn reviewed the status of studies on Precambrian sedimentary rocks in his last two publications on Precambrian rocks (1970, 1972). Altogether, he published more than a dozen papers each on Archean and Proterozoic sedimentation.

Nearly all basin-analysis techniques that are applied to the study of Phanerozoic basins are used in the study of Precambrian basins. A prominent exception has been biostratigraphy, although correlation methods based on stromatolites (Walter 1976) and acritarchs (Vidal and Knoll 1983) are in the process of development. This lack of fossils, however, has its positive side as well, for detailed sedimentary structures in Precambrian sediments have escaped destruction by bioturbation.

Rare-earth element (REE) analysis has potential in provenance studies (e.g., Taylor and McClennan 1985), and major-element analysis is useful in interpreting protoliths for high-grade metamorphic rocks in which all sedimentary structures have been destroyed. Dating of zircons, even individual zircons, has provided new data on cyclicity in Archean volcanic-sedimentary terranes (e.g., Nunes and Thurston 1980), and dating of individual boulders in conglomerates (e.g., Nunes and Wood 1980) has established minimum ages for basin fill and has helped to delineate source rocks. Magnetostratigraphy has not been markedly successful in older Precambrian rock units, but has been used in the Middle Proterozoic volcanic and sedimentary rocks of the mid-continent "rift zone" of the Lake Superior region, both in correlation and in plotting polar-wander curves (Halls and Pesonen 1982). Detailed facies analysis has been applied to lower Proterozoic glacial rock units (Miall 1983). Some subsurface data are available for several Precambrian basins because of drilling related to mineral exploration, but most stratigraphy is based on outcrops; seismic stratigraphic methods are not easily applied to metamorphosed and deformed sequences.

In the first of the three papers in this section, Eriksson, Kidd, and Krapez reinforce the view that standard techniques used in the analysis of younger siliciclastic basins can be applied to metavolcanic-metasedimentary sequences as well as to ancient high-grade gneissic terranes that lack sedimentary structures. Petrography, REE patterns, trace-element abundance, major-element analyses, sedimentary facies, and paleocurrent analyses are used to ascertain the compositions and relative locations of the source rocks, recycling histories, and depositional environments. Integration of various data allows them to interpret the tectonic settings of the basins, with results comparable to Phanerozoic passive-margin sequences, rifts, and foreland basins.

Grotzinger and McCormick show that the Kilohigok Basin, northwest Canadian Shield, and its sedimentary fill can be interpreted as the product of tectonic flexure of an initial platform caused by tectonic loading (thrust nappes). The basin is well suited for basin analysis in that a continuous cross section of the basin fill and correlative units on the adjacent craton (i.e., arch or peripheral bulge), and the included unconformities, are well exposed. Tectonic flexure of the platform is indicated by several relationships, including synchronous exposure (indicated by regolith development and erosion) and drowning (indicated by deposition and onlap) on the platform, as well as by large-scale slide deposits that were probably triggered by tectonic steepening of slopes. A number of unconformities within the shelf portion of the foredeep-basin fill pass into the conformable sequence in the axial portion of the basin; these seem to be valid sequence boundaries that are used as approximate chronostratigraphic correlation markers. Recycling within the column is interpreted as cannibalization related to the erosion of successive thrust wedges.

In my own contribution, I summarize the evidence for a synchronous Early Proterozoic glaciation on two cratons. (Interestingly, one of the North American sequences noted here was already identified in 1943[b] by Pettijohn as glaciogenic.) I suggest that glaciations, termed uncommon "mega-events," produce distinctive rock units that can be utilized in "mega-event stratigraphy," and that they have an application in chronostratigraphic correlations between basins and between continents as well. Because widespread glaciations are controlled by global climatic changes, they may be particularly valuable in correlations of Precambrian rocks in which chronostratigraphic markers are rare or nonexistent. Pleistocene and Late Paleozoic Gondwana glaciations are used as younger analogs.

References

HALLS, H.C. and PESONEN, L.J. (1982) Paleomagnetism of Keweenawan rocks. In: Wold, R.J. and Hinze, W.J. (eds) Geology and Tectonics of the Lake Superior basin. Geological Society America Memoir 156, pp. 173–202.

MIALL, A.D. (1983) Glaciomarine sedimentation in the Gowganda Formation (Huronian), Northern Ontario. Journal Sedimentary Petrology 53:477–492.

NUNES, P.D. and THURSTON, P.C. (1980) Two hundred and twenty million years of Archean evolution: A zircon U-Pb age stratigraphic study of Uchi–Confederation Lakes Greenstone belt, Northwestern Ontario. Canadian Journal Earth Sciences 17:710–721.

NUNES, P.D. and WOOD, J. (1980) Geochronology of the North Spirit Lake area, district of Kenora–Progress Report. In: Pye, E.G. (ed) Summary of Geochronology Studies 1977–1979. Ontario Geological Survey Miscellaneous Paper 92, pp. 7–14.

PETTIJOHN, F.J. (1943a) Archean sedimentation. Geological Society American Bulletin 54:925–972.

PETTIJOHN, F.J. (1943b) Basal Huronian conglomerates of Menominee and Calumet districts, Michigan. Journal Geology 51:387–395.

PETTIJOHN, F.J. (1970) The Canadian Shield – A status report, 1970. In: Baar, A.J. (ed) Symposium on Basins and Geosynclines of the Canadian Shield. Geological Survey Canada Paper 70–40, pp. 239–255.

PETTIJOHN, F.J. (1972) The Archean of the Canadian Shield: A resume. In: Doe, B.R. and Smith, D.K. (eds) Studies in Mineralogy and Precambrian Geology. Geological Society America Memoir 135, pp. 131–149.

TAYLOR, S.R. and McCLENNAN, S.M. (1985) The Continental Crust: Its Composition and Evolution (An Examination of the Geochemical Record Preserved in Sedimentary Rocks). Oxford: Blackwell Scientific Publications, 312 p.

VIDAL, G. and KNOLL, A.H. (1983) Proterozoic plankton. In: Medaris, L.G. Jr., Byers, C.W., Mickelson, D.M., and Shanks, W.C. (eds) Proterozoic Geology: Selected Papers from an International Proterozoic Symposium. Geological Society America Memoir 161, pp. 265–277.

WALKER, R.J. and PETTIJOHN, F.J. (1971) Archean sedimentation: Analysis of the Minnitaki Basin, northwestern Ontario, Canada. Geological Society America Bulletin 82:2099–2129.

WALTER, M.R. (1976) (ed) Stromatolites. Amsterdam: Elsevier, 790 p.

19

Basin Analysis in Regionally Metamorphosed and Deformed Early Archean Terrains: Examples from Southern Africa and Western Australia

KENNETH A. ERIKSSON, WILLIAM S.F. KIDD, and BRYAN KRAPEZ

Abstract

Early Archean (>3.0 Ga) sedimentary rocks are present in high-grade metamorphic terranes and greenstone belts in southern Africa and Western Australia. The viability and quality of basin analysis in these terranes improves with decreasing metamorphic grade and degree of structural complexity. Depending on the state of preservation of the rocks, it is possible to reconstruct some or all of: stratigraphy, source-area composition and age, sediment dispersal patterns and depositional environments, and basin configuration and tectonic setting.

Pre-greenstone (>3.6 Ga) gneisses in the high-grade Limpopo Province and Western Gneiss Terrain are envisaged as remnants of cratonic nucleii. These gneisses represent basement to cover rocks that are exclusively of sedimentary origin and accumulated between ca. 3.6 and 3.2 Ga. Limpopo cover rocks are devoid of primary sedimentary structures and consist of quartzite with detrital zircons (original quartz arenite), marble (limestone), metapelite (mudstone) and aluminous gneiss (possible wacke). This quartz arenite-carbonate association implies a stable tectonic setting and the best analogs may be younger cratonic-shelf deposits. Metapelites in the Limpopo Province have a complex rare earth element (REE) geochemistry indicating a mixed provenance that included differentiated continental crust. A thick sequence (ca. 2.5 km) of conglomerate and crossbedded aluminous gneiss (wacke) in the Western Gneiss Terrain is interpreted as an alluvial deposit and possibly accumulated in a rift setting. Rare earth element geochemistry of metapelites indicates that differentiated continental crust consisting of K-granites was the dominant component of the source terrain.

Predominantly mafic and ultramafic volcanism in the Barberton and Pilbara greenstone belts took place between 3.5 and 3.3 Ga in an oceanic environment distant from any continental influence. Intercalated sedimentary deposits indicate that volcanism took place at relatively shallow-water depths. The volcanic sequences are overlain by predominantly siliciclastic sedimentary intervals. In the Barberton greenstone belt, the distribution of facies, in conjunction with paleocurrent, petrographic, geochemical, and geochronological data indicate that the Fig Tree and overlying Moodies Groups (ca. 5 km thick) were derived by progressive unroofing of a southerly source terrain consisting of the older volcanic rocks with intercalated sedimentary deposits, and a 3.5 to 3.3 Ga gneiss complex. Sedimentation took place initially in a submarine-fan setting. Basin shoaling is indicated by the upward transition into braided-alluvial and shallow-marine sediments. The stratigraphic evolution of the Fig Tree and Moodies Groups is similar to that of Phanerozoic foreland or foredeep basins.

In the Pilbara Block the lower Gorge Creek Group is compositionally similar to the Moodies Group and was derived from a comparable provenance; the tectonic setting of the basin is unclear. The upper Gorge Creek Group unconformably overlies the lower subdivision and contains abundant quartzite clasts recycled from it. This ca. 3 km-thick stratigraphic sequence is exclusively of continental origin; depositional environments include alluvial fans, braided rivers, floodplains, and lakes. Stratigraphic and sedimentological evidence indicates that basin development was controlled by marginal strike-slip faulting.

Available age constraints indicate that sedimentary rocks in the high-grade terranes and greenstone belts developed contemporaneously. Cover rocks in the high-grade terranes accumulated on thick continental crust and reflect a stable cratonic setting. In contrast, volcanic intervals in the greenstone belts are considered to have developed in an oceanic setting, whereas the overlying siliciclastic intervals reflect active tectonic settings associated with crustal shortening.

Introduction

Sedimentary rocks older than 3 billion years are present in southern Africa and Western Australia in the high-grade Limpopo Province and Western Gneiss Terrain and in the greenstone belts of the

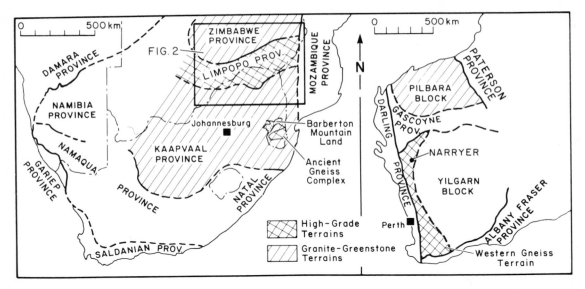

Fig. 19.1. Locality map of the high-grade Limpopo Province and Western Gneiss Terrain and the granite-greenstone Kaapvaal Province and Pilbara Block [adapted from Button (1976), Gee (1979), and Tankard *et al.* (1982)].

Kaapvaal Province and Pilbara Block, respectively (Fig. 19.1). This paper focuses on the predominantly siliciclastic sedimentary sequences in these two different types of crustal terranes. The sediments vary considerably in state of preservation of primary depositional features because of differences in degree of metamorphism and intensity of deformation. Despite the great antiquity of the sediments they are amenable to basin analysis employing many conventional sedimentological techniques (cf. Potter and Pettijohn 1977), as well as some less widely used techniques. The viability and quality of basin analysis improves with decreasing metamorphic grade and structural complexity; depending on the state of preservation of the rocks it is possible to reconstruct some or all of: stratigraphy, source-area composition and age, sediment dispersal patterns and depositional environments, and basin configuration and tectonic setting.

Analysis of these sedimentary rocks also allows a number of fundamental questions to be addressed that are relevant to the early evolution of the Earth's crust. These questions include land-sea proportions, crustal compositions, uniformitarian or nonuniformitarian nature of surface processes, and evidence for exposed land masses prior to 3 billion years ago. Of greater significance are the constraints that basin analysis can place on the temporal and spatial relationship between high-grade and greenstone terranes. High-grade terranes have variably been

considered as the metamorphosed roots of greenstone belts, as basement to greenstone belts, or as having developed contemporaneously with greenstone belts but in different tectonic settings (Percival and Card 1983; Windley 1984).

In this chapter we evaluate the application of conventional basin analysis techniques to these metamorphosed sediments before presenting a critical assessment of the problems of basin analysis in metamorphosed and deformed Archean terranes. In the final section of this paper we use the results of the basin analyses to address the broader questions outlined above.

Limpopo Province: Southern Africa

Geologic Setting

The high-grade Limpopo Province in southern Africa is located between the Zimbabwe and Kaapvaal Provinces that consist of arcuate greenstone belts surrounded by granites and gneisses (Fig. 19.2). Within the Limpopo Province a three-fold zonation is recognized; northern and southern marginal zones flank the central zone (Fig. 19.2). The marginal zones were derived by metamorphism and deformation of the adjoining granite-greenstone provinces, whereas the central zone displays a different history.

Fig. 19.2. Tectonic zones of the Limpopo Province and their spatial relation to the granite-greenstone terrane of the adjacent Zimbabwe and Kaapvaal Provinces [from Tankard *et al.* (1982)]. Location shown in Figure 19.1.

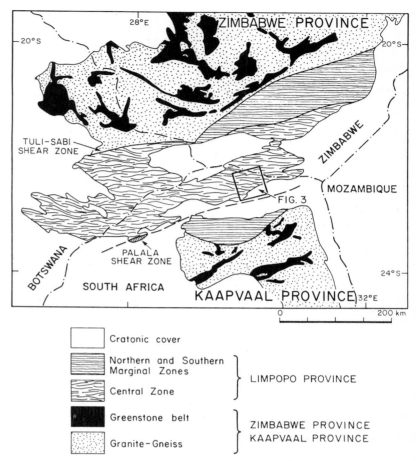

Within the central zone of the Limpopo Province a complex geological history is recognized dating from 3.8 to 2.4 Ga. The Sand River Gneisses with an age of 3,786 ± 61 Ma (Rb-Sr; Barton *et al.* 1983a) are intruded by 3,560 ± 100 Ma mafic dikes (Rb-Sr; Barton *et al.* 1977) and are recognized as basement to younger cover rocks represented by the Beitbridge Complex (Fig. 19.3; Fripp 1983). While the isotopic ages can be questioned, the basement-cover relationship is based on convincing field evidence that indicates that the oldest recognized structural fabrics are restricted to the Sand River Gneisses (Fripp 1983). In the Beitbridge-Messina-Tshipise region (Fig. 19.3), the Beitbridge Complex consists of four lithological associations, namely, quartzite-pyroxenitic amphibolite (Fig. 19.4); biotite-garnet-cordierite-sillimanite gneiss; magnetite quartzite, calc-silicate gneiss, and marble (Fig. 19.5); and gray gneiss. Intrusive into the cover rocks are the Messina Layered Intrusion (calcic anortho-

site suite) and the Singelele Granitoid (Fig. 19.3; Barton 1983a). The Messina Layered Intrusion is dated at 3,153 ± 47 Ma (Rb-Sr; Barton *et al.* 1979a) and 3,270$^{+105}_{-112}$ Ma (Pb-Pb; Barton 1983b). The latter age is considered to represent the time of emplacement and the lower value considered to record the earliest deformation and metamorphism of the layered intrusion along with the Beitbridge Complex and Singelele Granitoid (Barton 1983a; Watkeys *et al.* 1983) at pressures up to 10 kb and temperatures in excess of 800°C (Horrocks 1983). The ca. 3,150 Ma age of metamorphism is substantiated by mafic dikes of 3,000 to 3,100 Ma age (Rb-Sr; Barton *et al.* 1977, 1983b) that transect a major structural/metamorphic fabric in the basement, Messina Layered Intrusion, and Singelele Granitoid (Barton *et al.* 1983a; Watkeys *et al.* 1983). Subsequent deformation and a second high-grade metamorphism took place at 2.7 Ga (Van Reenen *et al.* 1987), with metamorphism continuing under conditions of

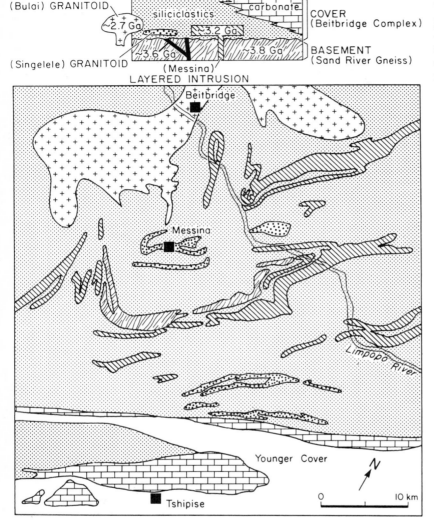

Fig. 19.3. Geological map of the central zone of the Limpopo Province in the Beitbridge-Messina-Tshipise area [modified after Tankard *et al.* (1982)]. Location shown in Figure 19.2.

progressive decompression until 2.2 Ga (Van Reenen 1986). The synkinematic Bulai Granitoid was emplaced early in this structural history (Watkeys *et al.* 1983) and is well dated at ca. 2,650 Ma (Fig. 19.3; Rb-Sr; Van Breemen and Dodson 1972; Rb-Sr and Pb-Pb; Barton *et al.* 1979b).

Stratigraphic Framework of the Beitbridge Complex

In the absence of primary structures such as cross-stratification, and because of great structural complexity, it is not possible to establish a stratigraphic sequence in the Beitbridge Complex. Ductile strain is ubiquitously shown in the central Limpopo Province by a layer-parallel foliation with strongly attenuated limbs of early isoclinal folds and wide-separation boudinage of some less ductile layers (Fig. 19.5). The effect of this strain has been to thin severely the original individual layers, now seen in outcrop at the 1 cm to 100 m scale. On a larger scale (km) it is, of course, likely that large-scale folding and thrust duplication have repeated and therefore thickened the overall original stratigraphic sections.

In this section an attempt is made to interpret primary lithologies and lithologic proportions that can be used to constrain other components of the basin analysis. One of the major problems in the Limpopo Province, as in many high-grade terranes,

Fig. 19.4. Interlayered quartzites (light) and amphibolites (dark) of the Beitbridge Complex. Outcrop located 10 km southwest of Messina. Scale bar = 1 m.

Fig. 19.5. Quarry outcrop of marble in the Beitbridge Complex. Quarry face is ca. 20 m high. Lower 5 m of outcrop is pink, dolomitic marble; upper 15 m is gray, calcitic marble. Dark attenuated layer with prominent boudins is calc-silicate rock. Outcrop located 20 km northwest of Alldays.

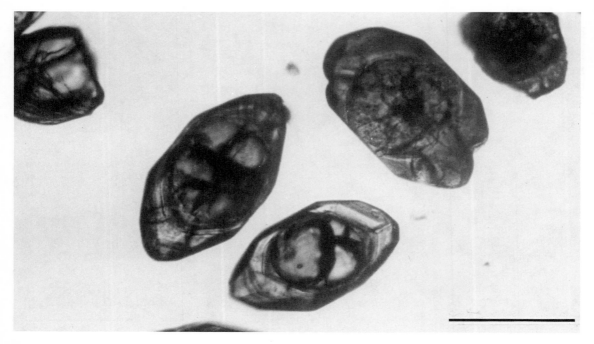

Fig. 19.6. Photomicrograph showing zircons with rounded cores and euhedral metamorphic overgrowths from quartzites in the Beitbridge Complex. Scale bar = .2 mm.

concerns the origin of the quartzite-pyroxenitic amphibolite association in the Beitbridge Complex. This association has conventionally been considered to represent mineralogically mature quartzose sandstone and mafic-ultramafic volcanic rock (Windley 1984). An alternative explanation for this association was proposed by Fripp (1983) who considered the quartzite as original chert and the amphibolite as original volcanic rock. A third possibility is that the amphibolites represent deformed and metamorphosed sills and/or dikes intrusive into quartzose sandstone or chert.

Quartzites in the Beitbridge Complex are presently up to 100 m thick in single occurrences (following severe tectonic thinning) and consist of up to 99.5% quartz. Metapelite partings are rare and no primary sedimentary structures are discernible. Layering defined by magnetite and fuchsite is common in the quartzites, and heavy minerals occur in higher concentrations than in most quartzose sandstones (J. Barton, personal communication 1983). Rutile and zircon are the most common heavy minerals with tourmaline also present; zircons average 0.4 mm in length and typically contain rounded cores with euhedral overgrowths (Fig. 19.6).

A number of the characteristics listed above favor an original detrital sedimentary origin for the quartzites. The high concentration of heavy minerals is incompatible with an interpretation of these rocks as metacherts. Zircon, rutile, and tourmaline are the most stable of the heavy minerals (Hubert 1962); they are typically concentrated in quartz arenite and are preserved despite prolonged source area weathering and reworking within the depositional basin. Other criteria that suggest a detrital sedimentary origin are thickness of the quartzites and the infrequent occurrence of argillaceous partings; thick chert sequences typically contain numerous shale interbeds (Nesbit and Price 1974; Lowe 1976). The presence of fuchsite may be used to suggest an exhalative origin for quartzites (cf. Schreyer et al. 1981), but fuchsite is common, and ascribed to the breakdown of detrital chromite, in quartz arenites from the Witwatersrand Supergroup (Antrobus and Whiteside 1964; Pretorius 1964). Although it is probably true that fuchsite is relatively more abundant in Archean sedimentary rocks, examples of prominent fuchsite in quartzose detrital rocks are known through much of the geological record (e.g., Eskola 1933; Padget 1956; Clifford 1957; Fabries and Latouche 1973). In all cases, the most plausible derivation of fuchsite is from detrital chromite of original ultramafic provenance. If a detrital origin for the quartzites is accepted they represent the oldest siliciclastic sedimentary rocks in southern Africa (cf. Tankard *et al.* 1982) and thus would represent first-cycle quartz arenite deposits.

Pyroxenitic amphibolites/granulites (for brevity termed amphibolites below) in the Beitbridge Complex are geochemically indistinguishable from average basaltic tholeiites of intrusive or extrusive origin (Fripp 1983). The amphibolites occur in sharp contact and alternate with quartzite on a scale of tens of metres to centimetres (Fig. 19.4). The cm-scale alternation is frequently developed through stratal thicknesses of up to 10 m. Amphibolites are homogeneous in composition and texture and, from our observations, are most abundant in proximity to the Messina Layered Intrusion (Fig. 19.3). Although most commonly associated with quartzite in the Messina area, amphibolite also occurs with all other lithologies in the Beitbridge Complex.

Although clearly of tholeiitic composition and therefore unlikely to be of sedimentary derivation, the extrusive or intrusive origin of the amphibolite protolith is enigmatic. However, we consider that they are most likely of intrusive origin. The greater abundance of amphibolite that we observed adjacent to the Messina Layered Intrusion suggests a genetic relationship between the two, and the geochemistry of the amphibolites reported by Fripp (1983) is consistent with this suggestion. The amphibolites and metasediments show evidence of significant ductile strain. It is difficult to be confident about the original thickness and scale of interlayering of the amphibolite and quartzite layers because no bulk strain markers are reported, and we observed none. If the strain is high, as we suspect from the qualitative indicators of early folds and boudinage, then the absence of cross-cutting intrusive relationships is understandable, since originally oblique features will have been rotated into parallelism. We observed a few possibly cross-cutting relationships, but they were not sufficiently convincing or oblique to distinguish from original lens-shaped sedimentary or volcanic layer terminations, or from sill edge terminations for that matter, given the likely ductile strain in the rocks. However, the lack of lithologies showing compositions indicative of mixing of quartzite and mafic protoliths, and the sharp contacts everywhere between the two lithologies, are features that

are more easily explained if the mafic protolith was largely or entirely intrusive. A tuffaceous protolith for the amphibolites is considered also unlikely. Ash beds do not display the degree of compositional homogeneity apparent in the amphibolite layers. In addition, ash-fall deposits are more commonly siliceous than mafic and typically are interbedded with pelagic sediments or mudstones rather than quartz arenites as in the Beitbridge Complex (cf. Fisher and Schmincke 1984; their Table 7-1). Also, reworking and mixing of the loose ash and arenite would be predicted; appropriate rocks are not seen. If we are incorrect, and the mafic layers are largely flows, then the assemblage of pure quartzites and mafics is one without a Phanerozoic analog. We see no reasons to assume, however, that extrusive mafic rocks were present in any significant quantity, and consider that the evidence at present, while less than completely satisfactory, favors an intrusive origin. We therefore prefer a sedimentary-intrusive origin for the quartzite-amphibolite association with the quartzite representing original quartz arenite and the amphibolite original sills and/or dikes.

The protoliths of two of the other lithological associations comprising the Beitbridge Complex are more easily defined. Marble in the calc-silicate gneiss-marble association attains thicknesses, for single outcrops, of at least 100 m (Fig. 19.5) and, in the Messina-Tshipise region, is developed in a belt parallel to and south of the quartzite occurrences (Fig. 19.3). Marbles display abundant compositional layering and are visualized as original sedimentary carbonate rocks. Calc-silicate gneiss is considered to have originally been calcareous pelite. Associated but subordinate magnetite quartzite may represent either metamorphosed iron-formation or iron-rich quartzite. Overall thicknesses of this association are uncertain. Most workers agree that biotite-garnet-cordierite-sillimanite gneisses represent metapelites and the major element geochemistry of those in the Beitbridge Complex closely resembles average "geosynclinal" shales (Brandl 1983; Fripp 1983). Taylor *et al.* (1986) argue that the mineralogy, and major, trace, and rare earth element (REE) geochemistry, and intimate association with recognizable quartz arenites and carbonates support a pelitic origin for these paragneisses.

The protolith of the fourth lithologic association, the gray gneisses, is highly problematic. On the basis of their geochemistry, Fripp (1983) considers that these gneisses may have been felsic to intermediate volcanic and/or terrigenous sedimentary rocks, whereas Brandl (1983) favors a sedimentary arkosic protolith.

Fripp (1983) estimated the proportions of various lithologies comprising the Beitbridge Complex in the Beitbridge-Messina-Tshipise region (Fig. 19.3). His estimates for rocks considered to be of original volcanic and sedimentary origin are: amphibolite (volcanic) – 30%; biotite-garnet-cordierite-sillimanite gneiss (metapelite) – 30%; quartzite (chert) – 10%; marble-calc-silicate gneiss (limestone-calcareous pelite) – 10%; gray gneiss (sediment or volcanic) – 10%; Singelele gneiss (felsic volcanic) – 10%. Based on these percentages Fripp (1983) considers the Beitbridge Complex to have contained a relatively high ratio of volcanic to sedimentary rocks. Barton (1983a) disputes this ratio arguing that the Singelele Gneiss was intrusive as evidenced by numerous xenoliths. He suggests a somewhat lower proportion of volcanic to nonvolcanic rocks. On the basis of the earlier discussion, it is possible that the Beitbridge Complex contained few, if any, volcanic rocks and our estimate of lithologic proportions of primary surficial deposits, assuming a sedimentary origin for the gray gneisses, is quartz arenite – 17%; limestone and calcareous mudstone – 17%; mudstone – 50%; arkose/feldspathic wacke – 16%.

Source-Area Composition

The Beitbridge Complex represents a mature assemblage of sedimentary rocks and their petrography thus provides little information on source-area composition. Heavy-mineral separates are similarly not informative as zircon, rutile, and rare tourmaline are the only detrital varieties present.

Rare-earth element (REE) patterns preserved in mudstones provide significant constraints on provenance composition because they provide an overall average of REE in upper crustal rocks exposed at the surface to erosion (Taylor and McLennan 1985). Rare-earth element curves for mafic rocks have low slopes, whereas felsic rocks are enriched in light rare-earth elements (LREE), such as La, Ce, Pr, and Nd, to give steep curves. Intermediate curves imply mixing of the two end members. In addition to the steepness of the curve, the presence or absence of a negative Eu anomaly is also informative. Upper crustal rocks lacking such anomalies are considered to represent mantle derivatives. In contrast, upper crustal rocks with negative Eu anomalies such as

granite and granodiorite are ascribed to intracrustal melting with retention of Eu in plagioclase in lower- or middle-crustal regions (Taylor and McLennan 1985). Archean mudstones characteristically lack negative Eu anomalies, whereas post-Archean mudstones display such anomalies. This contrast has been interpreted by Taylor and McLennan (1985) to record extensive crustal differentiation and craton development at the Archean-Proterozoic boundary, but their Archean sampling is heavily skewed to greenstone terranes, making the contrast equivocal.

Metapelite samples from the Beitbridge Complex show little evidence of modification by partial melting and hence their REE geochemistry is considered to reflect accurately the composition of the source terrain (Taylor et al. 1986). These data (Fig. 19.7) display variable slopes implying a mixed provenance. Samples LP20 and LP28 have typical Archean patterns with no negative Eu anomalies; sample LP20 was derived mainly from tonalitic granitoids or Na-rich felsic volcanic rocks and LP28 from a source with a larger mafic:felsic component. Rare-earth element curves for samples LP4, LP11, and LP30 resemble post-Archean shales. Enrichment of heavy REEs in sample LP30 is attributed to concentration of these elements in garnet as a result of metamorphic differentiation. The steep REE patterns and pronounced negative Eu anomalies displayed by these three samples are compatible with derivation from K-rich granites and granodiorites (Taylor et al. 1986).

Tectonic Setting

In the absence of primary sedimentary structures in metasedimentary rocks of the Beitbridge Complex, stratigraphic sequences, sedimentary facies, and sediment dispersal patterns cannot be reconstructed. Detailed paleoenvironmental analysis is thus impractical but the interpreted primary lithologies and lithologic proportions permit some general inferences on depositional environment and tectonic setting. In many respects the Beitbridge Complex is similar lithologically to the widely developed, cratonic-shelf quartz arenite-carbonate association present in younger, unmetamorphosed sequences dating back to 2.5 Ga. Examples include the lower part of the lower Proterozoic Transvaal Supergroup, South Africa (Tankard et al. 1982), the 1.9 Ga Epworth Group in Wopmay Orogen, Canada (Hoffman 1973, 1980), and, in particular, Cambro-

Fig. 19.7. Chondrite-normalized REE plot of metapelite samples in the Beitbridge Complex [compiled from data in Taylor *et al.* (1986)]. Heavy rare-earth element enrichment in sample LP30 is due to metamorphic differentiation and concentration of these elements in garnet. A primary sedimentary pattern would approximate the dashed line.

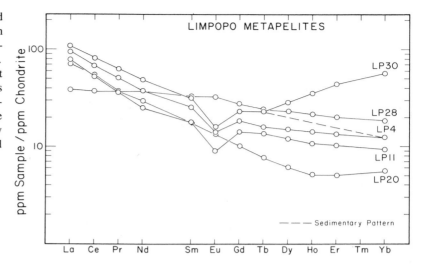

Ordovician sequences extending around much of North America (Stearn *et al.* 1984). The central Appalachian passive-margin sequence appears to be a particularly appropriate analog; it is notably rich in mudstone because of its near shelf-edge position in contrast to coeval platform sequences inboard to the west, which consist primarily of quartz arenite and carbonate. However, the data base for the Beitbridge Complex is inadequate to draw a direct analogy with a passive margin and the more general term *cratonic shelf* thus is preferred.

The REE geochemistry discussed above provides strong evidence for the exposure to erosion of an at least partly granitic terrain. We suggest that this terrain was continuous with the basement of the stable shelf on which the sedimentary protoliths of the Beitbridge Complex accumulated. Evidence for the erosion of products of intracrustal melting provided by the metapelite samples with steep patterns and negative Eu anomalies implies that 25 to 30 km thick continental crust existed prior to accumulation of the Beitbridge Complex. Taylor *et al.* (1986) consider that this granitic crust constituted small, stable continents or protocratons.

Discussion

The present configuration of the central zone of the Limpopo Province is a result of the following sequence of events: 1) Formation of continental crust greater than 25 to 30 km thick and probably represented by the Sand River Gneisses; 2) Intracrustal melting produced Eu-depleted felsic rocks of granitic and granodioritic composition; deformation of basement rocks took place prior to or after intracrustal melting; 3) Weathering and erosion accompanied/followed by sedimentation of quartz arenite, carbonate and mudstone on a stable continental shelf; 4) Intrusion of the Singelele and Bulai granitoids (Fig. 19.3); 5) Deformation and metamorphism at depths of 25 to 30 km; and 6) Uplift to the present surface.

The first three of these events and particularly the third are relatively well understood. The age of burial and metamorphism is constrained by the 2,650 to 2,700 Ma Bulai granitoid (Fig. 19.3), which was emplaced during or shortly after the peak of granulite metamorphism (Watkeys *et al.* 1983). The metasedimentary rocks of the Beitbridge Complex must therefore have been at mid-crustal depths at approximately 2.7 Ga to achieve the granulite facies assemblages seen at the present surface. Underthrusting of basement and cover rocks is attributed to Himalayan-style collisional tectonics at 2.7 Ga, and subsequent uplift to isostatic readjustment (Kidd 1985; Burke *et al.* 1986; Van Reenen *et al.* 1987).

Western Gneiss Terrain: Western Australia

Geologic Setting

The geologic framework of the Western Gneiss Terrain is similar in many respects to the central zone of the Limpopo Province. At Narryer in the north of

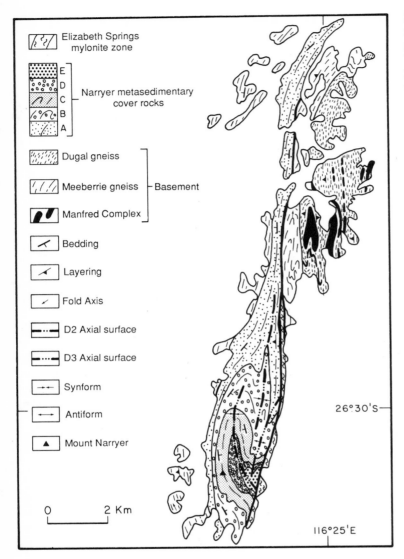

Fig. 19.8. Geological map of the Narryer region in the Western Gneiss Terrain [adapted from Myers and Williams (1985)]. See Figure 19.1 for location and Figure 19.10 for description of cover rocks.

the Western Gneiss Terrain (Fig. 19.1), basement rocks are identified on the basis of a pre-cover structural and metamorphic history (Myers and Williams 1985; Kinny *et al.* in press). Basement rocks (Fig. 19.8) consist primarily of the Meeberrie and Dugel Gneisses dated respectively at $3,630 \pm 40$ and $3,510 \pm 50$ Ma (Sm-Nd model; DeLaeter *et al.* 1981) and $3,688^{+33}_{-23}$ Ma and $3,416^{+82}_{-50}$ Ma (U-Pb zircon; Kinny *et al.* in press); the Dugel Gneiss contains xenoliths of layered gabbroic anorthosite (Manfred Complex; Fig. 19.8) that yield zircon ages of $3,750^{+72}_{-40}$ Ma (Kinny *et al.* in press). Zircon rim ages of $3,319^{+34}_{-16}$ Ma and a Rb-Sr whole rock isochron age of $3,350 \pm 43$ Ma are interpreted as metamorphic ages associated with gneiss formation (DeLaeter *et al.*

1981; Kinny *et al.* in press). Metasedimentary rocks, for which no detrital zircon ages are available, are present within the gneisses as either xenoliths or tectonically interleaved bodies of mainly pyroxene and sillimanite-mica quartzite and banded quartz-magnetite-pyroxene rocks (Myers and Williams 1985).

Cover rocks in the Narryer area (Fig. 19.8) consist of a ca. 2.5 km thick sequence of metaconglomerate, quartzite with well preserved crossbeds (Fig. 19.9), and subordinate paragneiss (Gee *et al.* 1981; Myers and Williams 1985). Cores of detrital zircons in the cover rocks are mainly ca. 3,750 Ma and ca. 3,500 Ma but zircons as young as 3,250 Ma and as old as 4,200 to 4,100 Ma are also present (Froude

Fig. 19.9. Crossbeds preserved in aluminous gneisses (original wackes) in Unit C in the Narryer cover rocks of the Western Gneiss Terrain. Foresets are defined by meta- morphic garnet. Note lack of tectonic modification of primary crossbed shapes.

et al. 1983). Metamorphic rims on detrital zircons cluster around 2,800 Ma (U-Pb zircon; Kinny 1986). The age of the cover rocks is thus constrained at less than 3,250 Ma, the age of the youngest detri- tal zircons and greater than 2,800 Ma, the age of metamorphic rims on detrital zircons. Sedimenta- tion is considered to have taken place closer to 3,250 Ma than 2,800 Ma (Taylor *et al.* 1986). The 2,800 Ma prograde metamorphism and accompanying deformation (D2) of the basement and cover rocks took place under granulite conditions (Blight and Barley 1981). Subsequent retrograde, amphibolite facies metamorphism was associated with a third phase of deformation (D3, Fig. 19.8; Myers and Williams 1985).

Stratigraphic Framework of the Narryer Cover Rocks

In marked contrast to the Limpopo Province, primary sedimentary structures and clast shapes are remarkably well preserved in the Narryer region despite the deformation and granulite facies meta- morphism. Crossbeds display no flattening or dis-

tortion of shape (Fig. 19.9) and the conglomerates show essentially no evidence of strain except for local rigid body rotation of some clasts into the plane of weak cleavage. Both early and late mylonites cut across the cover rocks. Primary struc- tures are destroyed in the mylonite zones but are preserved in lensoid bodies between the mylonite zones. Apparently strain partitioning into the mylonites spared most of the primary features from deformation.

The widespread occurrence of conglomerate and crossbedded quartzite provides unequivocal evi- dence that the Narryer cover rocks are of siliciclastic origin. Unlike the Beitbridge Complex in the Lim- popo Province, where no stratigraphic relationships can be deduced, a stratigraphic sequence can be recognized and mapped in the Narryer cover rocks. Primary sedimentary structures indicate that the strata are not overturned and a five-fold subdivision is recognizable (Figs. 19.8, 19.10). Each unit is lithologically distinct and it is possible that the ca. 2.5 km thickness closely approximates the deposi- tional thickness. However, in the absence of marker beds, structural interleaving within units and ductile attenuation cannot be ruled out. Older folded and

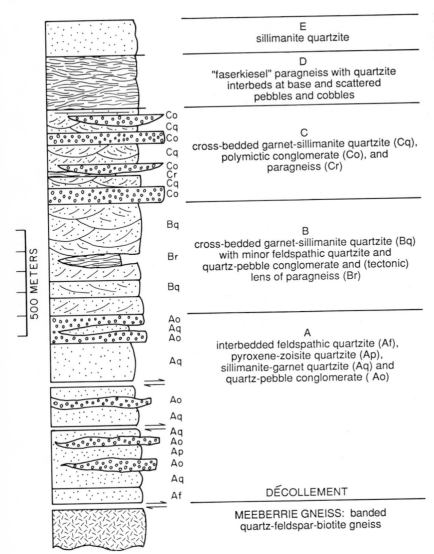

Fig. 19.10. Generalized stratigraphic column through the Narryer cover rocks. Thicknesses from maps of Williams and Myers (in press).

younger unfolded mylonitic shear zones within the cover rocks suggest possible structural duplication.

The base of the cover sequence is a mylonitic shear zone. Unit A at the base of the sequence is a succession of feldspathic, pyroxene-zoisite, and sillimanite-(garnet) quartzites with interbedded sheets of quartz-pebble conglomerate (Fig. 19.10). Unit B is composed of sillimanite-garnet(cordierite) quartzite with well developed crossbedding; also present are subordinate feldspathic quartzite, quartz-pebble conglomerate, and a single lens of paragneiss. Unit C also contains quartz-pebble conglomerate but is defined by the first appearance of polymictic conglomerate interbedded with cross-bedded garnet-sillimanite-(cordierite) quartzite. Cordierite-garnet-biotite-sillimanite gneiss and bimineralic quartz-garnet horizons are locally interbedded with the crossbedded quartzite. Conglomerate occurs as sheets and lenses containing clasts of banded magnetite-quartz-pyroxene-amphibole, vein quartz, quartzite and "garnetite" (garnet-quartz rock). Vein-quartz cobble conglomerate lenses with a sillimanite-rich matrix occur locally. Upper Unit C quartzites interfinger with paragneiss and conglomerate of Unit D that are characterized by the occurrence of distinctive "faserkiesel" of intergrown sillimanite and quartz. Unit D is abruptly overlain by sparsely sillimanitic quartzite of Unit E; primary stratification cannot be identified in either of Units D or E that occur in the core of the Narryer Syncline (Fig. 19.8).

The presence of diverse metamorphic minerals indicates that the original sandstones were texturally immature with variable but significant amounts of clay. Sillimanite is widespread and can constitute up to 20% of the quartzites. A lack of metamorphic K-feldspar in these quartzites indicates that the reaction muscovite + quartz ⇌ K-feldspar + sillimanite + H_2O did not take place (cf. Winkler 1974); this implies that the protolith was an aluminous, potassium-poor mineral such as kaolinite and not illite. The other widespread metamorphic mineral is the iron-aluminum garnet almandine, which can make up to 25% of the quartzite or as much as 50% of the intercalated garnet-quartz horizons. The protolith of the latter is enigmatic but ferruginous mudstones and siltstones, such as in the Witwatersrand Supergroup, South Africa (Fuller *et al.* 1981), are of similar bulk composition. By analogy, garnet within the crossbedded quartzites could have been derived from ferruginous clay. Pure Fe-Al clays, however, are uncommon, especially in association with kaolinite and the possibility exists that at least some of the iron in the siliciclastic sediments was present as iron oxide. If so, it is tempting to suggest that these metasedimentary rocks originally may have been redbeds in which case they would be the oldest occurrence in the geologic record.

The quartzite, vein-quartz, and banded quartz-magnetite-pyroxene-amphibole clasts are typically rounded; the banded clasts represent original iron formation. "Garnetite" clasts are both rounded and angular and are interpreted as original intraformational ferruginous siltstone-mudstone clasts. The biotite-garnet-sillimanite-cordierite paragneiss horizons represent original pelitic sedimentary rocks.

Source-Area Composition and Age

Information on the composition of the source area from which the Narryer cover rocks were derived can be obtained from clast types, heavy minerals, and REE data from metapelites. Banded quartz-magnetite-pyroxene-amphibole cobbles and boulders are the only diagnostic clasts in the conglomerates; these are identical to banded rocks in the sedimentary outliers within the basement Meeberrie Gneiss and may indicate that these outliers are

xenoliths, or older sedimentary rocks tectonically interleaved during gneiss formation at 3,350 Ma prior to deposition of the Narryer cover rocks, instead of tectonically interleaved cover rocks. Quartzite clasts may have been derived from associated lithologies in the outliers or could represent recycled cover rocks; the quartzites from these two locations are indistinguishable in the field.

As in the Limpopo Province, the detrital heavy mineral assemblage in the Narryer cover rocks is dominated by zircon and rutile with sphene and tourmaline comprising less than 1% of the population. Heavy minerals thus provide little information on source-area composition. Rare-earth element data from metapelites are more informative. Three REE curves have steep slopes and prominent negative Eu anomalies (Fig. 19.11; Taylor *et al.* 1986). There is no evidence of a mixed provenance of the type recognized in the Limpopo Province. Rather, these steep curves suggest that the cover rocks were derived from K-rich granitic or granodioritic plutons implying that thick, differentiated crust existed in the source area.

The predominant ages of detrital zircon cores of ca. 3,500 Ma are comparable to the Sm-Nd and U-Pb zircon ages of the basement Meeberrie and Dugel Gneisses. These gneisses are of granitic to granodioritic composition (Myers and Williams 1985) and, together with the sedimentary xenoliths, are a likely source for the cover rocks. The 3,750 Ma detrital zircons were probably derived from the Manfred Complex but the origin of the 4,200 to 4,100 Ma detrital zircons is unknown at present.

Sediment Dispersal Patterns and Depositional Environments

The excellent preservation of primary stratification in original sandstones and primary clast shapes in conglomerates, especially in Units B and C, permits a general facies analysis of the Narryer cover rocks. Three facies associations are present; stratified sandstone, clast-supported conglomerate, and matrix-supported conglomerate. Sandstone facies include trough crossbeds with subordinate tabular-planar crossbeds and horizontal stratification. The crossbeds indicate unimodal flow in a general southeasterly direction but with considerable variability in flow direction. Stratified sandstone is intimately associated with clast-supported conglom-

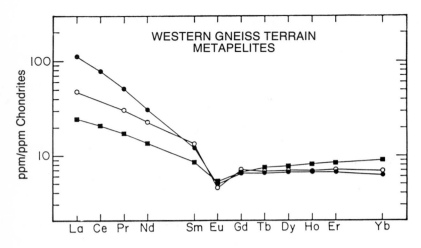

Fig. 19.11. Chondrite-normalized REE plot of metapelite samples from the Narryer cover rocks [from Taylor *et al.* (1986); reproduced with permission from Pergamon Press, Ltd.].

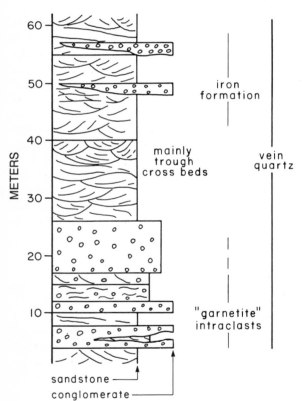

Fig. 19.12. Measured stratigraphic section through part of Unit C in the Narryer cover rocks illustrating the use of conventional sedimentology despite deformation and granulite facies metamorphism. Section located just north of Mt. Narryer.

erate in vertical sequence in Units A and C (Fig. 19.12). Detailed mapping of these two facies associations in Unit C demonstrates a primary lenticularity (Fig. 19.13) reflecting a channel and bar-type geometry typical of a high-energy, braided fluvial system (cf. Rust 1978). Unit A displays a similar interfingering of conglomerate and sandstone. Depositional environments are difficult to reconstruct for the crossbedded sandstones of Units B and E (Fig. 19.10). Possible depositional settings include braided river, tidal, eolian, and shoreface-shelf. Sediment-body geometries and stratification styles characteristic of tidal and eolian systems are not developed (cf. Hunter 1977; Homewood and Allen 1981; Kocurek 1981; Terwindt 1981). The presence of metamorphic sillimanite within the quartzites indicates that kaolinite was a significant constituent of the sandstone protolith. An environment of limited winnowing can thus be inferred; this inference is further evidence against a tidal or eolian origin and probably also against a shoreface-shelf origin for the sandstones. Furthermore, quartz-pebble conglomerates present in Unit B are not characteristic of tidal or eolian settings. Despite the absence of vertical sequences of facies and architectural elements typical of braided-alluvial systems (Miall 1977, 1985), a low-energy braided river is the preferred depositional environment of Units B and E.

The third facies consists of well rounded, vein-quartz cobble, matrix-supported conglomerates with a sillimanite-rich matrix. The matrix may compose up to 90% of the rock and from the earlier discussion

Fig. 19.13. Measured sections through part of Unit C. Top of a conglomerate bed was used as a datum for correlation of sections. Cross section illustrates syndepositional lenticularity of conglomerate within sandstone.

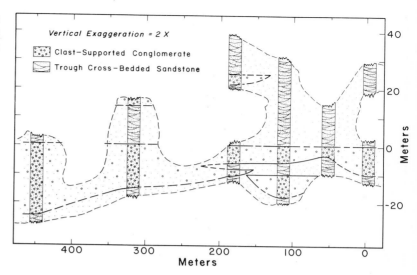

was probably a sandy kaolinitic mudstone. Matrix-supported conglomerates make up most of Unit D and occur locally in Unit C as broad lenticular bodies hundreds of metres wide and tens of metres thick; these bodies interfinger with clast-supported conglomerates and associated crossbedded sandstones. Based on their close spatial relationship with inferred braided-alluvial deposits, the matrix-supported conglomerates are considered to be of subaerial origin. Possible depositional processes include debris flows and mudflows.

Basin Configuration and Tectonic Setting

Basin configuration and paleogeographic relationships of the facies present in the Narryer cover rocks are unclear. The braided-alluvial sediments may be visualized as a trunk system that flowed in a general southerly direction, whereas the lenticularity, in north-south trending outcrops, of matrix-supported conglomerates in Unit C suggests transverse flow from the east or west. However, paleoflow azimuths are unreliable due to probable rotation of the cover sequence during late stages of deformation.

Rare-earth element geochemistry from the Narryer cover rocks (Fig. 19.11) provides evidence for thick continental crust in that area prior to 3.3 Ga. Other than requiring the existence of continental basement, the tectonic setting in which the cover rocks developed is problematic. The inferred debris-flow or mudflow deposits indicate at least moderate

relief, which in intracratonic settings is characteristically associated with faulting. Comparable thicknesses of debris flow or mudflow, and coarse-grained, braided-alluvial deposits have been interpreted as rift sequences (Miall 1981). However, in view of the poor constraints on the configuration of the Narryer basin, it is not possible to differentiate rift from nonmarine foreland basin settings similar to those described by Anderson and Picard (1974), DeCelles (1986), and DeCelles et al. (1987).

Discussion

The geological evolution of the Narryer region was similar to the central zone of the Limpopo Province and involved formation of 25 to 30 km thick continental crust, intracrustal melting to produce the K-rich protoliths of the Meeberrie and Dugel Gneisses, deformation and metamorphism of the basement and cover rocks at depths of 16 to 18 km, followed by uplift to the present surface. The age of burial and metamorphism is well constrained at 2.8 Ga but the mechanisms responsible for crustal thickening at this time are not understood. Gee et al. (1986) consider the possibility of collision of two unrelated segments of crust to explain the regional relationships between the high-grade Western Gneiss Terrain and the low-grade, granite-greenstone terrane of the Yilgarn Block to the east (Fig. 19.1); this proposed event might have been associated with the necessary crustal thickening in the Western Gneiss Terrain at ca. 2.8 Ga.

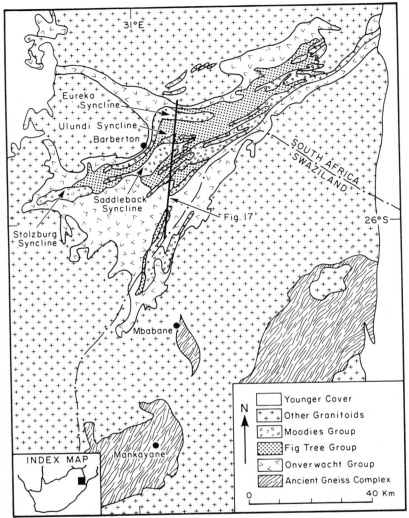

Fig. 19.14. Generalized geological map of the Barberton Greenstone Belt, Ancient Gneiss Complex, and surrounding granitoids [adapted from Tankard *et al.* (1982)]. Index map shows location in southern Africa.

Barberton Mountain Land: South Africa and Pilbara Block: Western Australia

Geologic Setting

The Barberton Mountain Land and Pilbara Block (Fig. 19.1) have long been considered as classical examples of Archean granite-greenstone terranes; in each terrane arcuate greenstone belts are surrounded by granite and gneiss (Figs. 19.14, 19.15). Metamorphism within the greenstone belts is greenschist grade or lower; in proximity to intrusive plutons amphibolite grades are locally attained. The Ancient Gneiss Complex southeast of the Barberton Moun-

tain Land (Fig. 19.14) consists of 3.5 Ga metavolcanic and metasedimentary rocks, a tonalitic-gneiss batholith, metabasite dikes, meta-anorthosites, and ca. 3.3 Ga undeformed granitoids (Jackson *et al.* 1987).

A common two-fold stratigraphic subdivision is recognized within the greenstone belts; namely, lower volcanic-dominated and upper siliciclastic-dominated intervals (Fig. 19.16; Viljoen and Viljoen 1970; Hickman 1981). The lower volcanic-dominated intervals, the Onverwacht and Warrawoona Groups (Fig. 19.16), consist mainly of mafic-ultramafic volcanic rocks that are characteristically pillowed and in the Barberton Mountain Land contain komatiites. Subordinate felsic volcanic rocks are present in both regions; these display calc-

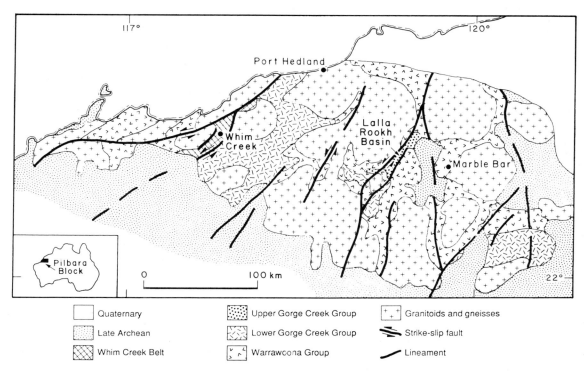

Fig. 19.15. Generalized geological map of a portion of the Pilbara Block [adapted from Hickman (1983)]. Index map shows location in Australia.

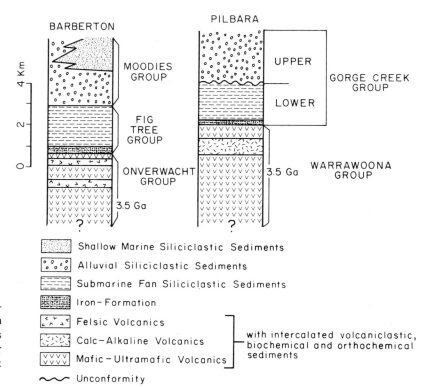

Fig. 19.16. Generalized stratigraphic columns of the Barberton and Pilbara greenstone belts [adapted from Anhaeusser (1973), Hickman (1983), Krapez (1984)].

alkaline affinities in the Pilbara Block (Barley *et al.* 1984). Lower Onverwacht mafic and felsic rocks together give a Sm-Nd isochron age of 3,540 ± 30 Ma (Hamilton *et al.* 1979), later revised to 3,530 ± 50 Ma (Hamilton *et al.* 1983). Warrawoona mafic volcanic rocks give a similar age of 3,560 ±32 Ma (Sm-Nd isochron; Hamilton *et al.* 1981). Minimum ages for the volcanic intervals are ca. 3,300 Ma (U-Pb zircon; Van Niekerk and Burger 1969; Pidgeon 1984).

Sedimentary rocks are minor but important components of the Onverwacht and Warrawoona Groups insofar as providing paleoenvironmental information. Lowe (1982) has recognized three types of sedimentary rocks in the volcanic sequences, which he categorizes as volcaniclastic, biochemical, and orthochemical. The sedimentary detritus is exclusively of intrabasinal derivation indicating an environment removed from continental influence. Furthermore, no continental basement to the volcanic intervals has been recognized in either region. These criteria have been used by Lowe (1982) to suggest that volcanism took place in an oceanic environment. Intercalated sedimentary rocks indicate that the preserved remnants of oceanic crust developed in shallow water (Barley *et al.* 1979; Lowe 1982). The mafic-ultramafic volcanism is considered to have formed relatively flat shield volcanoes with sporadic felsic cones that were often exposed to weathering and erosion (Lowe 1982).

The upper contact of the Onverwacht and Warrawoona Groups marks the base of the siliciclastic-dominated intervals and a dramatic change in tectonics as reflected by the Fig Tree-Moodies and Gorge Creek Groups (Fig. 19.16). Moodies Group sedimentation in the Barberton Mountain Land took place at less than 3,310 ± 10 Ma, the youngest age of granitoid cobbles in the basal conglomerate (U-Pb zircon; Tegtmeyer *et al.* 1981; Tegtmeyer and Kröner 1987), and before 3,200 ± 30 Ma, the age of an intrusive granite in the eastern mountain land (Fig. 19.14; Pb-Pb apatite; Oosthuysen and Burger 1973; recalculated by Cahen *et al.* 1984). The age of the Gorge Creek Group is less well constrained; direct geochronological evidence indicates an age between 3,300 Ma, the end of Warrawoona volcanism, and 2,768 ± 16 Ma, the age of volcanic rocks in an upper Archean sequence in the Pilbara Block (Fig. 19.15; U-Pb zircon; Pidgeon 1984). The upper Gorge Creek Group is considered to have accumulated at ca. 2,950 Ma (Krapez and Barley in press).

In the following sections, detailed basin analyses are presented for the Fig Tree and Moodies Groups in the Barberton Mountain Land (Figs. 19.14, 19.16) and the upper Gorge Creek Group in the Pilbara Block (Figs. 19.15, 19.16). These two sequences are well understood and provide an opportunity to evaluate the role of plate tectonics in the Archean. The lower Gorge Creek Group is not discussed; source-area composition and depositional environments were similar to the Fig Tree and Moodies Groups but sediment dispersal patterns, basin configuration, and tectonic setting are constrained poorly.

Stratigraphic Framework of the Fig Tree and Moodies Groups

Primary depositional textures and structures are well preserved throughout the Fig Tree and Moodies Groups, which crop out in a number of isolated synclines surrounded by Onverwacht volcanic rocks (Fig. 19.14). With the excellent preservation of "right-way-up" indicators in the form of primary sedimentary structures, a coherent stratigraphy can be mapped within each syncline. Correlation between synclines, however, is difficult in the Fig Tree Group because of the absence of marker or time horizons. In the southern half of the Barberton greenstone belt, the Fig Tree strata conformably overlie Onverwacht volcanic rocks and were involved in syndepositional folding and north-westerly directed, thrust-nappe tectonics prior to deposition of the unconformably overlying Moodies Group (Fig. 19.17; Lamb 1984; Lowe *et al.* 1985). The Fig Tree Group in this area commences with a 60 m-thick iron formation displaying evidence of synsedimentary deformation (Eriksson 1980a), overlain by an upward-coarsening sequence of sandstone and conglomerate. The overlying Moodies Group consists of a thick sequence of conglomerate with subordinate sandstone grading upward into sandstone (Fig. 19.17; Eriksson 1980a). In the northern half of the greenstone belt the Fig Tree and Moodies Groups are in conformable contact in the Saddleback, Stolzburg, and Eureka Synclines (Fig. 19.14). The Fig Tree Group consists primarily of ca. 2 km of interbedded wacke and mudstone with subordinate iron-formation (Eriksson 1980b), whereas the Moodies Group comprises one upward-fining sequence overlain by two upward-coarsening

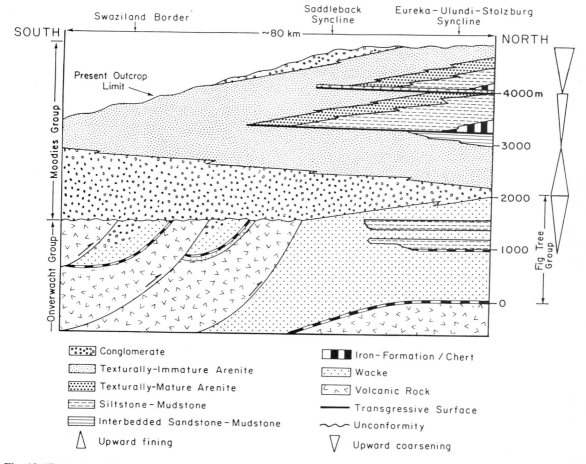

Fig. 19.17. Stratigraphic cross section of the Fig Tree and Moodies Groups and their relationship to the Onverwacht Group [based on data from Condie *et al.* (1970), Eriksson (1979), Lamb (1984), and Lowe *et al.* (1985)]. Location shown on Figure 19.14.

sequences; these sequences are ca. 1 km thick (Fig. 19.17; Eriksson 1979; Jackson *et al.* 1987). The base of the Moodies Group is defined by a conglomerate passing upward into a thick interval of sandstone followed by interbedded sandstone and mudstone. The upward-coarsening sequences are variable; in the Stolzburg and Eureka Synclines they commence with iron-formation grading into mudstone that contains increasing proportions of sandstone and siltstone upward and is capped by a thick sandstone. In the Stolzburg Syncline these sequences are coarser and consist primarily of sandstone and conglomerate (Fig. 19.17). Correlation of the Moodies Group among the Saddleback, Stolzburg, and Eureka Synclines is based on similarity of sequences and on an amygdaloidal lava horizon

at the base of the uppermost, upward-coarsening sequence (Fig. 19.17). In the absence of biostratigraphic control, the sequence boundaries and particularly the lava horizon are taken to represent time planes.

Source-Area Composition and Age

Clast types in conglomerates, sandstone petrography, and mudstone geochemistry provide information on the composition of the provenance for the Fig Tree and Moodies Groups. White, black, and banded chert are the predominant clast types; subordinate vein-quartz, iron-formation, and felsic- and mafic-volcanic clasts are also present (Eriksson

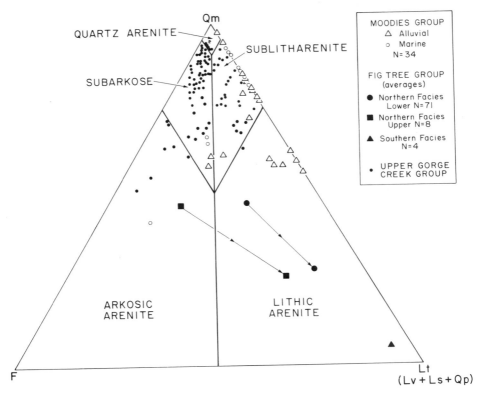

Fig. 19.18. Ternary plot of sandstone compositions in the Fig Tree, Moodies, and upper Gorge Creek Groups [data from Krapez (1984), Jackson *et al.* (1987), Eriksson (1980a), Herget (1966), and Reimer (1975) classification after Pettijohn *et al.* (1972)]. End members are monocrystalline quartz (Qm), total feldspar (F), and total lithics (Lt) incorporating volcanic lithics (Lv), sedimentary lithics (Ls), and polycrystalline quartz, mainly chert (Qp). For northern Fig Tree facies, average matrix component recalculated as Lt with shift towards lithic pole.

1978, 1980a). Granitoid cobbles and boulders occur in conglomerates of the southern Fig Tree outcrop belt and are abundant in the basal Moodies conglomerate in the northern Barberton greenstone belt (Gay 1969; Eriksson 1978; Heinrichs 1980). The granitoid clasts in the Moodies Group are massive or gneissic and vary from alkali granite and granite to granodiorite (Krupicka 1975; Reimer *et al.* 1985). Botryoidal quartz clasts and clasts of polymictic conglomerate are unique to the Moodies Group (Eriksson 1980a).

Sandstones in the Fig Tree and Moodies Groups become progressively more quartzose stratigraphically upward (Fig. 19.18). In the southern Fig Tree outcrop belt, sandstones are highly lithic, consisting of up to 90% metavolcanic rock fragments and angular chert (Eriksson 1980a). Sandstones in the northern outcrop belt are wackes with up to 35% matrix (Herget 1966; Reimer 1975); the matrix is considered to have been derived by diagenetic recrystalli-

zation of unstable lithic grains (Reimer 1972). If the matrix is recalculated as rock fragments, these sandstones are lithic arenites with chert and volcanic rock fragments the predominant components but with significant quantities of quartz and feldspar also present, in contrast to southern Fig Tree sandstones (Fig. 19.18). Lithic components include grains with perthitic intergrowth textures. Moodies Group sandstones are mainly sublitharenites with less common lithic, arkosic, and quartz arenites (Fig. 19.18; Jackson *et al.* 1987). The lithic character of most of the sandstones is due to their chert (Qp) component; volcanic rock fragments may constitute up to 10% of the arenites and, where present, orthoclase and microcline are the predominant feldspar varieties.

Clast types and sandstone petrography in the Fig Tree and Moodies Groups indicate a mixed provenance of Onverwacht volcanic rocks and associated chert, and granitoid and gneiss. Recycling of older

Fig. 19.19. Chondrite-normalized REE plot of averages of mudstone samples from the Fig Tree and Moodies Groups [compiled from data in McLennan *et al.* (1983)].

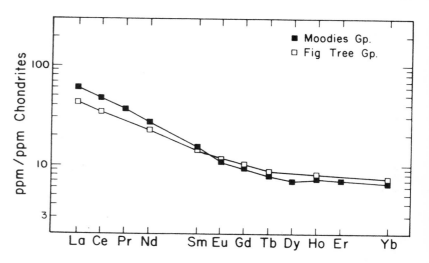

volcanic and sedimentary components is thus implied. Iron-formation clasts within all three groups suggest recycling within the predominantly siliciclastic intervals as iron formation is absent in the Onverwacht Group (Lowe 1982). Polymictic conglomerate clasts within the Moodies Group provide still stronger evidence of recycling.

Geochemical data from the Fig Tree and Moodies Groups substantiate the above provenance interpretation. Rare-earth element curves are relatively flat as reflected in low La/Yb ratios, and negative Eu anomalies are absent (Fig. 19.19; McLennan *et al.* 1983). Rare-earth element modelling implies a mixed provenance of mafic-ultramafic volcanic rocks and tonalites-felsic volcanic rocks; a significant plutonic contribution is indicated by the granitoid clasts in the Fig Tree and Moodies Groups and the grains of K-feldspar and perthitic intergrowth in several samples. The absence of negative Eu anomalies implies that the granitoid crust exposed to weathering and erosion had not undergone intracrustal melting, although granodiorites and monzonites without substantial negative Eu anomalies cannot be excluded (McLennan *et al.* 1983). In the Barberton Mountain Land, progressive unroofing of granitoids is reflected in the upward increase in quartz and feldspar (Fig. 19.18), Th, U (McLennan *et al.* 1983), and the steeper REE pattern in the Moodies versus the Fig Tree Group (Fig. 19.19). Rare-earth element patterns for the Fig Tree Group require a substantial mafic contribution, whereas light REE enrichment in the Moodies Group requires that felsic igneous rocks constituted up to 80% of the provenance but with less than 20% con-

sisting of granitoids with significant Eu anomalies (McLennan *et al.* 1983). This minor component of the source terrain could have supplied the K-granite clasts and K-feldspar in the Moodies Group. High Ni and Cr contents of the Fig Tree and Moodies Groups (Danchin 1967) support the contention that mafic-ultramafic rocks were a significant component of the source terrain but the high concentrations of these two elements are not compatible with the REE modelling (McLennan *et al.* 1983).

Available data indicate that the age of the provenance of the Fig Tree and Moodies Groups varied between ca. 3.3 and 3.5 Ga. Detrital zircons from the Fig Tree Group have a mean age of 3.52 Ga, whereas granitic and gneissic clasts in the basal Moodies Group vary in age from 3.3 to 3.47 Ga (U-Pb zircon; Tegtmeyer and Kröner 1987). The geochronological data also are consistent with derivation of the Fig Tree and Moodies Groups from the Onverwacht volcanics and Ancient Gneiss Complex.

Sediment Dispersal Patterns and Depositional Environments

A range of paleoenvironments has been recognized in the Fig Tree and Moodies Groups despite the antiquity of the rocks and the consequent lack of fossils. These paleoenvironments are the subjects of a number of papers and will be reviewed only briefly below with the objectives of: 1) illustrating that detailed facies analysis is feasible in Archean strata, and 2) providing the data base for the analysis of basin configuration and tectonic setting.

Fig Tree Group strata in the southern outcrop belts (Fig. 19.17) display many affinities with submarine-fan deposits (Eriksson 1980a), but the upward-coarsening sequences preserved in these outcrops are also compatible with a fan delta (Nocita and Lowe 1985). Above a basal chert horizon, the northern Fig Tree Group consists almost entirely of "classical" turbidites consisting of T_{ae}, complete and base-missing Bouma (1962) beds; flute casts indicate flow to the north. Specific subenvironments are not recognized but the overall stratigraphic sequence is interpreted as a lobe to slope transition (Fig. 19.20a; Eriksson 1980b); slope canyons are only inferred. Sandstones of the southern Fig Tree facies are considerably more lithic than those in the northern outcrop belt (Fig. 19.18). This characteristic, coupled with the presence of an unconformity between the Fig Tree and Moodies Groups in the south, has led Jackson *et al.* (1987) to suggest that the southern occurrences of the Fig Tree Group were deposited prior to those in the north and were being involved in northward-directed, fold-thrust tectonics at the time that the northern Fig Tree facies were accumulating (cf. Figs. 19.17, 19.20a). Submarine lobes of the northern Fig Tree Group were fed by a fluvial system represented by coeval braided-alluvial conglomerate and sandstone of the basal Moodies Group to the south (Figs. 19.17, 19.20a); paleocurrents from trough crossbeds indicate flow exclusively to the north.

Overlying conglomerates and texturally immature arenites in the Moodies Group (Fig. 19.17) are also of braided-alluvial origin. Within these sediments representatives are recognized of the Scott and Donjek models (Eriksson 1978; cf. Miall 1977). Northerly paleocurrent vectors, coupled with the northward transition into texturally mature arenite, siltstone, mudstone, and iron formation of inferred shallow-marine origin (Fig. 19.17; Eriksson 1979), indicate that a northerly paleoslope prevailed throughout Fig Tree and Moodies Group sedimentation.

Shallow-marine facies are particularly common in the northern outcrop belt of the Moodies Group

where they comprise the two upward-coarsening sequences in the Eureka and Stolzburg Synclines and the lowermost upward-coarsening sequence in the Saddleback Syncline (Fig. 19.17). Both sequences in the Eureka and Stolzburg Synclines are interpreted as progradational barrier-beach deposits in which distal-shelf iron formation passes upward (landward) into outer-shelf siltstone and mudstone followed by texturally mature, shoreface arenites (Eriksson 1979). Foreshore deposits are characterized by swash lamination with low-angle discordances and abundant wave ripples. The lowermost upward-coarsening sequence in the Saddleback Syncline (Fig. 19.17) is interpreted as a tide-dominated delta that developed coevally with barrier-beach sedimentation to the east and west (Fig. 19.20b). Because tide-dominated deltaic deposits are rare in the geologic record, this sequence is discussed in some detail.

Delta-front deposits at the base of the upward-coarsening sequence are composed of coarse-tail graded sandstone laminae capped by mudstone drapes, which contain numerous small-scale flame structures. In addition, dewatering sheets cut across up to 10 m of section. Overlying texturally mature quartz arenites make up large-scale, lenticular bodies oriented perpendicular to the paleoshoreline and are interpreted as linear sandstone shoals (Fig. 19.20b). The interior of the shoals consists of northerly or ebb-oriented crossbeds with reactivation surfaces, whereas bimodal-bipolar crossbeds are present on the shoal margins. Similar asymmetrical flow exists on Holocene tidal sand shoals such as are present along the north coast of Australia. Emergent shoal deposits in the Moodies Group are represented by coarse-grained sandstone with horizontal stratification and small-scale crossbeds, and mudstone laminae and drapes that characteristically display evidence of exposure. An uppermost interval consists of interlaminated sandstone and desiccated mudstone enclosing ca. 2 m thick tidal-channel deposits and is interpreted as a delta-plain sequence. Basal lags in the channel deposits include extrabasinal as well as intrabasinal mudstone clasts and are

►

Fig. 19.20a. a: Paleogeographic model for the Fig Tree Group in the Barberton Mountain Land illustrating coeval uplift of the Ancient Gneiss Complex and Onverwacht Group [modified after Tankard *et al.* (1982)]. b: Paleogeo-graphic model for the lowermost progradational sequence in the Moodies Group, Barberton Mountain Land [from Tankard *et al.* (1982)].

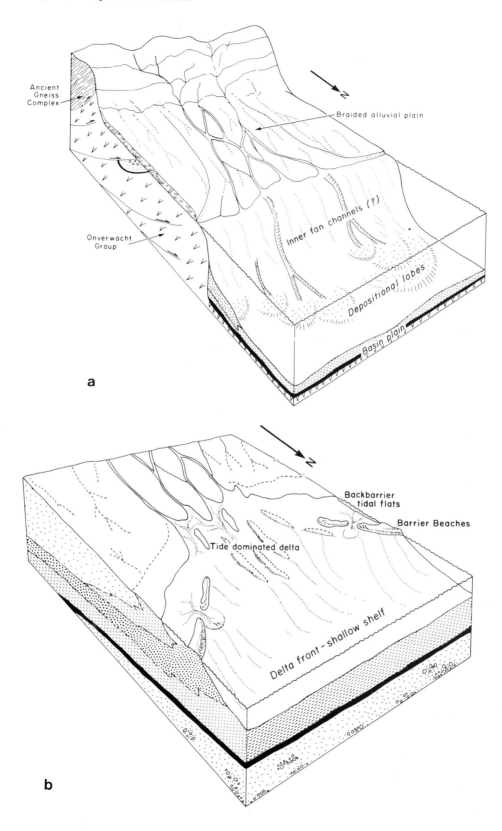

a

b

overlain by crossbedded sandstones with reactivation surfaces and mudstone drapes on foresets and between crossbed sets. Horizontally stratified sandstone laminae with numerous desiccated mudstone partings cap the sequences. This deltaic sequence is considered to have accumulated under macrotidal conditions that existed in an embayed reach of the coastline similar to that in the German Bight of the North Sea (Fig. 19.20b; Eriksson 1979).

Basin Configuration and Tectonic Setting

Dispersal patterns coupled with the spatial distribution of facies (Fig. 19.17) indicate that the Fig Tree and Moodies sediments were derived exclusively from the south. Sedimentation took place along a general east to west-trending shoreline (Fig. 19.20). Influx of extrabasinal detritus to the previously oceanic domain represented by the Onverwacht volcanics has been related to uplift of the Ancient Gneiss Complex to the south (Fig. 19.14; Jackson et al. 1987). Structures in the complex record progressive crustal shortening beginning at ca. 3.4 Ga. Massive uplift at ca. 3.3 Ga, the maximum age of the Moodies Group, was associated with large-scale, heterogeneous simple shear upwards and to the north. Deepening of the basin at the onset of Fig Tree sedimentation is attributed to crustal loading associated with this thrusting (Fig. 19.20a). The Fig Tree and Moodies Groups thus represent synorogenic deposits and their stratigraphic succession records progressive infilling of the basin under deepwater followed by shallow-water conditions. Jackson et al. (1987) have interpreted the above stratigraphic and structural evidence as favoring a foreland or foredeep-basin setting for the Fig Tree and Moodies Groups. The basin was floored by oceanic volcanics that by the time of siliciclastic sedimentation, had been underplated by tonalitic granitoids varying in age from 3.5 to 3.3 Ga (Barton 1983c). Additional evidence in support of a foreland or foredeep-basin setting for the Fig Tree and Moodies Groups includes: 1) the abundance of recycled Onverwacht volcanic and sedimentary components within the Fig Tree and Moodies Groups suggesting that the Onverwacht Group was telescoped onto the Ancient Gneiss Complex during northward folding and shearing of the latter; 2) the presence of older components of the Fig Tree Group within younger units of the siliciclastic interval; and 3) the occurrence of features indicative of syndepositional crustal short-

ening within the Fig Tree and Moodies Groups such as synsedimentary folding and brecciation of iron-formation, large growth folds, and unconformities above thrust faults verging in the direction of sediment influx (Fig. 19.17).

The wave of crustal shortening that uplifted the Ancient Gneiss Complex advanced progressively northward with time to uplift firstly the Onverwacht Group followed by the Onverwacht Group together with older components of the siliciclastic interval and finally involved the whole greenstone belt in folding and thrusting. Jackson et al. (1987) argue that the greenstone belt was sheared off along its base onto tonalities of similar age to the upper Onverwacht Group. The allochthon is considered to have acted as a thermal blanket and dense layer to initiate diapirism, which produced the present upright structural style and arcuate outcrop pattern of the Barberton greenstone belt (Fig. 19.14).

Stratigraphic Framework of the Upper Gorge Creek Group

The upper Gorge Creek Group is confined almost entirely to the Lalla Rookh Basin (Fig. 19.15) where it unconformably overlies shallow-marine quartz arenites, basinal iron formations, and turbiditic conglomerates, sandstones, and mudstones of the lower Gorge Creek Group (Krapez 1984). This upper subdivision of the Gorge Creek Group consists mainly of conglomerate and sandstone with subordinate mudstone. Basin-fill thickness is asymmetric; maximum thicknesses of ca. 3 km are developed adjacent to the northwestern boundary fault and decrease to ca. 1 km towards the southeastern basin margin (Fig. 19.15). Stratigraphic relationships within the basin are complex and are discussed later in this section.

Source-Area Composition

Clasts in the upper Gorge Creek Group consist of black, white and banded chert, iron formation, various quartzites, and quartz-mica schist. Sandstones are predominantly subarkoses but display considerable variability in composition (Fig. 19.18). In general, the sandstones display an upward change from lithic to quartzose to feldspathic arenites (Krapez 1984). Microcline is the dominant feldspar in the sandstones. Clast types and sandstone petrography indicate a mixed provenance consisting of Warrawoona volcanics, lower Gorge Creek sediments, and granitoids. The upward trend in sandstone petrography reflects progressive exposure and erosion of K-rich granites.

Heavy minerals in the upper Gorge Creek Group support the above provenance interpretation. Pyrite and chromite are the predominant heavy minerals; various radioactive placer minerals are also present (Krapez 1984). Rounded pyrite grains are massive, concentrically zoned, laminated, framboidal or porous; euhedral pyrite also occurs. The range of pyrite morphologies is similar to that of syngenetic pyrite in Archean shales and it is probable that the pyrite was derived from sulfide-facies iron formations in the lower Gorge Creek Group. Chromite grains are up to 1 mm in diameter; the most likely source was from layered ultramafic intrusions within the Warrawoona Group as chromite in the extrusives is finer grained. Radioactive heavy minerals reflect the K-granite component of the provenance.

Sediment Dispersal Patterns and Depositional Environments

The upper Gorge Creek Group in the Lalla Rookh Basin consists exclusively of continental deposits; Krapez (1984) recognizes four major depositional environments. Alluvial-fan deposits are represented by breccias of talus-slope origin and clast-supported and matrix-supported conglomerates deposited by debris flows and mudflows. These deposits are confined to the margins of the Lalla Rookh Syncline (Fig. 19.15). Braided-alluvial strata are dominated by two facies associations. Conglomeratic assemblages consist of lenses of clast-supported pebbles and boulders representing longitudinal-bar deposits. Associated trough crossbedded, horizontally stratified and wedge-planar crossbedded sandstone facies are interpreted as adjacent channel, and bar-top and bar-margin deposits, respectively (cf. Miall 1977). Sandstone assemblages consist of trough crossbedded cosets with interbedded sets of tabular-planar crossbeds; this association is comparable to the South Saskatchewan model of Cant and Walker (1978) in which trough crossbeds are produced by aggradation of mid-channel bars and tabular-planar crossbeds develop in response to accretion off bar margins. Paleocurrent data from trough axes suggest that source areas were present on most margins of the basin. The data reveal two dominant fluvial dispersal patterns; one along the basin axis parallel to the boundary faults and the other at a

high angle to the faults. Interbedded mudstone and fine-grained sandstone with horizontal lamination, ripple cross-lamination, and climbing ripples interfinger laterally with the braided-alluvial facies and are considered to represent floodplain deposits. Coarsening-upward sequences of mudstone grading into sandstone are visualized as lacustrine fan-delta deposits. The sandstones are trough crossbedded, or wave rippled and horizontally stratified and resemble channel and levee deposits, respectively. Associated lacustrine facies include sandstone with swash lamination and wave ripples, and lenticular bedding with desiccation cracks; these two associations are interpreted as lacustrine beach and mudflat deposits. Also present within the lacustrine association of facies are thinly bedded turbidites which accumulated below wave base.

Basin Configuration and Tectonic Setting

Sediment dispersal patterns and the spatial distribution of depositional environments have been used by Krapez (1984) to argue that the upper Gorge Creek Group accumulated in a small, enclosed, fault-bounded basin with roughly the same configuration as the present outcrop (Fig. 19.15). Alluvial fans along the high-gradient basin margins supplied terrigenous debris to a braided trunk system that was flanked by a floodplain and fed sediment to a lacustrine environment in the northern part of the outcrop belt (Fig. 19.21).

Krapez (1984) has argued that stratigraphic and sedimentological evidence in the upper Gorge Creek Group indicates that basin development was controlled by marginal strike-slip faulting. Criteria cited include: 1) the presence of intraformational low-angle unconformities; 2) the great stratigraphic thickness relative to the small basin area; 3) the elongate shape of the basin parallel to a northwestern boundary fault (Fig. 19.21) — the presence of numerous alluvial fans within the basin imply that this and other faults were active during sedimentation; 4) the asymmetry of basin fill and facies distribution; 5) the vertical stacking and limited lateral migration of depositional environments; and 6) the dominance of longitudinal infilling (cf. Reading 1980; Nilsen and McLaughlin 1985). The Lalla Rookh Basin is comparable in size, geometry and basin fill to small basins associated with strike-slip faulting in southern California (Crowell 1974; Nilsen and McLaughlin 1985), Norway (Steel and Gloppen

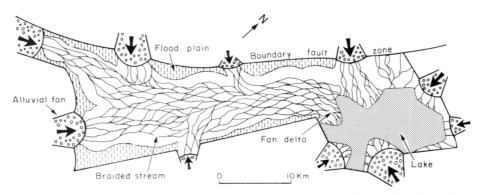

Fig. 19.21. Paleogeographic model for the upper Gorge Creek Group, Lalla Rookh Syncline [from Krapez (1984); reproduced with permission from Extension Division, University of Western Australia].

1980), and Turkey (Hempton *et al.* 1983). The Whim Creek Belt in the western Pilbara Block (Fig. 19.15) is envisaged similarly as a strike-slip basin (Krapez and Barley in press).

Problems and Perspectives

Techniques used in the analysis of the four basins are summarized in Table 19.1. This table clearly illustrates that the degree of sophistication of basin analysis increases with decreasing metamorphic grade and degree of structural complexity.

A stratigraphic framework is the starting point in a basin analysis. As for any Precambrian sequence, the lack of biostratigraphic control in the basins documented in this paper imposes limitations on correlation of stratigraphic units. In the Beitbridge Complex a lack of primary sedimentary structures prevents the recognition of a stratigraphic succession. In contrast, preservation of primary structures in the Narryer cover rocks as well as in the Fig Tree, Moodies, and upper Gorge Creek Groups allow reconstruction of stratigraphic sequences. At Narryer the possibility of structural duplication exists as mylonite zones are pervasive and marker horizons are absent. The stratigraphic succession in the Fig Tree and Moodies Groups is considered to be reliable, especially in the northern half of the Barberton Mountain Land. In this area marker beds are present and an amygdaloidal lava horizon that exists at the same stratigraphic level in a number of synclines can be used for stratigraphic correlation. Furthermore, deformational style is well understood. Despite com-

plex folding associated with transpression, detailed facies mapping in the upper Gorge Creek Group has provided a reliable stratigraphic framework.

Source-area compositions are relatively well constrained using mostly conventional techniques. Slopes and shapes of REE curves are not widely utilized in Phanerozoic sequences to interpret provenance compositions but have received wide acceptance among Precambrian workers as important signatures of crustal compositions through time (Taylor and McLennan 1985). Source-area ages are based on conventional U-Pb (Fig Tree and Moodies Groups), and single-grain ion probe (Narryer) dating techniques. The latter approach to determining source-area ages is relatively new and provides a means of discriminating different ages in a mixed population of detrital zircons.

Determination of sediment dispersal patterns in deformed terrains requires stereonet rotation of paleocurrent data (Potter and Pettijohn 1977). The sequence of deformational events at Narryer can be reconstructed but the amount of rotation of the cover rocks during the late stages of deformation cannot be determined; paleocurrent azimuths thus are unreliable. In the Barberton Mountain Land folds are isoclinal and plunging but no rotation of the siliciclastic intervals accompanied folding; stereonet rotation procedures are thus simple and paleocurrent azimuths are considered to be reliable. Folding in the upper Gorge Creek is complex and noncylindrical; only general, relative flow directions can thus be determined.

The lack of fossils in Precambrian sequences in general imposes certain limitations on paleoenvironmental interpretations, especially in discriminating

Table 19.1. Summary of techniques used in basin analyses.

	Beitbridge Complex Limpopo Province South Africa — Granulite facies metamorphism	Narryer Cover Rocks Western Gneiss Terrain Western Australia — Amphibolite/granulite facies metamorphism	Fig Tree and Moodies Groups Barberton Mountain Land South Africa — Greenschist facies metamorphism	Upper Gorge Creek Group Pilbara Block Western Australia — Greenschist facies metamorphism
Stratigraphic framework	Protolith identification Protolith proportions	Stratigraphic mapping Measured sections	Stratigraphic mapping Measured sections Correlation using lava horizon and sequence boundaries	Facies mapping Measured sections
Source-area composition	Slope and shape of REE curves Heavy minerals not informative	Clast types Slope and shape of REE curves Heavy minerals not informative	Clast types Sandstone petrography Slope and shape of REE curves	Clast types Sandstone petrography Heavy minerals
Source-area age	Data not available	Detrital zircons	Detrital zircons Granitoid clasts	Data not available
Sediment dispersal patterns	Cannot be reconstructed	Primary sedimentary structures	Primary sedimentary structures	Primary sedimentary structures
Depositional environments	Lithologic proportions	Facies interrelationships Facies modelling	Standard facies modelling	Standard facies modelling
Basin configuration including source-area location	Cannot be determined	Spatial interrelationships of depositional environments	Sediment dispersal patterns Spatial distribution of depositional environments	Sediment dispersal patterns Spatial interrelationships of depositional environments
Tectonic setting	Lithologic proportions REE data	REE data Basin configuration	Stratigraphic and structural relationships Stratigraphic evolution Basin configuration	Stratigraphic and sedimentological evolution Basin configuration

ambiguous marine from nonmarine deposits. Interpretations are based solely on facies modelling of physical criteria. In the absence of primary sedimentary structures in the Beitbridge Complex interpretation of depositional environments is equivocal, being based solely on proportions of interpreted protoliths.

Interpretations of basin configuration and tectonic setting are based largely on, and thus hinge on, the validity of the other components of the basin analyses. For example, the cratonic-shelf interpretation for the Beitbridge Complex may be invalid if the amphibolites were lavas or ash beds instead of sills or dikes as discussed earlier. If the amphibolite protolith were of extrusive origin, the Beitbridge Complex would represent an unique and unusual tectonic environment; no Holocene setting is known in which alternating quartz arenites and lavas or tuffs are accumulating. Furthermore, such an association is not known from less-deformed and metamorphosed sequences in the geological record. An unequivocal tectonic interpretation for the Narryer cover rocks is not possible because of post-depositional tectonic overprint and because basin configuration is not well constrained on the basis of only two depositional environments. In contrast, the interpretations of the Fig Tree and Moodies Groups as a foreland-basin sequence, and the upper Gorge Creek Group as a strike-slip basin deposit, are considered to be much less equivocal.

Discussion and Broader Implications

Available age constraints indicate that the high-grade terranes in southern Africa and Western Australia are of comparable age to the greenstone belts and developed between ca. 3.6 and 3.0 Ga, thus negating the possibility that these high-grade terranes represent basement to the greenstone belts. Primary lithological associations interpreted for cover rocks in the Limpopo Province and Western Gneiss Terrain are clearly different from lithological associations in the Barberton and Pilbara greenstone belts. The high-grade terranes in southern Africa and Western Australia thus do not represent the deep-seated metamorphosed roots of greenstone belts. Rather, the two terrane types discussed in this paper represent coeval but contrasting tectonic settings. Cover rocks in the high-grade terranes accumulated on thick continental crust and reflect a stable cra-

tonic setting. In contrast, volcanic intervals in the greenstone belts are widely considered to have developed in an oceanic setting (Lowe 1982; Hoffman 1984; DeWit 1986), whereas the overlying siliciclastic intervals reflect active tectonic settings associated with crustal shortening.

On a global scale, lower Archean sedimentary rocks of cratonic platform affinity are uncommon. In addition to the examples from the Limpopo Province and Western Gneiss Terrain, pre-3.0 Ga sedimentary associations of this type occur *locally* in Canada (Schau and Henderson 1983), India (Srinivasan and Ojakangas 1986; Argast and Donnelly 1986), the United States (Muller *et al.* 1982), and the Soviet Union (Kazansky and Moralev 1981). In view of the low susceptibility of cratonic platform sedimentary rocks to tectonic recycling (Veizer and Jansen 1985), their rare occurrence in the lower Archean rock record suggests that few deposits of this type developed prior to 3.0 Ga. This was probably due to the limited development of stable continental crust until 3.0 Ga in South Africa (Hunter 1974a; Anhaeusser and Robb 1981), 3.0 to 2.85 Ga in Western Australia (Blake and McNaughton 1984), and at the Archean-Proterozoic boundary at 2.5 Ga in most parts of the world (Eriksson and Donaldson 1986). Veizer and Jansen (1979) and Taylor *et al.* (1986) have proposed that small, stable continents comprised no more than 10% of the pre-3.0 Ga Earth. If continental crust was widespread prior to 3.0 Ga as suggested by Armstrong (1981) and Reymer and Schubert (1984), lower Archean cratonic platform sedimentary rocks should be more common in the rock record than they appear to be.

In contrast to the high-grade cratonic association, remnants of lower Archean oceanic crust and overlying active-margin siliciclastic sediments are relatively common on a global scale; most Archean cratons contain greenstone belts. Such volcanic and sedimentary rocks have a high potential for tectonic recycling (Veizer and Jansen 1985) and their occurrence suggests that greenstone assemblages were too voluminous to be entirely subducted and, in part, were involved in obduction as suggested for the Barberton greenstone belt by Jackson *et al.* (1987).

Data discussed in this paper suggest that between 3.6 and 3.3 Ga the surface of the Earth, at least in southern Africa and Western Australia, consisted of small nucleii of differentiated continental crust surrounded by extensive oceanic environments. The opposing view, that the volcanic intervals in the greenstone belts accumulated on continental crust (e.g., Hunter 1974b; Groves 1982; Kröner 1982), is not supported by field relationships or by geochronologic, isotopic, and sedimentologic evidence. No continental basement to the volcanic intervals in the greenstone belts has been recognized in either the Barberton Mountain Land or Pilbara Rock. Rather, geochronological data indicate that the volcanics are the oldest rocks in both areas and were underplated by tonalitic batholiths from 3.5 Ga (Barton 1983c; Barley *et al.* 1984; Jackson *et al.* 1987). Pre-3.6 Ga gneisses have been recognized locally (e.g., Kröner *et al.* 1986) but not in proximity to the greenstone belt. $^{87}Sr/^{86}Sr$ ratios from carbonates within the volcanic intervals indicate a high mantle:continental flux (Veizer *et al.* 1982) and continental detritus is absent from the volcanic intervals in both greenstone belts (Lowe 1982); these data are incompatible with the presence of extensive continental crust.

Granitoids may have been generated in the granite-greenstone terranes of the Kaapvaal Province and Pilbara Block (Fig. 19.1) as early as 3.5 Ga but were not exposed to erosion until ca. 3.3 Ga. These granitoids were uplifted together with the Onverwacht and Warrawoona volcanics (Fig. 19.16) and subordinate gneisses. The Fig Tree and Moodies Groups as well as the lower Gorge Creek Group were derived from this mixed provenance; the Fig Tree and Moodies Groups accumulated in a foredeep basin prior to 3.2 Ga. The upper Gorge Creek Group and coeval deposits developed in small strike-slip basins following stabilization of the Pilbara Block.

Siliciclastic sedimentary rocks in the Limpopo Province may be of comparable age to the oceanic volcanics in the greenstone belts. If so, this would imply that exposed land masses existed by 3.5 Ga and were subjected to weathering and erosion. In any case, the thick siliciclastic sequences in the greenstone belts provide unequivocal evidence that exposed land masses existed by 3.3 Ga. This evidence is at variance with the proposal of Hargraves (1976) that land masses became emergent only by 2.3 Ga. Geochemical data on mudstones indicate that the continental land masses exposed by 3.3 Ga were mainly tonalitic and that K-granites were a subordinate component of the pre-3.0 Ga continental crust. Early Archean (pre-3.0 Ga) sedimentation took place on the exposed land masses primarily in alluvial-fan and braided-alluvial environments as well as subaqueously in marginal-marine, shallow-marine, and deep-marine environments. Physical sedimentary processes and environments on the early Archean Earth were similar to those existing today.

Acknowledgments. Research in the Limpopo Province was supported by NASA Grants NAGW 488 to KAE and NAGW 487 to WSFK; in the Western Gneiss Terrain by NASA Grant NAG 9-95 to KAE; and in the Pilbara Block by NSF Grant EAR7842307 to KAE and an Anaconda Australia Inc. field grant to BK. The Council of the University of the Witwatersrand and the Council for Scientific and Industrial Research in South Africa supported fieldwork in the Barberton Mountain Land. Logistical assistance of the Geology Department, University of Western Australia and the Geological Survey of Western Australia is acknowledged. WSFK thanks the Lunar and Planetary Institute for assistance in manuscript preparation. We benefitted from discussions with J.M. Barton, M.J. Bickle, G. Brandl, R.D. Gee, D.I. Groves, C.W. Harris, M.P.A. Jackson, P.D. Kinny, S.M. McLennan, J.S. Myers, R.W. Ojakangas, G. Ross, S.R. Taylor, D.D. Van Reenen, J. Veizer, and I.R. Williams among others. We thank Llyn Sharp for photography, Ada Simmons for typing the manuscript, and Melody Wayne and Tom Wilson for drafting.

References

ANDERSON, D.W. and PICARD, M.D. (1974) Evolution of synorogenic clastic deposits in the intermontane Uinta Basin of Utah. In: Dickinson, W.R. (ed) Tectonics and Sedimentation. Society Economic Paleontologists Mineralogists Special Publication 22, pp. 167–189.

ANHAEUSSER, C.R. (1973) The evolution of the early Precambrian crust of southern Africa. Philosophical Transactions Royal Society London, Series A 273:359–388.

ANHAEUSSER, C.R. and ROBB, L.J. (1981) Magmatic cycles and the evolution of Archaean granitic crust in the eastern Transvaal and Swaziland. In: Glover, J.E. and Groves, D.I. (eds) Archaean Geology: International Symposium, Perth, 1980. Geological Society Australia Special Publication 7, pp. 457–467.

ANTROBUS, E.S.A. and WHITESIDE, H.C.M. (1964) The geology of certain mines in the East Rand. In: Haughton, S.H. (ed) The Geology of Some Ore Deposits in Southern Africa. Johannesburg: Geological Society South Africa 1:125–160.

ARGAST, S. and DONNELLY, T.W. (1986) Compositions and sources of metasediments in the upper Dharwar Supergroup, south India. Journal Geology 94:215–231.

ARMSTRONG, R.L. (1981) Radiogenic isotopes: The case for crustal recycling on a near-steady-state no-continental-growth Earth. Philosophical Transactions Royal Society London, Series A301:443–472.

BARLEY, M.E., DUNLOP, J.S.R., GLOVER, J.E., and GROVES, D.I. (1979) Sedimentary evidence for Archean shallow-water volcanic-sedimentary facies, eastern Pilbara Block, Western Australia. Earth and Planetary Science Letters 43:74–84.

BARLEY, M.E., GROVES, D.I., BORLEY, G.D., and ROGERS, N. (1984) Archaean calc-alkaline volcanism in the Pilbara Block, Western Australia. Precambrian Research 24:285–319.

BARTON, J.M. (1983a) Our understanding of the Limpopo Belt—a summary with proposals for future research. In: van Biljon, W.J. and Legg, J.H. (eds) The Limpopo Belt. Geological Society South Africa Special Publication 8, pp. 191–203.

BARTON, J.M. (1983b) Pb-isotopic evidence for the age of the Messina Layered Intrusion, central zone, Limpopo Mobile Belt. Geological Society South Africa Special Publication 8, pp. 39–41.

BARTON, J.M. (1983c) Isotopic constraints on possible tectonic models for crustal evolution in the Barberton granite-greenstone terrane, southern Africa. In: Anhaeusser, C.R. (ed) Contributions to the Geology of Barberton Mountain Land. Geological Society Australia Special Publication 9:73–79.

BARTON, J.M., FRIPP, R.E.P., HORROCKS, P., and McLEAN, N. (1979a) The geology, age and tectonic setting of the Messina Layered Intrusion, Limpopo Mobile Belt, southern Africa. American Journal Science 279:1108–1134.

BARTON, J.M., FRIPP, R.E.P., and RYAN, B. (1977) Rb/Sr ages and geological setting of ancient dikes in the Sand River area, Limpopo Mobile Belt, South Africa. Nature 267:487–490.

BARTON, J.M., RYAN, B., and FRIPP, R.E.P. (1983a) Rb-Sr and U-Th-Pb isotopic studies of the Sand River Gneisses, Central Zone, Limpopo Mobile Belt. In: van Biljon, W.J. and Legg, J.H. (eds) The Limpopo Belt. Geological Society South Africa Special Publication 8, pp. 9–18.

BARTON, J.M., RYAN, B., FRIPP, R.E.P., and HORROCKS, P. (1979b) Effects of metamorphism on the Rb-Sr and U-Pb systematics of the Singelele and Bulai gneisses, Limpopo Mobile Belt, southern Africa. Geological Society South Africa Transactions 82:259–269.

BARTON, J.M., FRIPP, R.E.P., and HORROCKS, P.C. (1983b) Rb-Sr ages and chemical composition of some deformed Archaean mafic dikes, central zone, Limpopo Mobile Belt, southern Africa. In: van Biljon, W.J. and Legg, J.H. (eds) The Limpopo Belt. Geological Society South Africa Special Publication 8, pp. 7–37.

BLAKE, T.S. and McNAUGHTON, N.J. (1984) A geochronological framework for the Pilbara Region. In: Muhling, J.R., Groves, D.I., and Blake, T.S. (eds) Archaean and Proterozoic Basins of the Pilbara, Western Australia: Evolution and Mineralization Potential. Perth: Geology Department and University Extension, The University of Western Australia Publication 9, pp. 1–22.

BLIGHT, D.F. and BARLEY, M. (1981) Estimated pressure and temperature conditions from some Western Australian Precambrian metamorphic terrains. Geological Survey Western Australia, Annual Report 1980, pp. 67–72.

BOUMA, A.H. (1962) Sedimentology of Some Flysch Deposits. Amsterdam: Elsevier, 169 p.

BRANDL, G. (1983) Geology and geochemistry of various supracrustal rocks of the Beit Bridge Complex east of Messina. In: van Biljon, W.J. and Legg, J.H. (eds) The Limpopo Belt. Geological Society South Africa Special Publication 8, pp. 103–112.

BURKE, K.C., KIDD, W.S.F., and KUSKY, T.M. (1986) Archean foreland basin tectonics in the Witwatersrand, South Africa. Tectonics 5:439–456.

BUTTON, A. (1976) Transvaal and Hamersley basins—review of basin development and mineral deposits. Minerals Science Engineering 8:262–293.

CAHEN, L., SNELLING, N.J., DELHAB, J., and VAIL, J.R. (1984) The geochronology and evolution of Africa. Oxford: Clarendon Press, 512 p.

CANT, D.J. and WALKER, R.G. (1978) Fluvial processes and facies sequences in the sandy braided South Saskatchewan River, Canada. Sedimentology 25:625–648.

CLIFFORD, T.N. (1957) Fuchsite from a Silurian quartz conglomerate, Acworth Township, New Hampshire. American Mineralogist 42:566–568.

CONDIE, K.C., MACKE, J.E., and REIMER, T.O. (1970) Petrology and geochemistry of Early Precambrian graywackes from the Fig Tree Group, South Africa. Geological Society America Bulletin 81:2759–2776.

CROWELL, J.C. (1974) Origin of Late Cenozoic basins in southern California. In: Dickinson, W.R. (ed) Tectonics and Sedimentation. Society Economic Paleontologists Mineralogists Special Publication 22, pp. 190–204.

DANCHIN, R.V. (1967) Chromium and nickel in the Fig Tree shale from South Africa. Science 158:261–262.

DeCELLES, P.G. (1986) Sedimentation in a tectonically partitioned, nonmarine foreland basin: The lower Cretaceous Kootenai Formation, southwestern Montana. Geological Society America Bulletin 97:911–931.

DeCELLES, P.G., TOLSON, R.B., GRAHAM, S.A., SMITH, G.A., INGERSOLL, R.V., WHITE, J., SCHMIDT, C.J., RICE, R., MOXON, I., LEMKE, L., HANDSCHY, J.W., FOLLO, M.F., EDWARDS, D.P., CAVAZZA, W., CALDWELL, M., and BARGAR, E. (1987) Laramide thrust-generated alluvial-fan sedimentation, Sphinx Conglomerate, southwestern Montana. American Association Petroleum Geologists Bulletin 71:135–155.

DeLAETER, J.R., FLETCHER, I.R., ROSMAN, K.J.R., WILLIAMS, I.R., GEE, R.D., and LIBBY,

W.G. (1981) Early Archaean gneisses from the Yilgarn Block, Western Australia. Nature 292:322–324.

DeWIT, M.J. (1986) A mid-Archean ophiolite complex, Barberton Mountain Land. Workshop on Tectonic Evolution of Greenstone Belts. Houston: Lunar and Planetary Institute Technical Report No. 86-10, pp. 86–88.

ERIKSSON, K.A. (1978) Alluvial and destructive beach facies from the Archean Moodies Group, Barberton Mountain Land, South Africa and Swaziland. In: Miall, A.D. (ed) Fluvial Sedimentology. Canadian Society Petroleum Geologists Memoir 5, pp. 287–311.

ERIKSSON, K.A. (1979) Marginal marine depositional processes from the Archean Moodies Group, Barberton Mountain Land, South Africa: Evidence and significance. Precambrian Research 8:153–182.

ERIKSSON, K.A. (1980a) Transitional sedimentation styles in the Fig Tree and Moodies Group, Barberton Mountain Land, South Africa: Evidence favouring an Atlantic- or Japan Sea-type Archean continental margin. Precambrian Research 12:141–160.

ERIKSSON, K.A. (1980b) Hydrodynamic and paleogeographic interpretation of turbidite deposits from the Archean Fig Tree Group of the Barberton Mountain Land, South Africa. Geological Society America Bulletin 91:21–26.

ERIKSSON, K.A. and DONALDSON, J.A. (1986) Basinal and shelf sedimentation in relation to the Archean-Proterozoic boundary. Precambrian Research 33:103–121.

ESKOLA, P. (1933) On the chrome minerals of Outokumpu. Comptes Rendus Societe Geologie Finlande 7:26–44.

FABRIES, J. and LATOUCHE, L. (1973) Presence de fuchsite dans les quartzites de la serie charnochitique des Grou Oumelalen. Societe Francaise Mineralogie Cristallographie Bulletin 96:148–149.

FISHER, R.V. and SCHMINCKE, H.-U. (1984) Pyroclastic Rocks. New York: Springer-Verlag, 427 p.

FRIPP, R.E.P. (1983) The Precambrian geology of the area around the Sand River near Messina, Central Zone, Limpopo Mobile Belt. In: van Biljon, W.J. and Legg, J.H. (eds) The Limpopo Belt. Geological Society South Africa Special Publication 8, pp. 89–102.

FROUDE, D.O., IRELAND, T.R., KINNY, P.D., WILLIAMS, I.S., COMPSTON, W., WILLIAMS, I.R., and MYERS, J.S. (1983) Ion microprobe identification of 4,100–4,200 Myr-old terrestrial zircons. Nature 304:616–618.

FULLER, A.O., CAMDEN-SMITH, P., SPRAGUE, A.R.G., WATERS, D.J., and WILLIS, J.P. (1981) Geochemical signature of shales from the Witwatersrand Supergroup. South African Journal Science 77:378–381.

GAY, N.C. (1969) The analysis of strain in the Barberton Mountain Land, eastern Transvaal, using deformed pebbles. Journal Geology 77:377–396.

GEE, R.D. (1979) Structure and tectonic style of the Western Australian Shield. Tectonophysics 58:327–369.

GEE, R.D., BAXTER, J.L., WILDE, S.A., and WILLIAMS, I.R. (1981) Crustal development in the Archaean Yilgarn Block, Western Australia. In: Glover, J.E. and Groves, D.I. (eds) Archaean Geology: Second International Symposium, Perth, 1980. Geological Society Australia Special Publication 7, pp. 43–56.

GEE, R.D., MYERS, J.S., and TRENDALL, A.F. (1986) Relation between Archean high-grade gneiss and granite-greenstone terrain in Western Australia. Precambrian Research 33:87–102.

GROVES, D.I. (1982) The Archean and earliest Proterozoic evolution and metallogeny of Australia. Revista Brasiliera Geosciencias 12:135–148.

HAMILTON, P.J., EVENSEN, N.M., O'NIONS, R.K., GLIKSON, A.Y., and HICKMAN, A.P. (1981) Sm-Nd dating of the Talga-Talga Subgroup, Warrawoona Group, Pilbara Block, Western Australia. In: Glover, J.E. and Groves, D.I. (eds) Archaean Geology: Second International Symposium, Perth, 1980. Geological Society Australia Special Publication 7, pp. 187–192.

HAMILTON, P.J., EVENSEN, N.M., O'NIONS, R.K., SMITH, H.S., and ERLANK, A.J. (1979) Sm-Nd dating of Onverwacht Group volcanics, southern Africa. Nature 279:298–300.

HAMILTON, P.J., O'NIONS, R.K., BRIDGEWATER, D., and NUTMAN, A. (1983) Sm-Nd studies of Archaean metasediments and metavolcanics from West Greenland and their implications for the Earth's early history. Earth Planetary Science Letters 62:263–272.

HARGRAVES, R.B. (1976) Precambrian geologic history. Science 193:363–370.

HEINRICHS, T. (1980) Lithostratigraphic untersuchungen in der Fig Tree Gruppe des Barberton Greenstone Belt zwischen Umsoli und Lomati (Sudafrika). Georg-August-Universitat Göttingen, West Germany: Göttingen Arbeiten Geologie Paläontologie Nr 22.

HEMPTON, M.R., DUNNE, L.A., and DEWEY, J.F. (1983) Sedimentation in an active strike-slip basin, southeastern Turkey. Journal Geology 91:401–412.

HERGET, G. (1966) Archaise sedimente und eruptive im Barberton Berg Transvaal-Sudafrika. Neues Jahrbuch Mineralogie Abhandlung 103:161–182.

HICKMAN, A.H. (1981) Crustal evolution of the Pilbara Block, Western Australia. In: Glover, J.E. and Groves, D.I. (eds) Archaean Geology: Second International Symposium, Perth, 1980. Geological Society Australia Special Publication 7, pp. 57–69.

HICKMAN, A.H. (1983) Geology of the Pilbara Block and its environs. Geological Survey Western Australia Bulletin, 127 p.

HOFFMAN, P.F. (1973) Evolution of an early Proterozoic continental margin: The Coronation geosyncline and associated aulacogens, northwest Canadian Shield. Philosophical Transactions Royal Society London, Series A273:547–581.

HOFFMAN, P.F. (1980) Wopmay Orogen: A Wilson cycle of early Proterozoic age in the northwest of the Canadian Shield. In: Strangway, D.W. (ed) The Continental Crust and its Mineral Deposits. Geological Association Canada Special Paper 20, pp. 523–549.

HOFFMAN, S. (1984) The 3.5 b.y. old Onverwacht Group: A remnant of ancient oceanic crust. Houston: Lunar and Planetary Institute Workshop on the Early Earth, pp. 31–33.

HOMEWOOD, P. and ALLEN, P. (1981) Wave-, tide-, and current-controlled sandbodies of Miocene Molasse, western Switzerland. American Association Petroleum Geologists Bulletin 65:2534–2545.

HORROCKS, P.C. (1983) A corundum and sapphirine paragenesis from the Limpopo Mobile Belt, southern Africa. Journal Metamorphic Petrology 1:13–230.

HUBERT, J.F. (1962) A zircon-tourmaline-rutile maturity index and the interdependence of composition of heavy mineral assemblages with the gross composition and texture of sandstones. Journal Sedimentary Petrology 50:489–496.

HUNTER, D.R. (1974a) Crustal development in the Kaapvaal craton, II: The Proterozoic. Precambrian Research 1:295–326.

HUNTER, D.R. (1974b) Crustal development in the Kaapvaal craton: I. The Archaean. Precambrian Research 1:259–294.

HUNTER, R.E. (1977) Basic types of stratification in small eolian dunes. Sedimentology 24:361–387.

JACKSON, M.P.A., ERIKSSON, K.A., and HARRIS, C.W. (1987) Early Archean foredeep sedimentation related to crustal shortening: A reinterpretation of the Barberton Sequence, southern Africa. Tectonophysics 136:197–221.

KAZANSKY, V.I. and MORALEV, V.M. (1981) Archaean geology and metallogeny of the Aldan Shield, USSR. In: Glover, J.E. and Groves, D.I. (eds) Archaean Geology: Second International Symposium, Perth, 1980. Geological Society Australia Special Publication 7, pp. 111–120.

KIDD, W.S.F. (1985) A review of tectonic aspects of the Limpopo belt and other Archean high-grade gneissic terranes. Houston: Lunar Planetary Science Institute Technical Report 85-01, pp. 48–49.

KINNY, P.D. (1986) Zircon ages from the Narryer metamorphic belt. Adelaide: Eighth Australian Geological Convention Abstracts 15, p. 107.

KINNY, P.D., WILLIAMS, I.S., FROUDE, D.O., IRELAND, T.R., and COMPSTON, W. (in press) Early Archaean zircon ages from orthogneisses and anorthosites at Mt. Narryer, Western Australia. Precambrian Research.

KOCUREK, G. (1981) Significance of interdune deposits

and bounding surfaces in aeolian dune sands. Sedimentology 28:753–780.

KRAPEZ, B. (1984) Sedimentation in a small, fault-bounded basin: The Lalla Rookh Sandstone, East Pilbara Block. In: Muhling, J.R., Groves, D.I., and Blake, T.S. (eds) Archaean and Proterozoic Basins of the Pilbara, Western Australia: Evolution and Mineralization Potential. Perth: Geology Department University Extension, University Western Australia Publication 9, pp. 89–110.

KRAPEZ, B. and BARLEY, M.E. (in press) Archean strike-slip faulting and related ensialic basins: Evidence from the Pilbara Block, Australia. Geological Magazine.

KRÖNER, A. (1982) Archean to early Proterozoic tectonics and crustal evolution: A review. Revista Brasileira Geosciencias 12:15–31.

KRÖNER, A., COMPSTON, W., and WILLIAMS, I.S. (1986) Evolution of early Archean gneiss-greenstone terrain in Swaziland, southern Africa, as revealed by ion microprobe zircon dating. Terra Cognita 6:124.

KRUPICKA, J. (1975) Early Precambrian rocks of granitic composition. Canadian Journal Earth Sciences 12:1307–1315.

LAMB, S.H. (1984) Structures on the eastern margin of the Archean Barberton greenstone belt, northwest Swaziland. In: Kröner, A. and Greiling, R. (eds) Precambrian Tectonics Illustrated. Stuttgart, Germany: Schweizerbartsche Verlagsbuchhandlung, pp. 19–39.

LOWE, D.R. (1976) Nonglacial varves in lower member of Arkansas Novaculite (Devonian), Arkansas and Oklahoma. American Association Petroleum Geologists Bulletin 30:213–216.

LOWE, D.R. (1982) Comparative sedimentology of the principal volcanic sequences of Archean greenstone belts in South Africa, Western Australia, and Canada: Implications for crustal evolution. Precambrian Research 17:1–29.

LOWE, D.R., BYERLY, G.R., RANSOM, B.L., and NOCITA, B.W. (1985) Stratigraphic and sedimentological evidence bearing on structural repetition in early Archean rocks of the Barberton Greenstone Belt, South Africa. Precambrian Research 27:165–186.

McLENNAN, S.M., TAYLOR, S.R., and KRÖNER, A. (1983) Geochemical evolution of Archean shales from South Africa, I: the Swaziland and Pongola Supergroups. Precambrian Research 22:93–124.

MIALL, A.D. (1977) A review of the braided river depositional environment. Earth-Science Reviews 13:1–62.

MIALL, A.D. (1981) Alluvial sedimentary basins: Tectonic setting and basin architecture. In: Miall, A.D. (ed) Sedimentation and Tectonics in Alluvial Basins. Geological Association Canada Special Paper 23, pp. 1–33.

MIALL, A.D. (1985) Architectural element analysis: A

new method of facies analysis applied to fluvial deposits. Earth-Science Reviews 22:261–308.

MULLER, P.A., WOODEN, J.L., and BOWES, D.R. (1982) Precambrian evolution of the Beartooth Mountains, Montana-Wyoming, U.S.A. Revista Brasileira Geosciencias 12:215–222.

MYERS, J.S. and WILLIAMS, I.R. (1985) Early Precambrian crustal evolution at Mount Narryer, Western Australia. Precambrian Research 27:153–163.

NESBIT, E.G. and PRICE, I. (1974) Siliceous turbidites: Bedded cherts as redeposited ocean ridge derived sediments. In: Hsü, K.J. and Jenkins, H.C. (eds) Pelagic Sediments: On Land and Under the Sea. International Association Sedimentologists Special Publication 1, pp. 351–366.

NILSEN, T.H. and McLAUGHLIN, R.J. (1985) Comparison of tectonic framework and depositional patterns of the Hornelen strike-slip basin of Norway and the Ridge and Little Sulphur Creek strike-slip basins of California. In: Biddle, K.T. and Christie-Blick, N. (eds) Strike-slip Deformation, Basin Formation, and Sedimentation. Society Economic Paleontologists Mineralogists Special Publication 37, pp. 79–103.

NOCITA, B.W. and LOWE, D.R. (1985) A fan-delta sequence in the Archean Fig Tree Group, Barberton Greenstone Belt, South Africa. Geological Society America Abstracts with Programs 17:678.

OOSTHUYSEN, E.J. and BURGER, A.J. (1973) The suitability of apatite as an age indicator by the uranium-lead isotope method. Earth Planetary Science Letters 18:29–36.

PADGET, P. (1956) The pre-Cambrian geology of west Finnmark. Norsk Geologisk Tidsskrift 36:80.

PERCIVAL, J.A. and CARD, K.D. (1983) Archean crust as revealed in the Kapuskasing uplift, Superior Province, Canada. Geology 11:323–326.

PETTIJOHN, F.J., POTTER, P.E., and SIEVER, R. (1972) Sand and Sandstone. New York: Springer-Verlag, 618 p.

PIDGEON, R.T. (1984) Geochronological constraints on early volcanic evolution of the Pilbara Block, Western Australia. Australian Journal Earth Sciences 31:237–242.

POTTER, P.E. and PETTIJOHN, F.J. (1977) Paleocurrents and Basin Analysis. New York: Springer-Verlag, 425 p.

PRETORIUS, D.A. (1964) The geology of the South Rand goldfield. In: Haughton, S.H. (ed) The Geology of Some Ore Deposits in Southern Africa. Johannesburg: Geological Society South Africa 1:219–282.

READING, H.G. (1980) Characteristics and recognition of strike-slip fault systems. In: Ballance, P.F. and Reading, H.G. (eds) Sedimentation in Oblique-Slip Mobile Zones. International Association Sedimentologists Special Publication 4, pp. 7–26.

REIMER, T.O. (1972) Diagenetic reactions in early Pre-cambrian graywackes of the Barberton Mountain Land (South Africa). Sedimentary Geology 7:263–282.

REIMER, T.O. (1975) Untersuchungen uber Abtragung, Sedimentation und Diagenese im fruhen Prakambrium am Beispiel der Sheba-Formation (Sudafrika). Geologisches Jahrbuch Reihe B., Heft 17, 108 p.

REIMER, T.O., CONDIE, K.C., SCHNEIDER, G., and GEORGI, A. (1985) Petrography and geochemistry of granitoid and metamorphic pebbles from the early Archaean Moodies Group, Barberton Mountain Land, South Africa. Precambrian Research 29:383–404.

REYMER, A. and SCHUBERT, G. (1984) Phanerozoic addition rates to the continental crust and crustal growth. Tectonics 3:63–77.

RUST, B.R. (1978) Depositional models for braided alluvium. In: Miall, A.D. (ed) Fluvial Sedimentology. Canadian Society Petroleum Geologists Memoir 5, pp. 605–625.

SCHAU, M. and HENDERSON, J.B. (1983) Archean chemical weathering in three localities in the Canadian shield. Precambrian Research 20:189–224.

SCHREYER, W., WERDING, G., and ABRAHAM, K. (1981) Corundum-fuchsite rocks in greenstone belts of South Africa: Petrology, geochemistry and possible origin. Journal Petrology 22:191–231.

SRINIVASAN, R. and OJAKANGAS, R.W. (1986) Sedimentology of quartz-pebble conglomerates and quartzites of the Archean Bababudan Group, Dharwar craton, South India: Evidence for early crustal stability. Journal Geology 94:199–214.

STEARN, C.W., CARROLL, R.L., and CLARK, T.H. (1984) Geological Evolution of North America. New Jersey: John Wiley and Sons, 566 p.

STEEL, R.J. and GLOPPEN, T.G. (1980) Late Caledonian basin formation, western Norway: Signs of strike-slip tectonics during infilling. In: Ballance, P.F. and Reading, H.G. (eds) Sedimentation in Oblique-Slip Mobile Zones. International Association Sedimentologists Special Publication 4, pp. 79–103.

TANKARD, A.J., JACKSON, M.P.A., ERIKSSON, K.A., HOBDAY, D.K., HUNTER, D.R., and MINTER, W.E.L. (1982) Crustal Evolution of Southern Africa: 3.8 Billion Years of Earth History. New York: Springer-Verlag, 523 p.

TAYLOR, S.R. and McLENNAN, S.M. (1985) The Continental Crust: Its Composition and Evolution. Oxford: Blackwell Scientific Publications, 312 p.

TAYLOR, S.R., RUDNICK, R.L., McLENNAN, S.M., and ERIKSSON, K.A. (1986) Rare earth element patterns in Archean high-grade metasediments and their tectonic significance. Geochimica et Cosmochimica Acta 50:2267–2279.

TEGTMEYER, A., LANCELOT, J.R., and KRÖNER, A. (1981) Zircon U-Pb dating on granitic boulders from the Moodies conglomerate (Barberton Mountain Land). Terra Cognita, Special Issue, Spring 1981, p. R23.

TEGTMEYER, A.R. and KRÖNER, A. (1987) U-Pb zircon ages bearing on the nature of early Archean greenstone belt evolution, Barberton Mountain Land, South Africa. Precambrian Research 36:1–20.

TERWINDT, J.H.J. (1981) Origin and sequences of structures in inshore mesotidal deposits of the North Sea. In: Nio, S.D., Schuettenhelm, R.T.E., and van Weering, T.C.E. (eds) Holocene Marine Sedimentation in the North Sea Basin. International Association Sedimentologists Special Publication 5, pp. 4–26.

VAN BREEMEN, O. and DODSON, M.H. (1972) Metamorphic chronology of the Limpopo belt, southern Africa. Geological Society America Bulletin 83:2005–2018.

VAN NIEKERK, C.B. and BURGER, A.J. (1969) A note on the minimum age of the acid lavas on the Onverwacht Series of the Swaziland System. Geological Society South Africa Transactions 72:9–21.

VAN REENEN, D.D. (1986) Hydration of cordierite and hypersthene and a description of the retrograde orthoamphibole isograd in the Limpopo belt, South Africa. American Mineralogist 71:896–911.

VAN REENEN, D.D., BARTON, J.M., ROERING, C., SMITH, C.A., and VAN SCHALKWYK, J.F. (1987) Deep crustal response to continental collision: The Limpopo Belt of southern Africa. Geology 15:11–14.

VEIZER, J., COMPSTON, W., HOEFS, J., and NIELSEN, H. (1982) Mantle buffering of the early ocean. Naturwissenschaften 69:173–180.

VEIZER, J. and JANSEN, S.L. (1979) Basement and sedimentary recycling and continental evolution. Journal Geology 87:341–370.

VEIZER, J. and JANSEN, S.L. (1985) Basement and sedimentary recycling. 2. Time dimension to global tectonics. Journal Geology 93:625–643.

VILJOEN, M.J. and VILJOEN, R.P. (1970) Archaean volcanicity and continental evolution in the Barberton region, Transvaal. In: Clifford, T.N. and Gass, I.G. (eds) African Magmatism and Tectonics. Edinburgh: Oliver and Boyd, pp. 27–49.

WATKEYS, M.L., LIGHT, M.P.R., and BRODERICK, T.J. (1983) A retrospective view of the central zone of the Limpopo Belt, Zimbabwe. In: van Biljon, W.J. and Legg, J.H. (eds) The Limpopo Belt. Geological Society South Africa Special Publication 8, pp. 65–80.

WILLIAMS, I.R. and MYERS, J.S. (in press) Archean geology of the Mount Narryer region, Western Gneiss Terrain of the Yilgarn Block, Western Australia. Geological Survey Western Australia Professional Paper.

WINDLEY, B.F. (1984) The Evolving Continents. New York: John Wiley and Sons, 399 p.

WINKLER, H.G.F. (1974) Petrogenesis of Metamorphic Rocks. Heidelberg: Springer-Verlag, 320 p.

20

Flexure of the Early Proterozoic Lithosphere and the Evolution of Kilohigok Basin (1.9 Ga), Northwest Canadian Shield

JOHN P. GROTZINGER and DAVID S. MCCORMICK

Abstract

The lower Goulburn Supergroup of Kilohigok Basin (1.9 Ga), N.W.T., Canada, contains three successively overlying tectono-stratigraphic sedimentary sequences of regional extent: a basal shallow-water siliciclastic/carbonate platform, overlain by deep-water flysch, in turn overlain by shallow-marine and fluvial molasse. The platform probably represents the coastal-plain wedge of an initial passive margin that foundered at a young age. As it collapsed, the outer part of the platform subsided rapidly while its interior arched and was subaerially exposed. Shelf drowning represents the onset of flexural subsidence subparallel to the trend of a major orogenic belt, the Thelon Tectonic Zone, and the establishment of a foreland basin. Arching and subsidence were perpendicular to the tectonic transport direction of intrabasinal nappes that probably root in the Thelon Tectonic Zone, indicating that convergence and uplift along the Thelon Tectonic Zone were probably responsible for flexural subsidence within the Kilohigok Basin. Following drowning, the platform was buried by deep-water trench deposits (flysch); with progressive uplift and basin filling, the foredeep entered the molasse phase and fluvial sediments prograded toward the foreland. Several features of basin evolution are consistent with its development on thermally young (warm) lithosphere, including: 1) the apparently short duration of the passive-margin platform; 2) the restricted development of the erosional unconformity along the top of the passive-margin platform; 3) the abrupt lateral transition from axial trench deposits to shelf deposits; and 4) the relatively high grade of metamorphism (upper anchizone) of the basin.

Introduction

As a valuable method in understanding how the Earth's lithosphere has evolved, it is important to contrast the evolution of Precambrian sedimentary basins with younger ones. Those basins originating primarily in response to flexure of the lithosphere during emplacement of loads (foredeeps) are particularly well suited in that their geometry is a function of the mechanical properties and thermal structure of the lithosphere (Beaumont 1981; Watts *et al.* 1982; Kominz 1986). The current understanding of lithospheric controls on the development of foreland basins has arisen primarily from studies involving geophysical models that quantitatively evaluate the geological response to changes in various lithospheric parameters such as its elastic thickness, the relaxation time for a viscoelastic plate, and the mass and location of a load (Beaumont 1981; Watts *et al.* 1982; Quinlan and Beaumont 1984; Kominz 1986). These factors determine foreland-basin geometry and influence internal basin stratigraphy. Consequently, the architecture of foreland-basin stratigraphy and the distribution of related unconformities can potentially be used to extract much information on the nature of continental lithosphere. Thus, long-term secular evolution of any parameters that affect lithospheric rheology may be manifested as systematic changes in the evolution of foreland basins.

The quantitative models that have been used to interpret the rheology of the lithosphere beneath foreland basins contain time-dependent variables that are generally constrained by biostratigraphic ages of sequences within the basin. Most well studied foreland basins are Mesozoic to Cenozoic in age. Unfortunately, the current lack of suitable age constraints for Precambrian sequences precludes the application of detailed quantitative modelling to understanding the development of Precambrian foreland basins, and therefore prohibits detailed comparison of Precambrian and Phanerozoic lithosphere. Nevertheless, as a first-order attempt at such an analysis, it is useful to compare qualitatively the general features of Precambrian and Phanerozoic foreland basins to recognize common elements in basin development that may have persisted through time, and that can be attributed to the establishment of a modern-type lithosphere in Precambrian time. Then, having established that certain previous assumptions concerning the behavior of the Precambrian lithosphere are indeed valid (e.g., that it flexes in response to loading), it may ultimately be possible to develop more quantitative models for selected Precambrian basins that could reveal important, but perhaps subtle, differences between Precambrian and Phanerozoic lithosphere.

This contribution discusses the stratigraphy and evolution of a foreland basin (Kilohigok Basin) of early Proterozoic age (1.9 Ga), emphasizing the evidence for lithospheric flexure during basin subsidence. The basin is superbly exposed and is uniquely suited for this analysis in that it has been cross-folded perpendicular to basin strike thereby exposing a cross section of the basin fill and adjacent cratonic cover that is continuous for nearly 350 km. This allows accurate reconstruction of basin stratigraphy and related unconformities as they pertain to the various stages of basin evolution. This study provides a detailed qualitative model for an early Proterozoic foreland basin, and develops a framework for more quantitative analysis in the future.

For clarity, certain terms are defined below as they pertain to the description of features discussed in the text. Definitions are taken from Bates and Jackson (1980) and include: 1) foreland—A stable area marginal to an orogenic belt, toward which the rocks of the belt were thrust or overfolded; 2) foredeep/ Foreland Basin—These terms are used interchangeably to describe an elongate depression of the crust bordering an orogenic belt; 3) flysch—Basinal submarine fan and related sediments deposited in the foredeep bordering advancing thrusts rising from the ocean floor; and 4) molasse—Paralic, marginal marine to fluvial siliciclastic sediments derived from erosion of rising mountains in orogenic belts and deposited in the adjacent foredeep.

Regional Setting and Previous Interpretations

The early Proterozoic Kilohigok Basin (1.9 Ga) is located in the northwestern corner of the Canadian Shield, along the northeastern margin of the

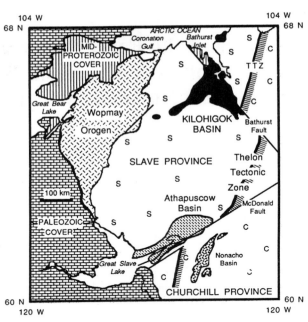

Fig. 20.1. Major tectonic elements of northwest Canadian Shield. Kilohigok Basin shown in black. C = Churchill Province; S = Slave Province; TTZ = Thelon Tectonic Zone.

Archean Slave Province (Fig. 20.1). It is bounded to the west by gently infolded Archean basement rocks of the cratonic foreland, and to the east by both basement rocks and contemporaneous high-grade rocks of the early Proterozoic Thelon Tectonic Zone (hinterland). The Thelon Tectonic Zone (1.98–1.92 Ga) is a regionally continuous belt of high-grade metamorphic rocks and plutonic rocks of possible arc affinity, that is also in part characterized by a major syn-plutonic dextral shear zone (Hanmer and Lucas 1985; Thompson et al. 1985; Hoffman et al. 1986; Van Breeman et al. 1986). It is a crustal-scale feature that may represent a suture between the Slave and western Churchill Provinces (Gibb 1978; Hoffman et al. 1986), and is the likely root zone for several northwest-directed thrust-nappes that occur in the southeastern corner of the Kilohigok Basin (Tirrul 1985). Flexural subsidence of the basin is most likely related to convergence along the Thelon Tectonic Zone and emplacement of related thrust nappes. Following sedimentation and thrust-related deformation, the basin underwent much younger episodes of sinistral strike-slip faulting and gentle basement-involved crossfolding so that the present configuration of the "basin" is related to selective structural preservation of infolded/faulted remnants, rather than its original shape, which was probably that of a prism trending parallel to the Thelon Tectonic Zone. Fortuitously, this selective preservation produced a cross section of the original basin, continuously exposed for nearly 350 km perpendicular to both facies belts and the original basin strike.

Hoffman (1973, 1980) interpreted Kilohigok Basin to be one of two aulacogens (Fig. 20.2a) related to rifting in the Wopmay Orogen. This interpretation was based on the linear trend of the basin, its association with a high-angle fault system, and a thick succession of sediments that were recognized as correlative to sequences in the Athapuscow Aulacogen and the Wopmay Orogen. However, citing the lack of evidence for extensive growth faulting and volcanism, Campbell and Cecile (1981) interpreted the basin as a northwest-trending intracratonic trough, developed as a splay off of a hypothetical aulacogen to the north of Kilohigok Basin (Fig. 20.2b); the hypothetical aulacogen was believed to open westward into the passive margin of Wopmay Orogen. The splay (Kilohigok Basin) was believed to have formed during eastward propagation of the inferred aulacogen. The driving mechanism of splay subsidence was unspecified. More recently, Grotzinger and Gall (1986) proposed a flexural (foredeep) model for the basin, based on detailed stratigraphic work in parts of the basin. Subsidence of the basin was interpreted to have been independent of events in the Wopmay Orogen, and related to convergence of the Slave and Churchill provinces along the Thelon Tectonic Zone (Fig. 20.2c). This model is strengthened independently by the documentation of thrust-nappes in the basin (Tirrul 1985), and evidence for extensive unroofing of the Thelon Tectonic Zone, which contains metamorphic assemblages (staurolite-kyanite) indicating 12 to 15 km of syn-tectonic to post-tectonic uplift (Thompson et al. 1985; D. Thompson personal com-

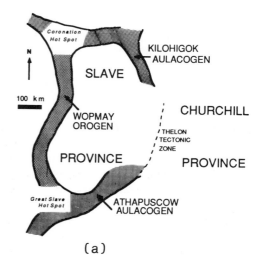

(a)

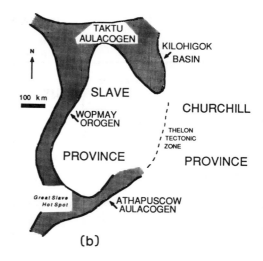

(b)

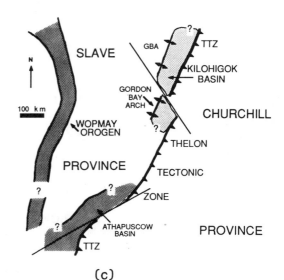

(c)

Fig. 20.2. Different models proposed for origin of Kilohigok Basin. A = Kilohigok Basin as one of two aulacogens related to break-up of Slave Craton and passive-margin development in Wopmay Orogen (Hoffman 1973, 1980). B = Kilohigok Basin as a "splay" from the hypothetical Taktu Aulacogen (Campbell and Cecile 1981). C = Kilohigok Basin as a foredeep related to convergence along Thelon Tectonic Zone and underthrusting of Slave Province beneath Churchill Province (Grotzinger and Gall 1986; Tirrul 1985). GBA = Gordon Bay Arch; TTZ = Thelon Tectonic Zone.

munication 1987). Furthermore, recently dated tuff beds in the basin (1.95 Ga; S.A. Bowring, personal communication 1987) demonstrate that subsidence of the basin was temporally linked to magmatic, metamorphic, and structural events in the Thelon Tectonic Zone (1.98–1.92 Ga; Van Breeman *et al.* 1986).

Foreland basins are characteristically elongate, rapidly subsiding troughs that form as "moats" adjacent to orogenic belts, where crustal loading causes downwarping of the adjacent lithosphere (Beaumont 1981; Watts *et al.* 1982). Facies are arranged asymmetrically so that shallow-water facies near the cratonic side of the basin pass abruptly into deep-water facies filling the basin axis. Many foreland basins display three fundamental stages of evolution: 1)

initial, rapid submergence or drowning of an earlier, slowly subsiding platform (e.g., passive-margin carbonates); 2) deep-marine sedimentation in a narrow axial trough – submarine-fan and related deposits are common and show longitudinal sediment dispersal patterns; and 3) shallow-marine to fluvial sedimentation associated with final filling of the basin, with sediment dispersal directed towards the foreland. Variations on this general theme occur depending on plate rigidity, extent of convergence, and erosion and sediment supply (Beaumont 1981; Quinlan and Beaumont 1984; Flemings and Jordan 1987; Jordan *et al.* 1988: this volume). Initial subsidence may be preceded by arching and subaerial exposure of older platform sediments, generating an unconformity along the exposure surface. This

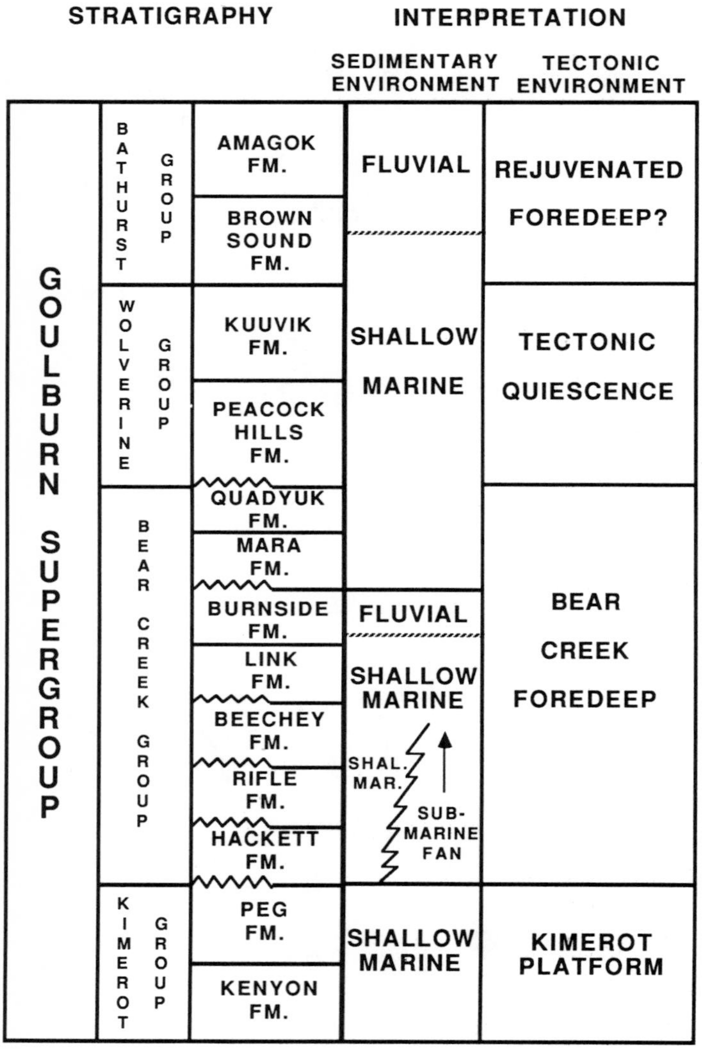

Fig. 20.3. Stratigraphic nomenclature for Goulburn Supergroup, Kilohigok Basin (Grotzinger *et al.* 1987).

arching is a consequence of the flexural origin of foredeep and the elastic properties of the lithosphere (Beaumont 1981; Watts *et al.* 1982). Finally, the trend of foreland basins and related arches commonly is parallel to the trend of the load (generally stacked thrusts) that generates the basin. Consequently, in ancient orogenic belts, foreland basins and arches should be approximately parallel to related thrust-fold belts, and perpendicular to tectonic transport indicators (e.g., stretching lineations).

Data presented here strongly suggest that the Kilohigok Basin formed by flexure of the Archean Slave craton during early Proterozoic convergence along its eastern edge, now represented by the Thelon Tectonic Zone. Evidence for this includes identification of a possible passive-margin terrace (Kimerot Platform, 500 m thick; Fig. 20.3), which underwent drowning of its outer edge and arching within its interior to produce a karstic unconformity before it was deeply submerged. Overlying sediments (Hackett and Rifle Formations, 2 to 2.5 km thick in axial zone) are deep-water submarine fan and interfan deposits, with paleocurrents indicating sediment dispersal parallel (axial) to the trends of both the flexural arch (Gordon Bay Arch) and the load (Thelon Tectonic Zone). Overlying deposits (Beechey and Link Formations, 1 to 2 km thick in axial zone) are shallow-marine shelf sediments with complex sediment dispersal patterns, succeeded by a major, southeastward-thickening (Burnside Formation, up to 2 km) wedge of fluvial coarse sandstones and conglomerates. Conglomerates contain clasts of most underlying lithologies, supplementing a dominant population of intraformational clasts. Such a mixture is common in migrating foredeeps that ultimately cannibalize themselves in advanced stages of development. Paleocurrents within fluvial sediments indicate sediment dispersal across the foreland-basin axis and arch, toward the craton. The Mara Formation (dominantly marine siltstone, 200 m thick) and Quadyuk Formation (marine carbonates, 50 m thick) indicate cessation of uplift and erosion in the hinterland and the termination of Bear Creek Foredeep. Units of the Wolverine Group (600 m thick) and Bathurst Group (3 km thick) may comprise a second cycle of shelf and foredeep sedimentation and are not discussed in detail.

The flexural basin model outlined here accounts for stratigraphic features and phases of basin evolution not observed during earlier work or satisfactorily explained in previous interpretations. It has the added strength of encompassing the predictable geodynamic effects of spatially and temporally related tectonic events along the eastern margin of Slave Craton. Furthermore, the model is consistent with geophysical data suggesting that the Thelon Tectonic Zone is a suture between the Slave and Churchill provinces (Gibb 1978; Gibb *et al.* 1983).

Kimerot Group:
Initial Platform Sequence

The Kimerot Group occurs in the southeastern part of Kilohigok Basin (restored for strike-slip faulting), where it is 0 to 500 m thick (Figs. 20.4, 20.5). Platform sediments are also preserved in several fault-bounded outliers, located east of Kilohigok Basin (Fig. 20.4). The Kimerot Group contains a transgressive siliciclastic unit (Kenyon Formation; 0–250 m thick), overlain by carbonate unit (Peg Formation; 0–250 m thick). Paleocurrent data and changes in thickness and facies within the preserved outcrop belt (Fig. 20.4) indicate that the platform probably faced southwest, south or southeast, and that the preserved cross section is at a high angle to the trend of facies belts. Note onlap of stratigraphic units to the northwest in the direction of decreasing subsidence. Facies relationships within the Kimerot are summarized below. Detailed descriptions and interpretations are presented in Grotzinger and Gall (1986).

Kimerot Siliciclastics (Kenyon Formation)

Facies Description

The Kenyon Formation (Figs. 20.3, 20.5) contains a basal transgressive lag overlain by a unit of trough crossbedded, coarse-grained to pebbly sandstone, overlain by a unit of fine-grained to medium-grained hummocky cross-stratified sandstone with minor dolomites, in turn overlain by a siltstone/dolomite unit with small-scale hummocky cross-stratification and interbedded stromatolitic and clastic-textured dolomites. Generally, units systematically overlap each other to the northwest.

Trough crossbedded sandstones (0–80 m thick) contain sets of troughs 0.05 to 0.2 cm thick, that form laterally continuous (>50 m) beds with scoured bounding surfaces (Fig. 20.6a). Sandstones are medium to very coarse, locally pebbly, sub-

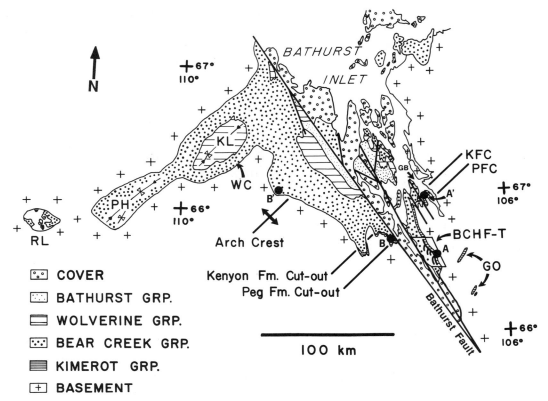

Fig. 20.4. Reconstruction of Kilohigok Basin correcting for 135 km of sinistral slip on Bathurst Fault (Campbell and Cecile 1981; Tirrul 1985), but not accounting for approximately 20 km of northwest-directed shortening in Bear Creek Hills Fold-Thrust Belt (Tirrul 1985). Locations A, A′, B, and B′ mark positions of stratigraphic cross sections shown in Figures 20.5 and 20.9. BCHF-T = Bear Creek Hills Fold-Thrust Belt; GO = Goulburn Supergroup outliers; GB = Gordon Bay; WC = Wolverine Canyon; RL = Rockinghorse Lake; PH = Peacock Hills; KL = Kuuvik Lakes.

rounded to well rounded, and contain 10 to 15% feldspar. The lowermost part of the unit shows unimodal sediment transport to the southeast (Fig. 20.7a), lacks glauconite and contains asymmetrical current ripples. In contrast, the upper part contains polymodal paleocurrent distributions, abundant glauconite, and symmetrical wave ripples.

Hummocky cross-stratified sandstones form a unit 0 to 140 m thick that is typically fine- to medium-grained, and thinly bedded. Beds contain low-angle truncation surfaces defining hummocks and uncommon swales, with wavelengths of 1 to 3 m and amplitudes of 0.1 to 0.3 m (Fig. 20.6b). Thicker beds up to 0.5 m commonly have scoured, trough crossbedded bases that flatten upward into hummocky strata. Hummocky cross-stratified sandstones pass upward into laminated siltstones.

The siltstone/dolomite unit (0 to 130 m thick) contains abundant small-scale hummocky cross-stratification (wavelengths a few dm, amplitudes <0.1 m), abundant clastic-textured and biohermal dolomites and uncommon hummocky cross-stratified, medium-grained sandstones. Dolomite beds are 0.3 to 2 m thick, and increase in abundance upward. Stromatolitic units are biohermal, but individual clastic-textured beds containing cross-bedding and hummocky cross-stratification are sheets that extend for tens of kilometres.

Facies Interpretation

The Kenyon Formation bears many features typical of transgressive shoreface deposits (Spearing 1975; Walker 1984). The sequence progresses upward

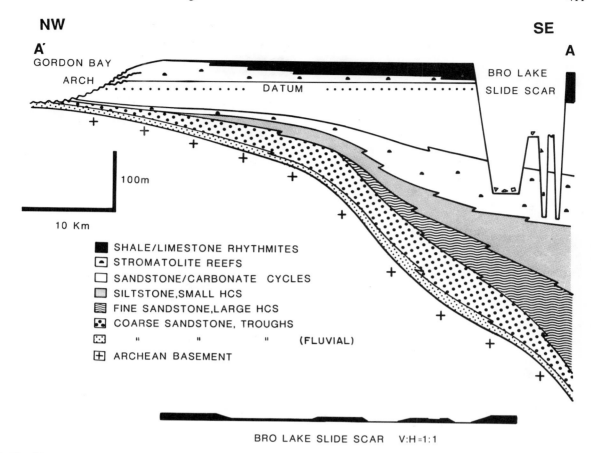

Fig. 20.5. Stratigraphic cross section of Kimerot Platform (Kimerot Group). Location of cross section shown in Figure 20.4. Kenyon Formation includes coarse sandstone, fine sandstone, and siltstone units. Peg Formation includes cyclic, reefal, and reef/rhythmite units. Note onlap of stratigraphic units and general thinning (decreased subsidence) to northwest.

from trough crossbedded coarse and pebbly sandstones, through medium sandstones with large-scale hummocky cross-stratification, up to siltstones with small-scale hummocky cross-stratification and abundant dolomites. This is consistent with a gradual northward onlap of depositional environments from fluvial and upper shoreface through lower shoreface to the offshore. The lower part of the trough crossbedded unit is interpreted to be fluvial on the basis of its unimodal paleocurrent pattern, lack of glauconite, and pervasive asymmetrical current ripples; the upper part is interpreted as marine and has polymodal paleocurrent patterns, abundant glauconite, and symmetrical wave ripples. The onset of carbonate sedimentation would have resulted from decreased siliciclastic influx, as a product of continued onlap of exposed basement. Eventually,

continued onlap and/or decreased siliciclastic influx allowed a major stromatolitic reef complex to develop as the lower unit in the overlying Peg Formation.

Kimerot Carbonates (Peg Formation)

Facies Description

The Peg Formation contains three units (Fig. 20.5): a lower reef unit (0–80 m thick), a middle cyclic unit (0–130 m thick), and an upper reef/rhythmite unit (0–30 m). The lower reef unit thins northwestward and may, in part, pass laterally into siltstones and dolomites of the underlying siliciclastic part of the platform (Fig. 20.5). The overlying cyclic unit also thins northward and the lower part of the unit may be

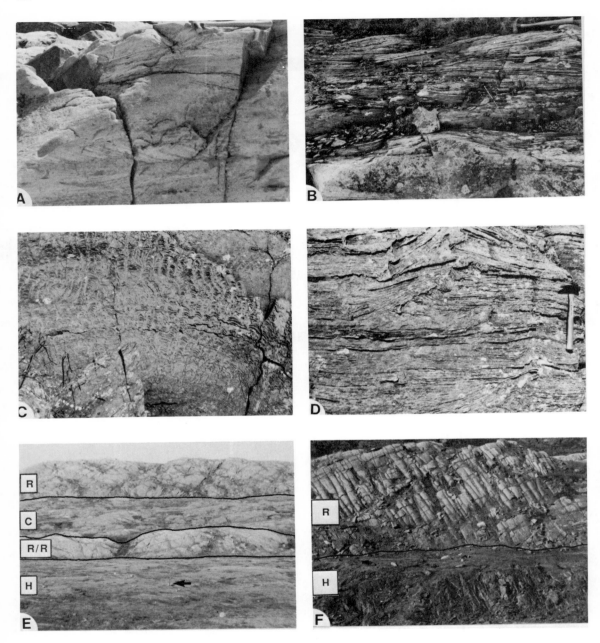

Fig. 20.6. Facies of Kimerot Platform. A = Trough cross-bedded pebbly sandstone of basal fluvial unit, Kenyon Formation. B = Hummocky cross-stratified siltstone, Kenyon Formation. C = Cross section through elongate stromatolite mound (compare with Fig. 20.6f), reefal unit, Peg Formation. Columns are about 0.05 m wide. D = Tepee structures in laminar tufas and cryptalgalaminites capping shallowing-upward cycle in cyclic unit. E = Peg Formation dipping steeply towards viewpoint (R = Reefal unit; C = cyclic unit; R/R = reef/rhyth-mite unit; H = deep-water shales of Hackett Formation). About 250 m of section are shown, as is a person for scale in foreground (arrow). F = Stratigraphic deletion of upper Peg Formation related to Bro Lake Slide. Dark shales of Hackett Formation (H) overlie elongate stromatolite mounds (R) of reefal unit, lower Peg Formation. Intervening units are missing and breccia blocks of deleted units line the slide scar (compare with Fig. 20.6e). Elongate mounds are about 10–15 m long.

laterally equivalent to the upper part of the underlying reefal unit. In the upper reef/rhythmite unit,
reefal facies thin southward, and pass laterally into
probably time-equivalent rhythmite facies (Fig.
20.5). Northward, reefal facies contain a few layers
of tepee and tidal-flat facies including cryptalgalaminites and tufas. The datum used in reconstructing Figure 20.5 is the top of a distinctive
shallowing-upward cycle that can be correlated
across the outcrop belt. Such contacts are very
nearly isochronous in both Phanerozoic and Proterozoic cyclic sequences (Fischer 1964; Anderson
et al. 1984; Grotzinger 1986a,b; Read et al. 1986).

The upper part of Kimerot Platform is conformable over much of its width, but becomes a major
erosional disconformity northwestward toward
Gordon Bay, where underlying units are progressively cut out all the way down to Archean basement
(Fig. 20.8a,b). Locally the unconformity is karstic,
with residual topographic relief of several metres.
Depressions between erosional remnants commonly
are filled with rounded pebbles and cobbles of
eroded lithologies. Similar relationships are
observed in the southeastern part of the outcrop belt
west of Bathurst Fault, where carbonates and basal
clastics of Kimerot Platform are successively cut out
to the northwest beneath the same disconformity.
These truncations define a northeast trend after
restoration of 130–140 km of sinistral displacement
on the Bathurst Fault (Campbell and Cecile 1981;
Tirrul 1985). The positive area corresponding to this
trend is known as the Gordon Bay Arch (Figs. 20.4,
20.5).

The Kimerot Platform is everywhere mantled by
dark laminated mudstone of the overlying Hackett

▶

Fig. 20.7. Paleocurrent data for selected units in Kimerot
and Bear Creek Groups. A = Fluvial unit at base of
Kenyon Formation. B = Cyclic unit, middle Peg Formation. C = Turbidites in Hackett and Rifle formations,
Axial Sequence. D = Shelf sandstones in upper Rifle Formation, Shelf Sequence. E = Fluvial sandstones of Burnside Formation. Outlines show mean ± 1 standard
deviation. All data consist of measured trough *axes* and
tabular-planar foresets except for the turbidites where data
were obtained exclusively from flute casts and climbing-
ripple foresets. Paleocurrents from the units of A, C, and
E provide critical information on the evolution of basin
geometry through time, and their vector means are plotted
on a base map in Figure 20.10 relative to major tectonic
elements of the region.

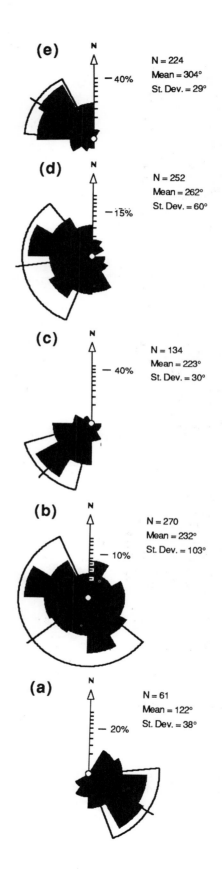

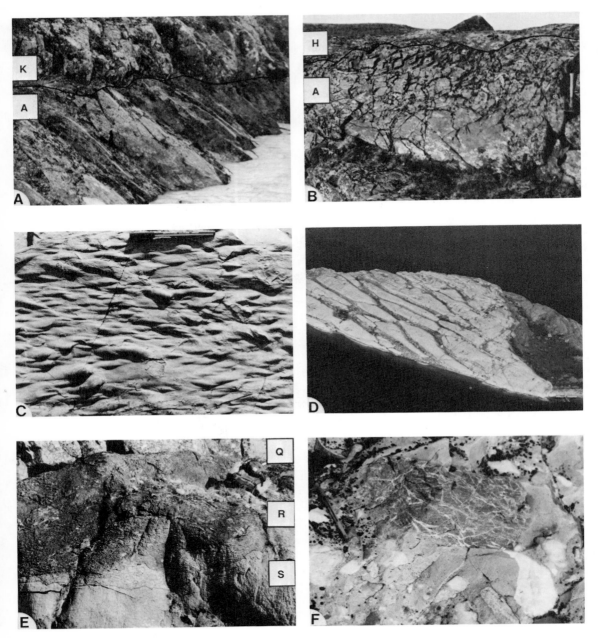

Fig. 20.8. Significant facies in the evolution of Bear Creek Foredeep. A = Contact between Archean basement (A) and fluvial unit of Kenyon Formation (K). Contact typically lacks well developed regolith. Compare with Figure 20.8b. Pack for scale at center right. B = Contact between Archean basement (A) and Hackett Formation (H) over the crest of Gordon Bay Arch where the Kimerot platform has been eroded down to the level of basement. Note well developed carbonate-matrix regolith in basement, which is typical of areas where the Kimerot platform has been removed by erosion. Hammer for scale at center right. C = Flute casts on base of turbidite bed, axial sequence of foredeep. D = Elongate stromatolite reefs of Beechey Formation, shelf sequence of foredeep. Note aircraft for scale in lower right corner. E = Disconformity-related regolith (R) developed on top of stromatolite unit (S) in Beechey Formation. Regolith is overlain by shelf quartzite (Q). Note scalloped solution surfaces at base of regolith. F = Intraformational cobble of Burnside sandstone in conglomerate unit at top of Burnside Formation. Note pervasive quartz-filled fractures in clast that do not extend beyond its perimeter.

Formation (Bear Creek Group). Locally, units at the top of the platform are missing, and megabreccias line the flanks and bases of possible slide scars (Fig. 20.5).

The lower reefal unit of the Peg Formation (Fig. 20.5) consists of large, strongly elongate stromatolite mounds (Fig. 20.6c). To the southeast, penetrative strain is high (Tirrul 1985), preventing accurate facies reconstruction. It appears, however, that the reefal unit passes gradually into fine-grained siliciclastics without an abrupt change into deeper-water facies; no breccias of shelf and slope facies are present.

The overlying cyclic unit (Fig. 20.5) consists of bout 70 shallowing-upward cycles, 0.5 to 2 m thick. Lower parts of most cycles contain trough crossbedded siliciclastic/carbonate sandstones with bimodal paleocurrents (Fig. 20.7b) and upper parts of cycles contain cryptalgalaminites and tufas; stromatolites are uncommonly developed as transitional facies. Cycle caps are commonly erosional, with well developed tepee structures and locally developed pisolite (Fig. 20.6d). To the southeast, trough cross-stratified sandstones at the bases of cycles pass laterally into hummocky cross-stratified fine sandstones, and cryptalgalaminites and tufas of cycle caps pass laterally into stromatolites. Although most of the cyclic unit lacks distinctive marker beds, several occur near the top and can be correlated for over 50 km perpendicular to depositional strike.

Reefal facies of the upper-reef/rhythmite unit comprise large, strongly elongate mounds (1–2 m wide, 0.5–1 m high), cored by smaller, branching columnar stromatolites, also elongate. Adjacent to Gordon Bay Arch, mounds contain brecciated horizons, associated with tidal flat facies including cryptalgalaminites, tufas, and tepees. To the southeast, reefs pass laterally into probably time-equivalent rhythmite facies that consist of thickly interlaminated dark mudstone and limestone (Fig. 20.5). Where rhythmites overlie reefal facies, the contact often contains 0.5 to 1.0 m of mounded edgewise conglomerate; clasts are derived from adjacent stromatolites.

Facies Interpretation

The lower reefal member was probably a subtidal ramp, deepening to the southeast, and may have formed during continued coastal onlap, which would have reduced siliciclastic influx. Reefal mounds decrease in size to the north-northwest, suggesting

a decrease in water depths where subsidence rates were lower.

Development of the cyclic unit represents stabilization of the platform. Pinch-out of tidal-flat facies within cycles to the south-southeast suggests that cycles formed by progradation of flats in that direction. Cycles probably formed in response to small changes in sea level (a few metres), indicated by brecciated, erosional cycle caps (cf. Grotzinger 1986a; Read et al. 1986).

The transition from the cyclic unit to the overlying reef/rhythmite unit is marked by a sharp lower contact, parallel to underlying cycle boundaries (Fig. 20.5). On this basis, it is inferred to be approximately chronostratigraphic. The contact represents an abrupt change to greater water depths where open-marine reefs and/or rhythmites were deposited, and heralds the establishment of a new basin-wide regime of subsidence and sedimentation.

Tectonic Significance of Platform Foundering

The nearly isochronous transition at the top of the Kimerot is identical to the transition between cyclic sediments at the top of the early Proterozoic Rocknest Platform and its overlying reefal veneer, located in Wopmay Orogen and interpreted as a passive margin-to-foredeep transition (Grotzinger 1986c). Initial increase in water depth over the Kimerot Platform was followed by diachronous drowning of the platform, as shown by progressive northwest onlap of rhythmite facies over reefal facies (Fig. 20.5). During drowning of the outer platform, uplift of the inner platform over Gordon Bay Arch occurred (Fig. 20.5). Arching was probably synchronous with drowning because both drowning and arching postdate deposition of the cyclic unit, and pre-date deposition of overlying deep-water dark mudstones (Hackett Formation), which cover the platform in all areas. Furthermore, reefs interfinger with tidal-flat facies and brecciated surfaces to the northwest, suggesting shoaling onto the Gordon Bay Arch. These relationships are important and indicate that drowning was caused by tectonic flexure of the platform, rather than a eustatic rise in sea level. Eustatic drowning is not likely because it cannot account for simultaneous exposure of the inner shelf and drowning of the outer shelf. Finally, flexure is required to produce systematic northwestward erosion and truncation of Kimerot units down to the level of basement. Support for a hypothesis of flexural tectonic drowning and simultaneous uplift of the platform is

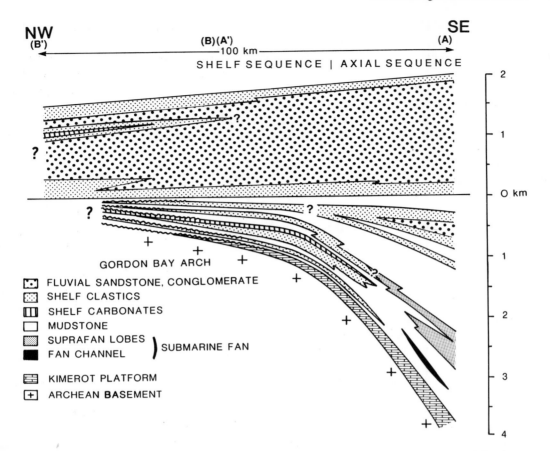

Fig. 20.9. Stratigraphic cross section of Bear Creek Foredeep (Bear Creek Group). Location of cross section shown in Figure 20.4. Note narrow, deep geometry of basin and the close proximity of trench deposits (submarine-fan sequences) to distal shelf deposits. Fluvial unit at top of foredeep is Burnside Formation. Shelf carbonate unit is Beechey Platform.

found in the orthogonal relationship between the trend of Gordon Bay Arch, and the trend of finite extension lineations within nappes (Tirrul 1985). Most likely, the northeast trend of Gordon Bay Arch is a product of northwest tectonic transport of thrust nappes, which would have emplaced a northeast-trending load on the lithosphere. It should be emphasized that rifting as a cause of drowning, with lateral heat flow accounting for the synchronous uplift and erosion of adjacent areas, is negated by the absence of any normal faults, which would have displaced the basement-Kimerot Platform contact.

Further evidence for flexural foundering of the platform is the Bro Lake Slide Scar, which scalloped the upper platform during drowning (Fig. 20.6e,f). Sliding and drowning probably were synchronous because both post-date deposition of

the cyclic unit, and predate deposition of overlying deep-water dark mudstones. Sliding of platform sediments is known to occur along the shelf break of rimmed platforms, simply due to gravitational instability (Mullins and Neumann 1979). The Bro Lake Slide Scar, however, is located well back on the platform, away from the probable shelf-to-slope transition zone. Furthermore the platform was probably a low-gradient ramp and lacked slopes that would allow sliding of early-cemented carbonate facies to occur. However, sliding might be expected to occur in response to steepening of the platform during down-bending of the lithosphere. Such an environment is not only characterized by slope steepening, but is also inherently the site of extensional stress (Chapple and Forsyth 1979; Bradley and Kusky 1986) and thus seismically induced mass wasting.

In sum, flexural (tectonic) rather than eustatic drowning of the Kimerot Platform is supported by the coincidence of rapid outer platform submergence with flexural arching of the platform interior, and simultaneous sliding of parts of the platform down post-depositionally steepened slopes. Diachronous southeast-to-northwest drowning of the platform is probably related to decreasing rates of tectonic subsidence (and thus relative sea-level rise) toward the Gordon Bay Arch; such a systematic decrease in subsidence rate toward the arch (i.e., pivot line) would be a direct consequence of flexural downwarping. Drowning of the Kimerot Platform marks the onset of foredeep subsidence and sedimentation in Kilohigok Basin.

Bear Creek Group: Establishment of Foredeep

The "Bear Creek Foredeep" includes sediments of the Bear Creek Group that overlie the Kimerot Platform; where the platform has been erosionally truncated, Bear Creek sediments lie directly on Archean basement (e.g., over the Gordon Bay Arch; Figs. 20.5, 20.8b). The foredeep sequence is subdivided spatially into a thicker axial sequence (3.0–5.5 km) developed above the outer part of Kimerot Platform, and a thinner shelf sequence (1.5–3.0 km) developed over the Gordon Bay Arch (Fig. 20.9). The axial sequence consists of submarine fan and related deposits up to 1.5 km thick, which pass up into continuous mudstones (1 km thick), in turn overlain by an upper sequence of fluvial sandstones and conglomerates (2 km thick).

The shelf sequence was developed synchronously with the axial sequence, but generally under much lower subsidence rates. Shelf sediments (0.5–2 km thick) are dominated by shallow-marine sandstones and mudstones, with minor carbonates, and pass upward into fluvial sandstones and conglomerates up to 1 km thick. These sediments include formations that are bounded by disconformities in the northwest. Disconformities pass into conformable contacts southeastward, toward the basin axis (Fig. 20.9). Shelf sandstones are laterally separated from deep-water axial sandstones by a facies belt of thick, continuous mudstones. This zone probably represents the regional paleoslope developed along the cratonic margin of the basin.

Axial Sequence, Lower Part

Facies Description

The lower part of the axial sequence (Hackett, Rifle, and Beechey Formations) consists of interbedded dark mudstones and fine to coarse sandstones (10–15% feldspar), locally containing carbonate blocks. Sandstones occur as large lenses, tens of kilometres wide and hundreds of metres thick. The lowermost lens consists of amalgamated sandstone layers, separated by intermittent mudstones. Sandstones are massive or normally graded, rarely trough crossbedded, and occasionally contain large blocks of clastic-textured or stromatolitic dolomite and limestone. Where present, beds containing blocks are inversely graded.

Overlying sandstone lenses are composed of thinly to thickly interbedded sandstones and dark mudstones. These may be arranged in thinning-and-fining-upward sequences, 30 to 70 m thick. Sandstones are fine to very coarse grained, and beds commonly are trough crossbedded or massive, and pinch out within 50 to 100 m along strike. Rarely, sandstone beds contain classical Bouma sequences, and many beds consist of a single set of trough foresets. Beds are 0.1 to 3 m thick, with scoured bases and flute casts (Fig. 20.8c). Thin mudstones may separate sandstone beds that are otherwise amalgamated. Paleocurrent measurements of crossbed foresets and flutes show longitudinal transport to the southwest, parallel to the basin axis and Gordon Bay Arch (Fig. 20.7c).

Facies Interpretation

A deep-water submarine-fan origin for sandstone lenses in the lower part of the axial sequence is supported by several features that compare with other well documented submarine-fan sequences (Howell and Normark 1982). These features are: 1) large sandstone lenses on the scale of tens of kilometres wide by hundreds of metres thick, completely enclosed in finely laminated dark mudstones; 2) individual sand beds that are channelized and laterally discontinuous over 50 to 100 m; 3) sandstone beds form thinning-and-fining-upward sequences; 4) sandstone beds contain trough crossbedding, graded bedding, massive bedding, rare Bouma sequences, and have scoured or channeled bases with flute casts; and 5) sediment transport is strongly unimodal, consistent with transport in a narrow trough. In the lower sandstone lens, coarse, amalgamated sandstones containing inversely graded beds of carbonate blocks may reflect a setting on the inner-fan channel or channelized portion of suprafan lobes (Eriksson 1982; Howell and Normark 1982). Dolomite blocks appear to have been derived from the siltstone/dolomite unit of the siliciclastic part of Kimerot Platform, but limestone blocks are of uncertain origin. In the upper sandstone lenses, discontinuous sandstone beds form thinning-and-fining-upward sequences that may reflect deposition on the channelized or braided portions of suprafan lobes.

In the stratigraphically highest sandstone "lens," apparent submarine-fan facies in the axial sequence overstep the zone of bypassing, and pass laterally into shelf sandstones of the shelf sequence. This may relate to the evolving geometry of the foredeep allowing lateral migration of fan deposits toward the foreland, as a result of widening and shallowing of the basin axis as it migrated cratonward. Filling of the basin would result from both sediment accumulation, and also rise of the basin floor (i.e., turbidite base level), as the mass of the load was decreased by erosion and the lithosphere underwent rebound.

Paleocurrents in sand turbidite beds show remarkably unimodal sediment dispersal, parallel to the basin axis and Gordon Bay Arch (Fig. 20.7c). Longitudinal sediment dispersal is a classical feature of foredeeps (McBride 1962; Potter and Pettijohn 1977; Mutti 1985), and is a product of the narrowly confined and linear geometry of the basin.

Axial Sequence, Middle Part

Facies Description

The middle part of the axial sequence is dominated by shelf sandstones and mudstones of the Link Formation. Sandstone units are 70 to 600 m thick, and laterally continuous for many tens of kilometres, unlike underlying fan deposits. Intervening mudstones are 30 to 300 m thick. Internally, sandstone units of marine origin contain sequences dominated by trough-crossbedded sandstone sheets with polymodal paleocurrent patterns (Fig. 20.7d) that pass laterally and vertically into units dominated by hummocky cross-stratification, and then into units dominated by mudstone with uncommon thin sandstone layers. A restricted wedge of maroon sandstone is developed in the southeast corner of the basin; it is characterized by

unimodal crossbedding, with abundant mud chips within foresets, and desiccation cracks along intervening mud layers.

Facies Interpretation

The middle part of the axial sequence represents filling and shallowing of the foredeep, and the onset of widespread shallow shelf sedimentation. Fluctuations in relative sea level produced alternation of shallow-shelf sandstones (with hummocky crossbedding and trough crossbedding) with deeper-shelf mudstones. The relative sea-level fluctuations may have resulted from either eustatic or flexural (tectonic) causes. The wedge of maroon sandstone containing unimodal paleocurrents, mud chips within foresets, and desiccation cracks is interpreted to be fluvial. Sediments of this wedge are southeasterly derived and represent an initial pulse of molasse sedimentation that preceded the major fluvial sequence (Burnside Formation), which dominates the upper part of Bear Creek Foredeep.

Shelf Sequence

Facies Description

The shelf sequence in Bear Creek Foredeep is dominated by shelf mudstone and sandstone sequences that comprise several units (Hackett, Rifle, Beechey, and Link Formations). Sequences consist of sandstone units (40–100 m thick) and mudstone units (10–80 m thick), with uncommon carbonate units (1–30 m thick). Shelf sandstones are sheets, containing hummocky cross-stratification, trough crossbedding, and tabular-planar crossbedding. Sandstones of the Beechey Formation are associated with 1 to 30 m of stromatolitic carbonate (Beechey Platform; Fig. 20.9). The Beechey Platform contains large stromatolite mounds that are locally up to 3 m wide, 1 to 2 m high, and 100 m long (Fig. 20.8d).

The upper contacts of the Beechey Formation and many sandstone units within the foredeep shelf sequence have erosionally disconformable upper contacts (Figs. 20.8e, 20.9). Unconformities are marked by brecciated or scalloped solution surfaces overlain by layers containing pebbles or cobbles of underlying sandstones. Thick sequences of iron-rich pisolite may be developed, and surfaces are commonly coated by layers of stromatolitic tufa. Siliciclastic mudstones and siltstones may contain layers of patchy carbonate cementation and poorly developed tepee profiles. Locally, zones of complex brecciation several metres thick are developed below disconformities, particularly where clastic carbonate was deposited in addition to siliciclastic sand. The intensity of regolith development is greatest over the Gordon Bay Arch.

Facies Interpretation

Shelf sandstones and mudstones were deposited over the Gordon Bay Arch, where subsidence rates were greatly reduced relative to the axial part of Bear Creek Foredeep. Consequently, water depths were generally much shallower, as shown by the abundance of hummocky cross-stratification, wave ripples, and crossbedding with complex polymodal paleocurrent patterns.

The presence of disconformities capping shelf sandstone and carbonate units indicates large fluctuations in relative sea level. Because these sequences are, in part, located over the Gordon Bay Arch, it is possible that relative sea-level oscillations were tectonically or eustatically induced. A eustatic origin is suggested by their lateral extent beyond the influence of arching (Grotzinger et al. 1987), as well as the direct juxtaposition of deep-water mudstone facies on many of the surfaces, without intervening paralic sediments. However, a flexural (tectonic) control on disconformity development is indicated by the increased intensity of regolith development over the Gordon Bay Arch (Grotzinger et al. 1987).

The Bear Creek unconformities separate offlap-onlap sequences, and therefore are valid sequence boundaries and probably relative time markers in the sense that the unit below the surface is everywhere older than the unit above the surface (Vail et al. 1977). This has important implications for the Bear Creek Foredeep as well as other Precambrian siliciclastic sequences where, in the absence of fossils, only tuff beds (easily reworked in shelf settings) are presently available for time correlation (Christie-Blick et al. in press). Although the Bear Creek sequences are probably somewhat thinner than those that can be resolved on seismic lines, the method is based on the recognition of geometric relationships between stratigraphic sequences and their bounding unconformities, and is not dependent on the scale at which the sequences occur (Christie-Blick et al. in press).

Axial and Shelf Sequences, Upper Parts

Facies Description

The upper parts of both the axial and the shelf sequences are dominated by a single unit (Burnside Formation), which generally consists of a thick ($>2,000$ m) northwestward-tapering wedge of maroon to tan sandstone and minor conglomerate. This unit comprises a series of sequences each several hundred metres thick consisting of: 1) a lower transitional unit found only in the axial part of the basin; 2) a middle sequence, present across the basin, of fining-upward cycles (100s m thick) beginning with minor conglomerate horizons that thin upward; and 3) an upper sequence of fining-upward cycles, also present across the basin, in which conglomerates contain intraformational clasts. All of these units interfinger northwest toward Wopmay Orogen with siltstones and silty sandstones that contain symmetrical wave ripples and hummocky cross-stratification, and lack desiccation features.

The basal transitional sequence coarsens upward from underlying cm-scale wave-rippled siltstone and sandstone to dm-scale erosionally based trough, tangentially, and horizontally bedded sandstone. Basal contacts are commonly broadly scoured over several metres; basal crossbed sets may contain argillite clasts. Poorly developed fining-upward cycles (0.3 to 3.0 m thick) are present; thin ($<$ few cm) silty sandstone caps are rarely preserved. The conglomeratic beds in the middle and upper sequences are thin sheets commonly only a few clasts thick and rarely thicker than 0.1 m. The size, frequency, and thickness of conglomeratic units decreases to the northwest, especially over the shelf area. Paleocurrent trends are northwesterly directed with low dispersion, indicating sediment transport across the Gordon Bay Arch (Fig. 20.7e). The only exception is a southwesterly directed mode, parallel to the Gordon Bay Arch, present only in the lower transitional unit toward the southeasternmost part of the basin.

Conglomerate lithologies are dominated by vein quartz, plus common intrabasinal Kimerot Group and Bear Creek Group clasts of quartzite and jasper, and some basement-derived metasediment, banded iron formation, and mylonite. Most intrabasinal clasts are derived from pre-Burnside Formation units. However, intraformational clasts of the Burnside Formation itself appear in the upper few hundred metres of measured sections; these clasts commonly contain quartz-filled fractures not present in the matrix (Fig. 20.8f).

Facies Interpretation

The upper parts of the axial and shelf sequences have been interpreted as braided fluvial deposits (Campbell and Cecile 1981; Grotzinger *et al.* 1987). In the foredeep model presented here, they represent the final basin filling, with transport of sediment across the basin. The transitional unit present only in the axial zone of the basin represents the progradation of fluvial facies across a marine basin. The basin-parallel paleocurrent trends in the transitional unit suggest that the Gordon Bay Arch may have acted as a dam in the initial stages of transitional marine to nonmarine sedimentation, effectively confining sediment transport parallel to the basin axis. The middle and upper coarse-grained braided fluvial facies overtopped the arch and were shed across the basin axis in response to continued uplift and erosion of the hinterland. Fluvial deposits of the Burnside Formation pass northwestward toward Wopmay Orogen into siltstones and silty sandstones interpreted as shallow-marine shelf deposits.

The composition of Burnside conglomerates is consistent with cratonward migration of the basin axis. Because foredeeps propagate in front of advancing thrust-fold belts (Bally *et al.* 1966; Beaumont 1981; Stockmal *et al.* 1986), initial foredeep deposits may be incorporated into thrust wedges advancing toward the foreland. Synorogenic sediments are delaminated by propagating thrusts, uplifted, eroded, and recycled into the foredeep, commonly at a high stratigraphic position. Such cannibalization of the Jurassic-Cretaceous foredeep adjacent to the Cordilleran thrust wedge occurred during deposition of the Paleocene Paskapoo Formation (Bally *et al.* 1966). In the Bear Creek Foredeep, conglomerates in the middle sequence of the Burnside Formation indicate that initially underlying Goulburn Group units and basement lithologies were being reworked. Vein quartz probably formed during fracturing of lithified sediments associated with thrusting. This is supported by the observation that intraformational quartzite clasts contain quartz-filled fractures not present in the matrix, suggesting fracturing at depth prior to resedimentation. The coarser upper conglomerates with intraformational clasts indicate that the deformation front had migrated toward the foreland and that fluvial synorogenic sediments, in addition to previous litholo-

gies, were uplifted, eroded, and recycled back into the foredeep.

Upper Bear Creek Group: Termination of Foredeep

Facies Description and Interpretation

The uppermost two units of the Bear Creek Group are the Mara and Quadyuk Formations (Fig. 20.3). These units have not been studied in detail and only a brief summary, based largely on previous work by Campbell and Cecile (1981), is presented here.

The Mara Formation is a sequence of marine siltstones and fine sandstones that disconformably overlies fluvial deposits of the Burnside Formation in northwest exposures (Campbell and Cecile 1981); the contact is conformable to the southeast and fluvial sandstones of the Burnside Formation grade up into marine siltstones of the Mara Formation (Campbell and Cecile 1981; our unpublished data). The Mara Formation grades up into the overlying Quadyuk Formation, which is dominated by stromatolitic and clastic-textured dolomites. Both the Mara and Quadyuk Formations pass laterally southeastward into probably time-equivalent fluvial sandstones in the southeasternmost part of the basin. To the northwest, Quadyuk carbonates are missing, and a thick pisolitic regolith is developed on the upper surface of the Mara Formation (Campbell and Cecile 1981). The Mara/Quadyuk sequence is overlain by the Wolverine Group (Fig. 20.3), characterized by a regionally extensive carbonate platform (Campbell and Cecile 1981).

Tectonic Significance

The transition from fluvial deposits of the Burnside Formation into marine siliciclastics and carbonates of the Mara and Quadyuk Formations signifies waning of sediment influx into the basin and the onset of tectonic quiescence in the hinterland. This type of transition from molasse to carbonate sedimentation is typical of other foreland basins. One example is the Taconic foredeep of the central Appalachians where fluvial deposits of the Juniata Formation are succeeded by shallow-marine siliciclastic deposits of the Tuscarora Formation and ultimately carbonates of the Helderberg Group (Rodgers 1971).

Truncation of the Quadyuk Formation to the northwest and development of a regional disconformity on top of the Mara Formation may be significant in that the zone of regolith development is spatially consistent with the Gordon Bay Arch. However, little is known concerning the origin of this feature and investigations are currently in progress.

Possible Foredeep Reactivation

Following Bear Creek Group time, there was a regionally extensive transgression of Slave Craton, and widespread deposition of carbonates (Wolverine Group, Fig. 20.3; Campbell and Cecile 1981). Carbonate sedimentation was abruptly terminated at the end of Wolverine time by offlapping shallow-marine sandstones followed by fluvial sediments and conglomerates of the Bathurst Group (Fig. 20.3). Fluvial sediments of the Bathurst Group were dispersed to the southwest, west, and northwest (Campbell and Cecile 1981) and are more feldspathic than sandstones of the Burnside Formation. Sedimentation was also accompanied by the extrusion of minor basaltic lava flows.

The similarity of sediment-dispersal trends between the Burnside Formation and Bathurst Group suggest that the Bathurst Group was deposited during renewed uplift and unroofing of the Thelon Tectonic Zone. This may have accompanied a second phase of convergence and overthrusting in the Thelon Tectonic Zone.

Discussion

Owing to a combination of poor time control, inadequate exposure, and/or incomplete preservation, documentation of subsidence mechanisms of Precambrian basins is difficult. The Kilohigok Basin is unique in this regard; it is superbly exposed as a result of Quaternary glacial scouring of bedrock, and although only partially preserved as a structural remnant, the basin and its cratonic extension are entirely exposed in cross section. This is fortunate because geodynamic models of basin formation rely heavily on data extracted from stratigraphic cross sections that pinpoint critical facies relationships, erosional surfaces, and stratigraphic pinch-outs (cf. Steckler and Watts 1982; Quinlan and Beaumont 1984). Only the current lack of time constraints prohibits more quantitative analysis of Kilohigok Basin. However, zircon-bearing tuff beds have been recently discovered at critical positions in the basin

(e.g., immediately above Kimerot Platform) and initial attempts at U-Pb dating of the zircons suggest that accurate dates will be available soon (S.A. Bowring personal communication 1987). Furthermore, efforts are being made using sequence stratigraphy (cf. Vail *et al.* 1977) to correlate stratigraphic intervals in Kilohigok Basin with tuff-bearing equivalents in Wopmay Orogen. Initial results of the sequence analysis are also promising.

Despite the current lack of extensive time control, this study has generated several converging lines of evidence that allow some basic aspects of basin development to be clarified. These are: 1) evaluation of the evidence for the foreland basin versus alternative rift or strike-slip basin models; 2) the relationship of the Kimerot Group to other passive-margin platforms including its location on the margin and duration of growth; and 3) the thermal structure of the lithosphere at the time of convergence and overthrusting.

Evidence for Foredeep Subsidence

The Bear Creek Group shares many of the distinctive characteristics of other well documented foreland-basin deposits (Bally *et al.* 1966; Benedict and Walker 1978; Allen and Homewood 1986). Deepwater axially deposited turbidites overlie shallow-water platform carbonates, and grade up into shallow marine and then fluvial molasse deposits that were shed toward the foreland. Fluvial deposits culminate in conglomerates dominated by intraformational clasts derived by tectonic reworking of earlier deposited lithofacies equivalents, thus indicating migration of the basin toward the foreland with time. The Kimerot Platform was probably drowned by flexure of the lithosphere as indicated by drowning and mass sliding of portions of its outer part, and synchronous uplift and erosion of its inner part. The resulting Gordon Bay Arch was intermittently reactivated throughout deposition of the Bear Creek Group, generating exposure surfaces on shelf sequences and strongly influencing sedimentation patterns. The trend of the Gordon Bay Arch, the depositional strike of the Bear Creek Foredeep (trench axis), and the trend of the Thelon Tectonic Zone are all co-parallel (Fig. 20.10). Similarly they are approximately perpendicular to the trend of finite-stretching lineations in thrust nappes. These relationships support a model of basin development in which foredeep subsidence and activation of the

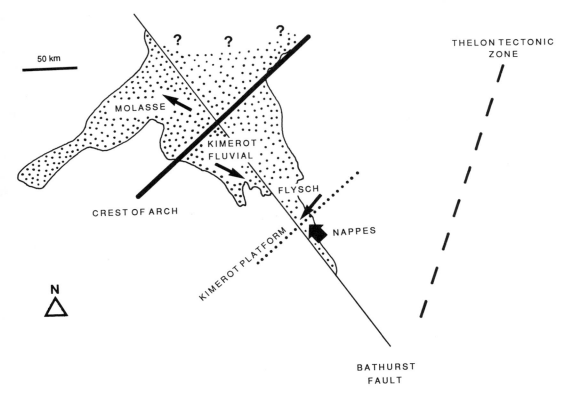

Fig. 20.10. Spatial relationships between critical elements in evolution of Kilohigok Basin. The strike of Kimerot Platform and transport direction of basal fluvial deposits help constrain the initial passive margin as southeast-facing. Development of the subsequent foredeep is supported by co-parallel trends of Gordon Bay Arch (heavy black line) and the basin axis (indicated by paleocurrents in turbiditic flysch). The cause of foredeep subsidence is inferred to be crustal-scale overthrusting in Thelon Tectonic Zone, which trends subparallel to the arch and basin axis, and probably caused down-bending (flexure) of the adjacent lithosphere. This is consistent with the northwest transport direction of intrabasinal thrust nappes (wide arrow). The vector means of paleocurrent data (thin arrows) from selected units (see Fig. 20.7) help constrain the geometry of the basin through time. Kimerot fluvial deposits (Fig. 20.7a) reflect the initial southeast gradient of the passive margin; flysch paleocurrents (submarine-fan deposits of the Bear Creek Group; Fig. 20.7c) reflect the southwest-northeast trend of the foredeep axis; molasse paleocurrents of the Burnside Formation (Fig. 20.7e) reflect the final basin fill, reversal of initial (passive-margin) paleoslope, and progradation of a major alluvial complex over the Slave Craton.

Gordon Bay Arch occurred by flexure of the lithosphere during emplacement of thrust nappes rooted in the Thelon Tectonic Zone (Tirrul 1985).

Flexure, as the controlling subsidence mechanism during Bear Creek time, is not characteristic of either rift or strike-slip basins, which are inherently fault controlled. To begin with, no evidence exists for basement-involved syn-sedimentary faulting of any type in the axial zone of the basin, where abrupt major thickening occurs. The Archean/Proterozoic unconformity ("floor" of basin) is continuously exposed perpendicular to depositional strike and any normal faults or stratigraphic thickening related to basement-involved faulting would be obvious given

the orientation of the exposed basin cross section. Furthermore, younger strike-slip faults that now preserve slices of the original basin are oriented at a high angle to the original basin strike and stratigraphic shelf edge, and therefore do not represent reactivated older growth faults that might have controlled basin subsidence but are now unexposed.

The restricted spatial and temporal distribution of the Bear Creek molasse (Burnside Formation) is also uncharacteristic of rift or strike-slip sedimentation. Both rift and strike-slip basins contain thick successions of coarse fluvial deposits that are formed at various times throughout the life of the basin, derived from the basin margins and transported to the basin center following episodic faulting events (Miall 1981; Christie-Blick and Biddle 1985). Fluvial deposits of the Bear Creek Group (Burnside Formation) are stratigraphically restricted to its uppermost part, completing a distinct evolutionary sequence that starts with platform sedimentation, progresses through deep-water turbidite sedimentation, and culminates with fluvial sedimentation. Also, fluvial sediments are derived from a southeastern source area and transported into and then beyond the axial zone of the basin, extending for over 100 km cratonward of the northwest margin of the basin (Gordon Bay Arch). In order for this to occur, the fluvial plain must aggrade at rates higher than the axial zone could subside so as to overstep the basin margin and build out beyond it. This is not known to occur in rift or strike-slip basins where sediments are trapped in down-dropped blocks adjacent to fault scarps (Christie-Blick and Biddle 1985); nor does it occur in simple sag-type intracratonic basins where sediment dispersal is centripetal (Sloss 1963). Conversely, alluvial sedimentation that extends far beyond the basin margin is characteristic of the culminating phase in the cycle of foreland-basin development. It reflects the attainment of balance between the weight of the load and the elastic strength of the lithosphere (Beaumont 1981; Quinlan and Beaumont 1984). At that point, basin subsidence is slow relative to sedimentation and erosional debris progrades beyond the basin margin and onto the craton. Ancient examples of this include the three separate collisional events in the U.S. Appalachians that produced the Queenston Delta (Taconic Orogeny; Rodgers 1971), the Catskill Delta (Acadian Orogeny; Allen and Friend 1968; Woodrow and Sevon 1985), and the Pocono-Mauch Chunk alluvial complex (Alleghenian Orogeny; Dunbar and Rodgers 1958). Modern progradation of collision-generated molasse is occurring by southward migration of the Indo-Gangetic Foredeep over the Indian Craton (Burbank *et al.* 1986).

Significance of Kimerot Platform

Several features of the Kimerot Platform suggest that it may have been a platform that formed during passive-margin subsidence. First, the platform was southeast facing, directly opposed to the transport direction of subsequent thrust nappes and parallel to the trend of overlying foreland basin, similar to many ancient passive-margin sequences (Dewey

and Bird 1970). Initial fluvial deposits at the base of the platform formed on a low-gradient slope that was inclined to the southeast. This was followed by northwest-directed onlap of various marine facies that pinch out systematically to the northwest (Fig. 20.5). Significantly, overlying carbonates of the Peg Formation retain their thickness, and overstep most of the siliciclastics of the underlying Kenyon Formation. Such stratigraphic onlap is typical of many Phanerozoic passive margins (Vail *et al.* 1977; Hiscott *et al.* 1984) and is controlled by the increasing flexural rigidity of the lithosphere associated with cooling of heated crust (Watts *et al.* 1982). Additionally, the Kimerot Platform consists of the classic combination of transgressive marine sandstones overlain by cyclic carbonates that characterizes passive-margin sequences of all ages (Read 1982). Bond *et al.* (1987) have suggested that this transition is related to the combined effects of long-term sea-level rise following continental break-up, and to the increasing flexural rigidity of the lithosphere due to cooling of heated lithosphere. The predicted effects of this are that younger cyclic carbonates will overstep older siliciclastics and extend farther onto the craton.

Cyclic carbonates of the Peg Formation present an opportunity to approximate roughly the amount of time represented by the preserved Kimerot Group as well as rates of platform subsidence. A similar analysis of the early Proterozoic Rocknest Platform (passive-margin carbonates) has been developed recently by Grotzinger (1986a), where the periodicity of the small-scale shallowing-upward cycles was used as a measure of time in subsidence analysis. Various observations now available indicate that for various periods in Phanerozoic time, small-scale (few m) shallowing-upward cycles of many carbonate platforms have average periods within the range of 20,000 to 100,000 years (e.g., Fischer 1964; Bova 1982; Koerschner 1983; Matthews 1984; Goodwin and Anderson 1985; Read *et al.* 1986; Goldhammer *et al.* in press). This range in average cycle period is also recorded in Precambrian cyclic carbonates and average cycle periods of 20,000 to 100,000 years have been documented in the early Proterozoic Rocknest Formation (Grotzinger 1986a). Platform cyclicity has on these time scales been interpreted to record a Milankovitch climatic signal by the workers cited above.

Regardless of the actual cause of the cyclicity, the range of average cycle period for any given sequence may be used to constrain the amount of time

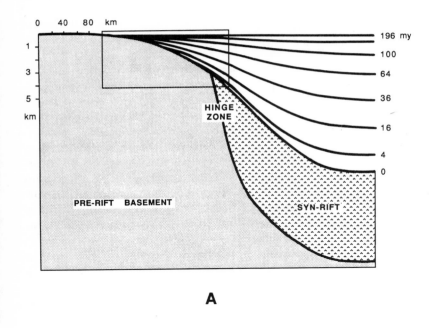

A

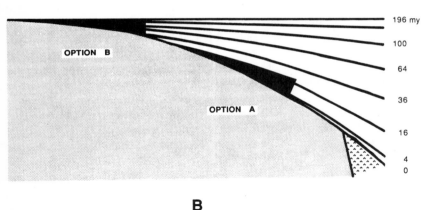

B

Fig. 20.11. A = Synthetic passive margin constructed from a two-dimensional, one-layer (uniform) stretching model. The coastal plain wedge, developed landward of the hinge zone, forms by flexural subsidence of the lithosphere due to sediment loading over the stretched part of the margin. Box outlines enlarged area shown in B. B = Comparison of model coastal plain with scaled reductions of Kimerot Platform (black). In Option A the platform is shown developed on basement at an early stage in the history of the hypothetical margin and is thus of short duration (approximately 20 million years). In Option B the platform is shown developed on basement at a late stage and is thus of much longer duration (approximately 150 million years).

represented by that sequence. Assuming that Peg Formation cycles had average periods between 20,000 and 100,000 years, and because there are approximately 70 cycles in the Peg Formation, its duration is bracketed between 1 and 7 million years. Estimated total subsidence rates (interval thickness/interval duration) are on the order of 0.02 to 0.08 m/10³ yr. Applying this rate to the underlying Kenyon Formation gives a total duration of the preserved Kimerot Platform on the order of 5 to 20 million years. Although there are errors in such approximations, specific values are probably correct within a factor of 2 or 3 (Grotzinger 1986a,c), and these permit the first-order conclusion that the Kimerot Platform probably had a short duration.

However, by itself this estimate does not allow the inference that the margin was also of short duration; it must also be specified on what part of the margin the platform was developed.

It is useful to compare the geometry of the Kimerot Platform wedge with a geophysical model of a passive-margin sequence to obtain more information on the subsidence history of the Kimerot Platform and on what part of the margin it developed. A synthetic passive margin (Fig. 20.11a) was constructed using the two-dimensional, forward extension model of Steckler (1981) that implements finite-element solutions for vertical and lateral heat flow and for flexure caused by both sediment and water loads, and the increase in rigidity as heated lithosphere cools and contracts. The model was run to generate the post-rift stratigraphy for a mature passive margin of almost 200 million years (Fig. 20.11a). Input variables for this model define an equilibrium, thermal lithospheric-plate thickness of 125 km with an elastic thickness determined by the depth to the 450°C isotherm, and an initial (pre-rift) crustal thickness of 31 km that is stretched and thinned laterally over 200 km to a value equivalent to ocean crust. The sedimentary infill has a density of 2,500 kg/m³, and other parameters (e.g., coefficient of thermal expansion, etc.) are as specified in McKenzie (1978).

Thermal-mechanical modelling of modern (e.g., Watts *et al.* 1982) and ancient (Bond and Kominz 1984) passive margins has shown that a significant amount of flexurally induced subsidence can occur landward of the hinge zone due to accumulation of sediment within the margin and the increase in flexural rigidity as heated lithosphere cools. The models predict that for a mature margin, the resulting sedimentary deposit landward of the hinge zone, or *coastal plain wedge*, can reach a maximum thickness of almost 3.5 km and a width of almost 200 km (Steckler 1981). Because the mass of the sedimentary load and the lithospheric rigidity both increase with time, the coastal-plain wedge contains diachronous onlapping units that thin and young away from the hinge zone. A representative coastal-plain wedge is shown in Figure 20.11a.

The results of the model are significant and illustrate that the Kimerot Platform could have formed as part of a coastal plain wedge, landward rather than seaward of the hinge zone. This interpretation is supported by geological evidence that demonstrates that the basement-platform contact is unfaulted in all areas where the platform is developed, indicating that brittle stretching of the upper crust was not involved in platform subsidence. This is in agreement with forward extension models, which do not predict faulting landward of the hinge zone (Steckler 1981; Steckler and Watts 1982). However, in two-layer, nonuniform extension models (Royden and Keen 1980) the amount by which the upper lithosphere is stretched may be substantially less than the lower lithosphere; a geological consequence of this is that faulting seaward of the hinge zone may be reduced (Royden and Keen 1980). Nevertheless, faulting is not eliminated and should be seen in a geological cross section such as that constructed for the Kimerot Platform, which is in part based on detailed mapping of the well exposed basement-

platform contact. On these grounds it is concluded that the Kimerot Platform was formed landward of the hinge zone and represents a coastal-plain wedge.

Two possible end-member interpretations of the Kimerot Platform as a coastal-plain wedge are illustrated in Figure 20.11b, and emphasize the potential disparity in duration of the platform as a function of its lateral position. The Kimerot Platform has been redrawn to scale and superimposed (black) on an enlarged replica of the model coastal-plain wedge shown in Figure 20.11a. Both options are geometrically consistent with the reconstructed shape of the Kimerot Platform, which is defined by the basement/platform contact and the datum (time line) at the top of the cyclic member. Time lines between the basement/platform contact and the datum at the top of the cyclic member are not required to make this comparison.

The major difference between the two end-member interpretations lies in the amount of time represented by the platform (shown in black for each case) depending on whether the Kimerot Platform was formed early or late in the history of coastal-plain subsidence. For Option A, the platform is developed early, following soon after the termination of rifting and the onset of cooling of the margin (Fig. 20.11b). At this time, the rate of thermal subsidence is high and consequently the rate of sediment loading is high, which generates a high rate of flexural subsidence for the coastal plain. Also, lithospheric rigidity is low at this time due to high heat flow, causing the coastal plain to subside more deeply than at later times after heat flow decays and rigidity is correspondingly increased. In contrast to Option A, Option B shows the platform as having developed at a later time, after the margin has cooled significantly and the rates of thermal subsidence and sediment loading are much lower (Fig. 20.11b). Consequently the rate of flexural subsidence of the coastal plain is low. This effect is enhanced by the tendency for the increased flexural rigidity of the plate to allow the coastal plain to subside less deeply as a result of continued cooling.

The decay of flexural coastal-plain subsidence mimics that of the cooling part of the margin where loading occurs, and is thus exponential in form (Steckler and Watts 1982; Bond and Kominz 1984). Thus, the predominant difference between the two end-members is that a platform developed early (option A) represents much less time than a platform developed late in the subsidence history of the margin (option B). For a coastal-plain wedge the size of the Kimerot Platform, developed on basement, this represents a difference between a platform duration of approximately 20 million years (option A) and a platform duration of approximately 150 million years (option B). Estimates of the duration of Kimerot Platform (5 to 20 million years) compare well with Option A of the model coastal-plain wedge (Fig. 20.11b) and suggest that the Kimerot probably formed early in the subsidence history of the margin, perhaps just landward of the hinge zone.

We emphasize, however, that other factors can affect the subsidence rate of the coastal plain and thus influence the duration of a sequence of unit thickness. For example, forward extension models show that if the plate thickness or stretching factor (β = initial thickness/final thickness) is changed, it will have a corresponding change in the subsidence rate of the margin, thereby modifying the amount of time represented by any given stratigraphic unit. Passive-margin subsidence will be reduced for thinner lithospheric plates and smaller values of β; the opposite is true for thicker plates and greater values of β (Bond et al. 1988: this volume). However, the general form of the subsidence remains unaffected (Bond et al. 1988: this volume), and the amount of time in a stratigraphic unit formed early on in the subsidence of a coastal-plain wedge will always be less than the amount of time represented by a unit of similar thickness but formed later on and farther away from the hinge zone. The model illustrated in Figure 20.11a was constructed using an average equilibrium thermal plate thickness (125 km) but stretching fully to ocean-crust values in order to maximize the rate of post-rift subsidence and the extent of the coastal-plain wedge. A maximum β value lower than this, or a thinner lithospheric plate, might yield slower subsidence rates than those apparently exhibited by the Kimerot Platform; only a thicker plate could yield faster post-rift subsidence rates. The model (Fig. 20.11b) is considered a maximized end member given that thicker-than-average plates (>125 km) are not consistent with the inferred higher heat flow during early Proterozoic time (cf. Bickle 1978), and that possibly plates may have been thinner. Consequently, the model is considered to be appropriate for comparison with the Kimerot Platform, and suggests that the margin may have been young at the time the platform was tectonically drowned.

Development of Bear Creek Foredeep on Thermally Young Lithosphere

Several features of Bear Creek Foredeep evolution are consistent with its development on thermally young (warm) lithosphere, relative to the end of inferred passive-margin extension. These include: 1) the apparent short duration and high subsidence rates of the Kimerot Platform as discussed above; 2) the spatially restricted occurrence of the erosional unconformity at the top of Kimerot Platform; 3) the narrow transition zone between the thick axial zone sequence and thin-shelf sequence; and 4) pervasive low-grade metamorphism of basin sediments and high ductile strain in thrust nappes.

The unconformity at the top of Kimerot Platform is developed only in the northwest (Figs. 20.5, 20.9), where it abruptly cuts down through the Kimerot Group until it intersects basement. No evidence for erosion of strata at the top of the platform exists southeast of this region. This can be interpreted in two ways. First, during convergence and overthrusting load emplacement was close enough to the platform that the outer part of the platform was within the flexural wavelength of the lithosphere (distance between load and peripheral bulge). Thus, at the onset of loading the outer part would descend and the inner part would rise over the peripheral bulge. It follows then that the basin would have been narrow, opening and closing within the flexural wavelength of the lithosphere. This is consistent with the apparent short duration of the passive margin.

A second possibility is that a bulge migrated over the entire platform but caused subaerial exposure of only the inner part. It is important to note that both inner and outer parts of Kimerot Platform were at sea level, with tidal-flat carbonates forming over most of its surface just prior to the onset of drowning and arching, and that variations in water depth created by initial platform slope cannot be invoked to explain the lack of subaerial exposure of its outer part (i.e., outer part too deep to be exposed during bulge migration). A possible explanation may relate to the apparent youthfulness of the margin at the time of loading. Stockmal et al. (1986) have shown that the thermal age of a rifted margin is important in the early stages of foredeep development. Given that the margin may have been young at the time of basin closure (perhaps only a few tens of million years), it would be expected that the lithosphere outboard of the hinge zone would be warm and sub-siding rapidly; inboard of the hinge zone the litho-sphere would have been cooler and more rigid (Stockmal *et al.* 1986). Such a scenario might predict that the transition from a conformable to erosional contact should occur in the vicinity of the hinge zone. The model of Stockmal *et al.* (1986) predicts that the amplitude of a peripheral bulge developed over a thermally young margin (15 to 20 million years) is on the order of only a few metres. By comparison, the amplitude of a bulge developed over an older margin (120 to 150 million years) is on the order of several hundred metres. Superficially, this observation apparently contradicts the docu-mented relationship that bulge amplitude is greater for a warmer, less rigid plate than for a cooler, more rigid plate (Beaumont 1981). However, bulge ampli-tude is also strongly influenced by the mass of the load (Beaumont 1981; Watts *et al.* 1982), increasing for greater loads and decreasing for smaller loads. Because a younger margin subsides less than an older margin prior to thrusting, the space available to accommodate the submarine portion of a load is lower for a younger margin than an older margin. Consequently, under conditions of constant subaerial topography for a load impinging on the edge of a rifted margin, the mass emplaced on an older margin will be much greater than for a younger margin simply because more material can be accom-modated within the space available below sea level; as the mass of the load increases, so does bulge amplitude (Stockmal *et al.* 1986). The model also indicates that as the load propagates toward the fore-land with time, its mass increases by continued tec-tonic shortening and thickening, producing a corresponding amplification of bulge height. Such a scenario is consistent with the observation that the amount of erosion at the top of the Kimerot Platform increases toward the craton.

Other evidence is compatible with development of the Bear Creek Foredeep on warm lithosphere. Shelf deposits pass abruptly into submarine-fan deposits of the axial sequence across a narrow transi-tional zone, interpreted as the foreland paleoslope. The width of this transitional zone would in part be a function of the thermal age of the lithosphere at the time of flexure, the zone being narrower for ther-mally younger, less rigid lithosphere and broader for thermally older, more rigid lithosphere. In the Bear Creek Foredeep shelf deposits occur within 15 to 20 km of submarine-fan deposits, suggesting a narrow

transitional zone (paleoslope). This contrasts with similar stratigraphic relationships in the Taconic foredeep, which was developed on a passive margin that had cooled for approximately 90 million years prior to tectonic loading (Bond *et al.* 1986). In the New York Taconic foredeep, shelf deposits of the Trenton Group pass laterally into a starved transi-tional zone (Utica Shale) 30 to 50 km wide, and then into trench-axis turbidites of the Schenectady and Austin Glen Formations (Fisher 1977; Bradley and Kusky 1986). Similarly, in the Tennessee Taconic foredeep, shelf deposits of the Lenoir, Hol-ston, and Chapman Ridge Formations pass laterally into a starved transitional zone (Blockhouse and western Sevier Formations) 30 to 60 km wide, and then into trench-axis turbidites of the Tellico For-mation and eastern Sevier Formation (Benedict and Walker 1978; Shanmugam and Walker 1978). Because for a given load the width of the transitional zone (paleoslope) is a measure of plate rigidity at the time of flexure, the Kimerot margin is interpreted to have been warmer (less rigid) than the Appalachian margin at the time of tectonic loading and margin collapse. However, foreland-basin geometry is also influenced by the load mass and distribution, and therefore the comparison described above may be invalid if the extent of loading was greatly different between hinterlands adjacent to each foredeep.

Finally, the occurrence of basin-wide low-grade metamorphism is consistent with elevated geother-mal gradients during burial of sediments over ther-mally young (warm) lithosphere. Thompson and Frey (1984) have reported values of illite crystal-linity for Goulburn Supergroup rocks that suggest temperatures of up to 300–375 °C. Tirrul (1985) similarly noted that movement and deformation of thrust nappes in the southeast part of the basin took place under temperatures sufficiently high to pro-duce pervasive ductile strain in dolomites. There-fore, it is possible that these data may record elevated geothermal gradients in the basin due to its formation over thermally warm, and thus possibly young, lithosphere. However, other interpretations for elevated basin temperatures are possible given that the timing of this metamorphism is poorly con-strained (Thompson and Frey 1984). Also, even if the timing is correct for burial metamorphism, the data could be alternatively interpreted to record an elevated average geotherm for the early Proterozoic continental lithosphere. Davies (1979), however, provides arguments against elevated Precambrian continental heat flux and suggests a model that promotes thermal buffering at the base of the Precambrian crust to shield against high mantle temperatures. This model is consistent with the recent findings of Lerner-Lam and Jordan (in press), that the continents may have thick root zones capable of such buffering.

Acknowledgments. We gratefully acknowledge the Geological Survey of Canada for logistical support and the opportunity to work in Kilohigok Basin. S. Bohan, B. Cadigan, Q. Gall, G. Luffman, and S. Pelechaty are thanked for field assistance. This paper has been improved through discussions with G. Bond, D. Bradley, P. Hoffman, M. Kominz, and R. Tirrul. G. Karner and D. Ojakangas are thanked for reading the final manuscript and providing helpful comments and suggestions. Financial support was provided by the Geological Survey of Canada and National Science Foundation Grant EAR 86-14670.

References

ALLEN, P.A. and HOMEWOOD, P. (eds) (1986) Foreland Basins. International Association Sedimentologists Special Publication 8, 453 p.

ALLEN, J.R.L. and FRIEND, P.F. (1968) Deposition of the Catskill Facies, Appalachian Region: With some notes on some other Old Red Sandstone basins. In: Klein, G. deV. (ed) Late Paleozoic and Mesozoic Continental Sedimentation, Northeastern North America. Geological Society America Special Paper 106:21–73.

ANDERSON, A.K., GOODWIN, P., and SOBIESKI, T. (1984) Episodic accumulation and the origin of formation boundaries in the Helderberg Group of New York State. Geology 12:120–123.

BALLY, A.W., GORDY, P.L., and STEWART, G.A. (1966) Structure, seismic data and orogenic evolution of southern Canadian Rockies. Bulletin Canadian Petroleum Geology 14:337–381.

BATES, R.L. and JACKSON, J.A. (1980) Glossary of Geology. Falls Church, Virginia: American Geological Institute, 749 p.

BEAUMONT, C. (1981) Foreland basins. Royal Astronomical Society, Geophysical Journal 65:291–329.

BENEDICT, G.L. and WALKER, K.R. (1978) Paleobathymetric analysis in Paleozoic sequences and its geodynamic significance. American Journal Science 278:579–607.

BICKLE, M.J. (1978) Heat loss from the Earth: A constraint on Archaean tectonics from the relation between geothermal gradients and the rate of plate production. Earth Planetary Science Letters 40:301–315.

BOND, G.C., GROTZINGER, J.P., and KOMINZ, M.A. (1987) The role of thermal subsidence, flexure and eustasy in the evolution of early Paleozoic carbonate platforms in the Appalachian and Cordilleran miogeoclines (Abstract). American Association Petroleum Geologists Bulletin 71:532.

BOND, G.C. and KOMINZ, M.A. (1984) Construction of tectonic subsidence curves for the early Paleozoic miogeocline, southern Canadian Rocky Mountains: Implications for subsidence mechanisms, age of breakup, and crustal thinning. Geological Society America Bulletin 95:155–173.

BOND, G.C., KOMINZ, M.A., and GROTZINGER, J.P. (1988) Cambro-Ordovician eustasy: Evidence from geophysical modelling of subsidence in Cordilleran and Appalachian passive margins. In: Kleinspehn, K.L., and Paola, C. (eds) New Perspectives in Basin Analysis, New York: Springer-Verlag, pp. 129–160.

BOND, G.C., KOMINZ, M.A., and NICKESON, P.A. (1986) Magnitudes of error in tectonic subsidence curves for ancient passive margins with examples from early Paleozoic of Appalachian-Caledonide Orogen, North America and Greenland. American Association Petroleum Geologist Bulletin 70:567–568.

BOVA, J.A. (1982) Peritidal Cyclic and Incipiently Drowned Platform Sequences: Lower Ordovician Chepultepec Formation, Virginia [unpublished Master's thesis], Virginia Polytechnic Institute: Blacksburg, 184 p.

BRADLEY, D.W. and KUSKY, T.M. (1986) Geologic evidence for rate of plate convergence during the Taconic arc-continent collision. Journal Geology 94:667–681.

BURBANK, D.W., RAYNOLDS, R.G.H., and JOHNSON, G.D. (1986) Late Cenozoic tectonics and sedimentation in the north-western Himalayan foredeep: II. Eastern limb of the Northwest Syntaxis and regional synthesis. In: Allen, P.A. and Homewood, P. (eds) Foreland Basins. International Association Sedimentologists Special Publication 8, pp. 293–308.

CAMPBELL, F.H.A. and CECILE, M.P. (1981) Evolution of the early Proterozoic Kilohigok Basin, Bathurst Inlet-Victoria Island, Northwest Territories. In: Campbell, F.H.A. (ed) Proterozoic Basins of Canada. Geological Survey of Canada Paper 81-10, pp. 103–131.

CHAPPLE, W.M., and FORSYTH, D.W. (1979) Earthquakes and bending of plates at trenches. Journal Geophysical Research 84:6729–6749.

CHRISTIE-BLICK, N. and BIDDLE, K.T. (1985) Deformation and basin formation along strike-slip faults. In: Biddle, K.T. and Christie-Blick, N. (eds) Strike-slip Deformation, Basin Formation, and Sedimentation. Society Economic Paleontologists Mineralogists Special Publication 37:1–34.

CHRISTIE-BLICK, N., GROTZINGER, J.P., and VON DER BORCH, C.C. (in press) Sequence stratigraphy in Proterozoic successions. Geology.

DAVIES, G.F. (1979) Thickness and thermal history of continental crust and root zones. Earth Planetary Science Letters 44:231–238.

DEWEY, J.F. and BIRD, J.M. (1970) Mountain belts and the new global tectonics. Journal Geophysical Research 75:2625–2647.

DUNBAR, C.O. and RODGERS, J. (1958) Principles of Stratigraphy. New York: John Wiley & Sons, Inc., 356 p.

ERIKSSON, K.A. (1982) Geometry and internal characteristics of Archean submarine channel deposits, Pilbara block, Western Australia. Journal Sedimentary Petrology 52:383–393.

FISCHER, A.G. (1964) The Lofer cyclothems of the Alpine Triassic. In: Merriam, D.F. (ed) Symposium on Cyclic Sedimentation. State Geological Survey Kansas Bulletin 169(1):107–149.

FISHER, D.W. (1977) Correlation of the Middle and Upper Ordovician rocks in New York State. New York State Museum Map and Chart Series 25, Plate 4.

FLEMINGS, P.B. and JORDAN, T.E. (1987) Synthetic stratigraphy of foreland basins. EOS 68(16):419.

GIBB, R.A. (1978) Slave-Churchill collision tectonics. Nature 271:50–52.

GIBB, R.A., THOMAS, M.D., LAPOINTE, P.L., and MUKHOPADHYAY, M. (1983) Geophysics of proposed sutures in Canada. Precambrian Research 19: 349–385.

GOLDHAMMER, R.K., DUNN, P.A., and HARDIE, L.A. (in press) High frequency glacioeustatic sea-level oscillations with Milankovitch characteristics recorded in Middle Triassic platform carbonates in N. Italy. American Journal Science.

GOODWIN, P.W. and ANDERSON, E.J. (1985) Punctuated aggradational cycles: A general hypothesis of episodic stratigraphic accumulation. Journal Geology 93:515–533.

GROTZINGER, J.P. (1986a) Cyclicity and paleoenvironmental dynamics, Rocknest Platform, northwest Canada. Geological Society America Bulletin 97:1208–1231.

GROTZINGER, J.P. (1986b) Upward shallowing platform cycles: A response to 2.2 billion years of low-amplitude, high-frequency (Milankovitch band) sea level oscillations. Paleoceanography 1:403–416.

GROTZINGER, J.P. (1986c) Evolution of early Proterozoic passive-margin carbonate platform, Rocknest Formation, Wopmay Orogen, N.W.T., Canada. Journal Sedimentary Petrology 56:831–847.

GROTZINGER, J.P. and GALL, Q. (1986) Preliminary investigations of early Proterozoic Western River and Burnside River formations: Evidence for foredeep origin of Kilohigok Basin, N.W.T., Canada. Current Research, Part A, Geological Survey Canada Paper 86-1A:95–106.

GROTZINGER, J.P., McCORMICK, D.S., and PELECHATY, S.M. (1987) Progress report on the stratigraphy, sedimentology, and significance of the Kimerot and Bear Creek groups, Kilohigok Basin, District of Mackenzie. Current Research, Part A, Geological Survey Canada Paper 87-1A:219–238.

HANMER, S. and LUCAS, S.B. (1985) Anatomy of a ductile transcurrent shear: The Great Slave Lake Shear Zone, District of Mackenzie, N.W.T. (preliminary report). Current Research, Part B, Geological Survey Canada Paper 85-1B:7–22.

HISCOTT, R.N., JAMES, N.P., and PEMBERTON, S.G. (1984) Sedimentology and ichnology of the Lower Cambrian Bradore Formation, coastal Labrador: Fluvial to shallow-marine transgressive sequence. Bulletin Canadian Petroleum Geologists 32:11–26.

HOFFMAN, P.F. (1973) Evolution of an early Proterozoic continental margin: The Coronation geosyncline and associated aulacogens of the northwestern Canadian Shield. Royal Society London, Philosophical Transactions A 273:547–581.

HOFFMAN, P.F. (1980) Wopmay Orogen: A Wilson cycle of early Proterozoic age in the northwest of the Canadian Shield. In: Strangeway, D.W. (ed) The Continental Crust and its Mineral Deposits. Geological Association Canada Special Paper 20, pp. 523–549.

HOFFMAN, P.F., CULSHAW, N.G., HANMER, S.K., LECHEMINANT, A.N., McGRATH, P.H., TIRRUL, R., VAN BREEMAN, O., BOWRING, S.A. and GROTZINGER, J.P. (1986) Is the Thelon Front (NWT) a suture? Program With Abstracts, Geological Association Canada 11:82.

HOWELL, D.G. and NORMARK, W.R. (1982) Sedimentology of submarine fans. In: Scholle, P.A. and Spearing, D. (eds) Sandstone Depositional Environments. American Association Petroleum Geologists Memoir 31:365–404.

JORDAN, T.E., FLEMINGS, P.B., and BEER, J.A. (1988) Dating thrust-fault activity by use of foreland-basin strata. In: Kleinspehn, K.L. and Paola, C. (eds) New Perspectives in Basin Analysis. New York: Springer-Verlag, pp. 307–330.

KOERSCHNER, W.F., III (1983) Cyclic Peritidal Facies of a Cambrian Aggraded Shelf: Elbrook and Conococheague Formations, Virginia Appalachians [unpublished Master's thesis], Virginia Polytechnic Institute, Blacksburg: 184 p.

KOMINZ, M.A. (1986) Geophysical Modeling Studies. Chapter III: Geophysical modeling of the thermal history of foreland basins (part 1) and thermal modeling of foreland basins (part 2). Unpublished Ph.D. dissertation, Columbia University, New York: 49–142.

LERNER-LAM, A.L. and JORDAN, T.H. (in press) How thick are the continents? Journal Geophysical Research.

MATTHEWS, R.K. (1984) Dynamic Stratigraphy: An Introduction to Sedimentation and Stratigraphy, Englewood Cliffs, New Jersey: Prentice-Hall, Inc., 489 p.

McBRIDE, E.F. (1962) Flysch and associated beds in the Martinsburg Formation (Ordovician), Central Appalachians. Journal Sedimentary Petrology 32:39–91.

McKENZIE, D. (1978) Some remarks on the development of sedimentary basins. Earth Planetary Science Letters 40:25–32.

MIALL, A.D. (1981) Alluvial sedimentary basins: Tectonic setting and basin architecture. In: Miall, A.D. (ed) Sedimentation and Tectonics in Alluvial Basins. Geological Association Canada Special Paper 23, pp. 1–34.

MULLINS, H.T. and NEUMANN, A.C. (1979) Carbonate slopes along open seas and seaways in the northern Bahamas. In: Doyle, L. and Pilkey, O.H. (eds) Geology of Continental Slopes. Society Economic Paleontologists Mineralogists Special Publication 27:165–192.

MUTTI, E. (1985) Turbidite systems and their relations to depositional sequences. In: Zuffa, G.G. (ed) Provenance of Arenites. NATO Advanced Science Institute Series 129:65–93.

POTTER, P.E. and PETTIJOHN, F.J. (1977) Paleocurrents and Basin Analysis (second edition). New York: Springer-Verlag, 425 p.

QUINLAN, G.M. and BEAUMONT, C. (1984) Appalachian thrusting, lithospheric flexure, and the Paleozoic stratigraphy of the Eastern Interior of North America. Canadian Journal Earth Sciences 22:973–996.

READ, J.F. (1982) Carbonate platforms of passive (extensional) continental margins – types, characteristics, and evolution. Tectonophysics 81:195–212.

READ, J.F., GROTZINGER, J.P., BOVA, J.A., and KOERSCHNER, W.F. (1986) Models for generation of carbonate cycles. Geology 14:107–110.

RODGERS, J. (1971) The Taconic Orogeny. Geological Society America Bulletin 82:1141–1178.

ROYDEN, L. and KEEN, C.E. (1980) Rifting process and thermal evolution of the continental margin of eastern Canada determined from subsidence curves. Earth Planetary Science Letters 51:343–361.

SHANMUGAM, G. and WALKER, K.R. (1978) Tectonic control of distal turbidites in the Middle Ordovician Blockhouse and Lower Sevier formations in East Tennessee. American Journal Science 278:551–578.

SLOSS, L.L. (1963) Sequences in the cratonic interior of North America. Geological Society America Bulletin 74:93–114.

SPEARING, D.R. (1975) Shallow marine sands. In: Harms, J.C., Southard, J.B., Spearing, D.R., and Walker, R.G. (eds) Depositional Environments as Interpreted from Primary Sedimentary Structures and Stratification Sequences. Society Economic Paleontologists Mineralogists Short Course 2:103–132.

STECKLER, M.S. (1981) Thermal and Mechanical Evolution of Atlantic-Type Margins [unpublished Ph.D. thesis]: New York, Columbia University: 161 p.

STECKLER, M.S. and WATTS, A.B. (1982) Subsidence history and tectonic evolution of Atlantic-type continental margins. In: Scrutton, R.A. (ed) Dynamics of Passive Margins. American Geophysical Union, Geodynamics Series 6:184–196.

STOCKMAL, G.S., BEAUMONT, C., and BOUTILIER, R. (1986) Geodynamic models of convergent margin tectonics: Transition from rifted margin to overthrust belt and consequences for foreland-basin development. American Association Petroleum Geologists Bulletin 70:181–190.

THOMPSON, P.H. and FREY, M. (1984) Illite "crystallinity" in the Western River Formation and its significance regarding the regional metamorphism of the early Proterozoic Goulburn Group, District of Mackenzie. Current Research, Part A, Geological Survey Canada Paper 84-1A:409–414.

THOMPSON, P.H., CULSHAW, N., THOMPSON, D.L., and BUCHANAN, J.R. (1985) Geology across the western boundary of the Thelon Tectonic Zone in the Tinney Hills-Overby Lake (west half) map area, District of Mackenzie. Current Research, Part A, Geological Survey Canada Paper 85-1A:555–572.

TIRRUL, R. (1985) Nappes in the Kilohigok Basin, and their relation to the Thelon Tectonic Zone, District of Mackenzie. Current Research, Part A, Geological Survey Canada Paper 85-1A:407–420.

VAIL, P.R., MITCHUM, R.M, and THOMPSON, S., III (1977) Seismic stratigraphy and global changes of sea level, Part 3: Relative changes of sea level from coastal onlap. In: Payton, C.E. (ed) Seismic Stratigraphy – Applications to Hydrocarbon Exploration. American Association Petroleum Geologists Memoir 26, pp. 63–81.

VAN BREEMAN, O., HENDERSON, J.B., LOVERIDGE, W.W., and THOMPSON, P.H. (1986) Archean-Aphebian geochronology along the Thelon Tectonic Zone, Healey Lake area, NWT. Program With Abstracts. Geological Association of Canada 11:139.

WALKER, R.G. (1984) Shelf and shallow marine sands. In: Walker, R.G. (ed) Facies Models (second edition). Geoscience Canada Reprint Series 1:141–170.

WATTS, A.B., KARNER, G.D., and STECKLER, M.S. (1982) Lithospheric flexure and the evolution of sedimentary basins. Royal Society London, Philosophical Transactions A 305:249–281.

WOODROW, D.L. and SEVON, W.D. (1985) The Catskill Delta. Geological Society America Special Paper 201, 246 p.

21

Glaciation: An Uncommon "Mega-Event" as a Key to Intracontinental and Intercontinental Correlation of Early Proterozoic Basin Fill, North American and Baltic Cratons

Richard W. Ojakangas

Abstract

Glaciation was an uncommon "mega-event" during early Proterozoic time, with a glaciation well documented in North America. Glaciogenic units, including diamictites and finely laminated sediments with dropstones, are present in the Huronian Supergroup of Ontario, in the Marquette Range Supergroup of Michigan, in the Snowy Pass Supergroup of Wyoming, in northern Quebec, and in the Northwest Territories.

Glaciogenic deposits have been recognized recently on the Baltic Shield in the Sariolian Group of the Karelian Supergroup of Finland. Since they were first described, they have been noted at the same stratigraphic horizon at several other Finnish localities. To the east of Finland in Karelia, USSR, rocks at several localities that were earlier interpreted as tectono-sedimentary in origin have been reinterpreted as glaciogenic; these occurrences, sixteen in number, are located in the Sariolian Group as in Finland. The Finnish and Karelian occurrences are scattered over an area of 500 km by 300 km, suggestive of a large glaciated terrain.

The aforementioned supergroups have many similarities in lithology and sequence. Similarity of sequence may be a reflection only of similar environments of deposition in basins with a similar tectonic evolution, but when the sequence includes products of a widespread, uncommon sedimentary event (a "mega-event"), in this case a glaciation, then the case for chronocorrelation is strengthened. The chronocorrelation between the Huronian and Karelian Supergroups is further strengthened by the observation that both are cut by ca. 2,150 Ma mafic dikes and rest upon ca. 2,450 Ma igneous rocks.

This application of "mega-event correlation" is an example of "mega-event stratigraphy" in which glaciations can be used to subdivide a sedimentary sequence whose age is otherwise constrained only within broad limits. If the age of the overall sequence is constrained by radiometric data, the sequence can be subdivided on the basis of the mega-event(s) into pre-mega-event, mega-event, and post-mega-event strata. The recognition of depositional sequences within the overall sequence may make possible the interpretation of eustatic sea-level changes related to glaciation.

Introduction

Most sedimentary rock types are relatively common and cannot be used to correlate strata over long distances. However, *uncommon* sedimentary rocks, such as glaciogenic rocks (i.e., the diamictite and dropstone unit association), are products of *uncommon* sedimentary events, and theoretically should be of value in correlation within and between basins, even on an intercontinental scale.

Correlation using markers of "instantaneous" geologic events such as ash layers, has been called "event correlation" (e.g., Boggs 1987, p. 685). I propose here that the use in correlation of the products of uncommon sedimentary events, such as glaciogenic units, constitutes a larger scale and longer term event correlation that can be termed "mega-event correlation," and is a valid basis for "mega-event stratigraphy."

An example of such correlation is the Late Paleozoic Gondwana glaciation of the southern hemisphere. Late Paleozoic glacial deposits have long been used as evidence for the existence of Pangea and its post-glacial fragmentation, although it appears that the glaciation spanned late Devonian to late Permian time (about 90 million years), with glacial centers in different locations dependent upon the position of Pangea relative to the South Pole (Caputo and Crowell 1985). Intercontinental correlation of Pleistocene glacial deposits is also well established.

Chumakov (1981), in a survey of Late Precambrian glacial deposits, termed glaciogenic deposits "glaciocomplexes," which he further subdivided into regional "glaciohorizons." The latter are useful in regional correlations, but are not reliable for inter-regional or intercontinental correlations if biostratigraphic or radiometric controls are not available. The base of the Laplandian glaciohorizon, for example, is used as the lower boundary of the Vendian in the official USSR time scale. Chumakov (1981) further suggested that eustatic sea-level changes might be manifested as unconformities in correlative stratigraphic columns that lack glaciogenic deposits.

Evidence is presented herein for intracontinental and intercontinental correlation of Early Proterozoic rock units that lack fossils for correlation purposes. In Precambrian rocks, without fossils, the rock types are of an even greater importance in chronostratigraphic correlation than in most Phanerozoic sequences.

The recognition of ancient glaciogenic deposits is often difficult; the same rocks may be interpreted as glacial by some workers and as sedimentary-tectonic by others. Definitive criteria have been reviewed by several authors (e.g., Crowell 1974; Schermerhorn 1974; Andrews and Matsch 1983; Anderson 1983; Ojakangas 1985). The most reliable criterion for a glacial history for deformed and metamorphosed ancient sequences is the diamictite/dropstone-unit association (Ojakangas 1985). Whereas diamictites can be formed in several ways, dropstones (oversized clasts in fine-grained sediment that have penetrated or disturbed fine laminae in the sediment) must have had an origin related to ice if they were deposited before the evolution of organisms capable of transporting clasts. The ice may have been present as floating glaciers, icebergs, or shore ice.

Glaciogenic formations may contain several rock types, such as diamictites and dropstone units, that were formed by glaciers and/or rain-out from ice. However, other rock types, which by strict definition might not be called glaciogenic, are also commonly present. Examples are diamictites deposited by debris flows, common along margins of ablating glaciers ("flow tills"), and glaciofluvial sands and gravels deposited by glacial meltwater.

The preservation potential of glacial sediments deposited on land is low. For example, the Pleistocene continental-glacier deposits of North America are already being rapidly removed by erosion. Prob-

ably the best-documented ancient terrestrial glacial deposits in the geologic record are those of Late Ordovician age of north and west Africa (e.g., Deynoux and Trompette 1981). However, glaciogenic sediments deposited on a subsiding continental margin or in a subsiding intracratonic basin may be buried by younger sediment and preserved, and as a result most of the thick ancient glaciogenic formations are glaciomarine.

Other products of Early Proterozoic uncommon mega-events that might be used for correlation under the right circumstances are siliceous iron formations and uraniferous quartz-pebble conglomerates. Two other products, of greater temporal distribution, are prominent regoliths and sabkha carbonate-evaporite deposits. While the aforementioned four rock types are probably more common and widely distributed, and therefore less unique than glacial deposits, their value in correlation is greatest when they occur in a similar order in a rock sequence containing glaciogenic deposits.

Early Proterozoic Glaciation, North America

Lower Proterozoic glaciogenic deposits have been documented at several places in North America (Fig. 21.1), including Ontario north of Lake Huron, northern Michigan, southeastern Wyoming, central Quebec at Chibougamau, and the Northwest Territory just west of Hudson Bay. These occurrences are briefly described below.

Huronian Supergroup, Ontario

Three diamictite and dropstone-bearing formations (the Ramsay Lake, the Bruce, and the Gowganda Formations) are included in the Huronian Supergroup, which is as thick as 12,000 m (Fig 21.2). The supergroup includes three cycles of sedimentation, each beginning with diamictite (the above-named formations) and passing upward through fine-grained sedimentary rocks and finally into fluvial sandstone (Roscoe 1969). There is no evidence of repetition by thrusting.

The Ramsay Lake Formation contains as much as 150 m of diamictite, and is overlain by a basal dropstone unit of the Pecors Formation (Young 1981a). The Bruce Formation, up to 300 m thick, also

Fig. 21.1. Map showing distribution of localities with Lower Proterozoic glaciogenic rocks in Ontario (ON), Michigan (MI), Wyoming (WY), Northwest Territories (NWT), and Quebec (Q). Also shown is the diamictite locality of the Black Hills, South Dakota (SD). Area enclosed by dashed line was suggested as the area of Early Proterozoic glaciation by Young (1973).

contains diamictites and thin dropstone units (Young 1981a).

The Gowganda Formation pinches out to the north because of both depositional factors and erosion, but it may be as thick as 1,000 m in the south. It may have covered more than 120,000 km² prior to erosion. Massive diamictites are abundant (Fig. 21.3), striated clasts are present, and laminated dropstone units are common (Fig. 21.4). Diamictites overlie frost-shattered(?) Archean basement near Cobalt, Ontario (Donaldson and Munro 1982; Mustard and Donaldson 1987). Also common are graded mass-flow deposits, sandstones interpreted as deltaic deposits, and units consisting of laminated, dropstone-free shale and graywacke with normal grading. Most lithologic units appear to be lenticular on scales of tens of metres to kilometres (Young 1981b; Miall 1985).

The Gowganda Formation contains abundant evidence of a glacial history, as first documented by Coleman (1907). A glacial origin for the formation at different localities has been advocated by many workers, including Schenk (1965), Lindsey (1969), and Miall (1983), with some differences of opinion concerning the extent of continental glacial versus glaciomarine deposits. Nonglacial origins for parts

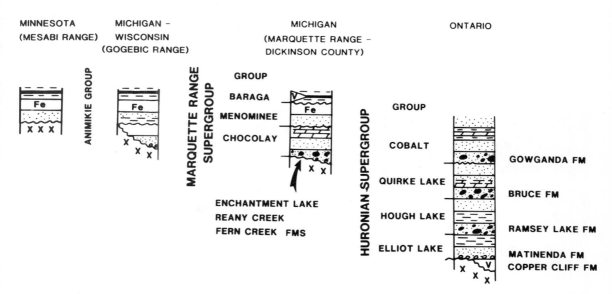

Fig. 21.2. Generalized stratigraphic columns of the Huronian Supergroup in Ontario, the Marquette Range Supergroup in Michigan, and the Animikie Group in Michigan, Wisconsin, and Minnesota. Only formations mentioned in the text are named in this illustration. The black conglomeratic pattern represents glaciogenic units, Fe represents iron formations, and dashes with solid black lines represent turbidite sequences. Other patterns are standard geologic patterns for volcanic rocks, conglomerate, sandstone, shale, and dolomite. Straight lines are conformable contacts and wavy lines are unconformities. Xs represent Archean basement. Not to scale.

Fig. 21.3. Diamictite in the Gowganda Formation, Ontario.

of the formation have also been proposed (Frarey and Roscoe 1970; Card *et al.* 1977); proglacial (non-ice-contact) components are certainly present.

Marquette Range Supergroup, Michigan

The Marquette Range Supergroup, 5,800 to 8,300 m thick, lies 200 km west of present exposures of the Huronian Supergroup. Basal units of the Marquette Range Supergroup in three different areas of the Upper Peninsula of Michigan are the Fern Creek, Enchantment Lake, and Reany Creek Formations (Fig. 21.2); these are 80 m, 150 m, and 1,000 m thick, respectively. Each is underlain by Archean basement and each contains diamictites (Pettijohn 1943; Puffett 1969; Young 1973; Gair 1981) and dropstone units (Ojakangas 1982). At one locality beneath the Reany Creek, the Archean granitic basement appears to be frost shattered. Figures 21.5 and 21.6 show a diamictite and a dropstone unit of the Reany Creek Formation.

Whereas the above-cited authors all advocate a glacial origin for the three formations, Mattson and

Fig. 21.4. Dropstone unit in the Gowganda Formation, Ontario.

Fig. 21.5. Diamictite in the Reany Creek Formation, Michigan.

Cambray (1983) interpreted the Reany Creek as the product of mass-flow mechanisms not related to glaciation, and Bayley *et al.* (1966), Gair and Thaden (1968), and Larue (1981) interpreted the Fern Creek and the Enchantment Lake Formations as alluvial-fan deposits. Certainly mass-flow and alluvial processes were active during deposition of these units, but the association of diamictites with dropstone units suggests a relationship to either a nonmarine glacially influenced environment or a glacially influenced continental margin.

Snowy Pass Supergroup, Southern Wyoming

The Snowy Pass Supergroup of the Medicine Bow Mountains and correlative units in the Sierra Madre Mountains of southern Wyoming include three diamictite-bearing units: the Campbell Lake, Vagner, and Headquarters Formations (Houston *et al.* 1981; Karlstrom *et al.* 1983). These are part of a 10,000 m thick, deformed, and metamorphosed Lower Proterozoic sedimentary sequence that includes thrust faults (Fig. 21.7). The diamictite

Fig. 21.6. Dropstone unit in the Reany Creek Formation, Michigan.

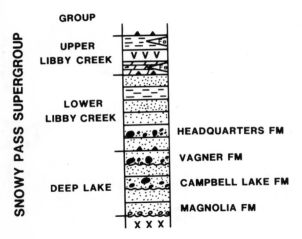

Fig. 21.7. Generalized stratigraphic column of the Snowy Pass Supergroup in the Medicine Bow Mountains, Wyoming. Only formations mentioned in the text are named in this illustration. Symbols as in Figure 21.2; thrust faults shown by small filled triangles. Not to scale. After Karlstrom *et al.* 1983.

members of the three formations are 12 m, 300 m, and 60 m thick, respectively. Each formation contains thick phyllitic members, and dropstone units have been found in the latter two formations. Houston *et al.* (1981) interpret the Vagner and Headquarters Formations to be glaciogenic. As in the Huronian Supergroup, three cycles of sedimentation, each beginning with a glaciogenic unit, have been proposed (Graff 1979), but thrust faults may have complicated the relationships of these units.

Chibougamau Formation, Quebec

The middle member of the Chibougamau Formation in central Quebec is more than 100 m thick and contains numerous diamictites and dropstone units as well as siltstone, conglomerate, and sandstone with normal grading. It has been interpreted as glaciogenic by Long (1974, 1981).

Padlei Formation, Horwitz Group, Northwest Territories

The basal diamictite of the Padlei Formation rests on Archean rocks or on older Proterozoic rocks (Young and McLennan 1981). The basal diamictite member is as thick as 400 m, and the overlying laminated unit with dropstones is as thick as 200 m. Young (1973)

and Young and McLennan (1981) have interpreted the Padlei Formation as glacial.

Summary, North America

The five units described above possess the essential characteristic of a glacial origin (the diamictite-dropstone unit association), as well as other attributes of a glacial origin, and can be interpreted as lithostratigraphically correlative.

Correlations between the Huronian Supergroup of Ontario and the Snowy Pass Supergroup of Wyoming have been proposed by several workers (Young 1970, 1973; Graff 1979; Houston *et al.* 1981). Strengthening that correlation, which is based in part on the three glacial units within the three cycles of sedimentation, are associated uraniferous quartz-pebble conglomerate units—the Matinenda Formation of Ontario (Fig. 21.2) and the Magnolia Formation of Wyoming (Fig. 21.7). The entire lower Proterozoic sequence is directly correlatable at the formation level (Karlstrom and Houston 1979), and the similarity of the two sequences is indeed striking. Although the Ontario and Wyoming localities are now widely separated, the Wyoming strata may have been moved several hundred kilometres westward relative to the Michigan-Ontario region by sinistral movement along the major Precambrian Mullen Creek-Nash Fork shear zone (Houston *et al.* 1981) in southeastern Wyoming.

Correlation of the glacial units in Michigan with each other and with the Gowganda Formation (Fig. 21.2) in Ontario has been proposed by several workers (e.g., Young 1973, 1983; Ojakangas 1982, 1985). Because only one glaciogenic horizon is present in Michigan, it has been correlated with the uppermost glaciogenic unit (the Gowganda) in Ontario 200 km to the east (e.g., Ojakangas 1982; Young 1983); however, it is theoretically possible that other glaciogenic units were removed by the erosion that produced two major unconformities stratigraphically higher in Michigan. Both the Chibougamau Formation and the Padlei Formation of the Northwest Territories have been correlated with the Gowganda Formation, largely on the basis of glaciogenic units (Long 1981; Young and McLennan 1981).

Young (1973) correlated all these glaciogenic formations and proposed that all were products of one major glacial period including minor advances and retreats, with the last advance of the ice sheet having been the most widespread (Fig. 21.1). Another locality may eventually fit in; diamictite of Early Proterozoic age is present in the Black Hills of South Dakota (Kurtz 1981).

The stratigraphic position of Proterozoic siliceous iron formation may be significant for correlation purposes. In the three sequences where iron formation is present [the Marquette Range Supergroup of Michigan (Fig. 21.2), the Snowy Pass Supergroup of Wyoming (Fig. 21.7), and the Mistassini Group, which is probably younger than the Chibougamau Formation (Long 1981)], it occurs within the sequence overlying the glaciogenic deposits.

The ages of these North American glaciogenic units are not tightly constrained, and are generally listed as between 2,500 Ma (a general age for the Archean basement) and about 1,900–1,700 Ma (the age of plutonism and metamorphism). The Huronian Supergroup is the best dated, bracketed by Nipissing diabases ($2,150–2,219.4^{+3.6}_{-3.5}$ Ma) that cross-cut the entire sequence (Van Schmus 1965; Corfu and Andrews 1986), and rhyolitic volcanics of the Copper Cliff Formation low in the sequence dated at $2,450^{+25}_{-10}$ Ma (Fig. 21.2; Krogh *et al.* 1984).

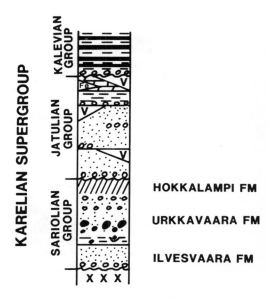

Fig. 21.8. Generalized stratigraphic column of the Karelian Supergroup of eastern Finland and adjacent USSR. Only formations mentioned in the text are named in this illustration. Symbols as in Figure 21.2. Meta-regolith shown by diagonal lines. Not to scale.

Early Proterozoic Glaciation, Baltic Shield

Sariolian Group, Finland

The Urkkavaara Formation of the Sariolian Group, Karelian Supergroup, of eastern Finland (Fig. 21.8) has been assigned a glacial origin by Marmo and Ojakangas (1983, 1984); they delineated three members within a dropstone unit that is gradational upward into diamictite (Fig. 21.9). Subsequent work has located a glaciogenic unit at the same stratigraphic horizon at four other localities as far away as 300 km, and the Urkkavaara Formation, originally described as 30–60 m thick, has been expanded to 300 m with the addition of overlying sandstones and conglomerates that are interpreted as proglacial (glaciofluvial) sediments (Marmo, in press).

A lower Proterozoic regolith as thick as 100 m is present within the Karelian Supergroup at the top of the Sariolian Group. Various members of the Urkkavaara Formation have been kaolinized and subsequently metamorphosed to kyanite, andalusite, and other aluminous minerals (Marmo, in press). Over-

lying this weathered horizon is the quartzose Jatulian Group.

East Karelia, USSR

In East Karelia, the part of the USSR adjacent to Finland, and in the Kola Peninsula farther north, the diamictite-dropstone unit association is known from sixteen localities, apparently stratigraphically equivalent to the Sariolian Group of the Karelian Supergroup (Negrutsa and Negrutsa 1981a, b, c). However, Negrutsa and Negrutsa did not favor a glacial origin for those rocks, interpreting them as tectono-sedimentary and volcano-sedimentary. These rocks have been subsequently reinterpreted as glaciogenic by Salop (1983, p. 138). This interpretation is supported by my observations of two diamictite-dropstone sequences in East Karelia, both associated with volcanic rocks; one is overlain by a regolith that is succeeded by the quartzitic Jatulian Group (K. Heiskanen, personal communication 1985). Elsewhere in East Karelia a prominent regolith is also present beneath the Jatulian Group (Salop 1983, p. 138; Sokolov *et al.* 1984), as it is in Finland.

Fig. 21.9. Dropstone unit passing upward into massive diamictite, Urkkavaara Formation, Finland. The transition zone is beneath the hammer handle, which points stratigraphically upward.

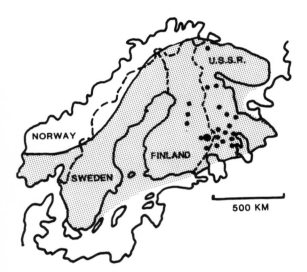

Fig. 21.10. Generalized map of the Baltic Shield. Precambrian rocks are shaded; Phanerozoic rocks are white. Large black dot in eastern Finland is the discovery site of glaciogenic deposits by Marmo and Ojakangas (1983); smaller dots in Finland represent other localities with undescribed glaciogenic rocks. Dots in USSR are diamictite localities described by Negrutsa and Negrutsa (1981 a,b,c).

Summary, Baltic Shield

The studied Finnish diamictite/dropstone-unit localities and those in the USSR that are described in the literature are shown on Figure 21.10. The total area involved is 150,000 to 200,000 km², suggesting a continental-scale glaciation on the Baltic Shield. Although some Baltic localities are isolated and difficult to correlate precisely, it seems that most of the aforementioned diamictite-dropstone units occur at approximately the same stratigraphic horizon in the Sariolian Group. They are cut by 2,180 to 2,160 Ma diabase dikes (Sakko 1971) and overlie igneous rocks dated at 2,450 to 2,430 Ma (Krats *et al.* 1976 in Salop 1983, p. 136; Alapieti 1982).

Stratigraphically below the Sariolian Group in East Karelia, USSR, is the Tunguda-Nadvoitsa or Sumian Group (not present in eastern Finland), which contains pyritiferous uranium-bearing quartz-pebble conglomerates (Salop 1977, p. 130). Equivalent units may be present in Finnish Lapland in the Lapponian Group. Iron formation is present in the Onega Group (equivalent to the upper Jatulian of Finland) in East Karelia and in the Kola Peninsula (Salop 1983, p. 139). Iron formation is also present in Finland in the upper part of the Jatulian Group, the so-called Marine Jatulian (Fig. 21.8); the single iron-rich horizon has been dated at 2,080 Ma (Laajoki and Saikkonen 1977). The uppermost Lower Proterozoic glaciogenic rocks on the Baltic Shield were deeply weathered in Early Proterozoic time (Marmo, in press).

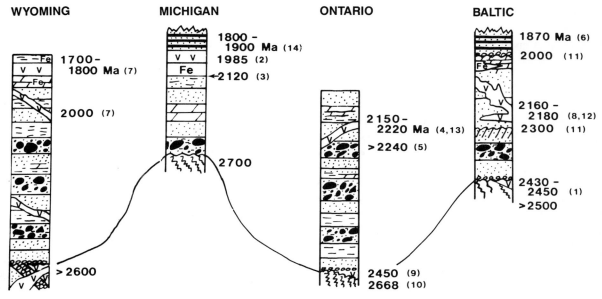

Fig. 21.11. Generalized correlation chart for Lower Proterozoic columns of North America and the Baltic Shield (Finland and Karelia, USSR). Symbols, except for basement, as in Figure 21.2. Regolith in Baltic column is shown by diagonal lines. Not to scale. Numbers are radiometric dates based on the following references: (1) Alapieti 1982; (2) Banks and Van Schmus 1971; (3) Beck and Murthy 1982; (4) Corfu and Andrews 1986; (5) Fairbairn *et al.* 1969; (6) Huhma 1986; (7) Karlstrom *et al.* 1983; (8) Krats *et al.* 1976, in Salop 1983; (9) Krogh *et al.* 1984; (10) Krogh and Turek 1982; (11) Meriläinen 1980; (12) Sakko 1971; (13) Van Schmus 1965; (14) Van Schmus 1980.

Intracontinental Correlation

Intracontinental correlation of Lower Proterozoic rocks on both the North American and the Baltic cratons, based on glaciogenic deposits and radiometric ages of underlying and cross-cutting igneous rocks, has been summarized above. In addition to the glaciogenic marker units, several other relatively uncommon lithologies occur in the same stratigraphic order within the five cited North American stratigraphic successions (Fig. 21.11). These lithologies, from the base up are as follows: uraniferous quartz-pebble conglomerate, glaciogenic rocks, sabkha-type carbonates, and iron formation.

A correlation between the Wyoming and Huronian successions was made by Blackwelder as early as 1926. Thus the correlation proposed herein for five areas on the North American craton is not new, but provides a re-emphasis on the long-established stratigraphic principle that certain unique sequences of rocks (especially glaciogenic units) that are products of long-term but uncommon sedimentary events are valid marker units.

Correlations among units of the Karelian Supergroup on the Baltic Shield are well established (e.g., Silvennoinen 1972). A more detailed analysis of the basin in which the supergroup was deposited (Sokolov and Heiskanen 1985) together with this study further strengthens these correlations.

Intercontinental Correlations

The intercontinental correlation between the Huronian and the Karelian Supergroups was, interestingly, proposed as early as 1876 by Wiik (in Laajoki and Saikkonen 1977, p. 128), and reiterated by Laajoki and Saikkonen (1977), but was based on the general similarity of the lithostratigraphic sequences without recognition of the presence of glaciogenic units to strengthen the correlation. The recent discovery of Lower Proterozoic glaciogenic units on the Baltic Shield that appear to be of the same age as those in Lower Proterozoic sequences in North America is a basis for intercontinental lithostratigraphic correlation of basins.

In addition to the glaciogenic deposits, the same general sequence of other relatively uncommon lithologies that is present in North America is seen in Finland and the adjacent USSR (Fig. 21.11).

The presence of a regolith in North America at a similar stratigraphic position to that discovered in Finland and Karelia was suggested prior to the discovery of the well developed regolith in Finland, on the basis of aluminous minerals within the overlying sandstone units (Chandler *et al.* 1969; Young 1973). This may indicate that the regolith was largely recycled into the overlying quartz-rich sandstones. However, if preserved regolith can be found by further searching at this stratigraphic horizon at several localities, perhaps another mega-event (i.e., a climate-controlled deep-weathering event) could be documented and the case for the proposed intercontinental correlation would be even stronger.

The radiometric ages appear to support chronostratigraphic correlations, as well as lithostratigraphic correlations. Not only more data but more precise radiometric ages and paleomagnetic pole positions are required to strengthen such intercontinental correlation.

Paleomagnetic data on Lower Proterozoic sequences in general are few and somewhat ambiguous, but the regions in which the Huronian and Karelian Supergroups occur may have been located at about 50°N latitude at approximately 2,400 Ma, as summarized by Pesonen and Neuvonen (1981). Piper (1976, 1983) placed the North American and Baltic cratons adjacent to each other based on a reconstruction of the continents. The presence of the glaciogenic deposits on both cratons is additional evidence that the cratons were at relatively high latitudes in Early Proterozoic time, and it is tempting to speculate that both cratons were glaciated by the same major continental ice mass.

There are also fairly well documented glacial deposits in the Lower Proterozoic Transvaal and Griqualand Supergroups of South Africa that are between 2,224 ± 21 Ma and 2,300 ± 100 Ma (Visser 1981). Clearly the age of these glaciogenic deposits is comparable to those of similar deposits on the North American and the Baltic cratons. If it can be ascertained that the South African deposits are the result of the same glacial mega-event, then it would seem that the location of the African continent at that time had to be at a relatively high latitude. But was this high latitude in the northern or southern hemisphere? Diamictites, without associated dropstone units and therefore not substantiated as glacial sediments, have been described from India and Australia as well.

Discussion

Mega-event stratigraphy is possible in rock sequences that contain products of relatively uncommon mega-events, such as glaciations of continental scale. Chronostratigraphic correlation on this basis is especially useful in rock sequences for which radiometric dating is not available or is imprecise. It should be especially useful in basins for which only the maximum and minimum absolute ages of the basin fill are known for it makes possible further chronologic subdivision of the stratigraphic sequences.

It would not be useful, for example, to attempt to correlate two 1,000 m thick arkosic sandstone units from two widely separated basins. Although such sandstones are the products of mega-events, each is likely to have been controlled by nonunique local tectonism and need not be synchronous. Similarly, two widely separated basins with flysch to molasse sequences are likely the results of a similar succession of sedimentary environments controlled by similar tectonic histories that are not likely to be synchronous. On the other hand, if lithostratigraphic units in widely separated basins are composed of distinctive lithologies, such as the products of a unique, global climate-controlled event that is constrained by radiometric dates, then the units in the two basins can be correlated with some confidence.

This approach to chronocorrelation is akin to, yet different from, correlation of instantaneous events (e.g., Boggs 1987, p. 685), such as a single volcanic eruption that is manifest in the rock record as a widespread ash bed. Mega-event correlation is different in that a mega event is of much longer duration and may be diachronous.

The glaciogenic units in the Huronian Supergroup of Ontario, based upon the radiometric ages cited above, were deposited between 2,475 and 2,150 Ma, a maximum time interval of approximately 325 million years. Similarly, the glaciogenic Urkkavaara Formation of Finland was deposited between 2,450 and 2,160 Ma, an interval of about 290 million years. These intervals represent maxima and it is likely that deposition took place over a small fraction of these time intervals.

If it is assumed that the above radiometric data are valid, and if it is assumed that a single ice age (a mega-event) formed these glaciogenic deposits on both the North American and Baltic Shields, the

Lower Proterozoic sedimentary successions on those shields between the dated basal flows and the cross-cutting dikes can be divided into: 1) pre-mega-event, 2) mega-event, and 3) post-mega-event (i.e., pre-glacial, glacial, and post-glacial). Unfortunately, absolute dates cannot be placed on these three subdivisions, but the subdivisions nevertheless should be useful in general chronocorrelation as well as in more detailed studies of stratigraphy and environments of deposition.

A depositional sequence, as defined by Mitchum *et al.* (1977), is a stratigraphic unit composed of relatively conformable, genetically related strata, and bounded at its top and base by unconformities that may pass basinward into conformable relationships. According to this definition, the Lower Proterozoic Urkkavaara Formation on the Baltic Shield constitutes a depositional sequence with its lower erosional surface being the Archean-Proterozoic contact (or slightly higher, above the Ilvesvaara Formation), and its upper erosional surface the top of the regolith developed upon the Urkkavaara Formation (Fig. 21.11). It is not yet clear whether the glaciogenic units in North America, and associated units, are depositional sequences as defined above. However, several, such as the units in Michigan and Ontario, rest unconformably upon Archean basement (Fig. 21.2). Depositional sequences within stratigraphic sequences that contain evidence for glacial mega-events, if documented, might be related to eustatic sea-level changes and hence to marine regressions and transgressions. Such unconformities have been documented for the Pleistocene of Baffin Island by Miller *et al.* (1977). Even in sedimentary sequences lacking glacial deposits on other continents, regressions and transgressions may be related to the glaciation, as documented by Veevers and Powell (1987) in upper Paleozoic deposits. However, there appears to be a lack of preserved sedimentary sequences of Early Proterozoic age that do not contain glaciogenic units.

Summary

Glaciogenic units in widely separated basins, such as those described herein in the Lower Proterozoic successions of North America and the Baltic Craton, provide a good example of correlation of the deposits of mega-events (i.e., mega-event stratigraphy).

Glaciogenic units of continental rather than valley-glacier scale are the products of global climatically controlled mega-events. Because such mega-events are not directly related to regional tectonic events, they are useful for intercontinental as well as intracontinental chronocorrelation. The Pleistocene glaciation is a well documented younger example of the use of a mega-event for correlation purposes, as is the Late Paleozoic glaciation of Gondwanaland. These younger analogs indicate clearly the diachronous nature of glacial mega-events. While this diachroneity affects the precision of correlation, it does not negate the usefulness of glacial mega-events in improving upon the resolution of the relative ages of sedimentary units within Precambrian sequences for which limited radiometric age data are available.

Acknowledgments. I wish to thank the following persons for their constructive reviews of this paper: Bruce Simonson, Andrew Miall, Karen Kleinspehn, and Chris Paola. Avis Hedin typed the manuscript.

References

ALAPIETI, T. (1982) The Koillismaa layered igneous complex, Finland – its structure, mineralogy and geochemistry, with emphasis on the distribution of chromium. Geological Survey Finland Bulletin 319, 116 p.

ANDERSON, J.B. (1983) Ancient glacial-marine deposits: Their spatial and temporal distributions. In: Molnia, B.F. (ed) Glacial-Marine Sedimentation. New York: Plenum Press, pp. 3–92.

ANDREWS, J.T. and MATSCH, C.L. (1983) Glacial Marine Sediments and Sedimentation, an Annotated Bibliography. Geo Abstracts, Bibliography No. 11, Norwich, England, 227 p.

BANKS, P.O. and VAN SCHMUS, W.R. (1971) Chronology of Precambrian rocks of Iron and Dickinson Counties, Michigan (Abstract). Duluth, Minn.: Proceedings, 17th Annual Institute Lake Superior Geology, p. 9.

BAYLEY, R.W., DUTTON, C.E., and LAMEY, C.A. (1966) Geology of the Menominee Iron-Bearing District, Dickinson County, Michigan, and Florence and Marinette Counties, Wisconsin. United States Geological Survey Professional Paper 513, 96 p.

BECK, W. and MURTHY, V.R. (1982) Rb-Sr and Sm-Nd isotopic studies of Proterozoic mafic dikes in northeastern Minnesota (Abstract). International Falls, Minn.: Proceedings, 28th Annual Institute Lake Superior Geology, p. 5.

BLACKWELDER, E. (1926) Precambrian geology of the Medicine Bow Mountains. Geological Society America Bulletin 37:615–658.

BOGGS, S., Jr. (1987) Principles of Sedimentology and Stratigraphy. Columbus, Ohio: Merrill Publishing Company, 784 p.

CAPUTO, M.V. and CROWELL, J.C. (1985) Migration of glacial centers across Gondwana during Paleozoic era. Geological Society America Bulletin 96:1020–1036.

CARD, K.D., INNES, D.G., and DEBICKI, R.L. (1977) Stratigraphy, Sedimentology, and Petrology of the Huronian Supergroup in the Sudbury-Espanola Area. Ontario Division of Mines Geoscience Study 16, 99 p.

CHANDLER, F.J., YOUNG, G.M., and WOOD, J. (1969) Diaspore in Early Proterozoic quartzites (Lorrain Formation) of Ontario. Canadian Journal Earth Sciences 6:337–340.

CHUMAKOV, N.M. (1981) Upper Proterozoic glaciogenic rocks and their stratigraphic significance. Precambrian Research 15:373–395.

COLEMAN, A.P. (1907) A lower Huronian ice age. American Journal Science 23:187–192.

CORFU, F. and ANDREWS, A.J. (1986) A U-Pb age for mineralized Nipissing diabase, Gowganda, Ontario. Canadian Journal Earth Sciences 23:107–109.

CROWELL, J.C. (1974) Climatic significance of sedimentary deposits containing dispersed megaclasts. In: Nairn, A.E.M. (ed) Problems in Paleoclimatology. New York: Wiley Interscience, pp. 86–99.

DEYNOUX, M. and TROMPETTE, R. (1981) Late Ordovician tillites of the Taovdeni Basin, West Africa. In: Hambrey, M.J. and Harland, W.B. (eds) Earth's Pre-Pleistocene Glacial Record. Cambridge, England: Cambridge University Press, pp. 89–96.

DONALDSON, J.A. and MUNRO, I. (1982) Guidebook for field excursion 16B: Precambrian geology of the Cobalt area, northern Ontario. Eleventh International Congress on Sedimentology, McMaster University, Hamilton, Ontario, pp. 7–54.

FAIRBAIRN, H.W., HURLEY, P.M., CARD, K.D., and KNIGHT, C.J. (1969) Correlation of radiometric age of Nipissing diabase and metasediments with Proterozoic orogenic events in Ontario. Canadian Journal Earth Sciences 6:489–497.

FRAREY, M.J. and ROSCOE, S.M. (1970) The Huronian Supergroup north of Lake Huron. In: Baer, A.J. (ed) Symposium on Basins and Geosynclines of the Canadian Shield. Geological Survey Canada Paper 70–40, pp. 143–157.

GAIR, J.E. (1981) Lower Proterozoic glacial deposits of northern Michigan, U.S.A. In: Hambrey, M.J. and Harland, W.B. (eds) Earth's Pre-Pleistocene Glacial Record. Cambridge, England: Cambridge University Press, pp. 803–806.

GAIR, J.E. and THADEN, R.E. (1968) Geology of the Marquette and Sands Quadrangles, Marquette County, Michigan. United States Geological Survey Professional Paper 397, 77 p.

GRAFF, P. (1979) A review of the stratigraphy and uranium potential of Early Proterozoic (Precambrian X) metasediments in the Sierra Madre, Wyoming. University Wyoming Contributions to Geology 17:149–157.

HOUSTON, R.S., LANTHIER, L.R., KARLSTROM, K.E., and SYLVESTER, G. (1981) Early Proterozoic diamictite of southern Wyoming. In: Hambrey, M.J. and Harland, W.B. (eds) Earth's Pre-Pleistocene Glacial Record. Cambridge, England: Cambridge University Press, pp. 795–799.

HUHMA, H. (1986) Sm-Nd, U-Pb and Pb-Pb isotopic evidence for the origin of the Early Proterozoic Svecokarelian crust in Finland. Geological Survey Finland Bulletin 337, 48 p.

KARLSTROM, K.E., FLURKEY, A.J., and HOUSTON, R.S. (1983) Stratigraphy and depositional setting of the Proterozoic Snowy Pass Supergroup, southeastern Wyoming: Record of an early Proterozoic Atlantic-type cratonic margin. Geological Society America Bulletin 94:1257–1274.

KARLSTROM, K.E. and HOUSTON, R.S. (1979) Stratigraphy and Uranium Potential of Early Proterozoic Metasedimentary Rocks in the Medicine Bow Mountains, Wyoming. Geological Survey Wyoming Reports and Investigations 13, 45 p.

KRATS, K.O., LEVCHINKOV, D.A., OUCHINNIKOV, L.B., SHULESHKO, I.K., YAKOVLEVA, S.Z., MAKEYEV, A.F., and KOMAROV, A.N. (1976) Vozrastnyye granitsy yatuliyskogo kompleksa Karelii (Age boundaries of the Yatulian complex of Karelia.) Akademiya Nauk SSSR Doklady; Svodnyi Vypusk 231 (5), pp. 1191–1194.

KROGH, T.E., DAVIS, D.W., and CORFU, F. (1984) Precise U-Pb zircon and baddeleyite ages for the Sudbury area. In: Pye, E.G., Naldrett, A.J., and Giblin, P.E. (eds) The Geology and Ore Deposits of the Sudbury Structure. Ontario Geological Survey Special Volume 1, pp. 431–446.

KROGH, T.E. and TUREK, A. (1982) Precise U-Pb zircon ages from the Gamitagama Greenstone Belt, Southern Superior Province. Canadian Journal Earth Sciences 19:859–867.

KURTZ, D.D. (1981) Early Proterozoic diamictites of the Black Hills, South Dakota. In: Hambrey, M.J. and Harland, W.B. (eds) Earth's Pre-Pleistocene Glacial Record. Cambridge, England: Cambridge University Press, pp. 800–802.

LAAJOKI, K. and SAIKKONEN, R. (1977) On the Geology and Geochemistry of the Precambrian Iron Formation in Väyrylänkylä, South Puolanka Area, Finland. Geological Survey Finland Bulletin 292, 137 p.

LARUE, D.K. (1981) The Chocolay Group, Lake Superior region, U.S.A.: Sedimentologic evidence for deposition in basinal and platformal settings on an early

Proterozoic craton. Geological Society America Bulletin 92:417–435.

LINDSEY, D.A. (1969) Glacial marine sediments in the Precambrian Gowganda Formation at Whitefish Falls, Ontario (Canada). Palaeogeography, Palaeoclimatology, Palaeoecology 9:7–26.

LONG, D.G.F. (1974) Glacial and paraglacial genesis of conglomeratic rocks of the Chibougamau Formation (Aphebian), Chibougamau, Quebec. Canadian Journal Earth Sciences 11:1236–1252.

LONG, D.G.F. (1981) Glaciogenic rocks in the early Proterozoic Chibougamau Formation of northern Quebec. In: Hambrey, M.J. and Harland, W.B. (eds) Earth's Pre-Pleistocene Glacial Record. Cambridge, England: Cambridge University Press, pp. 817–820.

MARMO, J.S. (in press) Sariolian stratigraphy and sedimentation in the Koli-Kaltimo area, North Karelia, Eastern Finland. In: Shokolov, V. (ed) Symposium on the Lower Proterozoic Formations on the Eastern Part of the Baltic Shield. Proceedings of a Finnish-Soviet Symposium, Petrozavodsk, 1985.

MARMO, J.S. and OJAKANGAS, R.W. (1983) Varhaisproterotsooinen Urkkavaara-muodostuma Kontiolahdella glasigeeninen metasedimenttisarja Sariolaryhman ylaosassa (Early Proterozoic Urkkavaara Formation in Kontiolahti, North Karelia—a glaciogenic meta-sedimentary sequence in the Upper Sariolan Group). Geologi (Finnish Geological Society) 35:3–6.

MARMO, J.S. and OJAKANGAS, R.W. (1984) Lower Proterozoic glaciogenic deposits, eastern Finland. Geological Society America Bulletin 95:1055–1062.

MATTSON, S.R. and CAMBRAY, F.W. (1983) The Reany Creek Formation: A mass-flow deposit of possible post-Menominee age (Abstract). Houghton, Mich.: Proceedings 29th Annual Institute on Lake Superior Geology, p. 27.

MERILÄINEN, K. (1980) On the stratigraphy of the Karelian formations. In: Silvennoinen, A. (ed) Jatulian Geology in the Eastern Part of the Baltic Shield. Proceedings of a Finnish-Soviet Symposium, Finland, 1979, pp. 17–112.

MIALL, A.D. (1983) Glaciomarine sedimentation in the Gowganda Formation (Huronian), Northern Ontario. Journal Sedimentary Petrology 53:477–492.

MIALL, A.D. (1985) Sedimentation on an early Proterozoic continental margin under glacial influence: The Gowganda Formation (Huronian), Elliot Lake area, Ontario, Canada. Sedimentology 32:763–788.

MILLER, G.H., ANDREWS, J.T., and SHORT, S.K. (1977) The last interglacial/glacial cycle, Clyde foreland, Baffin Island, N.W.T.: Stratigraphy, biostratigraphy, and chronology. Canadian Journal Earth Sciences 14:2824–2857.

MITCHUM, R.M. Jr., VAIL, P.R., and THOMPSON, S. III. (1977) The depositional sequence as a basic unit for stratigraphic analysis. In: Payton, C.E. (ed) Seismic Stratigraphy—Applications to Hydrocarbon Exploration. American Association Petroleum Geologists Memoir 26, pp. 53–62.

MUSTARD, P.S. and DONALDSON, J.A. (1987) Early Proterozoic ice-proximal glaciomarine deposition: Lower Gowganda Formation at Cobalt, Ontario, Canada. Geological Society America Bulletin 98:373–387.

NEGRUTSA, T.F. and NEGRUTSA, V.Z. (1981a) Early Proterozoic Lammos tilloids of the Kola Peninsula, U.S.S.R. In: Hambrey, M.J. and Harland, W.B. (eds) Earth's Pre-Pleistocene Glacial Record. Cambridge, England: Cambridge University Press, pp. 678–680.

NEGRUTSA, T.F. and NEGRUTSA, V.Z. (1981b) Early Proterozoic Yanis-Yarvi tilloids, South Karelia, U.S.S.R. In: Hambrey, J.J. and Harland, W.B. (eds) Earth's Pre-Pleistocene Glacial Record. Cambridge, England: Cambridge University Press, pp. 681–682.

NEGRUTSA, T.F. and NEGRUTSA, V.Z. (1981c) Early Proterozoic Sarioli tilloids in the eastern part of the Baltic Shield, U.S.S.R. In: Hambrey, M.J. and Harland, W.B. (eds) Earth's Pre-Pleistocene Glacial Record. Cambridge, England: Cambridge University Press, pp. 683–686.

OJAKANGAS, R.W. (1982) Lower Proterozoic glaciogenic formations, Marquette Supergroup, Upper Peninsula, Michigan, U.S.A. (Abstract). International Association of Sedimentologists Abstracts of Papers, Eleventh International Congress on Sedimentology, McMaster University, Hamilton, ON, p. 76.

OJAKANGAS, R.W. (1985) Evidence for Early Proterozoic glaciation: The dropstone unit—diamictite association. In: K. Laajoki and J. Paakkola (eds) Proterozoic Exogenic Processes and Related Metallogeny. Geological Survey Finland Bulletin 331:51–72.

PESONEN, L.J. and NEUVONEN, K.J. (1981) Paleomagnetism of the Baltic Shield—Implications for Precambrian tectonics. In: Kroner, A. (ed) Precambrian Plate Tectonics. Amsterdam: Elsevier, pp. 623–648.

PETTIJOHN, F.J. (1943) Basal Huronian conglomerates of Menominee and Calumet districts, Michigan. Journal Geology 51:387–397.

PIPER, J.D.A. (1976) Paleomagnetic evidence for a Proterozoic supercontinent. Philosophical Transactions Royal Society London 280:469–490.

PIPER, J.D.A. (1983) Dynamics of the continental crust in Proterozoic times. In: Medaris, L.G., Jr., Byers, C.W., Mickelson, D.M., and Shanks, W.C. (eds) Proterozoic Geology. Geological Society America Memoir 161, pp. 11–34.

PUFFETT, W.P. (1969) The Reany Creek Formation, Marquette County, Michigan. United States Geological Survey Bulletin 1274-F: pp. F-1–F-25.

ROSCOE, S.M. (1969) Huronian Rocks and Uraniferous Conglomerates in the Canadian Shield. Geological Survey Canada Paper 68-40, 205 p.

SAKKO, M. (1971) Varhais-Karjalaisten metadiabassien radiometrisia: Zirconi-ikia (with English summary): Radiometric zircon ages of the Early Karelian metadiabases. Geologi (Finnish Geological Society) 23:117–118.

SALOP, L.J. (1977) Precambrian of the Northern Hemisphere. New York: Elsevier, 378 p.

SALOP, L.J. (1983) Geological Evolution of the Earth during the Precambrian. Berlin: Springer-Verlag, 459 p.

SCHENK, P.E. (1965) Depositional environment of the Gowganda Formation (Precambrian) at the south end of Lake Timagami, Ontario. Journal Sedimentary Petrology 35:309–318.

SCHERMERHORN, L.J.A. (1974) Late Precambrian mixtites: Glacial and/or nonglacial? American Journal Science 274:673–824.

SILVENNOINEN, A. (1972) On the Stratigraphic and Structural Geology of the Rukatunturi Area, Northeastern Finland. Geological Survey Finland Bulletin 257, 48 p.

SOKOLOV, V.A. and HEISKANEN, K. (1985) Evolution of Precambrian volcanogenic-sedimentary lithogenesis in the southeastern part of the Baltic Shield. In: Laajoki, K. and Paakola, J. (eds) Proterozoic Exogenic Processes and Related Metallogeny. Geological Survey Finland Bulletin 331:91–106.

SOKOLOV, V.A., KULIKOV, V.S., and GOLUBEV, A.I. (1984) Guidebook, Geological Field Trips in Karelia. Institute Geology Academy Sciences USSR, Karelian Branch, Petrozavodsk: 104 p.

VAN SCHMUS, R. (1965) The geochronology of the Blind River–Bruce Mines area, Ontario, Canada. Journal Geology 73:755–780.

VAN SCHMUS, R. (1980) Geochronology of igneous rocks associated with the Penokean orogeny in Wisconsin. In: Morey, G.B. and Hanson, G.N. (eds) Selected Studies of Archean Gneisses and Lower Proterozoic Rocks, Southern Canadian Shield. Geological Society America Special Paper 182, pp. 159–168.

VEEVERS, J.J. and POWELL, C.M. (1987) Late Paleozoic glacial episodes in Gondwanaland reflected in transgressive-regressive depositional sequences in Euramerica. Geological Society America Bulletin 98:475–487.

VISSER, J.N.J. (1981) The mid-Precambrian tillite in the Griqualand West and Transvaal Basins, South Africa. In: Hambrey, M.J. and Harland, W.R. (eds) Earth's Pre-Pleistocene Glacial Record. Cambridge, England: Cambridge University Press, pp. 180–184.

YOUNG, G.M. (1970) An extensive early Proterozoic glaciation in North America? Palaeogeography, Palaeoclimatology, Palaeoecology 7:85–101.

YOUNG, G.M. (1973) Tillites and aluminous quartzites as possible time markers for Middle Precambrian (Aphebian) rocks of North America. In: Young, G.M. (ed) Huronian Stratigraphy and Sedimentation. Geological Association Canada Special Paper 12, pp. 97–128.

YOUNG, G.M. (1981a) Diamictites of the early Proterozoic Ramsay Lake and Bruce Formations, north shore of Lake Huron, Ontario, Canada. In: Hambrey, M.J. and Harland, W.B. (eds) Earth's Pre-Pleistocene Glacial Record. Cambridge, England: Cambridge University Press, pp. 813–816.

YOUNG, G.M. (1981b) The Early Proterozoic Gowganda Formation, Ontario, Canada. In: Hambrey, M.J. and Harland, W.B. (eds) Earth's Pre-Pleistocene Glacial Record. Cambridge, England: Cambridge University Press, pp. 807–812.

YOUNG, G.M. (1983) Tectono-sedimentary history of early Proterozoic rocks of the northern Great Lakes region. In: Medaris, L.G., Jr. (ed) Early Proterozoic Geology of the Great Lakes Region. Geological Society America Memoir 160, pp. 15–32.

YOUNG, G.M. and MCLENNAN, S.M. (1981) Early Proterozoic Padlei Formation, Northwest Territories, Canada. In: Hambrey, M.J. and Harland, W.B. (eds) Earth's Pre-Pleistocene Glacial Record. Cambridge, England: Cambridge University Press, pp. 790–794.

Index